AF566996

Methods in Cell Biology

VOLUME 58

Green Fluorescent Proteins

Series Editors

Leslie Wilson
Department of Biological Sciences
University of California, Santa Barbara
Santa Barbara, California

Paul Matsudaira
Whitehead Institute for Biomedical Research and
Department of Biology
Massachusetts Institute of Technology
Cambridge, Massachusetts

Methods in Cell Biology

Prepared under the Auspices of the American Society for Cell Biology

VOLUME 58

Green Fluorescent Proteins

Edited by

Kevin F. Sullivan

Department of Cell Biology
The Scripps Research Institute
La Jolla, California

Steve A. Kay

Department of Cell Biology
The Scripps Research Institute
La Jolla, California

A Harcourt Science and Technology Company

San Diego San Francisco New York Boston London Sydney Tokyo

Cover photo credit (paperback edition only): Mitosis in human cells is revealed by fluorescence microscopy showing GFP-labeled chromosomes that have incorporated a histone H2B-GFP fusion protein (green) coupled with immunofluorescence imaging of microtubules (blue) and centromeres (orange). The image was acquired using a Bio-Rad MRC-1024 confocal microscope. Provided by Kevin F. Sullivan.

This book is printed on acid-free paper. 

Copyright © 1999 by ACADEMIC PRESS

All Rights Reserved.

No part of this publication may be reproduced or transmitted in any form or by any means, electronic or mechanical, including photocopy, recording, or any information storage and retrieval system, without permission in writing from the Publisher.

The appearance of the code at the bottom of the first page of a chapter in this book indicates the Publisher's consent that copies of the chapter may be made for personal or internal use of specific clients. This consent is given on the condition, however, that the copier pay the stated per copy fee through the Copyright Clearance Center, Inc. (222 Rosewood Drive, Danvers, Massachusetts 01923), for copying beyond that permitted by Sections 107 or 108 of the U.S. Copyright Law. This consent does not extend to other kinds of copying, such as copying for general distribution, for advertising or promotional purposes, for creating new collective works, or for resale. Copy fees for pre-1999 chapters are as shown on the title pages. If no fee code appears on the title page, the copy fee is the same as for current chapters.
0091-679X/99 $25.00

Academic Press
A Harcourt Science and Technology Company
525 B Street, Suite 1900, San Diego, California 92101-4495, USA
http://www.academicpress.com

Academic Press
Harcourt Place, 32 Jamestown Road, London NW1 7BY, UK
http://www.academicpress.com

International Standard Book Number: 0-12-544160-6 (case)
International Standard Book Number: 0-12-676075-6 (pb)

PRINTED IN THE UNITED STATES OF AMERICA
00 01 02 03 EB 9 8 7 6 5 4 3

CONTENTS

CONTRIBUTORS

Numbers in parentheses indicate the pages on which the authors' contributions begin.

Zimran Ahmed (75), Department of Biomedical Sciences and CNR, Center for the Study of Biomembranes, University of Padua, 35121 Padua, Italy

Joseph C. Ayoob (239), Department of Cell and Developmental Biology, University of Pennsylvania School of Medicine, Philadelphia, Pennsylvania 19104

Carlo Bastianutto (75), Department of Biomedical Sciences and CNR, Center for the Study of Biomembranes, University of Padua, 35121 Padua, Italy

Andrew S. Belmont (203), Department of Cell and Structural Biology, University of Illinois, Urbana-Champaign, Urbana, Illinois 61801. E-mail: asbel@uiuc.edu

Maureen Blomberg-Wirschell (223), Program in Molecular Medicine, University of Massachusetts Medical Center, Worcester, Massachusetts 01605

Andrea Brand (165), Wellcome/CRC Institute and Department of Genetics, Cambridge University, Cambridge CB2 1QR, United Kingdom. E-mail: ahb@mole.bio.cam.ac.uk

Marisa Brini (75), Department of Biomedical Sciences and CNR, Center for the Study of Biomembranes, University of Padua, 35121 Padua, Italy

Janet L. Carminati (87), Department of Biological Sciences, Stanford University, Stanford, California 94305. E-mail: stearns@leland.stanford.edu

Walter Carrington (223), Biomedical Imaging Group, University of Massachusetts Medical Center, Worcester, Massachusetts 01605

Andrew B. Cubitt (19), Aurora Biosciences Corporation, La Jolla, California 92037

Guissou A. Dabiri (239), Department of Cell and Developmental Biology, University of Pennsylvania School of Medicine, Philadelphia, Pennsylvania 19104

Richard N. Day (293), Department of Medicine and Cell Biology, NSF Center for Biological Timing, University of Virginia Health Sciences Center, Charlottesville, Virginia 22908

Francesca De Giorgi (75), Department of Biomedical Sciences and CNR, Center for the Study of Biomembranes, University of Padua, 35121 Padua, Italy

Stephen J. Doxsey (223), Program in Molecular Medicine, University of Massachusetts Medical Center, Worcester, Massachusetts 01605. E-mail: stephen.doxsey@ummed.edu

J. Ellenberg (261), Cell Biology and Metabolism Branch, NICHD, NIH, Bethesda, Maryland 20982

Sharyn A. Endow (153), Department of Microbiology, Duke University Medical Center, Durham, North Carolina 27710. E-mail: endow@galactose.mc.duke.edu

Paul Ferrigno (107), Department of Biological Chemistry and Molecular Pharmacology, Harvard Medical School and the Dana Farber Cancer Institute, Boston, Massachusetts 02115

David W. Galbraith (315), Department of Plant Sciences, University of Arizona, Tucson, Arizona 85721. E-mail: galbraith@arizona.edu

Paul C. Goodwin (343), Applied Precision, Inc., Issaquah, Washington 98027. E-mail: pgood@api.com

James Haseloff (139), MRC Laboratory of Molecular Biology, Cambridge CB2 2QH, England. E-mail: jph@mrc-lmb.cam.ac.uk

Roger Heim (19), Aurora Biosciences Corporation, La Jolla, California 92037

K. Hirschberg (261), Cell Biology and Metabolism Branch, NICHD, NIH, Bethesda, Maryland 20982

Laurence Sophie Jouaville (75), Department of Biomedical Sciences and CNR, Center for the Study of Biomembranes, University of Padua, 35121 Padua, Italy

Steve A. Kay (283), Department of Cell Biology, The Scripps Research Institute, La Jolla, California 92037. E-mail: stevek@scripps.edu

Susan M. Knobel (31), Department of Molecular Physiology and Biophysics, Vanderbilt University, Nashville, Tennessee 37232

Gang Li (203), Department of Cell and Structural Biology, University of Illinois, Urbana-Champaign, Urbana, Illinois 61801

J. Lippincott-Schwartz (261), Cell Biology and Metabolism Branch, NICHD, NIH, Bethesda, Maryland 20982. E-mail: jlippen@helix.nih.gov

Robert Marsault (75), Department of Biomedical Sciences and CNR, Center for the Study of Biomembranes, University of Padua, 35121 Padua, Italy

C. D. Miller (261), Cell Biology and Metabolism Branch, NICHD, NIH, Bethesda, Maryland 20982

Marta Murgia (75), Department of Biomedical Sciences and CNR, Center for the Study of Biomembranes, University of Padua, 35121 Padua, Italy

George H. Patterson (31), Department of Molecular Physiology and Biophysics, Vanderbilt University, Nashville, Tennessee 37232

Ammasi Periasamy (293), Department of Biology, Advanced Cellular Imaging Facility, University of Virginia, Charlottesville, Virginia 22903. E-mail: ap3t@virginia.edu

Daniel W. Pierce (49), Howard Hughes Medical Institute and Department of Pharmacology, University of California, San Francisco, California 94143

Paolo Pinton (75), Department of Biomedical Sciences and CNR, Center for the Study of Biomembranes, University of Padua, 35121 Padua, Italy

David W. Piston (31), Department of Molecular Physiology and Biophysics, Vanderbilt University, Nashville, Tennessee 37232. E-mail: dave.piston@mcmail.vanderbilt.edu

Jeffrey D. Plautz (283), Department of Cell Biology, The Scripps Research Institute, La Jolla, California 92037

Tullio Pozzan (75), Department of Biomedical Sciences and CNR, Center for the Study of Biomembranes, University of Padua, 35121 Padua, Italy

F. G. Prendergast (1), Department of Pharmacology, Mayo Graduate School, Rochester, Minnesota 55905

J. F. Presley (261), Cell Biology and Metabolism Branch, NICHD, NIH, Bethesda, Maryland 20982

Rosario Rizzuto (75), Department of Biomedical Sciences and CNR, Center for the Study of Biomembranes, University of Padua, 35121 Padua, Italy

Carmen Robinett (203), Department of Cell and Structural Biology, University of Illinois, Urbana-Champaign, Urbana, Illinois 61801

Jean M. Sanger (239), Department of Cell and Developmental Biology, University of Pennsylvania School of Medicine, Philadelphia, Pennsylvania 19104

Joseph W. Sanger (239), Department of Cell and Developmental Biology, University of Pennsylvania School of Medicine, Philadelphia, Pennsylvania 19104. E-mail: sangerj@mail.med.upenn.edu

Kenneth E. Sawin (123), Imperial Cancer Research Fund, London WC2A 3PX, United Kingdom

Richard D. Shelby (183), Department of Cell Biology, The Scripps Research Institute, La Jolla, California 92037. E-mail: rshelby@scripps.edu

Pamela A. Silver (107), Department of Biological Chemistry and Molecular Pharmacology, Harvard Medical School and the Dana Farber Cancer Institute, Boston, Massachusetts 02115. E-mail: pamela_silver@dfci.harvard.edu

Tim Stearns (87), Department of Biological Sciences, Stanford University, Stanford, California 94305

Gail Sudlow (203), Department of Cell and Structural Biology, University of Illinois, Urbana-Champaign, Urbana, Illinois 61801

Kevin F. Sullivan (183), Department of Cell Biology, The Scripps Research Institute, La Jolla, California 92037. E-mail: ksullivan@scripps.edu

Richard Tuft (223), Biomedical Imaging Group, University of Massachusetts Medical Center, Worcester, Massachusetts 01605

Kenan K. Turnacioglu (239), Department of Cell and Developmental Biology, University of Pennsylvania School of Medicine, Philadelphia, Pennsylvania 19104

Ronald D. Vale (49), Howard Hughes Medical Institute and Department of Pharmacology, University of California, San Francisco, California 94143. E-mail: valee@phy.ucsf.ed

Leslie A. Woollenweber (19), Aurora Biosciences Corporation, La Jolla, California 92037

Aaron Young (223), Program in Molecular Medicine, University of Massachusetts Medical Center, Worcester, Massachusetts 01605

K. J. M. Zaal (261), Cell Biology and Metabolism Branch, NICHD, NIH, Bethesda, Maryland 20982

PREFACE

Fluorescence microscopy plays an essential role in cell biology by allowing visualization of specific molecular components of cells. The initial implementation through immunofluorescence techniques transformed the cytoplasm from a gelatinous bag of enzymes into the highly organized cytoarchitecture that we appreciate today. Methods for directly labeling proteins and successfully reintroducing them into cells added a dynamic element, allowing specific chemical reactions, such as polmerization/depolymerization, intracellular transport, and ligand binding, to be examined in living cells by time-lapse microscopy. The discovery of GFP, the green fluorescent protein of the jellyfish *Aequorea victoria,* has revolutionized experimental analysis of the dynamic molecular organization of cells by reducing the methods required for fluorescently labeling proteins to the techniques of molecular biology.

The beauty of GFP, aside from its brilliant green color, is that it has established a universal method for introducing a fluorescent tag into nearly any biological structure. GFP fusion protein technology is applicable to essentially any protein, beginning with a gene, to visualize it in living cells. At the cellular level, specific biological structures—membranes, spindles, chromosomes, organelles—can be imaged using nonperturbing GFP-based labels. Above the cellular level, GFP alone or as a fusion protein can be used to mark populations of cells for experimental manipulation, to visualize tissue organization, and to dissect cell lineage relationships. In short, GFP has proven to be a tremendously versatile biological tool that has worked its way into nearly all aspects of experimental biology in the scant four years since its introduction.

This volume of *Methods in Cell Biology* contains a set of protocols for work with GFP at several levels. Chapters 1–3 deal with GFP biophysics, GFP variants, and quantitative imaging of GFP and provide a foundation for thinking about GFP fluorescence and for building probes. The remainder of the volume, Chapters 4–16, provide specific examples of GFP expression in a variety of cell systems, including yeasts, plants, insects, and animal cells. Many chapters focus on development and use of specific biological assay systems and include details of the design and implementation of GFP-based probes. Thus, the rationale for fusion protein design and cloning methods, techniques for expression, microscopy and data collection, data analysis, and other methods are discussed in several contexts. Other chapters detail specific experimental techniques such as single-molecule fluorescence analysis, fluorescence resonance energy transfer, and flow cytometric applications of GFP. It is our hope that this volume will serve as a source of specific protocols for researchers wishing to implement the methods described here, as well as a guidebook that can help generate ideas on how to use GFP to solve new experimental problems.

Acknowledgments

The editors would like to thank the following organizations for their generous support of this volume:

Clontech Laboratories, Inc.
Hamamatsu Photonics
Chroma Technology Corporation
Olympus America, Inc., Precision Instrument Division

Kevin F. Sullivan
Steve A. Kay

CHAPTER 1

Biophysics of the Green Fluorescent Protein

F. G. Prendergast
Department of Pharmacology
Mayo Graduate School
Rochester, Minnesota 55905

I. Introduction

Green bioluminescence has been known as a feature of marine coelenterates for many years, but only in 1971 with the discovery by Morin and Hastings (1971a) did it become known that the green color derives from an intrinsically green fluorescent protein. Most of the early work to characterize this protein was then done in the laboratories of Frank Johnson and Osamu Shimomura (Shimomura and Johnson, 1975) and of John Blinks (Blinks *et al.,* 1978), who studied the bioluminescence of the jellyfish *Aequorea victoria* and characterized the tandem of proteins responsible for the light emission, namely aequorin and a green fluorescent protein. Subsequently, Ward and Cormier (1978) characterized another green bioluminescent system from the sea pansy, *Renilla reniformis.* Although both systems evinced green bioluminescence and both had proteins that were brilliantly green fluorescent evident on native polyacrylamide gels, distinct differences in the bioluminescent processes were discovered early on. In *Aequorea,* the bioluminescence *per se* derived from aequorin, a calcium-ion-activated photoprotein that required no other exogenous cofactors for light

METHODS IN CELL BIOLOGY, VOL. 58
Copyright © 1999 by Academic Press. All rights of reproduction in any form reserved.
0091-679X/99 $25.00

emission (for review see Morie *et al.*, 1976; Youvan and Larrick, 1996). Aequorin luminescence was distinctly blue, and because the organism itself clearly emitted green light only, there must be a process of energy transfer from aequorin to the green fluorescent protein, at least *in vivo.* Interestingly, all attempts to demonstrate such energy transfer *in vitro* have failed, as have attempts to show physical interaction between aequorin and the green fluorescent protein of *Aequorea.* The *Renilla* system was shown by the elegant work of Ward and Cormier (1978) to be different. Bioluminescence derived from the action of a protein, a luciferase, acting enzymatically upon a substrate, an imidazolopyrazinone, in the presence of molecular oxygen. The light emitted from luciferase was then transferred very efficiently via a Förster mechanism of resonance energy transfer (RET) to a green fluorescent protein, the latter bound tightly to the luciferase. Early on, the fluorescence emission spectra of both the *Renilla-* and *Aequorea-*derived green fluorescent proteins were shown to be essentially identical despite the differences in the bioluminescence processes of the parent organisms. Thus, although the abbreviation GFP generally refers, for reason of popularity and history, only to the protein from *Aequorea victoria,* on the basis of emission spectra obviously it can be applied equally accurately to the green fluorescent protein of *Renilla,* or for that matter to similarly fluorescent proteins from a host of other marine coelenterates such as *Mitrocomia* (*Halistaura*), *Phialidium,* and *Ptilosarcus.* However, there must be a caveat because there is not yet much evidence published showing the measure of similarity or difference among all of these green fluorescent proteins, especially with regard to their amino acid sequences or tertiary structures.

The widespread occurrence of similar, or at least analogous, bioluminescence processes in these marine organisms is peculiar. There are no simple explanations, but the fact that the "substrate" for the Ca^{2+}-activated photoproteins is chemically identical to the true enzymatic substrate for the enzyme luciferase (in *Renilla* and presumably also in closely related species such as *Ptilosarcus*) and the identity of the final green fluorescence emitted by these organisms is unlikely to be simply coincidental. Given the apparently similar fundamental biophysical processes underlying the fluorescence of GFP, comparisons of biophysical features among the various organisms will inevitably raise some interesting questions, particularly if there are substantive differences in amino acid sequences and molecular sizes for different natural green fluorescent proteins.

Shimomura (1979) was the first to show through a series of heroic experiments, that the chromophore in GFP was covalently adducted and to speculate that this chromophore probably arose from an unusual set of condensation and oxidation reactions. Subsequently, Ward *et al.* (1993) published the correct structure of the chromophore of BFP which largely corroborated Shimomura's earlier structure and his speculation regarding the origin of the chromophore. It was left to the molecular biology work of Prasher and co-workers (1992), however, to demonstrate unequivocally that the visible green fluorescence must be genetically encoded, the chromophore being derived from a -Ser-Tyr-Gly- sequence in the nascent protein. One has to suspect that those working earlier in the field had failed to follow through with actual expression of the protein from recombinant systems because they did not anticipate what was subsequently found. *A priori* it seemed inevitable that the jellyfish employed some unique enzyme to synthesize this unusual structure. Chalfie *et al.* (1994) did the seminal experiments to show that the expressed protein (apparently) spontaneously turned green fluorescent, an inference verified subsequently by Tsien and colleagues (Heim *et al.,* 1994) and (Kolb *et al.,* 1996). The evidence is now conclusive that the chromophore is formed autocatalytically by an unusual and unique series of reactions, all occurring intramolecularly, the only exogenous agent needed being molecular oxygen, which is required for a final oxidation step.

Many of these mechanistic details have been realized in the past few years only, and with the elucidation of the tertiary structure of *Aequorea* GFP the biophysical picture of this protein might be deemed largely complete. The principal purpose of this chapter is to show that this latter assumption is not justified; many interesting questions regarding the biophysics of the green fluorescent protein remain, questions which, if answered, may make this already remarkable useful cell and molecular biology tool even more valuable as a probe.

II. Protein Folding and the Generation of This Chromophore

The tertiary structure of wild-type (wt) GFP and of selected mutants bearing a limited number of amino acid substitutions (Ormö *et al.,* 1996; Brejc and Remington, 1997; Palm *et al.,* 1997; Yang *et al.,* 1996) have revealed a unique protein fold comprising primarily a tightly woven, 11-stranded β-barrel. There are several short and distorted helical segments on one end of the cylindrical structure and a central helical segment containing the imidazolidinone chromophore, the plane of the latter being oriented at an angle of about 60° to the symmetry axis of the cylinder (Ormö *et al.,* 1996; Brejc and Remington, 1997; Yang *et al.,* 1996). This unusual structure has been called a "β-lantern" by some, a "β-can" by ourselves. These seemingly trivial terms are actually usefully descriptive, particularly if we consider the can to be closed on either end, creating what Ormö *et al.* (1996) have termed a "completely encapsulated chromophore" and describing a structure overall that, *a priori,* should be physicochemically very stable either to the effects of denaturants or to the actions of proteases. Ward and co-workers, in a series of very careful and detailed studies conducted over many years, have shown that *Aequorea* GFP is indeed extremely stable to such conditions (Levine and Ward, 1982; Ward *et al.,* 1982; Ward and Bokman, 1982). Thermal denaturation results in essentially irreversible denaturation with a T_m of approximately (Ward and Bokman, 1982) but, as Ward and colleagues have shown, GFP will show reversible denaturation in response to pH extremes. Green fluorescence is completely lost upon protein unfolding and is regained only when the protein refolds. This extreme sensitivity of the green fluorescence to solvent exposure is clearly valuable as a signal of regained tertiary structure and hence a very useful indicator of tertiary structural perturbation of GFP itself. However, the unusual β-barrel motif of GFP should not be ignored inasmuch as it provides a useful, unique model of how proteins with this topology might fold.

Surprisingly little use has so far been made of GFP as a model for studying the denaturation and renaturation of a predominantly β-sheet protein. In fact, the vast majority of folding studies have been done on proteins that are primarily α-helical or of an α,β-motif. Such systems have substantially supported Kauzmann's suggestion (1954) that the principal driving force for protein folding is the thermodynamic advantage inherent in the sequestration of hydrophobic residues in the interior of the protein away from solvent (water), which occurs concomitantly with the equally advantageous thermodynamic circumstances of having *hydrophilic* residues exposed to water. In recent years, evidence has accumulated for the existence of folding "intermediates" in many, maybe most, of the proteins studied to date. A "folding intermediate" is in effect a partially folded structure intermediate between the presumably fully unfolded structure and the fully folded structure, the latter defined as being replete with tertiary interactions and solvent exclusion from the protein core (Kim and Baldwin, 1990; Sosnick *et al.,* 1994; Clark *et al.,* 1997). As Gierasch and colleagues have recently discussed (Clark *et al.,* 1997), the folding of a β-sheet protein poses a different problem, requiring as it does "concurrent formation of secondary structure and native-like topology, since distant segments of the chain must approach one another to form strand–strand contacts" (Clark *et al.,* 1997). These authors also highlighted some of the key remaining questions regarding the folding of β-sheet proteins, to wit: "when does the native topology develop in the folding pathway; do β-sheet proteins fold through 'molten globule' intermediates and, if so, what is their structural nature; and when do the specific interactions that stabilize the unique native state—such as hydrogen bonding—develop in the course of folding?" (Clark *et al.,* 1997). Although these questions are obviously generally important for the folding of any and all β-sheet proteins, they bear particular significance for the biophysics of GFP, because it is now apparent that the development of green fluorescence is critically dependent on the fidelity of protein folding.

Tsien and co-workers (Heim *et al.,* 1994) have broached a plausible two-step chemical mechanism for the intramolecular, autocatalytic formation of the mature chromophore. This mechanism is depicted in Fig. 1. The evidence is good and the chemical logic unchallengeable that the cyclization reaction occurs first. Although mechanistically the reaction shown in Fig. 1 is reasonable, one has to speculate the need for stringent steric relationships to facilitate attack of a poorly nucleophilic amido nitrogen (of Gly-67) on the equally poorly electrophilic peptidic carbonyl group of the Ser-65 moiety. Since either Ser-65 or Tyr-66 can be replaced with other amino acids, even amino acids bearing bulky side chains, with retention of the ability to cyclize, this implies that side chain (protein matrix) packing around the site of the reaction is not critical. For application of the

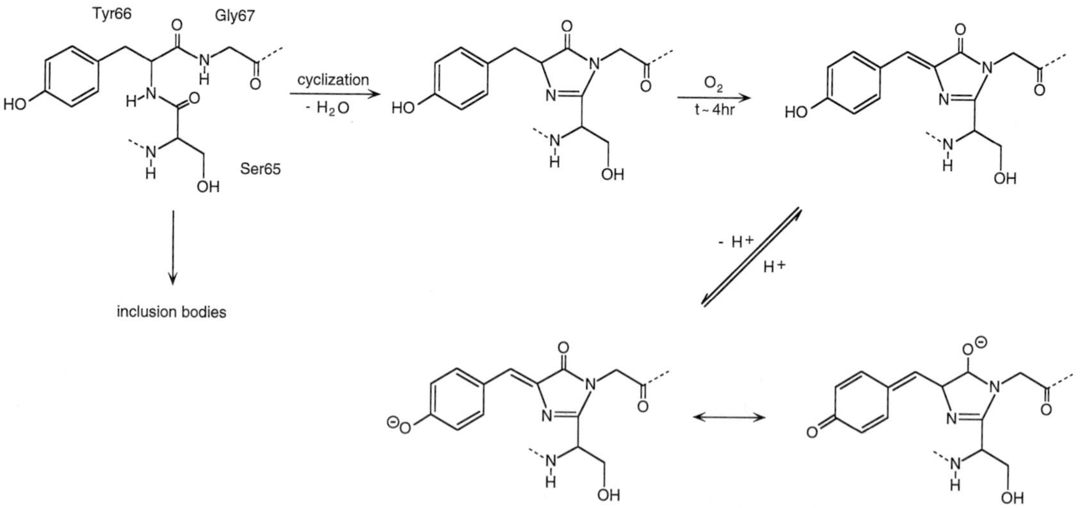

Fig. 1 Proposed scheme for the autocatalytic formation of the *p*-hydroxybenzylidene-imidozolidinone chromophore of *Aequorea* GFP taken from Heim *et al.* (1994).

reactants, however, the secondary fold *is* critical and the available evidence indicates strongly also that the nascent protein must complete at least an initial tertiary structure *before* cyclization can occur. Thereafter, the changes in configuration attending the formation of the imidazolidinone ring almost certainly require further secondary or tertiary structural accommodation, but there are as yet no experimental data to substantiate this speculation and such experimental evidence will be hard to acquire. These arguments gain credence from the fact that there are other proteins with the -Ser-Tyr-Gly- sequence that are not green fluorescent; *a priori* there is nothing inherently special about this sequence, which by itself can confer no unique conformation sufficient to promote the cyclization (condensation) reaction.

There are additional structural considerations. For example, the cyclization reaction portrayed in Fig. 1 would not proceed efficiently in the presence of water. As Remington and co-workers have discussed (Ormö *et al.,* 1996; Brejc and Remington, 1997), the X-ray structure of native GFP shows that the chromophore is inaccessible to bulk water. Admittedly, these assessments are based on the mature protein, but, as was noted earlier, it is improbable that the nascent, folded protein can have a substantially different structure. The water formed as a consequence of the condensation reaction either would then have to be expelled or conceivably could become one of the "structural" water molecules accommodated in the adjacent large cavity described by Ormö *et al.* (1996). Further, it is not known whether formation of the imidazolidinone requires catalytic assistance of as yet unidentified amino acid side chains as potential activating aides either for the amido nucleophile or for the (assumed) electrophilic carbonyl.

The conclusion may thus be drawn reasonably that the conformational circumstances promoting the (putative) first reaction in the generation of the GFP chromophore are at least unusual and probably unique. Unfortunately, spectroscopic probing of this reaction will be difficult because there is no simple or unique spectroscopic signal attending the reaction. The carbonyl functionality of the imidazolidinone would have a good IR signature, but that signal would almost certainly be buried by the much larger absorbance bands from the many other (peptidic) carbonyl moieties. However, in principle, NMR experiments could shed light on the process. In reality, it might be impossible to isolate the protein *prior* to formation of the imidazolinone or to slow the reaction sufficiently to allow spectroscopic analysis, and in any event the protein is close to the upper molecular weight limit for NMR analysis using the techniques in common usage today. Molecular biology manipulation of the amino acid sequence could provide some interesting approaches to the problem, especially if mutants that fold "normally" but are unable to promote the cyclization could be found. Such mutants could then be gainfully studied at leisure by use of crystallography and a variety of spectroscopic techniques.

Green fluorescent protein appears to be remarkably impervious to molecular oxygen. The $-C=N-$ double bond of the imidazolidinone moiety formed as a result of the condensation reaction would promote spontaneous dehydrogenation (oxidation) of the C_2–C_3 bond of the adjacent Tyr-66 moiety, a reaction that would be driven by the stability gained from the resulting conjugation. As Heim *et al.* (1994) have pointed out, imidazolidin-5-ones are known to "undergo autoxidative formation of double bonds at the 4 position (Kjaer, 1953; Kidwai and Devasia, 1962)." Molecular oxygen is known to be the requisite oxidant, and the reaction is known to be slow. (Because the product is green fluorescent, the kinetics of oxidation is easily followed.) This reaction with molecular oxygen appears to be the rate-limiting step for the appearance of green fluorescence both *in vitro* and *in vivo,* and the reaction *in vitro* is often incomplete as evidenced by the existence of unoxidized GFP molecules in purified preparations of the recombinant protein. Admittedly, the kinetics of oxidation as reflected in the appearance of green fluorescence could reflect the intrinsic slowness of the chemical reaction, not necessarily the rate of diffusion into the protein matrix. However, oxygen quenches the fluorescence of *Aequorea* GFP very poorly, the apparent bimolecular diffusion constant (k_q) for oxygen quenching of the green fluorescence being $<10^8\ s^{-1}$. Lakowicz and Weber (1973a,b) showed more than two decades ago that oxygen could quench the fluorescence of Trp residues in proteins, even for Trp residues whose side chains were apparently deeply buried in the protein matrix, seemingly inaccessible to water, often with k_q values $>10^9\ s^{-1}$ for oxygen quenching. The most straightforward interpretation of those data was the existence of nanosecond fluctuations in the protein matrix structure of sufficient amplitude to allow access of oxygen (probe radius 1 · 4 Å) directly to the fluorophore, whereupon there was collisional quenching of the fluorescence. There are complications to this interpretation occasioned by the paramagnetism of ground-state oxygen, which raises the possibility that actual *collision* between oxygen and the fluorophore may not be necessary for the quenching of Trp fluorescence in proteins, especially if oxygen adsorbs weakly to the protein matrix. That problem notwithstanding, *in general* the inference of Lakowicz and Weber regarding fluctuations in protein structure seems reasonable, whence our conclusion of the relative inaccessibility of the chromophore in GFP to molecular oxygen. The most straightforward inference is that this protein is conformationally rigid, whence the poor penetrance of molecular oxygen. It should be noted that in an extensive study of the ability of oxygen to quench fluorescence we ourselves have not found a single fluorophore whose fluorescence is insensitive to oxygen's effects (Prendergast, P. G. unpublished data).

Despite the logic of these arguments, they should not be taken too far. For example, the double bond of the dehydrotyrosine moiety *is* susceptible to reasonably facile reduction by agents such as dithionite, a relatively weak reductant, with loss of green fluorescence. The structural pathway for access of the reducing species to the reactive double bond can only be guessed at but is unlikely to involve any sort of global conformational changes in the protein. The reduced protein is also quite easily reoxidized, a finding that has been exploited for studying the kinetics of oxidation *in vitro*. It is also known that the kinetics of the "greening" process can also be accelerated, at least *in vivo,* by selective amino acid substitutions, which could alter "oxygen accessibility" to the incipient fluorophore and hence be the agent of accelerated oxidation. Palm *et al.* (1997), however, have offered a different explanation for how some mutations might affect the yield of soluble mature GFP, for example, for the F64L and V163A mutations. The crystal structure of wt GFP shows no close contacts for the side chains of either F64 or V163, which led Palm *et al.* to suggest that their replacement with sterically less bulky amino acids is advantageous, not so much for the mature protein but by promoting more rapid and efficient folding of the nascent protein or for crucial, but unidentified, conformational steps in the formation of the chromophore. Palm *et al.* suggest that the overall protein fold constrains residues 57–71 to facilitate fitting this segment into the β-barrel and to ensure the strained conformation of the -Ser-Tyr-Gly- sequence necessary for apposition of the reactive moieties needed for the cyclization reaction. Their argument goes further, namely, that the increased free volume afforded by the F64L and V163A mutations eases the closing of the β-barrel during protein folding. These speculations are credible and offer a very different model for how the kinetics of chromophore formation may be controlled substantially by protein folding and not be dominated by accessibility to oxygen. The hypothesis also suggests that whereas some bulky side chain substitutions are apparently readily accommodated (e.g., Y66W), others may not be.

All of these considerations suggest a host of experiments that might yield fruitful information on the biophysical properties of GFP. Despite GFP's relatively high molecular weight and what was said earlier about the use of multidimensional NMR, exciting new techniques (Ottiger *et al.,* 1997; Pervushin *et al.,*

1997) offer high promise for the easier application of NMR to solution structure determination of proteins even larger than GFP. Also, there is no substantial barrier to the use of NMR relaxation measurements for studying the dynamics of GFP. Moreover, from the host of recombinant mutants that have already been made, the protein is obviously conformationally very tolerant of sequence changes, suggesting a variety of experiments that could be done by a combination of selective amino acid substitutions and selective isotopic enrichment. Molecular dynamics simulations would also be useful, although the high molecular weight would make simulations computationally expensive. However, we could gain some insight into the conformational fluctuations occurring in the vicinity of the chromophore, particularly as these might relate to the dynamics of water into and out of the protein and possible pathways of oxygen access. Molecular mechanics and other simulations such as minimum perturbation mapping (Haydock, 19) Branchini *et al.*, (1997) might also be useful to probe likely conformational states of amino acid side chains believed to play roles in the proton transfers posited to be key to the fluorescence properties of wt GFP and GFP mutants.

One final comment: As noted earlier, several coelenterates emit essentially identical fluorescence from GFPs. We know that the chromophores of *Renilla* and *Aequorea* GFP are structurally identical. The temptation is to speculate that all green fluorescent proteins will be at least highly homologous in sequence given the structural stringencies required, first for autocatalytic production of the fluorophore and second for the fluorescence emission of the proteins. Yet, there is no published evidence to substantiate such a speculation, and our own admittedly incomplete work on the sequence of the GFP of a close relative of *Renilla,* namely *Ptilosarcus* GFP, suggests an amino acid sequence very different from that of *Aequorea* GFP (Homer, M., and Prendergast, F. G., unpublished data). The implications of such a result are clear and underscore the need for determination of the amino acid sequences of several GFPs.

III. The Biophysics of the Fluorescence of GFP

During the past few years, especially since 1994, a great deal has been learned and much has been written regarding the physicochemical mechanisms underlying the fluorescence of GFP. This veritable explosion of information has been catalyzed by the availability of the tertiary structure. The latter has been particularly valuable for those interested in structure-driven site-directed mutagenesis in their quest primarily for mutant proteins with altered excitation or emission spectra, or to find proteins that develop their green fluorescence more efficiently *in vivo* than does the wild type. In this brief review, some duplication of information with other articles in this volume is unavoidable. The principal objective here, however, is a critical review of what we believe we understand of the photophysics of the *p*-hydroxybenzylideneimidazolidinone chromophore of GFP. The intense green fluorescence of this imidazolidinone in the folded green fluorescent protein stands in stark contrast to the virtual absence of fluorescence in the denatured protein or for the isolated chromophore dissolved in any and every solvent so far tested. The loss of fluorescence for solutions of the chromophore occurs simultaneously with marked changes in its absorption spectrum, the latter being also strongly dependent on pH. Under acidic conditions the absorption wavelength maximum is approximately 384 nm, whereas upon base denaturation ($pH > 12$) the absorption maximum is at 447 nm. These absorption changes, ably described by Ward and Bokman (1982), are now known to presage the prominent role of proton equilibria in the photophysics of GFP. They can be understood generally from examination of the structure of the chromophore.

There are clearly two principal sites of potential pH sensitivity, namely the phenolic hydroxyl and the imino nitrogen of the imidazolidinone ring. Under the basic conditions described earlier an anionic form comprising a ground-state phenolate and an unprotonated nitrogen would exist. Under acidic conditions ($pH < 4$) the aromatic (phenyl) side chain would exist as the phenol and the iminonitrogen might be cationic. The tautomeric equilibria will depend, therefore, on both pH and the solvent employed. Quantum mechanical evaluation of transition energies and hence of the likely pH-induced absorption spectral shifts should be tractable, particularly for the (simulated) chromophore dissolved in polar solvents such as water. Calculations of this sort would be especially valuable for investigating how torsional motions of one ring about the other (i.e., motions that distort coplanarity might affect the absorption spectra and fluorescence emission and also what effect there might be from the simulated "solvent" environment created by the protein matrix. These are important questions because it is not evident, *a priori,* why the fluorescence quantum yield of the chromophore in solution is so low. The most tempting rationalization is rapid deactivation of the excited state by loss of planarity consequent on ring mobility, motion that is restricted in the folded protein by steric factors (packing) and (possibly) by electrostatic constraints. It would be interesting to study the spectral properties of synthetic analogs of the chromophore dissolved in highly viscous media or as inclusions in cyclodextrins to determine if fluorescence can in fact be evoked in solution by constraining ring mobility.

The fluorescence of native GFP understandably has been the object of intense interest. From a strictly spectroscopic perspective, two recent papers stand out, namely those of Chattoraj *et al.* (1996) and Lossau *et al.* (1996). These authors employed femtosecond and picosecond transient absorption and time-resolved fluorescence methodologies to probe the photophysical processes of wt GFP and mutants thereof. One of the principal objectives was to explain the appearance of the two absorption maxima of wt GFP at 398 and 478 nm, respectively, compared to mutants showing one or other of these absorption maxima but not both. As Chatteraj *et al.* (1996) argue, these two excitation bands could derive from transitions between a single ground state and first and second excited singlet states of a single chromophore or could indicate the existence of separate species in the ground state capable of interconverting in both ground and excited states.

Recalling the tautomeric possibilities discussed earlier, one may readily imagine the plausibility of the second possibility raised by Chattoraj *et al.* (1996).

Förster discovered the photolylic reaction depicted by the scheme

$$\mathrm{RH} \xrightarrow{h\nu} \mathrm{RH}^* \longrightarrow \mathrm{R}^{-*} + \mathrm{H}^+,$$

where RH represents the protonated ground state species, RH^* is its excited-state counterpart, and R^{-*} is the excited-state anion created by proton transfer in the excited state. As Lossau *et al.* (1996) have discussed, photolytic protein transfer is a classic feature of phenol photophysics. Phenolic hydroxyls are acidic because of the participation of one of the lone pairs on the oxygen (in a $2p_Z$ orbital) with the aromatic π-system. (This in part also explains the chemical reactivity of the *ortho* and *para* positions of the phenyl ring and the proclivity of phenols to be transformed into quinonoid structures.) Upon photoexcitation the π–π^* transition originating in the phenyl ring causes substantial loss of electron density on the phenolic oxygen, which translates into increased acidity of the phenol in the excited state compared to the ground state. If there are good proton acceptors in the surrounding solvent or in the protein matrix adjacent to the phenol, then formal proton transfer from the phenol is facilitated to create a cationic acceptor and the anionic phenolate. Fluorescence emission spectrum may thus occur from the protonated species, from the phenolate anion, or from both, depending on the kinetics of interconversion between the two forms and also on whether the phenolate can be stabilized long enough for emission to occur.

From first principles, therefore, there is every reason to believe that such processes would be likely for the GFP chromophore, providing that the protein matrix affords facile intramolecular proton transfer. The experiments of Chattoraj *et al.* (1996) and later of Lossau *et al.* (1996) show almost unequivocally that the excitation spectra of wt *Aequorea* can indeed be rationalized in terms of a ground-state phenol form absorbing maximally at 398 nm and a phenolate form with maximal absorbance at 377 nm. These authors have summarized the evidence particularly well. Their analysis is given next.

They noted first the sensitivity of the related intensities of the visible absorption maxima of GFP to a host of physicochemical conditions, including pH, temperature, ionic strength, protein concentration, and exposure to light. Changes in any of these properties evoke continuous shifts in the absorption spectra, with distinct isobestic points indicating at least two interconverting ground-state forms. As Chattoraj *et al.* (1996) point out, the ground-state energy barrier to interconversion of the two forms must be quite high given the slowness of reversion of the absorption spectrum to the usual spectrum subsequent to UV-photo excitation.

Picosecond spectroscopic measurements of the rise and decay of GFP fluorescence showed clearly that the electronic state created by excitation of the higher energy band at 398 nm undergoes nonradiative decay to a second electronic state whose emission is at lower energy and is subject to a substantial deuterium isotope effect. The final scheme proposed by these authors from studies at room temperature and at 77° K suggested a rather complicated set of unidentified intermediate forms to explain spectroscopic data obtained on the protein in a 77° K glass. The scheme proposed to explain the existence of these intermediates (Fig. 7 of Chattoraj *et al.*, 1996) is plausible, but experimental validation will be hard to come by, and one also has to be concerned that the data recovered from up-conversion spectroscopy could be tainted by the effects of the unavoidably high excitation powers used. Chattoraj *et al.* also proposed that excited-state solvation is a principal determinant of at least some of the kinetics observed, the "solvent" in this instance being defined as comprising elements of the protein matrix. These determinations of the kinetics of light emission in GFP corroborate the hypothesis of the key role of proton transfer in the fluorescence process.

The tertiary structure of GFP combined with the detailed fluorescence spectroscopic studies of Lossau *et al.* (1996) subsequently provided added insight into the role of both ground-state and excited-state proton transfer—abbreviated ESPT by Lossau *et al.* (1996)—on the fluorescence process. Their conclusions are summarized well by the Förster cycle depiction (Fig. 12 of Lossau *et al.*, 1996), which is duplicated in Fig. 2. Fundamentally, excitation into the high-

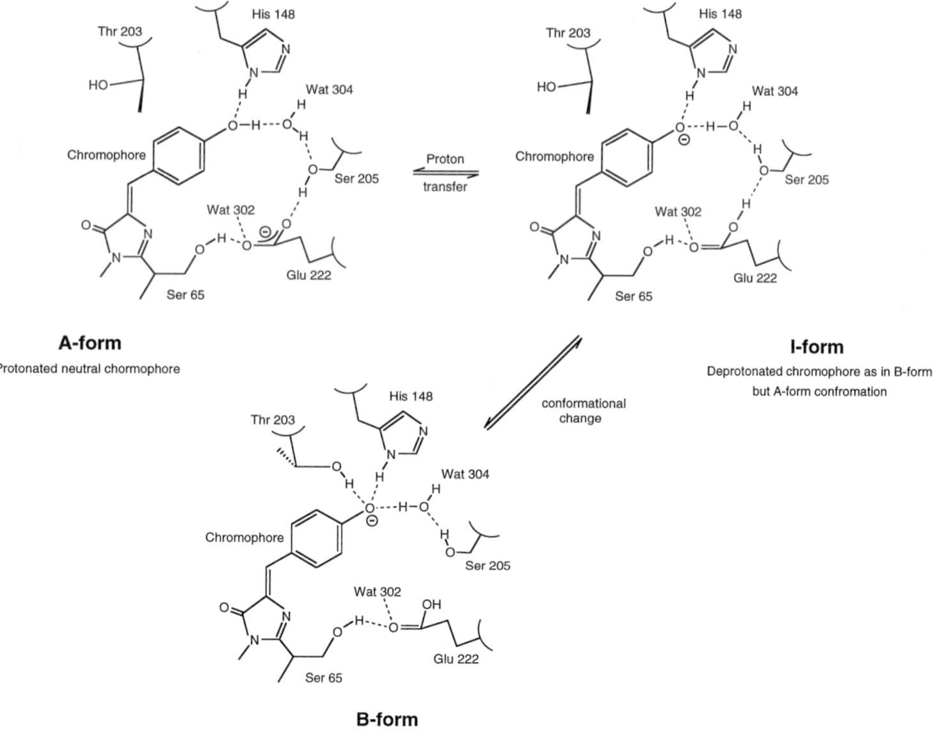

Fig. 2 Proposed protonation states and partial hydrogen bonding patterns for the chromophore in GFP—after Palm *et al.* (1997).

energy (398 nm) band promotes the chromophore into an excited state with initial retention of the phenolic proton (RH). The extensive H-bond network inferred from crystallographic data (Ormö *et al.,* 1996; Brejc and Remington, 1997), and particularly hydrogen bonding of the hydroxyl proton either to the nitrogen of His-148 or via bridges to Glu-222, facilitates proton transfer—within 6 ps of excitation—to yield the phenolate (anionic) excited state. This latter state emits light at 510 nm with a fluorescence lifetime of 3.3 ns, the typical fluorescence lifetime measured for GFP in crystals (Palm *et al.,* 1997) or in simple aqueous solution. Examination of the extensive hydrogen-bonding network in the vicinity of the chromophore shows that it comprises both elements of the protein matrix and water molecules embedded inside the protein. Given the inevitable existence of protein dynamics occurring on the same time scale as the initial proton-transfer events, the proton initially transferred from the phenol need not reside on any particular residue but could be moved around within the network, the excited state union being stabilized by a reorganization of the H-bonding pattern within the protein matrix and including water molecules. This is schematically depicted in Fig. 3, taken from Palm *et al.* (1997).

From such a scheme it should be apparent that mutation of one or more of the potential donors or acceptors could have a substantial effect, depending on how overall proton exchange within the network is affected, and also could alter the stability of the putative anionic form of the chromophore.

The complexity of the presumed network is striking, so much so that precise definition of the path taken by a proton abstracted from the excited state of the chromophore is not really tenable. The arguments of Brejc *et al.* (1997) for a primary role for the carboxylate of Glu-222 in the photophysics of GFP are intrinsically reasonable, but others, notably Lossau *et al.* (1996) and Palm *et al.* (1996) have equally strong arguments in favor of a key role for His-148, nor does it seem, *a priori,* that the question can be answered trivially by site-directed mutations given the seeming adaptability (malleability) of the network. The large deuterium isotope effect reported by Chattoraj *et al.* (1996) suggests strongly that there are multiple proton transfer steps, but the simple reality is that there

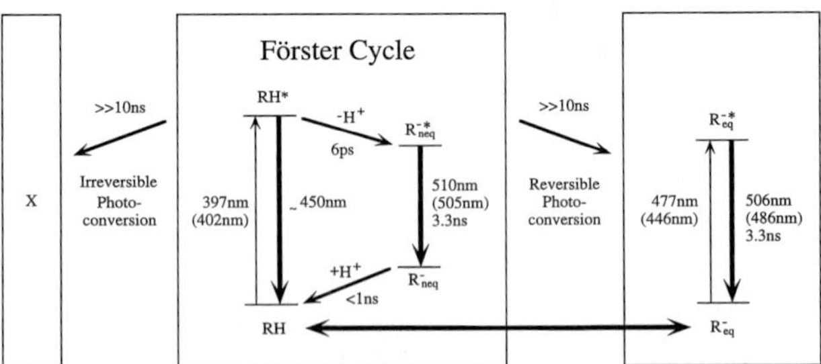

Fig. 3 Schematic depiction of Förster cycle proposed by Lossau *et al.* to rationalize GFP photophysics inferred from time-resolved fluorescence intensity decay data.

is as yet insufficient evidence to support the precise patterns of proton movement in the network.

These arguments are not meant to dissuade further study of the effects of mutations on both the excitation and emission spectra of GFP. Quite the contrary. Palm *et al.* (1996) have suggested several mutations that could yield considerable insight into the motions of protons in the pocket and regulation of the excited-state behavior of the GFP chromophore. For example, His-148N and E222Q mutations would retain H-bonding capability but negate proton transfer, whereas S205A or S205V mutations, they suggest, should prevent transformation of the protonated (neutral) chromophore to the deprotonated anionic chromophore. All of these possibilities focus only on the protonation of the phenolic moiety and generally ignore either the possible protonation of the imino nitrogen or the possible role of Gln-94 and Arg-96 in stabilizing the enolate (resonance) form of the chromophore (Fig. 4). It is important to realize that conjugation extends from the *p*-hydroxybenzylidene moiety to either the imino nitrogen or to the carbonyl group of the imidozolidinone rings and hence that it is trivially possible to delocalize charge from the phenolate to the carbonyl. This possibility was raised by Heim *et al.* (1994) and by Lossau *et al.* (1996) but was not discussed to any extent. This is a surprising omission because the enolate (quinonoid) form would have red-shifted absorption and yellow-green fluorescence, and could have its negative charge substantially stabilized by the cationic Arg-96 aided, through hydrogen bonding, by Glu-94. The effects of site-specific mutations involving both of these residues would therefore be interesting. We should also note here that a quinonoid structure is feasible for the Y66H mutant providing that there was either formal proton transfer from the histidyl moiety (e.g., to Glu-222) *or* some way to stabilize a cationic resonance form of the imidazole side chain, again possibly employing the Glu-222 carboxylate.

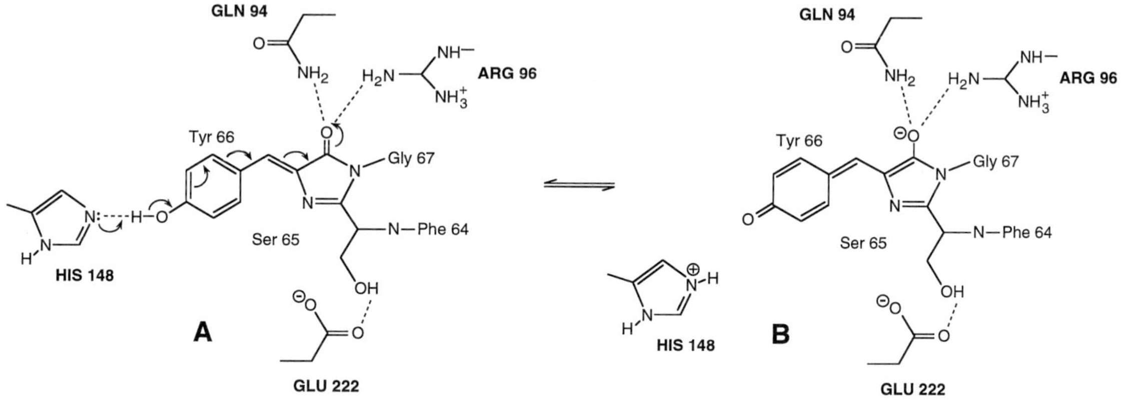

Fig. 4 Simplified schematic depicting possible mechanism for generation and stabilization of enolate form of GFP chromophore.

The arguments regarding the potential malleability of the H-bond network in the *kinetics* of proton transfer and our proposed indeterminacy of precise proton transfer paths notwithstanding, it *is* clear that several amino acid substitutions do profoundly affect the *stability* of the emitting state and could also ensure a "permanent" ground-state phenolate and thereby a shift of the excitation (ground-state or absorption) maximum to lower energy such as in the red-shifted mutants of Youvan and colleagues (Delagrave *et al.,* 1995). The results of Delagrave et al. (1995), for example, show that mutations of amino acid residues other than proton active species can also effect substantial red shifts in the absorption spectrum of GFP. One of the most striking is the mutant denoted RSGFP4 (Delagrave *et al.,* 1995) which contains the sequence for residues 64–69 MGYGVL, compared to FSYGVQ for *Aequorea* GFP, which has an absorption maximum of 498 nm, strikingly similar to the absorption maximum for *Renilla* GFP. However, not only is the absorption maximum red shifted, it now shows essentially mirror image symmetry with the fluorescence emission, an unusual finding for a protein-bound fluorophore. Such symmetry indicates that the nuclear configurations of the ground and excited state are identical, and implies strongly not only a well-stabilized ground-state form but also that the chromophore is held with both rings rigidly coplanar and is incapable of changing its physical form in its excited state. The most probable rationalization is that there is more dense packing around the chromophore in RSGFPU occasioned by the amino acid substitutions and a stabilized, static hydrogen-bonding network supporting the anionic form. Surprisingly, to this date the tertiary structure of this mutant has not been solved, nor unfortunately has either the sequence or the crystal structure of *Renilla* GFP been determined. Irrespective of the final sequence identity between FSGFP4 and *Renilla* GFP, one may predict reasonably confidently a strong similarity between the environments of the chromophores in the two proteins.

Finally, although we have obviously come a long way in our understanding of the basic processes underlying GFP fluorescence, there clearly are many unanswered questions and hence many opportunities for further study. Some of these have been either alluded to or described in the earlier discussions.

There is no easily discernible experimental approach to determining precisely how deuteriation affects GFP photophysics and what role may be played by the dynamics of protein matrix-embedded water molecules. It has already been noted above that of the water molecules evident in the X-ray crystal structure, several seem to be playing "structural" roles and hence will exchange very slowly; others form parts of the H-bond network but it is impossible *a priori* to predict local residency times or exchange rates either within the chromophore "pocket" or with bulk water. Absent such information, the role of water in any "solvent relaxation" is highly speculative, albeit a rational notion. Unfortunately, speculation based on water dynamics in other proteins is probably of little value. Water–protein interactions are best evaluated for each individual protein, generalizations being most tenable only for water molecules demonstrably capable of exchanging with bulk water, whence similar exchange rates can be reasonably predicted from one protein to the next. Computer simulations of water dynamics might provide useful insight, and one also can conceive of interesting NMR experiments that might reveal the accessibility of water to and assess semiquantitatively the dynamics of water in the chromophore binding region of GFP. Molecular dynamics simulations would also be interesting to probe for potential energy barriers to proton migration and to evaluate the validity of the speculation that the slow recovery of the 398-nm absorption band after light excitation is indeed due to a large energy barrier hindering rotation of the side chain of Thr-203. In our view, the tertiary structure by itself does not show a compelling reason why rotation of Thr-203 should present a barrier large enough to justify the very slow conversion shown by Chattoraj *et al.* (1996). It would obviously be useful to measure the height of this barrier. This could be achieved at least to first approximation by determining the temperature dependence of the spectral change.

Lossau *et al.* (1996) have discussed the nonexponential decay of the excited protonated state (RH^* in their notation). For the wt protein this can be reasonably explained by the high probability of variable proton transfer paths and rates mediated by the extensive hydrogen-bonding network, but this explanation does not work for the mutants with blue-shifted excitation. The view of Lossau *et al.* (1996) is that this phenomenon may reflect dipolar relaxation processes, but there is as yet little experimental evidence to support this suggestion.

IV. Resonance Energy Transfer Involving GFP

The most dramatic display of energy transfer involving GFP is *in vivo* green bioluminescence. For *Renilla* and bioluminescence, the careful experiments of Ward and Cormier (1978, showed unequivocally some form of resonance energy transfer. The high efficiency of the process in *Renilla* appears to be attributed, at least in part, to the high-affinity complex formed between the bioluminescing luciferase and the GFP. What is not clear is whether the energy transfer occurs via a Förster-type mechanism or through a Dexter-type process from the chemically excited luciferin to the GFP. By analogy, energy transfer *in vivo* in *Aequorea* is also resonance determined. It is not due to trivial transfer (i.e., not due to simple reabsorption of the blue chemiluminescence emitted by aequorin followed by reemission as green fluorescence). For the aequorin–*Aequorea* GFP tandem the overlap integral defined by the blue bioluminescence and the two absorption bands of the GFP is optimal. Interestingly, this overlap is less ideal in the *Renilla* bioluminescent tandem because of the relatively red-shifted excitation spectrum of *Renilla* GFP. Energy transfer could be made very efficient in both systems by perfect alignment of the respective energy donor and acceptor transition moments and by minimizing the separation distance between donor and acceptor chromophores. It would be interesting to construct an aequorin–*Aequorea* GFP fusion protein to determine whether bioluminescence energy transfer can be detected for this system *in vitro,* although there is no reason, *a priori,* to believe that such transfer would not occur given the *in vivo* result.

Aequorea GFP has a single tryptophan residue but displays no tryptophan fluorescence. The tertiary structure reveals that both chromophores are separated by about 15 Å and are essentially coplanar. Given the relatively tight packing around both the indole and the GFP fluorophore, hence the essential immobility of the two moieties, the orientation factor is optimized, and with absorption bands at 398 and 477 nm for wt *Aequorea* GFP, inevitably so is the overlap integral. In other words, conditions are ideal for fluorescence resonance energy transfer from the Trp to the green emitting chromophore, hence the total quenching of intrinsic Trp fluorescence in the native GFP. This is a very useful spectroscopic signature because appearance of Trp fluorescence must then imply either structural changes in the protein or substantial changes in the excitation absorption spectrum of the acceptor.

V. Summary

It is almost certainly a truism that interpretation of the fluorescence of a protein matrix-embedded chromophore in terms of the physicochemical character of its environment requires that the tertiary structure of the protein be known to high resolution. This reality derives from the complexity of the photophysics of most fluorescent molecules, complexity that reveals the imperfections of available theory. The accuracy of these dicta is highlighted by the biophysical properties of the green fluorescent protein now being so elegantly elucidated from the application of X-ray crystallography, ultrafast optical spectroscopy, and site-specific mutagenesis. Despite the mass of recent data, however, the physicochemical basis of the green fluorescence cannot be regarded as having been fully defined, nor has the role of protein folding in chromophore formation been solved. In addition, GFP consistently yields surprises typified by the recent experiments of Vanden Bout *et al.* (1997) on the fluorescence of single molecules mutant (T203F and T2034) GFPs, which showed unique reversible photobleaching and were whimsically termed "blinking molecules" by Moerner (1977). Given the apparent malleability of the GFP sequence and the sensitivity of the chromophore's photophysics to a broad spectrum of physicochemical factors, it is inevitable that additional useful and intriguing biophysical properties will emerge from the study of other mutants. Although on the surface it may seem mundane, determination of the amino acid sequence and tertiary structures of the GFPs from other coelenterates is quite likely to provide very useful insights into the biophysical bases of both protein folding and the green fluorescence *per se.* Finally, a broader set of spectroscopic techniques need to be applied to the study of GFPs, and future fluorescence examination should include measurements of transient absorption and fluorescence emission anisotropy decays.

Acknowledgments

I would like to acknowledge the efficient assistance of Lori VanderPlas, Jill Kappers, and Peter Callahan in the preparation of this manuscript. Supported by GM348347.

References

Blinks, J. R., Prendergast, F. G., and Allen, D. G. (1978). Photoproteins as biological calcium indicators. *Pharmacol. Rev.* **28,** 1–93.

Branchini, B. R., Lusins, J. O., and Zimmer, M. (1997). A molecular mechanics and database analysis of the structural preorganization and activation of the chromophore-containing hexapeptide fragment in green fluorescent protein. *J. Biomolecular Structure and Dynamics* **14,** 441–448.

Brejc, K., Sixma, T. K., Kitts, P. A., Kain, S. R., Tsien, R. Y., Ormö, M., and Remington, S. J. (1997). Structural basis for dual excitation and photoisomerization of the *Aequorea victoria* green fluorescent protein. *Proc. Natl. Acad. Sci. U. S. A.* **94,** 2306–2311.

Chalfie, M., Tu, Y., Euskirchen, G., Ward, W. W., and Prasher, D. C. (1994). Green fluorescent protein as a marker for gene expression. *Science* **263,** 802–804.

Chattoraj, M., King, B. A., Bublitz, G., and Boxer, S. G. (1996). Ultra-fast excited state dynamics in green fluorescent protein: multiple states and proton transfer. *Proc. Natl. Acad. Sci. U. S. A.* **93,** 8362–8367.

Clark, P. L., Liu, Z. P., Rizo, J., and Gierasch, L. M. (1997). Cavity formation before stable hydrogen bonding in the folding of a β-clam protein. *Nat. Struct. Biol.* **4,** 883–886.

Cody, C. W., Prasher, D. C., Westler, W. M., Prendergast, F. G., and Ward, W. W. (1993). Chemical structure of the hexapeptide chromophore of the *Aequorea* green fluorescent protein. *Biochemistry* **32,** 1212–1218.

Delagrave, S., Hautin, R. E., Silva, C. M., Yang, M. M., and Youvan, D. C. (1995). Red-shifted excitation mutants of the green fluorescent protein. *Biotechnology* **13,** 151–154.

Haydock, C. (1993). Protein side chain rotational isomerization: A minimum perturbation mapping study. *J. Chem. Phys.* **98,** 8199–8214.

Heim, R., Prasher, D. C., and Tsien, R. Y. (1994). Wavelength mutations and post-translational autoxidation of green fluorescent protein. *Proc. Natl. Acad. Sci. U. S. A.* **91,** 12501–12504.

Kauzmann, W. (1954). In "The Mechanism of Enzyme Action" (McElroy, W. D., and Glass, B., Eds.) pp. 70–120. Johns Hopkins Press, Baltimore, Maryland.

Kidwai, A. R., and Devasia, G. M. (1962). A new method for the synthesis of amino acids. Synthesis of amino acids and their derivatives through 2,4-disubstituted 2-imidazolin-5-ones. *J. Org. Chem.* **27,** 4527–4531.

Kim, P. S., and Baldwin, R. L. (1990). Intermediates in the folding reactions of small proteins. *Annu. Rev. Biochem.* **59,** 631–660.

Kjaer, A. (1953). Reactions between amino esters and α-amino acid esters Part III: 2-phenyl-5(4)Imidazolone and its reactions.

Kolb, V. A., Makeyev, E. V., Ward, W. W., and Spirin, A. S. (1996). Synthesis and maturation of green fluorescent protein in a cell-free translation system. *Biotechnology Letters* **18,** 1447–1452.

Lakowicz, J. R., and Weber, G. (1973a). Quenching of fluorescence by oxygen. A probe for structural fluctuations in macromolecules. *Biochemistry* **12,** 4161–4170.

Lakowicz, J. R., and Weber, G. (1973b). Quenching of protein fluorescence by oxygen. Detection of structural fluctuations in proteins on the nanosecond time scale. *Biochemistry* **12,** 4171–4179.

Levine, L. D., and Ward, W. W. (1982). Isolation and characterization of a photoprotein, "Phialidin," and a spectrally unique green fluorescent protein from the bioluminescent jellyfish, *Phialidium gregarium. Comp. Biochem. Physiol.* **72B,** 77–85.

Lossau, H., Kummer, A., Heinecke, R., Pöllinger-Dammer, F., Kompa, C., Bieser, G., Jonsson, T., Silva, C. M., Yang, M. M., Youvan, D. C., Michel-Beyerle, M. E. (1996). Time-resolved spectroscopy of wild-type and mutant green fluorescent proteins reveals excited state deprotonation consistent with fluorophore-protein interactions. *Chem. Phys.* **213,** 1–16.

Moerner, W. E. (1977). Those blinking molecules. *Science* **277,** 1059–1060.

Morise, H., Shimomura, O., Johnson, F. H., and Winant, J. (1996). Intermolecular energy transfer in the bioluminescent system of *Aequorea. Biochemistry* **13,** 2656–2662.

Morin, J. G., and Hastings, J. W. (1971a). Biochemistry of the bioluminescence of colonial hydroids and other coelenterates. *J. Cell. Physiol.* **77,** 305–312.

Morin, J. G., and Hastings, J. W. (1971b). Energy transfer in a bioluminescent system. *J. Cell Physiol.* **77,** 313–318.

Ormö, M., Cubitt, A. B., Kallio, K., Gross, L. A., Rsien, R. Y. and Remington, S. J. (1996). Crystal structure of the Aequorea victoria green fluorescent protein. *Science* **273,** 1392–1395.

Ottiger, M., Tjandra, N., and Bax, A. (1997). Magnetic field dependent amide ^{15}N chemical shifts in a protein-DNA complex resulting from magnetic ordering in solution. *J. Am. Chem. Soc.* **119,** 9825–9830.

Palm, G. J., Zdanov, A., Gaitanaris, G. A., Stauber, R., Pavlakis, G. N., and Wlodawer, A. (1997). The structural basis for spectral variations in green fluorescent protein. *Nat. Struct. Biol.* **4,** 361–365.

Perozzo, M. A., Ward, K. B., Thompson, R. B., and Ward, W. W. (1988). X-ray diffraction and time-resolved fluorescence analysis of *Aequorea* green fluorescent protein crystals. *J. Biol. Chem.* **263,** 7713–7716.

Peryushin, K., Reik, R., Wider, G., and Wuthrich, K. (1997). Attenuated T_2 relaxation by mutual cancellation of dipole-dipole coupling and chemical shift anisotropy indicates an avenue to NMR structures of very large biological macromolecules in solution. *Proc. Natl. Acad. Sci.* **94,** 12366–12371.

Prasher, D. C., Eckenrode, V. K., Ward, W. W., Prendergast, F. G., and Cormier, M. J. (1992). Primary structure of the *Aequorea victoria* green fluorescent protein. *Gene* **111,** 229–233.

Shimomura, O. (1979). Structure of the chromophore of *Aequorea* green fluorescent protein. *FEBS Lett.* **104,** 220–222.

Shimomura, O., and Johnson, F. H. (1975). Chemical nature of bioluminescence systems in coelenterates. *Proc. Natl. Acad. Sci. U. S. A.* **72,** 1546–1549.

Sosnick, T. R., Mayne, L., Hiller, R., Englander, S. W. (1994). The barriers in protein folding. *Nat. Struct. Biol.* **1,** 149–156.

Vander Bout, D. A., Yip, W.-T., Hu, D., Fu, D.-K., Swager, T. M., Barbara, P. F. (1997). Discrete intensity jumps and intermolecular electronic energy transfer in the spectroscopy of single conjugated polymer molecules. *Science* **277,** 1076–1997.

Ward, W. W., and Bekman, S. H. (1982). Reversible denaturation of Aequorea green fluorescent protein: physical separation and characterization of the renatured protein. *Biochemistry* **21,** 4535–4540.

Ward, W. W., and Cormier, M. J. (1978). Energy transfer via protein–protein interaction in *Renilla* bioluminescence. *Photochem. Photobiol.* **27,** 389–396.

Ward, W. W., and Cormier, M. J. (1979). An energy transfer protein in coelenterate bioluminescence characterization of the *Renilla* green fluorescent protein (GFP). *J. Biol. Chem.* **254,** 781–788.

Ward, W. W., Prentice, H. J., Roth, A. F., Cody, C. W., and Reever, S. C. (1982). Spectral perturbations of the *Aequorea* green fluorescent protein. *Photochem. Photobiol.* **35,** 803–808.

Yang, F., Moss, L. G., and Phillips, G. M. (1996). The molecular structure of green fluorescent protein. *Nat. Biotech.* **14,** 1246–1252.

Youvan, D. C., and Larrick, J. W. (1996). Fluorescent proteins and applications. *Gene* **173,** 1–117.

CHAPTER 2

Understanding Structure–Function Relationships in the *Aequorea victoria* Green Fluorescent Protein

Andrew B. Cubitt, Leslie A. Woollenweber, and Roger Heim
Aurora Biosciences Corporation
San Diego, California 92121

I. Introduction

The green fluorescent protein (GFP) from *Aequorea victoria* is one member of a small but important class of proteins that exhibit strong visible fluorescence without the requirement of cofactors or other enzymes. The *in vivo* function of GFP is to convert the blue light emitted by the photoprotein aequorin to green light. By doing so it is believed to increase both the efficiency of light output by aequorin and the transmission of the light in the ocean (Prasher, 1995). Learning more about the physiological role of GFP and its interaction with aequorin could help us to understand how to create mutants that exhibit efficiency energy transfer and how to effectively control dimerization.

The interaction of GFP with aequorin is readily reversible (Morise *et al.,* 1974) and is stabilized by high protein and salt concentrations, conditions likely to be encountered within the light-emitting organelles of *Aequorea victoria.* Both GFP

Copyright © 1999 by Academic Press. All rights of reproduction in any form reserved.
0091-679X/99 $25.00

(Yang *et al.,* 1996) and aequorin (Prendergast and Mann, 1978) can dimerize under appropriate conditions, and it is the dimerized forms that are believed to interact. The molecular details of the interaction of aequorin with GFP are unknown, although it has been suggested that a C-terminal hydrophobic patch, deriving from amino acids 206, 221, and 223 (Yang *et al.,* 1996), or a stretch of negative electrostatic potential (Brejc *et al.,* 1997) could be plausible interaction domains.

At least nine isoforms of GFP have been identified from preparations of purified protein prepared from mixed populations of jellyfish collected at Friday Harbor, Washington (Tsien and Prasher, 1995). In the case of aequorin, five isoforms have been identified from a single *Aequorea* circumoral ring, suggesting that heterogeneity also exists within individual jellyfish (Charbonneau *et al.,* 1985; Prasher *et al.,* 1987). Although the significance of the heterogeneity of both GFP and aequorin has largely been ignored, it is possible that it favors the correct association of GFP with aequorin. At least one site of heterogeneity in the *Aequorea*-derived GFP nucleotide residues occurs at a position involved in GFP dimerization (Yang *et al.,* 1996). Isoform variation at this position (172) was split between positively and negatively charged amino acids, which would favor the selective association of different isoforms.

II. Structure

The crystal structures of a number of GFP mutants have been described, including wild-type GFP (Yang *et al.,* 1996), S65T (Ormo *et al.,* 1996), BFP (Y66H, Y145F) (Wachter *et al.,* 1997), and T203Y, S65G, S72A, V68L (10C) (Remington, personal communication). The main structural features of GFP are very well conserved between these mutants, even though they have quite dissimilar spectral properties. GFP resembles a cylinder with a diameter of about 24 Å and a length of 42 Å, and has a tight, well-packed structure that lacks any large extended loop domains (Ormo *et al.,* 1996). The cylindrical fold of the protein is made up of 11 strands of β-sheet, which provide a regular pattern of hydrogen bonding, within which the chromophore is buried. The chromophore is located half way along the central stretch of irregular α-helix close to the geometric center of the cylinder (Fig. 1, see color plate). Small sections of α-helix also form caps on the top and bottom of the cylinder to directly protect the chromophore. The cavity containing the fluorophore has a number of charged residues in the immediate vicinity of the chromophore and also contains four water molecules that are important in establishing a hydrogen-bonding network around the chromophore. This network is probably critical in establishing the spectroscopic and dynamic photochemical properties of GFP. The β-sheets that make up the cylinder of GFP are separated slightly along the face of the cylinder in which the chromophore faces (comprising residues 145–150 and 164–169). The significance of this feature is not yet known, although it corresponds in part to the dimerization domain of wild-type GFP. Dimerization was also evident in the crystal structure of the T203Y mutant 10C and BFP, but not that of S65T. In the case of wild-type GFP, dimerization involved two main interaction areas, around amino acids 142–151 and 200–208; additionally, a number of more isolated contacts were identified (Y39, R168, N170, E172, L221, and F223) (Yang *et al.,* 1996).

Despite the identification of the dimerization contacts and direct verification of dimerization in crystals of GFP, we still have relatively little idea how easily different GFP mutants dimerize in free solution. Developing such an understanding is critical for the use of GFP mutants in protein localization and tagging experiments. Many mutations that improve the ability of GFP to fold efficiently at 37°C may do so by replacing hydrophobic protein association contacts, with less-hydrophobic substitutions potentially changing the ability of the protein to dimerize. For example, Yokoe and Meyer (1996) observed an almost 10-fold difference in translational diffusion between an optimized GFP mutant (F99S, M153T, V163A Crameri *et al.,* 1996) and wild-type GFP when expressed within mammalian cells. However, these authors did not measure the relative diffusional coefficients in buffer, so it is not possible to determine if the differences were due to dimerization or collisional interactions of GFP with macromolecular structures. An independent study determined the translational diffusion of wild-type GFP in aqueous buffer as 8.7×10^{-7} $cm^2 \cdot sec^{-1}$ (Terry *et al.,* 1995), a value identical to that determined for S65T (Swaminathan *et al.,* 1997). This mutant exhibited only a 1.5-fold change in translational diffusion coefficient when measured in saline and within the cytoplasm of cells, suggesting that translational diffusion may vary dramatically between mutants.

III. Chromophore Formation

The correct folding of GFP requires three distinct physical processes, the attainment of the correct three-dimensional structure, cyclization of the chromophore, and oxidation of the cyclized intermediate. Available evidence, based on kinetic (Heim *et al.,* 1994; Reid and Flynn, 1997) and mass spectra data (Cubitt *et al.,* 1995), is consistent with the idea that cyclization occurs prior to oxidation of the chromophore. The mechanism of fluorophore formation is speculative, but is believed to proceed through nucleophilic attack of the amino group of Gly-67 on the carbonyl carbon of Ser-65 to form a five-membered ring (imidazolinone) intermediate (Heim *et al.,* 1994). This process probably proceeds through a large number of cooperative interactions that help to distort the central α-helix, thereby moving Gly-67 close to the carbonyl of Ser-65. Ring closure, and the elimination of water, is probably activated by interaction of Arg-96 with the carbonyl oxygen of the imidazolidinone ring of the fluorophore (Ormo *et al.,* 1996). Oxidation of the hydroxybenzyl side chain of Tyr-66 by atmospheric oxygen produces the final fluorescent product *p*-hyroxybenzylideneimidazolinone (Cody *et al.,* 1993).

Studies designed to investigate fluorophore formation have used either the anaerobic production of GFP in bacteria (Heim *et al.,* 1994) or the use of denatured GFP to study the kinetics of refolding (Makino *et al.,* 1997; Reid and Flynn, 1997). By comparing the rate of renaturation of reduced and nonreduced GFP, Reid and Flynn concluded that folding and chromophore cyclization were relatively fast and that oxidation of the fluorophore was rate limiting.

The rate of chromophore oxidation seems to vary significantly between mutants in those cases where it has been measured. For example, the mutant S65T becomes fluorescent about five times faster than wild-type GFP when oxygen is readmitted to anaerobic bacterial extracts of the proteins (Heim *et al.,* 1995). In contrast, the mutant V163A, S175G (Siemering *et al.,* 1996) becomes fluorescent significantly slower than wild-type GFP when oxygen is readmitted to anaerobic cultures.

IV. Effects of Mutations on the Spectroscopic Properties of GFP

Since the original demonstration that GFP could be functionally expressed in *Caenorhabditis elegans,* GFP has been the subject of intensive study (Chalfie *et al.,* 1994). The need to create mutants that were brighter and could be spectrally resolved and folded more efficiently at 37°C was soon recognized by a number of groups who set out to develop enhanced mutants. Although mutants of GFP retain the overall structural organization of the wild-type protein, they often appear to fold less efficiently, exhibit greater sensitivity to environmental factors, and display enhanced rates of photobleaching. To overcome these problems, it has often been necessary to improve folding and stability of an identified mutant protein with novel spectral properties by further mutagenesis. Mutations that directly effect the spectroscopic properties of GFP do so either by directly effecting the covalent structure of the chromophore or by significantly changing its microenvironment. Presently known mutations that cause favorable changes in the spectroscopic or folding properties of GFP are located at about 20 separate positions within the molecule. Within this group all but three are conservative changes (the exceptions being F99S, S148P, and T203Y), implying that GFP is extremely intolerant to gross changes in structure. Many mutations result in a total loss of fluorescence, presumably because they prevent fluorophore formation. For the most part these mutations remain uncharacterized, even though they could potentially help to elucidate residues important in GFP folding and chromophore formation.

The best and some of the earliest examples of mutations with direct effects on the fluorophore structure are substitutions of Tyr-66 by other aromatic amino acids (Heim *et al.,* 1994). Changing the electronic properties of the side chain at this position has a significant impact on GFP because the fluorescent properties of the chromophore derive in large part from the π-electron-conjugated framework of Tyr-66. Substitution of Tyr-66 by amino acids with less electron-rich side chains produces proteins with blue-shifted excitation and emission spectra. Thus, phenylalanine, which has no electron donor group on its benzene ring side chain, gives rise to the shortest excitation wavelengths (358x, 442m). Histidine, with its imidazole group, has moderate electron-withdrawing ability and gives rise to slightly longer excitation and emission spectra (382x, 448m). Tryptophan, with its relatively large conjugated system and electron-rich indole, further red-shifts the excitation and emission spectra to 433x and 475m. The most electron-rich side chain is the charged (phenolate) form of tyrosine that is responsible for the 475-nm excitation peak in wild-type GFP. When the tyrosine is not charged, it is obviously much less electron rich and gives rise to the 395-nm excitation peak. Proteins that retain Tyr-66 are considerably more fluorescent than those with other aromatic substitutions at this position such as Y66H, Y66W, and Y66F (Table I). Of these mutations, the tryptophan mutants retain the best fluorescent properties, exhibiting overall quantum yields around 50–60% of typical tyrosine-based chromophores. These proteins also exhibit significantly reduced molar extinction coefficients, resulting in proteins that are about 20% as bright as the best GFP mutants. Improved versions of the tryptophan mutants have been developed that fold efficiently at 37°C, and some of these can be successfully resolved for multiple color analysis or for use in fluorescence resonance energy transfer (FRET)–based measurements. Y66H mutants, although well spectrally resolved from most other GFP mutants, have relatively low quantum yields (0.24–0.3) and extinction coefficients (21,000–26,000), making them relatively less fluorescent. In addition, most mutants of this type undergo rapid photobleaching (Rizzuto *et al.,* 1996; Patterson *et al.,* 1997). Mutants that exhibit improved folding at 37°C have been developed for the Y66H-based proteins, for example, P4-3, +F64L, and V163A, although they provide only modest improvements in spectral properties. The most blue-shifted mutants produced to date incorporate Y66F; however, they remain almost completely uncharacterized.

In contrast, substitutions at position 65 have been extensively studied, primarily because they significantly simplify and red-shift the excitation spectrum and improve the molar extinction compared to wild-type GFP (Heim *et al.,* 1995). While within the tripeptide core of the fluorophore, mutations at Ser-65 act through indirect environmental effects because their side chains are not conjugated to the double-bonded ring system of the chromophore. The best-characterized mutations are substitutions at Ser-65 by small aliphatic amino acids such as Ala, Cys, Leu, Thr, and Gly (Delagrave *et al.,* 1995; Heim *et al.,* 1995). These mutations are thought to act by preventing the ionization of Glu-222 by changing the hydrogen-bonding network around the fluorophore, thereby enabling the essentially complete ionization of Tyr-66 (Fig. 2, see color plate). Within the series of S65 mutants, all but S65A result in small red shifts of the 475-nm excitation peak in the wild-type protein due to a reduction in the energy required for the generation of the excited-state intermediate because of the elimination of electrostatic repulsion between Ser-65 and the excited state (Brejc

Table I
Spectroscopic and Biochemical Properties of Selected GFP Mutants

Mutations	Common name	Quantum yield and molar extinction	Excitation and emission max	Relative fluorescence at 37°C
i. S65T type				
S65T, S72A, N149K, M153T, I167T	Emerald	$\Phi = 0.68$ $\varepsilon = 57{,}500$	487 509	100
F64L, S65T, V153A		$\Phi = 0.58$ $\varepsilon = 42{,}000$	488 511	54
F64L, S65T (EGFP)	EGFP	$\Phi = 0.60$ $\varepsilon = 55{,}900$	488 507	20
S65T		$\Phi = 0.64$ $\varepsilon = 52{,}000$	489 511	12
ii. Y66H type				
F64L, Y66H, Y145F, V163A	P4-3E	$\Phi = 0.27$ $\varepsilon = 22{,}000$	384 448	100
F64L, Y66H, Y145F		$\Phi = 0.26$ $\varepsilon = 26{,}300$	383 447	82
Y66H, Y145F	P4-3	$\Phi = 0.3$ $\varepsilon = 22{,}300$	382 446	51
Y66H	BFP	$\Phi = 0.24$ $\varepsilon = 21{,}000$	384 448	15
iii. Y66W type				
S65A, Y66W, S72A, N146I, M153T, V163A	W1C	$\Phi = 0.39$ $\varepsilon = 21{,}200$	435 495	100
F64L, S65T, Y66W, N146I, M153T, V163A	W1B	$\Phi = 0.4$ $\varepsilon = 32{,}500$	434 452 476 (505)	80
Y66W, N146I, M153T, V163A	W7	$\Phi = 0.42$ $\varepsilon = 23{,}900$	434 452 476 (505)	61
Y66W			436 485	N.D.
iv. T203Y type				
S65G, S72A, K79R, T203Y	Topaz	$\Phi = 0.60$ $\varepsilon = 94{,}500$	514 527	100
S65G, V68L, S72A, T203Y	10C	$\Phi = 0.61$ $\varepsilon = 83{,}400$	514 527	58
S65G, V68L, Q69K, S72A, T203Y	10C Q69K	$\Phi = 0.71$ $\varepsilon = 62{,}000$	516 529	50
S65G, S72A, T203H		$\Phi = 0.78$ $\varepsilon = 48{,}500$	508 518	12
S65G, S72A, T203F		$\Phi = 0.70$ $\varepsilon = 65{,}500$	512 522	6
v. T203I type				
T203I, S72A, Y145F	H9-40	$\Phi = 0.64$ $\varepsilon = 29{,}000$	399 511	100
T203I	H9	$\Phi = 0.6$ $\varepsilon = 20{,}000$	399 511	13

Quantum yields were calculated by comparison to the standards fluorescein or 9-amino acridine. Molar extinction coefficients were calculated in Tris (10 m*M*)–EDTA (10 m*M*) buffer, pH 8.0, and represent the average of two determina-

Table I (*Continued*)

tions. Protein measurements were carried out using the BCA assay kit (BioRad) with BSA as the standard for all proteins at the same time to minimize assay-to-assay variation in protein determinations. It should be noted that the molar extinctions presented here are all equivalent, but significant day-to-day and assay-to-assay variations in protein assays can significantly influence apparent molar extinction coefficients. To reliably compare coefficients from different publications, at least one GFP of known extinction should be included with the unknown.

The ability of the GFP mutants to fold (relative fluorescence at 37°C) was determined by expression of the GFP in the bacterial expression cassette pRSET, under the control of the T7 promoter in bacteria BL21 (DE3) at 37°C overnight. Samples of bacteria grown at 37°C were measured for fluorescence at their peak excitation and emission wavelengths in a SPEX fluorimeter after resuspension in TE buffer, pH 8.0. To normalize for the amount of bacteria, the same samples were also measured for OD at 600 nm. Relative fluorescent intensities for each mutant were calculated by adjustment of the fluorescence measurement by OD at 600 nm. Relative fluorescence in the table was normalized to 100% for the most fluorescent mutant measured.

et al., 1997). Although the mutations S65A, S65C, and S65G have been extensively used both for dual-color imaging and FRET-based studies, the S65T-based mutants are likely the most commonly used, probably because they were initially well characterized, had reasonable fluorescent properties, and could be efficiently excited with the popular argon laser line at 488 nm. A number of improved mutants of S65T have been developed that exhibit improved folding at 37°C and retain or exceed the fluorescent properties of the original S65T mutant. Mutants such as Emerald (S65T, S72A, N149K, M153T, I167T) or EGFP (F64L, S65T) have relatively high quantum yields and molar extinction coefficients, making them highly detectable, and these are now commercially available.

Mutation of Ser-65 to Gly results in the most red-shifted mutants of this series. One of these (RSGPF4) is available commercially, although its spectroscopic properties have not been published (Delagrave *et al.,* 1995). The inclusion of additional mutations such as T203Y creates proteins with yet longer excitation and emission spectra (Ormo *et al.,* 1996). Aromatic mutations at Thr-203 achieve their effect through π-system interactions with Tyr-66, which reduce the energy required to excite the fluorophore from the ground state to the excited state. By doing so they enable lower energy light (redder light) to efficiently excite the chromophore. All aromatic amino acids tested caused a shift in excitation and emission spectra in the order Tyr > Phe > His > Trp, resulting in emission spectra that range from 512 nm to 527 nm. T203Y creates the biggest red shift of nonchromophore mutations yet described.

The mutant Topaz (S65G, S72A, K79R, T203Y) folds efficiently at 37°C and exhibits a high molar extinction coefficient (94,500) and quantum yield (0.6), making it the brightest GFP mutant developed to date. Probably the most interesting aspect of the T203Y mutants is their ability to undergo a reversible photochemical conversion process.

Like photoisomization in wild-type GFP, photoconversion of the T203 mutants seems to involve the gain and loss of protons on Tyr-66 (Chattoraj *et al.,* 1996;

Niwa *et al.*, 1996; Brejc *et al.*, 1997). In wild-type GFP, excitation with UV light causes proton transfer from the chromophore in its excited state to Glu-222, followed by hydrogen bond rearrangement and isomerization of Glu-222, which prevents the rapid reverse of the process after UV illumination has ceased (Fig. 2). By comparison, the T203Y mutants exhibit reversible photoconversion in which conversion to a dark state by illumination at 488 nm can be reversed by illumination at 400 nm (Dickson *et al.*, 1997). The greater spectral separation between photoconverted forms, increased brightness, and reversibility of the T203Y mutants should greatly improve the potential of GFP for studying spatial dynamics of tagged proteins compared to wild-type GFP (Yokoe and Meyer, 1996).

Mutation of the residues that are close to the chromophore such as the T203Y mutants can have effects on the spectral properties of GFP that are as significant as those mutations actually within it. For example, Delagrave (1995) identified a mutant (RSGFP1) that does not contain a substitution at Ser-65 but that nevertheless red-shifts the excitation and emission spectra. This mutant contains the mutation Q69L, which probably acts to disrupt the hydrogen bonding of Glu-222, thereby destabilizing its negatively charged form, hence favoring ionization of the fluorophore. Mutation of the same amino acid to lysine (Q69K) in the T203Y mutants results in a slight additional red shift, although it also significantly reduces the ability of the protein to fold at 37°C. Direct mutation of Glu-222 to Gly enables unrestricted ionization of Tyr-66 and results in a protein that lacks the 395-nm excitation peak and has a much larger 475-nm absorption peak (Ehrig *et al.*, 1995). Conversely, mutation of Thr-203 to Ile results in a stabilization of the neutral form of the chromophore, creating a protein that almost completely lacks the 475-nm excitation peak (Heim *et al.*, 1995). The emission spectrum of this protein is well separated from that of Y66H-based mutants, potentially making it a useful dual-color fluorescent tag for UV-based studies. Furthermore, a version of the T203I mutant has been developed that exhibits efficient folding at 37°C and has a superior quantum yield (0.62) and molar extinction comparable to that of other blue-shifted GFP mutants.

V. Effects of Mutations That Improve Thermosensitivity

Mutations that improve the thermostability of GFP ultimately do so by improving the efficiency of fluorophore formation. Although these mutations may improve the thermostability of GFP folding, they do not necessarily improve the spectroscopic or biochemically properties of the mature protein. In fact, in a number of cases absolute fluorescent brightness are decreased in the folding mutants compared to their less thermostable versions. A number of groups have addressed the problem of poor fluorophore formation in GFP when it is expressed at elevated temperatures and have independently identified a core set of mutations that are common to many enhanced mutants. This set includes the mutations F64L (Delagrave *et al.*, 1995; Cormack *et al.*, 1996), M153T (Crameri *et al.*, 1996;

Heim and Tsien, 1996) and V163A (Crameri *et al.,* 1996; Heim and Tsien, 1996; Siemering, *et al.,* 1996). Additionally, a number of mutations have been identified that seem to aid folding in specific combinations, and these are discussed later.

The precise mechanisms through which the folding mutations work are unknown; however, their distribution within GFP falls within four classes, those buried and physically close to the chromophore, those buried and physically far from the chromophore, and those with surface locations either close to the chromophore or distant from it.

Folding mutations close to the chromophore include those actually in the chromophore (S65A, -G, -C, -T, or -L) and two (F64L and S72A) that are close to it in the central α-helix. Additionally, the mutations Y145F, I167T, T203Y, and S205T are also physically close to the chromophore and partially or completely buried. These mutations are presumed to directly improve chromophore cyclization and oxidation but may also improve folding by other mechanisms. V163A is distant from the chromophore and in a buried location. The independent identification by three groups and its widespread occurrence in many different mutants suggest that V163A is probably critical in directing folding through more productive intermediates at elevated temperatures. The surface-located folding mutations include F99S, S147P, N149K, M153T, and S175G, (Crameri *et al.,* 1996; Kimata *et al.,* 1997; Siemering *et al.,* 1996) and these presumably act to reduce surface hydrophobicity. S147P and N149K also lie close to the chromophore in locations that could influence the hydrogen-bonding structure around the chromophore.

Studies to determine the optimum combinations of folding mutations (see later) suggest that not all combinations of mutations act cooperatively to produce progressively larger increases in folding efficiency. Rather, particular chromophore structures appear to require particular limited combinations of mutations for optimal folding. The fact that sets of some mutations act synergistically (F64L+V163A) whereas others (F64L+S72A) never appear to occur together suggests that they may act through common mechanisms.

Mutations are generally introduced into GFP to improve its folding or spectral properties, and their effects on the biochemical properties of GFP generally go unscreened. Although wild-type GFP is structurally resistant to urea, detergents, proteolytic attack, and thermal denaturation (Ward *et al.,* 1982; Ward and Bokman, 1982), it does exhibit significant spectral changes in response to mild changes in environmental conditions such as protein concentration, ionic strength, and pH (Ward, 1981; Ward *et al.,* 1982; Robart and Ward, 1990).

VI. The Development of Enhanced Mutants

Although we have made dramatic improvements in the fluorescent properties and folding of GFP, there is clearly a need for GFP mutants with enhanced properties for FRET and multiple-color analysis of protein distribution. Further improvements in the spectral properties of GFP are possible by applying what we presently know about the structure of GFP and reanalyzing known mutations that create desirable improvements in folding, fluorescence, or pH stability.

One approach to do this is to take all known point mutations that have been identified to improve folding, expression, or brightness of GFP, and to systematically test them to identify the optimal combination for different-colored fluorescent proteins. This can be achieved by designing the mutagenesis approach so that at each position either the mutation will be introduced or the position will be left unchanged. By carrying out the mutagenesis with pools of oligos, which potentially incorporate large combinations of mutants, it is possible to directly compare different combinations of mutants. For example, by using this approach it is possible to identify optimal pools of mutants, based on the ability to increase the brightness of GFP when expressed at 37°C. By repeated rescreening and mutagenesis of partially optimized mutants, it is possible to avoid the potential limitation of becoming trapped in favorable but not optimal sequences. It is also possible to determine if further improvements in folding could be achieved by the progressive accumulation of additional mutations.

This approach was used to develop better folding versions of the mutants H9, S65T, and 10C, and resulted in improvements of about eightfold for H9 and S65T, and 1.7-fold for 10C. The much lower enhancement of 10C is due to the relatively efficient folding of this protein prior to the addition of other mutations. Surprisingly, no additional mutations (other than the reversion of V68L) were identified that improved folding, despite our best efforts to introduce them. By comparison to general mutagenesis screens this approach is relatively simple, works well, and is quick. It suffers from poor diversity and is restricted to known mutations, which may not always be the most optimal.

Further improvements in GFP will probably require mutagenesis screens with large diversity libraries. The total number of mutants screened by different groups involved in GFP mutagenesis to date is probably significantly less than 1×10^9. By comparison, the total potential diversity of GFP is around 20^{238}, a value that is vastly larger, yet not only were better mutations found, many were independently identified by several groups. This observation is probably explained because in PCR-based mutagenesis screens, which most (but not all) groups used, the most common amino acid mutations introduced are single-point mutations at one position within a codon. The maximum number of alternative amino acids that can be encoded by variations at one position in a codon is three, severely restricting library diversity. Introducing two mutations per codon would dramatically increase the maximum potential diversity but at the cost of creating vast numbers of nonfluorescent or poorly fluorescent mutants. The development of improved mutagenesis screens requires the development and refinement of an expert system that can be used to restrict library diversity to functionally plausible mutants that appear to be important in determining the spectroscopic or folding properties of GFP. At the same time the library needs to contain sufficient diversity in key structural motifs around the chromophore and in turns to enable structural flexibility. To identify the relatively rare mutations that improve the spectroscopic properties of GFP, it is necessary to develop methods that can efficiently screen at least 10^7–10^8 mutants relatively rapidly and accurately enough to identify small changes in fluorescent properties. Such studies are highly worthwhile and will likely to lead to significantly improved fluorescent proteins.

References

Brejc, K., Sixma, T. K., Kitts, P. A., Kain, S. R., Tsien, R. Y., Ormo, M., and Remington, S. J. (1997). Structural basis for dual excitation and photoisomerization of *the Aequorea victoria* green fluorescent protein. *Proc. Natl. Acad. Sci. U. S. A.* **94,** 2306–2311.

Chalfie, M., Tu, Y., Euskirchen, G., Ward, W. W. and Prasher, D. C. (1994). Green fluorescent protein as a marker for gene expression. *Science* **263,** 802–805.

Charbonneau, H., Walsh, K. A., McCann, R. O., Prendergast, F. G., Cormier, M. J., and Vanaman, T. C. (1985) Amino acid sequence of the calcium-dependent photoprotein aequorin. *Biochemistry* **24,** 6762–6771.

Chattoraj, M., King, B. A., Bublitz, G. U., Boxer, S. G. (1996). Ultra-fast excited state dynamics in green fluorescent protein. Multiple states and proton transfer. *Proc. Natl. Acad. Sci. U. S. A.* **93,** 8362–8367.

Cody, C. W., Prasher, D. C., Westler, W. M., Prendergast, F. G. and Ward, W. W. (1993). Chemical structure of the hexapeptide chromophore of the *Aequorea* green fluorescent protein. *Biochemistry* **32,** 1212–1218.

Cormack, B. P., Valdivia, R. H., and Falkow, S. (1996). FACS-optimized mutants of the green fluorescent protein (GFP). *Gene* **173,** 33–38.

Crameri, A., Whitehorn, E. A., Tate, E., and Stemmer, W. P. C. (1996). Improved green fluorescent protein by molecular evolution using DNA shuffling. *Nat. Biotech.* **143,** 315–319.

Cubitt, A. B., Heim, R., Adams, S. R., Boyd, A. E., Gross, L. A., and Tsien, R. Y. (1995). Understanding, improving and using green fluorescent proteins. *TIBS* **20,** 448–455.

Delagrave, S., Hawtin, R. E., Silva, C. M., Yang, M. M., and Youvan, D. C. (1995). *Nat. Bio Tech.* **13,** 151–154.

Dickson, R. M., Cubitt, A. B., Tsien, R. Y, and Moerner, W. E. (1997). On/Off blinking and switching behavior of single green fluorescent protein molecules. *Nature* **388,** 355–358.

Ehrig, T., O'Kane, D. J., and Prendergast F. G. (1995). Green fluorescent protein mutants with altered fluorescent excitation spectra. *FEBS Lett.* **367,** 163–166.

Heim, R., Prasher, D. C., and Tsien, R. Y. (1994). Wavelength mutations and posttranslational autoxidation of green fluorescent protein. *Proc. Natl. Acad. Sci. U. S. A.* **91,** 12501–12504.

Heim, R., Cubitt, A. B., and Tsien, R. Y. (1995). Improved green fluorescence. *Nature* **373,** 663–664.

Heim, R., and Tsien, R. Y. (1996). Engineering green fluorescent protein for improved brightness, longer wavelengths and fluorescent resonance energy transfer. *Curr. Biol.* **6,** 178–182.

Kimata, Y., Iwaki, M., Lim, C. R., and Kohno, K. (1997). A novel mutation which enhances the fluorescence of green fluorescent protein at high temperatures. *Biochem. Biophys. Res. Comm.* **232,** 69–73.

Makino, Y., Amada, K., Taguchi, H., and Yoshida, M. (1997) Chaperonin-mediated folding of green fluorescent protein. *J. Biol. Chem.* **272,** 12468–12474.

Morise, H., Shimomura, O., Johnson, F. H., and Winant, J. (1974). Intermolecular energy transfer in the bioluminescent system of *Aequorea. Biochemistry* **13,** 2656–2662.

Niwa, H., Inouye, S., Hirano, T., Matsuno, T., Kojima, S., Kubota, M., Ohashi, M., and Tsuji, F. I. (1996). Chemical nature of the light emitter of the Aequorea green fluorescent protein. *Proc. Natl. Acad. Sci. U. S. A.* **93,** 13617–13622.

Ormo, M., Cubitt, A. B., Kallio, K., Gross, L. A., Tsien, R. Y., and Remington, S. J. (1996). Crystal structure of the *Aequorea victoria* green fluorescent protein. *Science* **273,** 1392–1395.

Patterson, G. H., Knobel, S. M., Sharif, W. D., Kain, S. R., and Piston, D. W. (1997). Use of the green fluorescent protein and its mutants in quantitative fluorescence microscopy. *Biophys J.* **73,** 2782–2790 (1997).

Prasher, D. C. (1995) Using GFP to see the light. *Trends Genet.* **11,** 320–323.

Prasher, D. C., McCann, R. O., Longiaru, M., Cormier, M. J. (1987). Sequence comparisons of complementary DNAs encoding aequorin isotypes. *Biochemistry* **26,** 1326–1332.

Prendergast, F. G., and Mann, K. G. (1978). Chemical and physical properties of aequorin and the green fluorescent protein isolated from *Aequorea forskalea. Biochemistry* **17,** 3448–3453.

Reid, B. R., and Flynn, G. C. (1997). Chromophore formation in green fluorescent protein. *Biochemistry* **36,** 6786–6791.

Rizzuto, R., Brini, R., DeGiorgi, F., Rossi, R., Heim, R., Tsien, R. Y., and Pozzan, T. (1996). Double labelling of subcellular structures with organelle-targeted GFP mutants *in vivo. Curr. Biol.* **6,** 183–188.

Robart, F. D., and Ward, W. W. (1990). Solvent perturbations of *Aequorea* green fluorescent protein. *Photochem. Photobiol.* **51,** 92s.

Romoser, V. A., Hinkle, P. M., and Persechini, A. (1997). Detection in living cells of Ca^{2+}-dependent changes in the fluorescent emission of an indicator composed of two green fluorescent protein variants linked by a calmodulin binding sequence. *J. Biol. Chem.* **272,** 13270–13274.

Siemering, K. R., Golbik, R., Sever, R., and Haseloff, J. (1996). Mutations that suppress the thermosensitivity of green fluorescent protein. *Curr. Biol.* **6,** 1653–1663.

Swaminathan, R., Hoang, C. P., and Verkman, A. S. (1997). Photobleaching recovery and anisotropy decay of green fluorescent protein GFP-S65T in solution and cells: cytoplasmic viscosity probed by green fluorescent protein translational and rotational diffusion. *Biophys. J.* **72,** 1900–1907.

Terry, B. R., Matthews, E. K. and Haseloff, J. (1995) Molecular characterization of recombinant green fluorescent protein by fluorescence correlation microscopy. *Biochem. Biophys. Res. Comm.* **217,** 21–27.

Tsien, R. Y., and Prasher, D. (1998). "Molecular Biology and Mutation of GFP" (Chalfie, M., and Kain, S., Eds.) Plenum, New York 97–118 (in press).

Wachter, R. M., King, B. A., Heim, R., Kallio, K., Tsien, R. Y., Boxer, S. G., and Remington, S. J. (1996) Crystal structure and photodynamic behavior of the blue emission variant of Y66H/Y145F of GFP. *Biochemistry* (in press). **36,** 9759–9765 (1997)

Ward, W. W. (1981) Properties of the coelenterate green fluorescent proteins. *In* "Bioluminescence and Chemiluminescence." Academic Press, New York. p. 235–242 Eds. DeLuca M & McElroy, W. D.

Ward, W. W., and Bokman, S. H. (1982) Reversible denaturation of Aequorea green fluorescent protein: physical separation and characterization of the renatured protein. *Biochemistry* **21,** 4535–4540.

Ward, W. W., Prentice, H. J., Roth, A. F., Cody, C. W., and Reeves, S. C. (1982) Spectral perturbations of the aequorea green fluorescent protein. *Photochem. Photobiol.* **35,** 803–808.

Yang, F., Moss, L. G., and Phillips, G. N. (1996) The Molecular structure of green fluorescent protein *Nat. Biotech.* **14,** 1246–1251.

Yokoe, H., and Meyer, T. (1996) Spatial dynamics of GFP-tagged proteins investigated by local fluorescence enhancement. *Nat. Biotech.* 14, 1252–1256.

CHAPTER 3

Quantitative Imaging of the Green Fluorescent Protein (GFP)

David W. Piston, George H. Patterson, and Susan M. Knobel
Department of Molecular Physiology and Biophysics
Vanderbilt University
Nashville, Tennessee 37232

I. Introduction

The advent of GFP as an expression marker in heterologous systems stands as one of the most significant advances for optical microscopy of living cells (Chalfie *et al.,* 1994). A major advantage of GFP is that it opens up the possibility of many experiments that are not possible using other existing fluorescence techniques. Among these new applications are the use of GFP as a real-time reporter gene in living systems, a dynamic marker for subcellular structures and organelles, and a tracer of intracellular protein trafficking. However, another application of GFP that has not yet been fully developed is its use as a quantitative reporter of polypeptide concentrations and dynamics. For example, immunoflu-

METHODS IN CELL BIOLOGY, VOL. 58
Copyright © 1999 by Academic Press. All rights of reproduction in any form reserved.
0091-679X/99 $25.00

orescence intensity can be measured accurately, but the absolute values of the target molecule are always uncertain due to limitations of the immunoreaction chemistry. On the other hand, it is possible to couple a single GFP to each target polypeptide, thus overcoming the uncertainties in fluorophore/target ratio or nonspecificity of labeling. Although genetic manipulations have generated GFP molecules with nearly ideal properties for fluorescence microscopy, GFP is not a panacea, and its use brings up a new set of potential problems that may affect any quantitative imaging.

In this chapter, we will discuss the properties of GFP that are important for quantitative imaging. We will also discuss the properties of the fluorescence microscope that are important in quantitative imaging, such as microscope components (objective lenses, fluorescence filters, etc.), signal-to-noise ratio, detection linearity, and fluorophore saturation. Due to the large number of GFP mutants and the variety of potential biological applications, a comprehensive description of all possible quantitative imaging situations is not possible. Thus, most descriptions of methods for quantitative imaging will be limited to the use of fluorescein-like GFP mutants (e.g., ones containing the S65T mutation) with laser scanning confocal microscopy.

II. Factors That Influence/Limit Quantitation of GFP in Fluorescence Microscopy

A. Fluorescence Properties of GFP That Are Important for Quantitative Imaging

Spectral and physical properties of GFP affect the accuracy and usefulness of any quantitative measurement. Many of these properties, such as extinction coefficient, quantum yield, photobleaching rate, and pH dependence, can be measured with purified GFP *in vitro*. However, other important properties, especially the time course of chromophore formation and protein degradation *in vivo,* cannot be easily determined. In general one chooses the brightest, most photostable GFP available (Heim and Tsien, 1996; Cormack *et al.,* 1996), which may make complicated corrections for background and photobleaching unnecessary in less demanding applications. In addition, improving the performance of GFP by optimization of codon usage and using temperature stabilization mutations is still helpful in quantitative experiments (Siemering *et al.,* 1996; Patterson *et al.* 1997). For more demanding imaging applications in which the amount of GFP is low (i.e., in which GFP is fused to a protein that is in low abundance), the intrinsic properties of GFP will likely be the limiting factor. Therefore, it is necessary to understand the properties described here and use appropriate controls to minimize artifacts arising from them.

Photobleaching (photoinduced destruction of the chromophore) of GFP is much slower than fluorescein under similar conditions (Patterson *et al.,* 1997). This resistance to photobleaching is likely due to the protection of the GFP chromophore by the tightly packed β-can structure (Ormo *et al.,* 1966; Yang *et al.,* 1996). Still, it is important to perform bleaching control experiments in any quantitative imaging experiment. Generally, one can simply acquire a time-lapse image series on GFP-labeled control cells (Niswender *et al.,* 1995) and measure the photobleaching time (photobleaching will generally give an exponential decay of intensity). This photobleaching time can then be used to correct the imaging experiments. If photobleaching becomes a limiting factor in the imaging of living cells, it is sometimes possible to reduce it by adding 10 μM Trolox (Rizzuto *et al.,* 1996). Using the S65T or F64L, S65T mutants of GFP, we find the photobleaching to be close to a single exponential decay with a half-time of >20 min for typical illumination intensities (Patterson *et al.,* 1997). However, it is very important to note that photobleaching of wild-type GFP is not characterized by a simple exponential decay. Wild-type GFP has two absorption peaks, 395 nm and 475 nm, which are related via an intramolecular rotation (isomerization) of the chromophore structure. The isomerization can be induced by irradiation of wild-type GFP with either 395- or 490-nm light, and the kinetics of this photoinduced reaction have recently been measured (Chattoraj *et al.,* 1996). Because both photoisomerization and photobleaching are occurring when wild-type GFP is irradiated at 488 nm, the temporal behavior is complicated: At first the photoisomerization dominates so the intensity increases, then, as the isomerization is complete and photobleaching begins to dominate, the intensity falls. Correcting a time-lapse series of images for such photoinduced behavior is difficult, so we have generally only used "red-shifted" fluorescein-like GFP mutants in our experiments.

Because the photobleaching of GFP is low, photodamage arising from GFP can also be expected to be low. Anecdotal evidence suggests that this is the case and indicates that GFP will be excellent for time-lapse and four-dimensional imaging. Although photodamage may be low, it is still important to perform viability controls during any imaging experiments on living cells. This is discussed in more detail in Section III.

The brightness of GFP and its mutants is also sensitive to pH. As an example, wild-type GFP shows relatively even brightness from pH 5 to pH 10 (Ward, 1981), whereas the S65T and F64L, S65T mutants are twofold brighter at pH 7 than at pH 6 (Patterson *et al.,* 1997). This behavior may affect any measurements in subcellular compartments such as lysosomes.

Finally, many physical parameters of GFP may dramatically affect its measurement but cannot be measured *in vitro.* The most important of these properties is the chromophore folding kinetics (or "turn-on" time). Because GFP chromophore formation requires oxidation and oxygen does not readily penetrate the GFP β-can structure, there can be a considerable delay (>1 h) between protein expression and the appearance of fluorescence. Unfortunately, it is not easy to distinguish between delays due to GFP expression (which one would like to measure) and those due to slow chromophore formation (which is an artifact of the GFP). In addition, the opposite process, GFP degradation, is also difficult to determine *in vivo,* although evidence from many studies indicates that GFP is very stable in cells (see, e.g., Hampton *et al.,* 1996).

B. Quantitative Imaging of Fluorescence in a Microscope

The fluorescence microscope is commonly used for visualizing biological specimens. Its basic elements are the excitation light source, objective lens, filter cube, and detector. The choice of each of these components can affect the accuracy of any image quantitation. Generally, a mercury lamp or argon-ion laser is used for GFP excitation; these two sources have bright blue lines that are nearly ideal for GFPs containing the S65T mutation. For use with blue-shifted BFPs, other light sources, such as a xenon arc lamp, may be preferable. Regardless, the only issue for quantitative imaging is that there be enough excitation light available to obtain the required fluorescence signal. The second element is the objective lens, which is defined by three parameters: magnification, numerical aperture (NA), and lens design type. Despite general belief to the contrary, magnification is the least important of these parameters. It determines how large an object will appear in the eyepiece or detector, but not the resolution (i.e., the smallest detail that can be clearly observed) or brightness (the amount of fluorescence signal that will be collected). Both of these parameters are a function of the NA, which is the *most important* criterion for selection of an objective. The NA defines the amount of light that will enter the objective lens. The larger the NA, the higher the resolution and the brighter the GFP signal will be. In general, the highest NA lens available should be used, although the working distance (the maximum focal depth into the sample) decreases with higher NA. The lens design is also important, and for GFP a Fluor or Plan-Neofluor design is preferable because it will be brighter (i.e., it will pass more of the fluorescence signal) than a Plan-Apochromat design.

The two most critical components for quantitative imaging are the emission filter and the detector. The excitation filter, dichroic mirror, and emission filter are usually combined in a "filter cube." The choice of an excitation filter depends on the light source to be used for the fluorescence excitation. For example, with a mercury lamp an excitation filter with transmission of 450–490 nm is typically used. Similarly, the choice of dichroic mirror is defined by the excitation and emission filters, which leaves only the choice of the emission filter.

The general rule of thumb for GFP visualization is that if the level of GFP expression is high enough, GFP can be easily imaged (or even observed by eye) using a standard fluorescein filter set. However, use of a filter set designed specifically for GFP will enhance GFP signal collection (Niswender *et al.,* 1995; Endow and Komma, 1996). A dedicated filter set for quantitative imaging of GFP should use a narrow bandpass filter to maximize the collection of GFP signal versus the collection of background (usually due to autofluorescence). With a narrow bandpass filter, the sharp green GFP signal is preferentially passed to the detector over the greenish yellow or greenish orange background of cellular autofluorescence (autofluorescence generally exhibits a very broad spectrum, whereas the GFP emission is relatively narrow). In principle, one would choose the narrowest bandpass filter to increase the discrimination of the GFP over other background signals. However, the optimal passband of the filter will be limited by the required signal-to-noise ratio (S/N), as discussed later. The optimal bandpass (i.e., the narrowest bandpass that allows a sufficient S/N) should be determined for each imaging situation, but we have had good luck using filters with ~30 nm bandpass (e.g., a 27-nm bandpass centered at 512 nm for wild-type GFP). It should be noted that for qualitative applications a wider bandpass filter will yield a "prettier" image. In addition, the specific filter chosen depends not only on the GFP signal strength, but also on the fluorescence spectrum of the particular GFP being used.

The use of a narrow bandpass is demonstrated in Fig. 1, which shows HeLa cells transfected with a glucokinase–GFP fusion plasmid (Niswender *et al.,* 1995). The autofluorescence background is considerably more pronounced using the longpass filter (panel A). If the level of GFP expression is high, then the choice of filter would make little difference (see the cells labeled "a" and "b"). However, for a cell that is expressing very low levels of GFP (see cell "c"), it is difficult to distinguish the GFP signal from autofluorescence with a longpass filter, although this cell clearly stands out with the narrow bandpass emission filter.

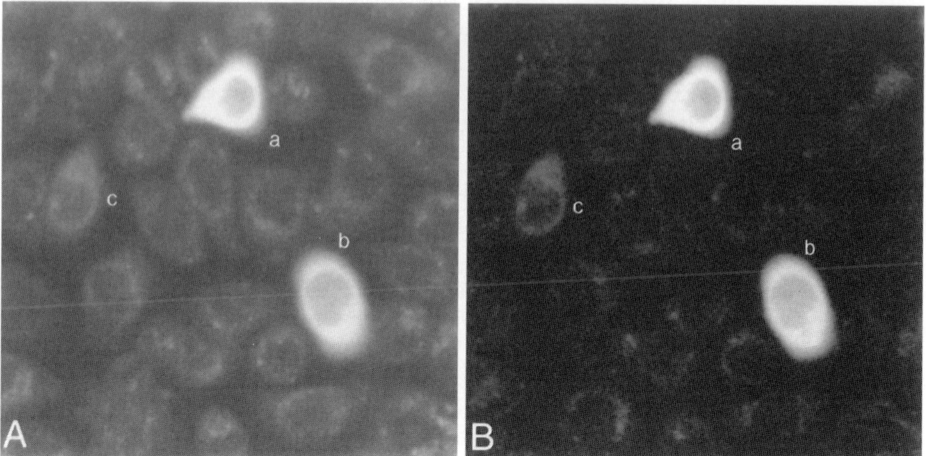

Fig. 1 Fluorescence images of HeLa cells transfected with a glucokinase–GFP fusion plasmid (Niswender *et al.,* 1995). Laser scanning confocal microscopy was performed using a Zeiss 410 microscope with a 40× Plan-Neofluar 1.3-NA oil-immersion objective. Excitation was by the 488-nm line of an argon–krypton-ion laser, and a Q498LP dichroic mirror was used. In panel A, a 515LP longpass emission filter was used, and in panel B a HQ512/27 emission filter designed specifically for GFP was used. Three GFP-positive cells (labeled a, b, c) can be seen in each panel.

The detectors that are typically used for quantitative microscopy are the photomultiplier tube (PMT) and the cooled charge-coupled device (CCD). PMTs are not imaging devices, so raster scanning (as is used in LSCM) must be performed to build up the image. Conversely, CCD cameras *are* imaging devices and can basically be used as a sensitive and linearly responding "eye." Other digital imaging cameras can be used for quantitative imaging as long as they yield a linear response to the fluorescence signals, but because film does not have a linear response, rigorous quantitation of photographs is not possible. One important design criterion for mounting the detector to the microscope is to avoid using an intermediate eyepiece lens. For example, an "iso C" mount connects the camera directly to the microscope with no extra lenses, which yields maximal optical efficiency. Although both PMT and CCD detectors have their own strengths and weaknesses, the usual bottom line is that PMTs are useful in microspectrofluorometry and laser scanning confocal microscopy (LSCM), whereas CCDs are more useful in widefield microscopy. Regardless of the detector used, basic principles such as the signal-to-noise ratio, detection linearity, and fluorophore saturation must be understood to ensure the accuracy of any quantitation.

The most important parameter for defining the usefulness of quantitative imaging is the signal-to-noise ratio (S/N). In modern commercial fluorescence microscopes the S/N is generally given by the shot noise, which is defined as the square root of the number of photons collected. Some systematic errors may be introduced in the detection system, but they are small compared to the shot noise for the two detectors, PMT and CCD camera, discussed in this chapter. For example, if one collects 100 photons per pixel the noise would be expected to be $\sqrt{100} = 10$, or 10% of the signal, and for 10,000 photons per pixel, $\sqrt{10{,}000} = 100$, or 1% of the signal. These examples reflect the typical signal levels in fluorescence microscopy, and errors introduced in the detection electronics are usually less than 1%. It would appear that one would always want to attain a maximal S/N, but in reality, factors such as photobleaching, fluorophore saturation, number of fluorophores in the sample, and spatial resolution limit the available S/N.

Photobleaching is often the ultimate limiting factor in fluorescence microscopy. Although the photobleaching rate of GFP is slow (as discussed earlier), photobleaching can still limit the amount of signal that can be obtained from a GFP-labeled sample. Another limiting factor, fluorophore saturation, does not occur in widefield fluorescence microscopy, but in LSCM, excitation intensities high enough to reach saturation are sometimes used. Saturation occurs when all the available fluorophores are constantly in the excited state, so that an increase in excitation light gives no increase in fluorescence. In this case, it is very difficult to calibrate fluorescence signals due to the introduced nonlinearity. Saturation can be avoided, though, by lowering the laser input power, and in general, the lowest excitation power that yields a sufficient S/N should be used. The number of fluorophores per pixel also limits the S/N. Each fluorophore can only generate one fluorescence photon at a time, so to obtain a high S/N, a large number of fluorophores is required. In addition, if a high degree of spatial resolution is required, then there will be fewer fluorophores in each pixel, which will limit the S/N.

Another consideration in the detection system is its linearity and offset. Both PMT and CCD detectors are highly linear, but it is sometimes possible to adjust the offset (sometimes called black level). This should be set so that the image reads ~0 (in analog-to-digital units, ADU) when there is no fluorescence coming from the sample. If there is a nonzero offset in the system, then it will not be possible to quantitate accurately differences between images. For example, if there is an offset of 10 ADU, then a cell with an average value of 30 ADU would actually contain twice as much GFP as one with an average value of 20 ADU ($30 - 10 = 20$, which is twice $20 - 10 = 10$). A negative offset causes even worse problems, because cells with the GFP signal may not even show up in the image. Details about how to set the correct offset in LSCM are given in Section III.

C. Microscopy Modes for Quantitative Fluorescence Imaging

Each technique that is used to measure fluorescence has advantages and limitations that require special consideration. Because many GFP mutants with altered spectral characteristics are now available, it is impossible to detail the optimal quantitative imaging criteria for each GFP mutant in every specimen with each type of instrument. The one basic principle that governs the visualization of GFP is to maximize the GFP signal over existing background. Because the background varies from specimen to specimen, optimization of the GFP signal over background can be accomplished by a combination of different means. These may include selecting an excitation wavelength that minimizes autofluorescence of the specimen, using a narrower bandpass emission filter, or using a GFP variant that is optimized for the specific organism under consideration. Specific considerations related to instrument selection and use are described in this section for four different types of microscopy: microspectrofluorometry, widefield (or conventional) fluorescence microscopy, confocal microscopy, and two-photon excitation microscopy.

1. Microspectrofluorometry

One of the simplest ways to measure total fluorescence from a cell population is with a spectrofluorometer. Although this technique is not imaging based, it is a very useful way to measure GFP spectra both *in vitro* and *in vivo,* quantitate GFP signals, and help calibrate results from quantitative imaging experiments. In a spectrofluorometer, fluorescence is measured directly from a solution (this could be a cell suspension, a whole cell extract, a purified GFP sample, etc.) placed in a cuvette, so very little special preparation is needed. For some mammalian cells that are particularly adherent, a stir bar must be used in the cuvette, and most commercial spectrofluorometers have accessories for stirring the sample. In general, samples should be diluted to an optical density (O.D.) of less than 1.0. We typically measure fluorescence preparations in the spectrofluorometer at an O.D. of 0.1 or less. A useful variation of the spectrofluorometer is the microspectrofluorometer, in which the sample is observed on an attached microscope stage instead of in a cuvette. Although it still uses cell preparations (mounted on slides, perfusion chambers, etc.), this system give a couple of advantages over imaging methods. First, because many cells are usually observed in the field of view, the fluorescence signal is much larger than in an imaging experiment. This increases the S/N and also allows a lower excitation intensity, which in turn can reduce any possible photodamage to the cells. Second, one can acquire a complete fluorescence spectra. This can yield separation of many different GFP variants and may prove extremely useful for FRET experiments based on multiple GFPs. Of course, the microspectrofluorometer has the major disadvantage that there is no subcellular resolution.

2. Conventional Microscopy Using a Cooled CCD Camera

A conventional fluorescence microscope can be used for quantitative imaging by equipping it with a low-light, linear-response camera. These cameras are typically based on a CCD chip, which is inherently extremely linear and introduces little offset. Thermoelectric cooling of a CCD detector greatly improves sensitivity by reducing electronic noise, thus yielding essentially zero background, high dynamic range (16 bits = 2^{16} gray scales), and high detection efficiency (40 to 80%). Because low illumination levels can be used with this type of imaging detector, fluorophore saturation is not a problem. In addition, it is possible to use a long integration time so that very low light levels can be detected, but in living samples, usable integration times may be limited by the time scale of the biological events under investigation.

The use of a cooled CCD permits quantitation of GFP signals, especially in thin samples such as cell monolayers. However, because the CCD is used on a conventional fluorescence microscope, there is no discrimination between in-focus and out-of-focus fluorescence. Therefore, image quantitation can be complicated due to nonuniformities in cell thickness (i.e., a thicker cell can easily be confused for a cell with higher GFP expression). In some cases in which out-of-focus background is a problem, image deconvolution (described later) can be used to allow accurate quantitation and subcellular resolution.

In practice, many of the factors that must be considered for quantitative imaging with a CCD camera are the same as those for LSCM (see Section III), especially for time-lapse imaging. Some issues, though, are specific for use with a CCD. One requirement is a computerized image acquisition system that is capable of controlling the camera, shutter, and focus motor. Synchronization of the shutter and camera so that the samples are irradiated with excitation illumination only when the camera is taking an exposure minimizes photodamage and photobleaching. Because of time constraints due to the dynamics of the process under observation, it is often not possible to optimize exposure times with respect to the camera's dynamic range for imaging live specimens. In these cases, it is sometimes possible to increase the image acquisition rate by increasing the illumination intensity, but often the optimal exposure times for acquiring images of good to acceptable quality over the observation time should be determined empirically. Usually, it is possible to reduce the illumination so that cells can be observed on the microscope stage for at least an hour (Endow and Komma, 1996; Moores *et al.,* 1996).

a. Image Deconvolution

Because a widefield microscope collects fluorescence from throughout the sample and the cells are not always of a uniform thickness, the volume that is contributing to the observed signal is unknown. This uncertainty greatly complicates any absolute quantitation. For this reason, we use confocal microscopy for our quantitative imaging applications. However, image deconvolution can allow absolute quantitation of widefield microscopy data by using a mathematical algorithm to deblur a three-dimensional data set (Agard *et al.,* 1989; Carrington *et al.,* 1990; Holmes *et al.,* 1995). The advent of fast desktop computers has allowed deconvolution to be useful to a wide variety of laboratories for many applications. However, deconvolution algorithms may introduce artifacts in the data set and are less useful on densely fluorescent samples. A "densely" fluorescent sample is one where fluorescence arises from many places in the cell or sample (i.e., not just the DNA or a single organelle). For instance, cytoplasmic staining or microtubule staining of an embryo may look like a blur in a widefield microscope, whereas a confocal image may show sharp detail. There is a point, for any given number of collected photons, at which deconvolution techniques can no longer find the details in the blurry image and deconvolve them. This point, however, has never been strictly determined, and each research team using deconvolution usually establishes its own rule of thumb about which samples are appropriate.

To use image deconvolution, three criteria must be met. First, a computer-controlled fluorescence microscope with a high-quality cooled CCD camera and an accurate motorized focus control is required. Optical sections are generally taken throughout the sample at 0.1- to 0.5-μm intervals, and from 4 to 128 sections are used to perform the deconvolution. To achieve the best results, the acquired data set must be optimized for every sample in terms of spacing and number of sections. Second, a powerful computer must be available for the image processing: Requirements depend mainly on the number of sections in the data set, but Pentium-based computers with sufficient memory can be used to deconvolve most data sets in under an hour. Third, the PSF (point spread function) of the microscope–how the microscope images a single point of fluorescence–must be carefully measured using small ($<$0.2-μm) fluorescent latex beads. Some new algorithms, however, do not require a priori knowledge of the PSF (Holmes *et al.*, 1995).

3. Confocal Microscopy

Although limitations on quantitative imaging with a cooled CCD camera can sometimes be overcome through image deconvolution, we have found that confocal microscopy offers significant advantages for fluorescence quantitation in tissues and optically thick samples. Because a well-defined optical section (typically ~1 μm thick) is acquired in confocal microscopy, nonuniformities in cell thickness are not a major problem. However, during imaging of thick samples such as tissue slices, the fluorescence signal may decrease as the focal plane moves deeper into the sample, making it more difficult to interpret the quantitative amount of GFP in the cells.

In a laser scanning confocal microscope, the illumination is scanned in a raster pattern to form the image and a PMT detector is used. In most confocal microscopes, PMTs usually give lower dynamic range (8 bits = 2^8 gray scales) and detection efficiency (15 to 30%) than a cooled CCD. Although it would be desirable to increase the detection efficiency, the 8-bit dynamic range is usually not a problem because generally fewer than 255 photons/pixel/scan are collected. Still, the excellent linearity of the PMT response, background rejection properties, and availability of appropriate lasers (particularly the 488-nm line of an argon-ion laser) make confocal microscopy near ideal for quantitative imaging of GFP fluorescence. The use of confocal microscopy for quantitative imaging of the fluorescein-like S65T mutants of GFP is detailed in Section III.

4. Two-Photon Excitation Microscopy

Two-photon excitation microscopy (TPEM) is a new alternative to confocal microscopy. Use of this technique eliminates out-of-focus photodamage, avoids chromatic aberrations, and provides high-resolution three-dimensional measurements. Two-photon excitation arises from the simultaneous absorption of two photons, each of which is half the energy required for the transition to the excited electronic state of the fluorophore. In practice, two-photon excitation is made possible by the very high local instantaneous intensity that is provided by a combination of diffraction-limited focusing of a single laser beam in the microscope and the temporal concentration of a subpicosecond mode-locked laser. Resultant instantaneous peak excitation intensities are 10^6 times greater than those that are typical in confocal microscopy, but the pulse duty cycle of 10^{-5} maintains the average input power at less than 10 mW, which is only slightly greater than the power that is typically used in conventional confocal microscopy. Currently, the cost of state-of-the-art lasers, which are required for TPEM, has limited this technique to only a few laboratories. However, the field of ultrafast lasers is undergoing rapid development, and as the price of lasers decreases over the next 10 years, TPEM may become a widely used method of optical sectioning microscopy.

Three properties of two-photon excitation provide the significant advantages over conventional optical sectioning microscopies for the study of UV excitable fluorophores in thick samples:

1. The excitation is limited to the focal volume due to the second-order dependence of the two-photon excitation on intensity and the decrease in intensity with the square of the distance from the focal plane. The excitation localization yields three-dimensional discrimination equivalent to an ideal confocal microscope without requiring a confocal spatial filter. Absence of the need to descan the fluorescence to pass a confocal aperture enhances fluorescence collection efficiency. Confinement of fluorophore excitation to the focal volume minimizes photobleaching and photodamage associated with UV illumination–the ultimate limiting factors in fluorescence microscopy of living cells and tissues.
2. Two-photon excitation allows imaging of UV fluorophores with conventional visible light optics in the scanning and detection systems because both the red excitation light ($\sim$700 nm) and the blue fluorescence ($>$400 nm) are within the visible spectrum.
3. Red light interacts much less strongly than UV light with most living cells and tissues (aside from plant cells) because fewer biological molecules absorb at longer wavelengths and red light is scattered less than shorter wavelengths. This nearly eliminates out-of-focus photodamage and background, and allows most of the input power to reach the focal plane. The relative transparency of biological specimens at 700 nm permits deeper sectioning than would be possible with UV excitation.

A limitation of TPEM is that the two-photon excitation spectrum may bear little resemblance to the one-photon absorption profile. GFP is easy to use in TPEM, however, because the two-photon excitation spectra of both wild-type GFP and the S65T mutant have been shown to overlap twice the one-photon absorption spectra, with the same relative absorption cross sections (Xu *et al.,* 1996). This overlap means that the major absorption peak of 395 nm for wild-type GFP is found at 2×395 nm (790 nm) with two-photon excitation. Preliminary results show a similar behavior for two-photon excitation of the F64L, S65T double mutant, and also that it is one of the strongest two-photon absorbers ever measured (C. Xu, M. Albolta, and W. W. Webb, personal communication). This is consistent with the extremely large one-photon absorption of this double mutant. Using wild-type GFP, useful two-photon-excited images have been obtained using $\sim$790- (Niswender *et al.,* 1995; Patterson *et al.,* 1997) and $\sim$900-nm (Potter *et al.,* 1996) excitation from a Ti:Sapphire laser.

The use of two-photon excitation microscopy is identical to that of confocal microscopy (as described later) except for the use of a different laser and elimination of the confocal pinhole. Currently, the laser of choice for TPEM is an ultrafast mode-locked Ti:Sapphire (e.g., Coherent Mira or Spectra-Physics Tsunami) pumped by either an argon-ion laser or an all-solid-state, frequency-doubled Nd:YAG laser that has recently achieved the power levels necessary for this application. The mode-locked Ti:Sapphire laser produces ~100-fs pulses at a repetition rate of ~100 MHz and can be tuned over a broad range of output wavelengths from 700 to 1100 nm, allowing two-photon excitation of most UV and visibly excited fluorophores. The ability to adjust laser wavelength, though, makes this laser considerably more complicated than the "turnkey" lasers used in confocal microscopy.

In the future, ultrafast lasers yielding a single wavelength or a small group of selected wavelengths may allow the development of easy-to-use and relatively inexpensive two-photon excitation microscopes. Most current two-photon excitation systems have been built around a confocal laser scanning microscope, but because the pinhole is not needed to obtain optical sectioning, it should be opened fully or removed when using two-photon excitation. The lack of a pinhole requirement also allows for various other detection schemes. For instance, one can use a CCD camera or full-field photomultiplier detector to increase the fluorescence collection efficiency and still acquire optical sections from thick samples. The lasers and detection schemes available have been described in detail elsewhere (Denk *et al.,* 1995).

III. Applications of LSCM for Quantitative Imaging of GFP

The basic strategy used to quantitate GFP images is to perform parallel spectroscopy and microscopy experiments on a standard solution of fluorescein (Molecular Probes #F-1300, extinction coefficient = 90,000 cm^{-1} M^{-1} at 488-nm excitation, quantum yield = 85%) and a purified GFP solution, and compare these results to images of GFP-labeled cells. We have used both GST–GFP and His_6–GFP (Patterson *et al.,* 1997) for fast and easy GFP purification (a useful protocol for His_6–GFP purification is given in Section IV), and in neither case did we find any photophysical differences between purified GFP and the purified fusion proteins. Once a purified GFP sample is obtained, its concentration can be determined by absorption and fluorescence spectroscopy using the known extinction coefficient and quantum yield of the particular form of GFP being used (Heim and Tsien, 1996; Patterson *et al.,* 1997). For example, the S65T mutant of GFP has an extinction coefficient of ~55,000 cm^{-1} M^{-1} at 488 nm (Patterson *et al.,* 1997). The concentration of the GFP sample is then calculated from the absorption relative to fluorescein, that is, if the absorption at 488 nm of the S65T sample is the same as that of a 1 μM fluorescein sample, the S65T concentration is [S65T] = [fluorescein] (extinction coefficient of fluorescein/extinction coefficient of S65T) = 1 $\mu M \times (90/55) = 1.64\ \mu M$. It is also helpful to compare any spectral determination of GFP concentration with a standard biochemical method, such as the BCA assay (Patterson *et al.,* 1997).

As described in Section II, it is important to set the correct gain and offset for the detection system. To begin, set the gain so that the image of a typical GFP-labeled cell is not saturating (i.e., no pixel values = 255). Once the correct gain is set, it should not be changed during the calibration or imaging steps. Next, the microscope offset must be set so that zero pixel value corresponds to zero fluorescence signal. This can be done using the fluorescein sample as follows: Image four concentrations (in a ratio of 1:2:3:4) of fluorescein that each give a good signal but for which the maximum pixel value is <255 at the gain used for the GFP imaging. Determine the mean pixel value of each image using the histogram command, and plot the mean pixel value versus the concentration. The resulting plot should be a straight line, and the y-intercept will go through zero when the black level is correctly set. If the y-intercept is positive the offset should be reduced, and if the y-intercept is negative the offset should be increased. Image the GFP-labeled cell again. It may require several trials to achieve the optimum gain/offset combination for your sample. Dilute or concentrate the GFP sample to match the signal obtained in the cell GFP image, and determine the concentrations of GFP in the cell over the linear range of the image data (pixel values = 0–255). This procedure is somewhat difficult and time-consuming, but it only needs to be done one time. Once the correct offset is obtained, it should not change unless there is a modification/repair to the detection system of the microscope.

The known GFP sample can now be compared with GFP-labeled cells imaged by the confocal microscope. This is done by comparing the mean pixel values of the regions of interest in images of GFP with the mean pixel values from known GFP standards. The GFP standards are simply various concentrations of purified GFP sealed in deep-well slides or embedded in polyacrylamide. The unknown GFP concentration is determined from the linear equation generated by the mean pixel values of the standards. In addition to estimating absolute levels of target proteins, the standards also give experimenters two important advantages. First, the standards provide a way to normalize imaging experiments performed under different conditions, such as day-to-day fluctuations in excitation power or different optical alignments. Second, even though GFP photobleaches slowly, the standards allow compensation of photobleaching that may occur over the course of a lengthy experiment.

A. Setting Up the Confocal Microscope

Many of the details here depend on the particular microscope used. We usually use the Zeiss LSM410 confocal microscope for our quantitative GFP experiments, but all currently available confocal microscopes are integrated with comprehensive software packages that allow time-lapse and three-dimensional image acquisition. In addition, most of the processing needed for accurate quantitative imaging can be done with the resident confocal software or NIH Image, which is available free from the National Institutes of Health Web page (http://hrsb.info.-

nih.gov/nih-image/). For experiments on living cells, it is almost always necessary to have a temperature-controlled sample chamber. For work on an inverted scope, it is important to have a sufficiently large opening to allow access to high-NA objectives. In addition, if perifusion of different solutions is used, we have found it extremely advantageous to heat the perfusate to the temperature of the cell bath before it reaches the sample chamber. All the protocols described here are normally performed using a 40×/1.3-NA Plan-Neofluar oil-immersion objective. For this 40x objective, we use a pinhole value of 60, which maximizes the signal-to-noise ratio in the confocal microscope (Sandison *et al.,* 1995). We use the 488-nm line of the argon-ion laser with custom filters from Chroma Technology (Brattleboro, VT), a Q498LP dichroic mirror, and an HQ512/27 bandpass filter.

When working with a new preparation of cells in a case in which we are not sure of the GFP intensity, we begin with the following procedure. First, place the cells on the microscope stage and focus on the cells using transmitted light. The use of phase contrast or Nomarski DIC is often helpful in locating the cells. To minimize photobleaching and photodamage, fluorescence should not be used to initially focus the microscope. If a temperature-controlled stage is used, the cells should be allowed to equilibrate in the sample chamber for ~10 min. Next, it is usually helpful to quickly observe the cells by eye with conventional epifluorescence to assure that there is a sufficient GFP signal and to gauge how strong this signal is. Preliminary examination is also helpful to find positive cells in transient transfection experiments in which only a small percentage of them may contain GFP. Once an appropriate cell has been found, let the confocal imaging begin. It is a good idea to start the laser scanning with a low laser power and a high PMT gain. For example, we start with laser power = 10 (out of 100), and contrast = 325. Conveniently, the correct offset on our LSM410 is brightness = 9800 for all contrast values between 200 and 400. The exact offset value is likely to be different for each LSM410, though, and for some microscopes, the correct offset may change depending on the gain used. It should be noted that on the Zeiss system, the F9 key will automatically adjust the contrast and brightness to give a low-noise image, but this will not give appropriate values for quantitative measurements. While scanning, the laser power should be increased until an image of the cell is visible on the display. Once an image has been acquired, the laser power and gain can be adjusted to optimize the image S/N, but this procedure should be performed as quickly as possible to minimize photobleaching and photodamage. For example, with brighter samples a lower gain can be used. If the GFP sample is extremely dim, the pinhole can be opened further to allow more light to be collected, but this will reduce the spatial discrimination of the confocal microscope and thus decrease the accuracy of any quantitation.

B. Preparation and Imaging of Standards

For most applications of this technique, a range of concentrations that bracket the expected levels in the region of interest should be prepared. If the concentration of GFP and/or its fusion partner is completely unknown, a good starting range is ~200 nm to 10 μM. The GFP standards for the deep-well slides consist of ~100-μl GFP solutions containing 100 μg/ml BSA to minimize adherence of the GFP to the glass surfaces. These are loaded into the slide well, covered with coverslips, and sealed with hot paraffin. Images from the middle of the homogeneous solutions will give fluorescent signals and thus mean pixel values indicative of that GFP concentration. Although this method works well and provides excellent quantitation, it requires the inconvenience of replacing the sample slide, culture dish, chamber, and so on with the standard slide. An alternative method is to embed the GFP in a polyacrylamide slab, place the slab with the sample, and image both simultaneously. The GFP is mixed with acrylamide–bisacrylamide and diluted with GFP elution buffer (see Section IV) to the appropriate concentrations for each. One μl of 10% ammonium persulfate (APS) and 0.5 μl of *N,N,N′,N′*-tetramethylethylenediamine (TEMED) per 100 μl solution are added and mixed thoroughly. Ten microliters are placed on a microscope slide, covered by a coverslip to form a flat GFP–polyacrylamide slab, and allowed to polymerize over 5 to 10 min at room temperature. Once the slab has polymerized, the coverslip can be removed, the slab can be sliced with a razor blade, and the pieces placed with samples for imaging experiments. An example of this method is shown in Fig. 2 (see color plate). Panel (A) shows the fluorescence, panel (B) shows the DIC, and (C) shows the overlay of GFP–polyacrylamide imaged simultaneously with yeast, *Saccharomyces cerevisiae,* growing on 1% agarose supplemented with medium. In regions close to the GFP–polyacrylamide slab, the DIC images of the yeast are slightly distorted, but the fluorescence images are not hampered at all.

C. Time-Lapse Imaging

One major strength of GFP is that, because it is used with living cells, temporal dynamics can be observed and measured using time-lapse imaging. Methods for time-lapse imaging are described in detail in Chapters 10 and 12, but we will list some of the considerations for such experiments. First, one must make sure that the sample does not move too much during the time lapse. This is very sample dependent; for example work with *Caenorhabditis elegans* is often performed on anesthetized animals. The sample can also appear to "move" if there is any drift in the microscope focus. Problems of focus drift can be overcome by manually refocusing during the time lapse, acquiring a full three-dimensional data set at each time point, or using a feedback-controlled focusing stage (these can maintain focus position to nanometer precision). Second, it is important to know that the sample remains viable throughout the imaging process. With embryos and whole organisms, a morphological viability assay can be used. With yeast cells, for instance, it is possible to observe the cells by eye using transmitted light and to confirm that cell division is continuing. For quantitative imaging, the only additional consideration is a correction for photobleaching effects. As described earlier, most photobleaching exhibits an exponential decay profile. In this case, it is straightforward to correct any quantitation by a simple normalization to the exponential decay curve. However, for long-time-lapse series of GFP-labeled cells, the absolute quantitation can be complicated by the fact that new GFP is being expressed throughout the experimental period.

IV. Preparation of Purified GFP Samples

To use many of the methods described here, a source of purified GFP is required. Because the addition of an affinity tag does not greatly perturb GFP fluorescence, purification of Histidine-tagged proteins from *E. coli* offers a quick and easily prepared source of highly purified GFPs. The protocol listed here is from an earlier study of GFP comparisons and should be easily adaptable for any GFP mutant (Patterson *et al.,* 1997).

A. Plasmid Construct

The 2.9-kb plasmid pRSET A (Invitrogen Corporation, Carlsbad, CA) was used for expression of Histidine-tagged protein. The cDNA of GFP was subcloned in frame with the hexa-Histidine tag sequence to produce an N-terminal His_6 fusion protein. GFP was prepared by PCR amplification using the N-terminal primer containing a *Bam*HI site and the C-terminal primer containing an *Eco*RI site and the TU#65 plasmid as template. The amplification product was gel purified, digested with the restriction endonucleases, *Bam*HI and *Eco*RI, and ligated into a similarly digested pRSET A to produce the GFP expression plasmid. The pRSET plasmid was transformed into the *Escherichia coli* strain BL21 pLysS for protein expression. In this system, high-level expression is driven by the T7 promoter in front of the His6-tagged proteins and the T7 RNA polymerase is provided by the host BL21 strain (Studier *et al.,* 1990). The pRSET and pLysS plasmids confer ampicillin and chloramphenicol resistance, respectively, to the expression strain.

B. Protein Expression and Purification

The His-tagged GFP protein was expressed in *E. coli* grown at 28°C. A 100-ml starter culture grown overnight in Terrific Broth (Sambrook *et al.,* 1989) containing 100 μg/ml ampicillin and 25 μg/ml chloramphenicol at 37°C was used to inoculate a 1-liter culture at 28°C. This culture was incubated for 2 h, induced with 0.1 m*M* isopropylthio-β-D-galactopyranoside (IPTG), and grown for 5 h before harvesting by centrifugation. The cells were resuspended in sonication buffer (50 m*M* Na_2HPO_4, 300 m*M* NaCl, pH 8.0) and stored at −70°C until purification was continued. After thawing, the cells were lysed by incubation with lysozyme followed by sonication. Insoluble debris were pelleted by centrifugation. The supernatant was incubated with Ni NTA agarose (Qiagen

Inc., Chatsworth, CA) in a 15-ml conical tube (Sarstedt, Newton, NC) for 1 h at room temperature on a rocker platform. The resin was pelleted by slow centrifugation, the supernatant was removed, and the resin was washed once with sonication buffer containing 10 m*M* imidazole. The wash was repeated once with sonication buffer containing 50 m*M* imidazole. The resin was packed into columns and the proteins were eluted from the Ni NTA column with 1 ml of elution buffer (50 m*M* Na_2HPO_4, 300 m*M* NaCl, 200 m*M* imidazole, pH 8.0). Protein concentration was determined by BCA assay, and the purification efficiency was determined by scanning densitometry of SDS gels stained with Coomassie brilliant blue. Only proteins that are purified to >95% homogeneity should be used for standards.

References

Agard, D. A., Hiraoka, Y., Shaw, P., and Sedat, J. W. (1989). Fluorescence microscopy in three dimensions. *Methods Cell Biol.* **30,** 353–377.

Carrington, W. A., Fogarty, K. E., and Fay, F. S. (1990). 3D fluorescence imaging of single cells using image restoration. *In* "Non-invasive Techniques in Cell Biology" (Foskett, J. K., and Grinstein, S., Eds.) pp. 53–72. Wiley-Liss. New York.

Chalfie, M., Tu, Y., Euskirchen, G., Ward, W. W., and Prasher, D. C. (1994). Green fluorescent protein as a marker for gene expression. *Science* **263,** 802–805.

Chattoraj, M., King, B. A., Bublitz, G. U., and Boxer, S. G. (1996). Ultra-fast excited state dynamics in green fluorescent protein: multiple states and proton transfer. *Proc. Natl Acad. Sci. U. S. A.* **93,** 8362–8367.

Cormack, B. P., Valdivia, R. H., and Falkow, S. (1996). FACS-optimized mutants of the green fluorescent protein (GFP). *Gene* **173,** 33–38.

Denk, W., Piston, D. W., and Webb, W. W. (1995). Two-photon excitation in laser scanning microscopy. *In* "The Handbook of Biological Confocal Microscopy" (Pawley, J., Ed.) pp. 445–458. Plenum Press, New York.

Endow, S. A., and Komma, D. J. (1996). Centrosome and spindle function of the *Drosophila* Ncd microtubule motor visualized in live embryos using Ncd-GFP fusion proteins. *J. Cell Sci.* **109,** 2429–2442.

Hampton, R. Y., Koning, A., Wright, R., and Rine, J. (1996). *In vivo* examination of membrane protein localization and degradation with green fluorescent protein. *Proc. Natl. Acad. Sci. U. S. A.* **93,** 828–833.

Heim, R., and Tsien, R. Y. (1996). Engineering green fluorescent protein for improved brightness, longer wavelengths and fluorescence resonance energy transfer. *Curr. Biol.* **6,** 178–182.

Holmes, T. J., Bhattacharyya, S., Cooper, J. A., Hanzel, D., Krishnamurthi, V., Lin, W., Roysam, B., Szarowski, D. H., and Turner, J. N. (1995). Light microscopic images reconstructed by maximum likelihood deconvolution. *In* "The Handbook of Biological Confocal Microscopy" (Pawley, J., Ed.) pp. 389–402. Plenum Press, New York.

Moores, S. L., Sabry, J. H., and Spudich, J. A. (1996). Myosin dynamics in live *Dictyostelium* cells. *Proc. Natl Acad. Sci. U. S. A.* **93,** 443–446.

Niswender, K. D., Blackman, S. M., Rohde, L., Magnuson, M. A., and Piston, D. W. (1995). Quantitative imaging of green fluorescent protein in cultured cells: comparison of microscopic techniques, use in fusion proteins and detection limits. *J. Microscopy* **180,** 109–116.

Ormo, M., Cubitt, A. B., Kallio, K., Gross, L. A., Tsien, R. Y., and Remington, S. J. (1996). Crystal structure of the *Aequorea victoria* green fluorescent protein. *Science* **273,** 1392–1395.

Patterson, G. H., Knobel, S. M., Sharif, W. D., Kain, S. R., and Piston, D. W. (1997). Use of the Green Fluorescent Protein (GFP) and its mutants in quantitative fluorescence microscopy. *Biophys. J.* **73,** 2782–2790.

Potter, S. M., Wang, C.-M., Garrity, P. A., and Fraser, S. E. (1996). Intravital imaging of green fluorescent protein using two-photon laser-scanning microscopy. *Gene* **173,** 25–31.

Rizzuto, R., Brini, M., De Giorgi, F., Rossi, R., Heim, R., Tsien, R. Y., and Pozzan, T. (1996). Double labelling of subcellular structures with organelle-targeted GFP mutants *in vivo. Curr. Biol.* **6,** 183–188.

Sambrook, J., Fritsch, E. F., and T. Maniatis. (1989). Appendix A: Bacterial media, antibiotics, and bacterial strains. *In* "Molecular cloning: a laboratory manual" (C. Nolan, Ed.) p. A. 2. Cold Spring Harbor Laboratory Press, Plainview, NY.

Sandison, D. R., Piston, D. W., Williams, R. M., and Webb, W. W. (1995). Resolution, background rejection, and signal-to-noise in widefield and confocal microscopy. *Appl. Optics* **34,** 3576–3588.

Siemering, K. R., Golbik, R., Sever, R., and Haseloff, J. (1996). Mutations that suppress the thermosensitivity of green fluorescent protein. *Curr. Biol.* **6,** 1653–1663.

Studier, F. W., Rosenberg, A. H., Dunn, J. J., and Dubendorff, J. W. (1990). Use of T7 RNA polymerase to direct expression of cloned genes. *In* "Methods of Enzymology" (D. V. Goeddel, Ed.), Vol. 185, pp. 60–89. Academic Press, San Diego.

Ward, W. W. (1981). Properties of coelenterate green fluorescent proteins. *In* "Bioluminescence and Chemiluminescence" (DeLuca, M. A., and McElroy, W. D., Eds.) pp. 225–234. Academic Press, San Diego.

Xu, C., Zipfel, W., Shear, J. B., Williams, R. M., and Webb, W. W. (1996). Multiphoton fluorescence excitation: new spectral windows for biological nonlinear microscopy. *Proc. Natl. Acad. Sci. U. S. A.* **93,** 10763–10768.

Yang, F., Moss, L. G., and Phillips, G. N., Jr. (1996). The molecular structure of green fluorescent protein. *Nat. BioTech.* **14,** 1246–1251.

CHAPTER 4

Single-Molecule Fluorescence Detection of Green Fluorescence Protein and Application to Single-Protein Dynamics

Daniel W. Pierce and Ronald D. Vale*

Howard Hughes Medical Institute and Dept. of Cellular and Molecular Pharmacology
* University of California
San Francisco, California 94143

I. Introduction

The first measurement of the activity of a single protein molecule was reported for an ion channel using patch-clamp methods in 1976 (Neher and Sakmann, 1976) and spawned the vast enterprise of ion channel biophysics that exists today. Thirteen years elapsed before the second single molecule measurement, which was detection of the movement induced by a single kinesin molecule in 1989 (Howard *et al.,* 1989; Block *et al.,* 1990). Single-molecule measurements on motor proteins have also proliferated (Svoboda *et al.,* 1993; Finer *et al.,* 1994; Ishijima *et al.,* 1994; Coppin *et al.,* 1995; Meyhofer and Howard, 1995). What both these systems have in common is that a unique property of the enzyme was exploited to perform the measurement. Ion channels are essentially amplifiers of their own

Copyright © 1999 by Academic Press. All rights of reproduction in any form reserved.
0091-679X/99 $25.00

activity, allowing picoampere ion currents to flow in response to changes in membrane potential or the presence or absence of ligands. Motor proteins elicit movement, and the forces they produce are sufficient to move objects visible in the light microscope.

Meanwhile, Moerner and Kador (1989) achieved single-molecule detection of a small organic molecule in 1989, again making using of a unique molecular property: fluorescence light emission. The capability of an efficient fluorophore to shift the wavelength of and reradiate thousands of photons per second allowed its detection in a manner analogous to that of ion channels. However, a fluorophore is a far more modular "amplifier" than an ion channel because it can be attached to almost any molecule of interest. In addition, various parameters of fluorescence emission can be measured. The most widespread measurements are those of position and intensity, in which case the fluorescence image of a sample is obtained. However, the spectrum, polarization, and lifetime can also provide valuable information on the environment of the probe. Here we restrict our attention to fluorescence imaging of single molecules.

Two recent developments have set the stage for more widespread use of single-molecule fluorescence detection in the study of protein dynamics. First, the development of a very low-background, high-sensitivity total internal reflection (TIR) fluorescence microscope by Funatsu and co-workers (1995) enabled single Cy3 molecules to be imaged at video rate (30 images per second). This allowed dynamic processes such as the turnover of ATP molecules by surface-bound myosin or the movement of fluorescently labeled kinesin (Vale *et al.,* 1996) to be observed in real time. Second, the cloning (Prasher *et al.,* 1992) and rapid improvement (Heim *et al.,* 1995) of *Aequoria victoria* green fluorescence protein (GFP) has allowed proteins to be labeled at known sites with hitherto unimaginable specificity while circumventing the need to purify the protein and avoiding the inactivation that often accompanies labeling by chemical means.

To be a potential candidate for single molecule imaging, a fluorophore should have an extinction coefficient of at least 10,000 M^{-1} cm^{-1} at a convenient laser wavelength and a high (>0.8) fluorescence quantum yield (Φ_f). However, the most important parameter is the quantum yield for photobleaching (Φ_b) because the ratio Φ_f/Φ_b gives the number of photons available from the fluorophore before photobleaching. This quantity determines for how long, and with what signal-to-noise ratio, a fluorophore can be imaged. In addition to Cy3, single-molecule detection of common fluorophores such as Cy5 (Funatsu *et al.,* 1995), Texas Red, FITC (Enderle *et al.,* 1997), tetramethylrhodamine (Ha *et al.,* 1997), and others has been reported.

Because the GFP mutant S65T has an extinction coefficient of ~39,200 at 488 nm (Heim *et al.,* 1995) and is believed to have a high fluorescence quantum yield, we anticipated that single-molecule GFP imaging would be possible. In the last year, our laboratory (Pierce *et al.,* 1997) and others (Dickson *et al.,* 1997; Iwane *et al.,* 1997) have demonstrated that single-molecule fluorescence detection of GFP is feasible. This method has already proven useful in observing the motion of single kinesin molecules (a motor protein) fused to GFP along microtubules (termed "processive movement") (Case *et al.,* 1997; Pierce *et al.,* 1997). GFP was also found to exhibit fluorescence behavior distinct from that of small organic fluorophores (Dickson *et al.,* 1997; Pierce *et al.,* 1997).

In this chapter, we discuss the features of fluorescence microscopes for single-molecule detection and describe the design and construction of the low-background TIR fluorescence microscope that we have used in our experiments. In addition, the unusual properties of GFP compared to small organic fluorophores are discussed, with particular attention to its behavior under single-molecule imaging conditions. Finally, applications of single-molecule GFP imaging are summarized. We have utilized single-molecule detection of kinesin–GFP, expressed by conventional means or by *in vitro* translation, to develop an assay for visualizing single-molecule motility of this motor protein. The assay is described in greater detail in a related contribution to *Methods in Enzymology* (Pierce and Vale, 1998). GFP fusions of many molecules of interest are already available, and more appear on a daily basis. Thus, it should be possible to extend these studies to other moving-enzyme systems such as helicases, polymerases, and ribosomes, and more generally to the quantitation of affinities and rate parameters of protein–protein interactions. Lastly, preliminary evidence suggests that, in some circumstances, single-molecule imaging of GFP fusion proteins in living cells may be possible.

II. Design Considerations for Fluorescence Microscopes for Single-Molecule Detection

What refinements to a conventional fluorescence microscope are necessary to detect the fluorescence of single molecules? With focused laser illumination and fluorescence collection using a modern 1.3- or 1.4-numerical-aperture (NA) oil-immersion objective lens, signal levels are adequate for detection with cooled CCD cameras and integration times in the range of tens of seconds, ICCD cameras at video rate, or nonimaging photon-counting detectors at kHz rates. The key technical problem is not boosting the signal levels, but reducing the background so that the single-molecule signal is not obscured.

There are two fundamental choices to be made when designing a fluorescence microscope: the illumination mode to be employed and the nature of the detector that will measure the resulting fluorescence. Single-molecule fluorescence detection has been achieved using all three illumination modes in current use, which are epi-illumination (Funatsu *et al.,* 1995; Ha *et al.,* 1996), near-field excitation (Betzig and Chichester, 1993; Ha *et al.,* 1996), and TIR illumination (Funatsu *et al.,* 1995; Macklin *et al.,* 1996; Dickson *et al.,* 1997; Lu and Xie, 1997; Pierce *et al.,* 1997). In epi-illumination, which is employed in all standard fluorescence microscopes, the excitation light is focused on the sample by the lens that is used to collect the fluorescence and therefore traverses the sample traveling away from the collection lens. In near-field excitation, an optical fiber with an extremely thin (~10-nm) tip is brought in close proximity to molecules in the sample, resulting in energy transfer from the fiber to the molecule. The advantage of this technique is that, because the excitation is confined to an area related to the diameter of the tip, the diffraction limit of resolution does not apply and images with a resolution of a few tens of nanometers may be obtained. In TIR illumination, the illumination totally internally reflects within a glass or quartz slide at the surface bounding the sample and therefore does not traverse the sample. Fluorophores in the sample near the interface are excited by an evanescent field that weakens exponentially with distance from the interface, with a falloff constant [~150 nm for visible light (Funatsu *et al.,* 1995) at a fused silica/water interface] related to the wavelength of the light, the angle of incidence, and the contrast in refractive index (Axelrod, 1989).

Detectors can be either cameras, in which case an image is obtained directly, or nonimaging devices such as avalanche photodiodes or photomultiplier tubes. In the latter case, images are built up by raster-scanning the sample; a plot of detector signal versus sample coordinates yields the image (Ha *et al.,* 1997). In terms of fundamental characteristics such as quantum efficiency, noise characteristics, and bandwidth, nonimaging detectors are superior. Because images must be built up by raster scanning, frame rates are slow (typically minutes to acquire an image with a surface area of ~100 μm^2), but if high-temporal-resolution data from a fixed position in the sample is desired, these systems can offer bandwidths in the kHz range. Imaging detectors allow acquisition of images with much higher frame rates [up to 30 Hz continuously in the case of intensified charge- coupled device (ICCD) or intensified silicon target (ISIT) cameras] but do not have as high quantum efficiency (50% for a Gen III+ intensified camera at the peak wavelength vs. 80% for a silicon avalanche photodiode) and suffer from intensifier-derived noise, both in the form of "snow" (bright pixels) resulting from thermal emission from the photocathode of the intensifier tube and from amplifier noise in the traditional sense of decreased signal-to-noise ratio of the output relative to the input signals.

In epi-illumination using a camera, the entire field of view is illuminated, whereas the illumination is focused to a diffraction-limited spot to minimize photodamage to the remainder of the sample when a nonimaging detector is employed. TIR illumination could in principle be used with either type of detector, but in practice is used with cameras because the grazing angle of incidence of the illuminating light on the sample interface renders the area illuminated much larger than the diffraction limit (Axelrod, 1989). Near-field illumination, in contrast, is only used in conjunction with nonimaging detectors because only sub-diffraction-limited areas of the sample are illuminated.

Which design is then optimal? If high-bandwidth data from single points in the sample is required, a nonimaging detector is necessary, in which case diffraction-limited epi-illumination or sub-diffraction-limited near-field excitation should be employed. If higher frame rates are required, a camera is necessary, in which case either epi-illumination or TIR illumination can be employed. Because the illumination traverses the entire sample in epi-illumination, background due to out-of-focus fluorescence is increased relative to TIR illumination, and TIR illumination is therefore preferred. However, this reduction in background comes at the cost of only being able to observe parts of the sample within ~150 nm of the surface.

Our experiments require use of an imaging detector with high frame rates, so the microscope constructed in this laboratory employs TIR illumination. However, there are two ways of introducing the illuminating laser beam into the area of the sample: through the slide on the far side of the sample from the objective, or through the objective lens (Axelrod, 1989) (Fig. 1). If the sample is illuminated from the far side of the objective, the short working distances and aberration correction characteristics of 1.4-NA oil-immersion objectives restrict the thickness of the samples that can be imaged to a few tens of microns, because the objective must focus across the sample to the slide surface. This restriction makes use of microscopic perfusion chambers (flow cells) impractical and examination of thicker samples such as tissue slices impossible. Through-the-objective or "prismless" TIR circumvents these limitations because the coverslip surface is imaged. A 1.4-NA oil immersion objective is an absolute requirement in this case to attain a high enough angle of incidence to obtain total internal reflection inside the coverslip. However, because all the illumination light passes through the objective lens, immersion oil, and coverslip, background autofluorescence from these materials decreases the signal-to-background ratio. High-NA objectives make use of high-index rare earth glasses, and substitutes using quartz are not available. Use of a low-fluorescence silicon oil substitute for standard immersion oil is recommended in this case (Funatsu *et al.,* 1995), but the background is still on the order of threefold higher when using 488- or 514-nm illumination as compared to TIR illumination from the far side of the objective (T. Funatsu, personal communication). Because we wished to optimize our system for single-molecule detection at video rate and did not envision imaging thick specimens, our instrument employs TIR illumination from the far side of the objective.

TIR illumination reduces background, because the illumination light is confined to a thin boundary layer between the sample and the slide or coverslip. However, even when TIR illumination is used, several technical points need to be optimized for good single-molecule viewing. First, illumination light scattered by optical imperfections at the slide/sample interface or sample objects resting on this interface is typically as much as six orders of magnitude brighter than the fluorescence of a single molecule. Thus, the dichroic mirrors and barrier filters used to isolate the fluorescence must be designed to attenuate the illumination wavelength by at least this amount while passing the fluorescence band as efficiently as possible. Second, care must be taken in choosing all the optics that the illumination light passes through to minimize autofluorescence, particularly in the immediate area of the sample. Ultraviolet-grade synthetic fused silica

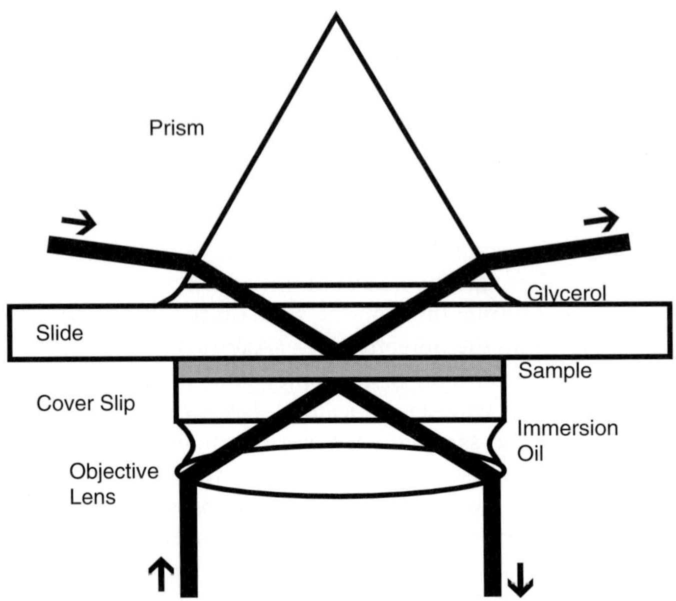

Fig. 1 Possible geometries for total internal reflection (TIR) illumination of the sample. The ray in the upper part of the figure shows the pathway for TIR from the far side of the objective. It enters a prism (preferably antireflection coated and fabricated from UV-grade fused silica) and exits the bottom (uncoated) surface into liquid glycerol (n = 1.47), which approximately matches the index of refraction of fused silica (1.43). The beam continues into the fused silica microscope slide, again experiencing minimal change of refractive index at the interface. At the lower surface of the slide, the index contrast between the fused silica and the aqueous sample leads to total internal reflection inside the slide if the angle of incidence is greater than ~66°, assuming the index of the sample is close to that of pure water (n = 1.33). Note that in this geometry, the surface of the slide, not the surface of the coverslip, is imaged. Because high-NA plan-apochromatic objective lenses are corrected to image objects near (within a few microns of) the surface of the coverslip, aberration correction and image brightness both decline for thicker samples, and the sample thickness that can be imaged at all is limited by the working distance of the objective. The ray in the lower part of the figure shows the path followed by the illumination for TIR through the objective. The laser beam is introduced into the back aperture of a NA 1.4 objective at the extreme edge of the back aperture propagating parallel to but displaced from the optic axis. Total internal reflection occurs at the interface between the coverslip and the sample, so sample thickness is not an issue. However, because the full-strength illumination passes through the objective lens, autofluorescence of the optical glasses and cements in the objective can lead to significant increases in background light levels.

(quartz) has the lowest level of fluorescent impurities of all the commonly available optical materials used for visible light and is therefore the material of choice. Third, care should be taken that the bandpass of the filters employed to isolate the fluorescence does not include the major ~3400 cm^{-1} Raman line of water (assuming aqueous samples will be imaged) to eliminate this background source. For example, for 488-nm excitation, the water Raman line will appear at ~584 nm, so the barrier filter should be chosen to attenuate this wavelength.

Dust was also minimized in the Funatsu *et al.* work (1995) by placing the entire microscope in a clean room, because the autofluorescence of dust particles is on the same order as the intensity of single fluorophore. However, we have not found it necessary to operate in a clean room environment to reduce dust background to negligible levels.

TIR optics can be set up on commercial upright or (preferably) inverted microscopes without great difficulty (Axelrod, 1989). The advantages of this approach are that the optical performance of the microscope with regard to aberration correction is preserved and that optical path lengths are generally short, yielding a compact instrument. The disadvantages are that commercial microscopes generally contain numerous optics between the sample and the detector, and each of these components causes an inevitable loss of light. In addition, the user has less flexibility in determining the overall magnification of the system. With a custom optical bench design, there is ready access to the sample area from above so that the TIR prism can be easily moved in and out of position (see Fig. 1), unnecessary folds of the image light path and attendant light losses can be avoided, and high-magnification images can be formed by the use of long optical paths, as opposed to the short paths and compound lens systems (and accompanying reflection losses) found in commercial microscopes. For these reasons, we elected to build a custom optical bench microscope. The optical layout of our instrument is given in Fig. 2. Details of the components used and the purposes they serve are given in Appendix I, which also includes a discussion of camera selection. The legend to Fig. 2 gives further details of the optics.

III. Characteristics of the Fluorescence from Single GFP Molecules

A TIR fluorescence image of a sample containing surface-adsorbed GFP at low density appears as bright spots on a dark background (Fig. 3). What characteristics allow the observer to conclude that the spots arise from the fluorescence of a single GFP molecule? The arguments are best understood in the context of similar observations on Cy3, as this fluorophore exhibits less complex behavior. In the original report by Funatsu and co-workers (1995) of Cy3 single-molecule detection, surface-adsorbed myosin labeled with Cy3 was imaged. In this situation, labeling at independent sites on the myosin is expected to lead to a mixture of single, double, and more highly labeled myosin molecules, so not all the spots were expected to contain only one Cy3 molecule. Three arguments based on the characteristics of the fluorescence led to the conclusion that the dimmer subpopulation of the spots contained single Cy3 molecules: first, the intensities of the spots were quantized (occurred in multiples of a fundamental value), as was expected for statistical labeling. Second, increasing the illumination intensity made the individual spots brighter, but did not reveal new spots that had pre-

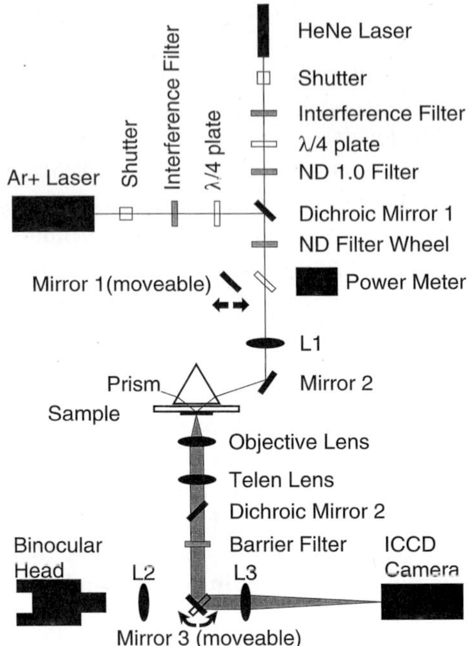

Fig. 2 Optical layout of the TIR microscope constructed in this laboratory. This diagram is intended for schematic purposes only, and not all the fold mirrors for the illumination pathway are shown; all optics present between the sample and the camera are indicated. The objective lens, telen lens, dichroic mirror 2, and barrier filter are housed within a Nikon nosepiece and epifluorescence attachment; the binocular head and lens L2 are from a Zeiss infinity-corrected inverted microscope. All other components are mounted in standard or custom optical bench mounts. The sample is held on a Leitz X–Y stage using a custom-made low-profile slide holder so as not to constrain motion of the prism relative to the slide; the prism, slide, and stage are moved as a unit by the focus micrometer (Newport ESA-CSA electrostrictive actuator). The lasers and ICCD camera are discussed in Appendix I. In brief, the other components are these:

Shutters: Newport 846 HP.

Interference filters: CVI Laser F10-490-4, F03514.5-4, and F10-632.8-4 for the 488-, 514-, and 632-nm laser lines respectively.

λ/4 plates: for 632 nm, CVI laser QWPM-632.8-05-4; for 488 and 514 nm, CVO QWP0-514.5-05-4 (this is a zero-order waveplate with sufficient bandwidth to circularly polarize both 488- and 514-nm light).

ND 1.0 filter: Melles–Griot, quartz substrate.

Dichroic Mirror 1: Chroma 530 DCLP.

ND Filter Wheel: New Focus 5214A.

Mirrors 1 and 2 (and other mirrors not shown): New Focus 5101-VIS.

Power Meter: Melles–Griot 13PEM001.

Lens L1: 100-mm doublet achromat.

Prism: CVI laser custom fabricated from UV-grade fused silica with CVI BBAR antireflection coating for 488–633 nm on the two diagonal surfaces only.

Fig. 3 Image of surface-adsorbed kinesin–GFP. This image is the average of four video frames. The excitation power is 20 mW. A monomeric kinesin–GFP construct at a concentration of 1 n*M* in a low-ionic-strength buffer containing 7.5 mg/ml bovine serum albumin was spotted onto a cleaned quartz slide, covered with a coverslip, sealed with rubber cement, and imaged by TIR fluorescence microscopy. A large fraction of the bright spots in the image correspond to single GFP molecules, although some may contain two GFP molecules whose positions are not resolved. The resolution of the image is diffraction-limited and the diameter of the spots is ~300 nm. Scale bar, 5 μm.

Fig. 2 (*Continued*)

Slides: fused silica, from Matsunami Trading Company, Japan.

Objective lens: Nikon PlanApo100/1.4.

Lens L3: CVI laser-fused silica PLCX-25.4-257.5-425-675, with the last two numbers indicating the approximate range of the antireflection coating. The magnification can be changed (without changing the objective) by substitution of this lens and movement of the camera (mounted on an optical rail) to the new focal plane.

The telen lens is supplied with the Nikon nosepiece. The dichroic mirror 2 and barrier filter sets for GFP, Cy3, and Cy5 are from Chroma (see text and Fig. 6). Mirror 3 is a custom optic from CVI coated with visible-enhanced aluminum.

viously been too dim to see, confirming that sources dimmer than the single-molecule spots were not present. Third, spots containing one quantum of intensity disappeared (photobleached) in an abrupt manner (Fig. 4, trace A), and spots containing two quanta of intensity photobleached in two abrupt steps (Fig. 4, trace B).

Similar data for GFP spots is shown in Fig. 5. Parallel arguments to all three mentioned above for Cy3 can be advanced. In this case, the labeling is not

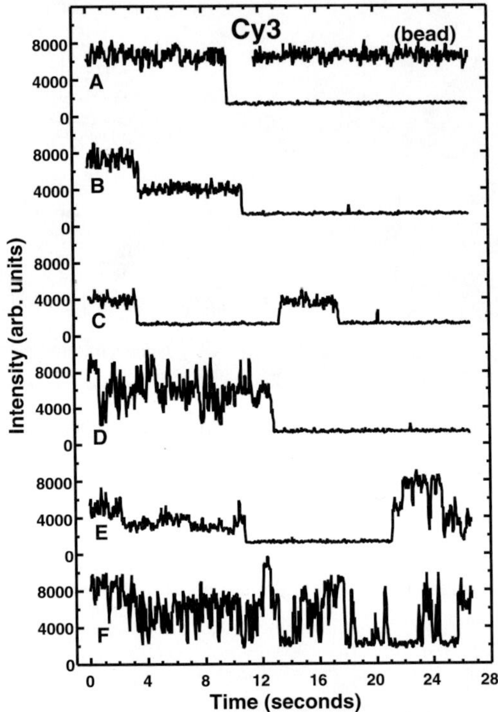

Fig. 4 Cy3 single molecule spot intensity records. Most of the records obtained resembled record A, with a stable fluorescence intensity followed by an abrupt photobleaching event. The signal measured from a 100-nm-diameter fluorescent microsphere using 10^4 lower illumination intensity is also shown. The noise in this trace is instrument limited and is due to photon statistics and the camera intensifier. Because the sample imaged was a kinesin motor chemically labeled with Cy3, some motors are expected to carry more than one Cy3 molecule, and the two-step photobleaching behavior shown in record B indicates that two Cy3 molecules were present in the corresponding spot in the image. The remainder of the records illustrate that, although uncommon, more complicated behavior is also observed with Cy3. Record C shows that the photobleaching is not always irreversible and photobleached molecules occasionally recover. Records D and E illustrate that not all Cy3 molecules give a stable intensity. The abrupt transitions in these records confirm that they are single-molecule fluorescence, as opposed to multiple molecules or an artifact such as autofluorescence of dust. Record F is an extreme example and illustrates the most complex behavior seen in hundreds of records.

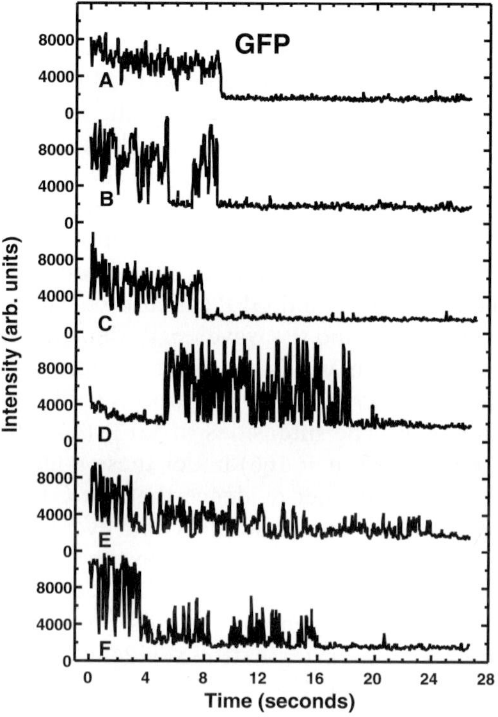

Fig. 5 GFP single-molecule spot intensity records. These records were selected to illustrate the types of behavior observed. Record A shows the classic behavior expected of a unitary photobleaching event of a single molecule; however, note that before bleaching, the noise level is higher than that observed for a Cy3 molecule or fluorescent bead at a similar intensity (Fig. 4). A majority of the records resemble one or the other of the examples shown. Record B shows a molecule that first exhibited "blinking" behavior, then photobleached for a ~2-s interval, recovered for another ~3 s, and finally photobleached. Record C was chosen to illustrate the high noise, rapid blinking, and decline in the average intensity that were often observed. Record D shows the extreme case of "blinking" behavior, with on the order of 100 trips to the baseline level occurring within the observation time. Records E and F show yet more complex behaviors, with declining fluorescence intensity, blinking, and the absence of a clearly defined, unitary photobleaching event. All the records were obtained from image data by summing up the intensity present in a 7- × 7-pixel window (corresponding to the diffraction-limited spot size) in each video frame over the 27-s time interval shown. Each record therefore contains 30 × 27 = 810 intensity measurements. Video data was obtained without frame averaging at an excitation power of ~1600 W/cm^2. Acceptable signal-to-noise images may be obtained at video rate with a fourfold lower excitation power, with a corresponding decrease in the rate of photobleaching.

statistical, so a quantized intensity histogram is not expected. However, an analogous argument can be made by comparing the intensities of monomeric and dimeric protein constructs (Pierce *et al.,* 1997). Two dimeric constructs were found to have an average spot intensity of 1.9 and 2.1 times that found for a monomeric construct under identical illumination conditions. Increasing the illumination intensity was found to increase the brightness but not the number of spots present in an image. Last, and most important, abrupt, quantal photobleaching events were often observed (Fig. 5, trace A).

In addition to these arguments, the intensities of single GFP spots and single Cy3 spots (obtained by labeling a kinesin construct at very low stoichiometry) can be quantitatively compared to see if they occur in the predicted ratio. The extinction coefficient of Cy3 at 514 nm is 67,500 M^{-1} cm^{-1}, whereas that of GFP S65T is 39,200 M^{-1} cm^{-1} at 488 nm, allowing calculation of the relative excitation rates of the two fluorophores at a given laser power. The overall efficiency of detection for Cy3 and GFP can also be calculated based on the filter set spectra shown in Fig. 6 and the wavelength-dependent quantum efficiency of the camera. If the quantum yields for fluorescence are the same for GFP and Cy3, the predicted GFP/Cy3 intensity ratio is 0.90 for twofold more intense illumination of the GFP. The intensities of GFP (44,056 $\pm$ 1302; n = 182) and Cy3 spots (51,043 $\pm$ 1592; n = 166) under these conditions give a ratio of 0.86. This close agreement is further evidence that the GFP spots are single molecules.

Lastly, reversible photobleaching events in which the fluorescence disappears and then reappears at the same level after some delay are occasionally observed for Cy3 (Fig. 4, trace C). For GFP, this behavior is more the rule than the exception, and it generally occurs on a faster time scale (Fig. 5, trace B and others). This behavior is strong evidence of the quantal nature of the fluorescence for both GFP and Cy3.

However, the GFP spot records differ in many respects from those obtained for Cy3. A large majority of the records observed for Cy3 follow the pattern of trace A in Fig. 4, where the intensity is stable and well defined until an abrupt photobleaching event. The noise level at this intensity is comparable to that obtained for a fluorescent bead illuminated with very dim laser light (~10^4 attenuation relative to the intensities used for single-molecule imaging). This indicates that the noise is dominated by photon statistics and intensifier noise from the camera as opposed to true fluctuations in the intrinsic brightness of the fluorophore. In contrast, none of the records for GFP show low, instrument- and photon-statistics-limited noise levels (Pierce *et al.*, 1997). Some of the records follow the pattern of trace A in Fig. 5, but a majority are more similar to one or the other of traces B to F. For a large majority of records, at least one clear example of an abrupt transition from fluorescent to photobleached (or vice versa) is present, and the hypothesis that this complex intensity behavior is due to the presence of multiple GFP molecules is not tenable. Although noisy intensity records are the rule for GFP, traces in D, E, and F in Fig. 4 show that they are also occasionally observed for Cy3.

In addition, the GFP intensity records are characterized by frequent, short-lived periods in which the spot appears to be photobleached but rapidly recovers, shown most dramatically in Fig. 5, trace C (Pierce *et al.*, 1997). Dickson and coworkers (1997) also reported repeated photobleaching and recovery cycles, which they termed "blinking," and additionally found that the long-lived photobleached

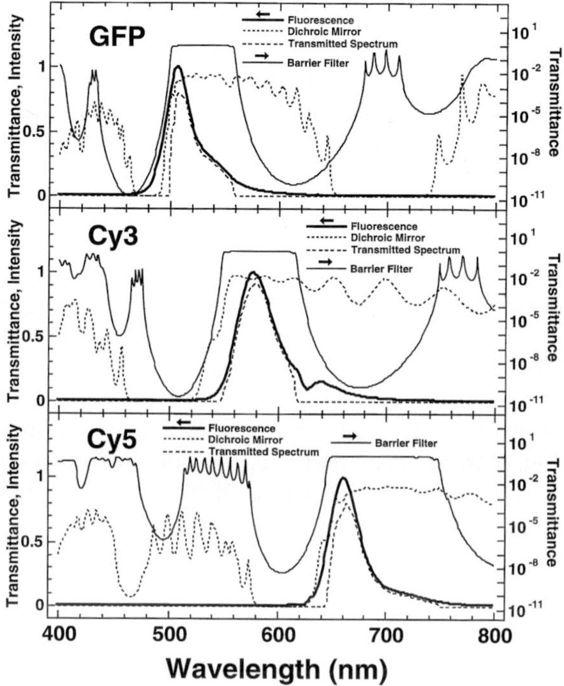

Fig. 6 Spectra of the chromophores and corresponding dichroic mirror/barrier filter sets. These filter sets were optimized for the maximum possible transmission of the fluorescence spectrum consistent with acceptable ($>10^6$) attenuation of the laser line used to excite each fluorophore. The fluorescence spectra are plotted on the left intensity axis and are not corrected for distortion due to the wavelength- and polarization-dependent sensitivity of the SLM fluorimeter employed. The filter sets were custom manufactured by Chroma and the spectral data were supplied by them. Transmittance spectra of the dichroic mirrors are for unpolarized light at 45° incidence and are plotted on the left transmittance axis. Transmittance spectra of the barrier filters are for normal incidence and are plotted on the logarithmic right axis. The transmitted spectrum (i.e., the spectrum one would measure after filtering) is the product of the preceding three spectra and is plotted on the left axis. The integrated areas of these transmitted spectra are 62, 75, and 72% for GFP, Cy3, and Cy5, respectively. The overall detection efficiency (as defined by the ratio of the number of photoelectrons generated at the intensifier tube in the camera to the number of photons collected and passed by the objective) may be calculated by multiplying the transmitted spectra by the quantum efficiency curve of the camera (not shown), and is 23, 30, and 30%, respectively.

state could be rendered fluorescent again by illumination at shorter wavelength. In their case, this wavelength was 405 nm, but it should be noted that this data was obtained on two yellow-shifted GFP triple mutants (S65A/S72A/T203F and S65A/S72A/T203Y), not with GFP S65T.

Lastly, we noted (Pierce *et al.,* 1997) that the fluorescence intensity of (unbleached) spots tends to decline with time. In order to quantitate this effect, the photobleaching rate was measured in two ways. First, 297 records of the type shown in Fig. 5 were averaged, the average of similar records from background areas in the image was subtracted, and the resulting decay was fitted to a single exponential model function, yielding a photobleaching rate constant of 0.48 s^{-1} under these illumination conditions (Fig. 7). Second, the number of spots present at various time delays after illumination was scored from the same image data, and the histogram of spot number versus time was again subjected to exponential fitting, yielding a rate constant of 0.39 s^{-1}. Both methods are insensitive to blinking, as a spot that disappears and then reappears is scored as present in the later image and contributes to the average intensity at that time. The difference is that the averaging method takes into account the quantitative intensity of the spots, whereas the histogram method is binary: spots are either present or absent. The difference in the two rate constants may represent a real, although somewhat subtle, effect, and gives an effective lifetime for this decline of ~10 s. This value is in qualitative agreement with the appearance of records A, C, E, and F in Fig. 5 and many other records not shown. In contrast, a similar analysis of Cy3 fluorescence gave lifetimes of 0.15 s^{-1} (averaging) and 0.16 s^{-1} (spot counting).

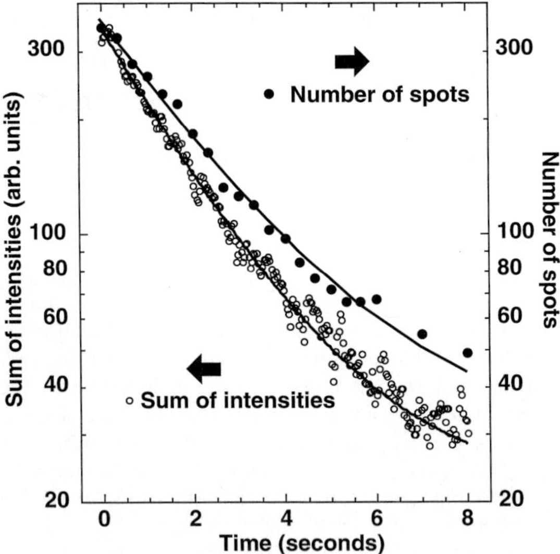

Fig. 7 Photobleaching of GFP spots quantitated by two different methods. In the first method, plotted against the right axis, 297 spot records such as those shown in Fig. 5 were summed, baseline subtracted using records from dark areas of the image, scaled, and fit to a single exponential model function (shown as a solid line through the data), yielding a photobleaching rate constant of 0.48 s^{-1} under these strong (20 mW) illumination conditions. The curvature in the fit is due to the presence of a baseline parameter. In the second method, the number of spots present as a function of time was scored from the same video data and similarly fit to an exponential model function, yielding a rate constant of 0.39 s^{-1}. Because the first method is sensitive to any changes in the average intensity of the spots, whereas the second is not, we interpret this discrepancy to indicate that the GFP spots decline in intensity in an average sense during illumination.

We have also investigated the effect of oxygen depletion and triplet-quenching reagents on the stability of GFP fluorescence. In contrast to the ~fivefold increase of the photobleaching rate of Cy3 when a coupled-enzyme oxygen scavenger system is omitted from the sample (Harada *et al.,* 1990; Pierce and Vale, 1997), GFP is not affected by either oxygen depletion or by addition of the vitamin E derivative Trolox (Aldrich Chemical Company). For both Cy3 and GFP, the photobleaching rate was found to be linear with excitation power over the 1- to 20-mW range tested, indicating that nonlinear phenomena such as excited-state absorption or two-photon absorption do not play a role in the observed photobleaching behavior.

What gives rise to the complex behavior observed for GFP? It is likely that several phenomena are occurring simultaneously. In wild-type GFP, structural (Brejc *et al.,* 1997) and spectroscopic (Chattoraj *et al.,* 1996) evidence suggests that the two observed absorption maxima at 396 and 475 nm (Heim *et al.,* 1994) correspond to the protonated and deprotonated states, respectively, of the hydroxyl group of Tyr-66, which forms part of the chromophore. Excitation at 396 nm results in rapid deprotonation and emission from the deprotonated form (Chattoraj *et al.,* 1996). Due to differences in the hydrogen-bonding pattern around the chromophore in GFP S65T, the 396-nm absorption is absent and the deprotonated state is observed in the crystal structure (Brejc *et al.,* 1997; Ormo *et al.,* 1996). Because the excited state is expected to be even less readily protonated at the phenolic oxygen, it seems unlikely that the blinking behavior is due to direct protonation of Tyr-66; it may result from protonation of the imidazolinone moiety of the chromophore. The long-lived photobleached state may be due to either further relaxation and stabilization of this species, its isomerization to the protonated form of the Tyr-66 side chain, or an independent process. We note that if the long-lived dark state involves protonation of Tyr-66, the resulting species must have an extremely low absorption at 488 nm or it would be rapidly re-ionized under the strong illumination conditions required for single-molecule imaging.

Proton transfers, if indeed they occur, do not yet account for the excess noise and possible intensity decrease we observe. The greater noise may not be a separate phenomenon but simply on/off blinking on a time scale too short to be resolved by our video rate apparatus (and hence the traces are not always observed to fall to the background level when the "off" state is generated), or it may result from shifts in the absorption spectrum relative to the laser wavelength and/or a shift of the fluorescence spectrum relative to the cutoff wavelengths of the filter set. Light-driven fluctuations in the GFP absorption and/or fluorescence spectrum, such as those observed for sulforhodamine 101 at a glass/polymethylmethacrylate interface (Lu and Xie, 1997), would also provide an explanation of the apparent intensity decrease, as high-intensity illumination would tend to drive chromophores to configurations that absorb the laser wavelength less strongly. Testing this possibility will require use of an apparatus capable of measuring transient spectra. Another possible source of intensity variations is rotation of the absorption and fluorescence transition dipoles. This could affect both the probability of absorption per unit time and the efficiency with which fluorescence is collected, and could be observable at video rate for surface-adsorbed GFP but unobservable on this time scale for small organic fluorophores due to their rapid motion. Because there is no plausible mechanism by which illumination could orient the sample, such rotations cannot explain the intensity decrease.

To what degree does this complex behavior compromise the utility of GFP for fluorescently tagging proteins for single-molecule studies? This will largely depend on the application. Measurement of phenomena on the same time scale as the blinking, or measurements such as resonance energy transfer that rely on quantitation of intensity differences, could be complicated by such fluctuations. However, in our experience, the blinking and intensity fluctuations do not create serious difficulties in tracking the movements of GFP-tagged kinesin motors, principally because the time scale of the intensity changes is short compared to the duration of a movement. The photobleaching behavior may turn out to be quite useful in certain situations. Because the fluorescence of the yellow-shifted mutants S65A/S72A/T203F and S65A/S72A/T203Y can be regenerated by exposure to 405-nm light, it may be possible to label a population of GFP molecules by photobleaching them, and subsequently reactivate them to examine changes in their distribution (Dickson *et al.,* 1997). If the fluorescence of GFP S65T can also be recovered by exposure to shorter wavelength light, simultaneous illumination at 488 nm and this second wavelength may allow single molecules to be continuously imaged for longer than would be possible with 488-nm illumination alone. Performing detailed side-by-side comparisons of the available GFP variants (Heim *et al.,* 1994; Cormack *et al.,* 1996; Dickson *et al.,* 1997) will be necessary to characterize the unique limitations and advantages of different GFPs for single molecule experiments.

IV. Advantages of Using GFP for Single-Molecule Detection

What are the advantages of fluorescent labeling with GFP relative to labeling by chemical modification with an organic fluorophore? In our experience with GFP fusions of kinesin motors, we have realized several advantages of this approach:

1. Many proteins are partially or completely inactivated by chemical modification. For instance, dye labeling of the *Drosophila* mitotic kinesin Ncd causes inactivation due to the presence of reactive cysteine groups in the motor domain. Even if labeling does not cause inactivation, time must be devoted to experiments demonstrating that this is the case.

2. Labeling reactions require quantities of highly purified protein in the hundreds of micrograms to milligram range, most of which is usually lost in the

separation of the dye-labeled protein from free dye and inactive aggregates (~90% of the starting material in the case of kinesin). For many interesting proteins, purification of the active, native forms is difficult or impossible, and fragments containing tags for affinity purification are often expressed instead. In such cases, GFP fusions can be generated with very little additional effort. Furthermore, purification of GFP fusion proteins is not necessary, as they can be tested in crude extracts. For example, we have shown that kinesin–GFP fusions can be expressed by *in vitro* translation in rabbit reticulocyte and wheat germ lysates, and, when diluted into the appropriate motility buffer, display motility *in vitro*.

3. When proteins are chemically labeled, a statistical mixture of proteins with differing numbers of fluorophores is obtained. This mixture must be characterized, particularly if single-molecule experiments are anticipated, because demonstration of single-molecule protein activity will require that the correspondence between numbers of fluorophores and numbers of protein molecules be established. GFP labeling is nonstatistical, with exactly one GFP molecule per polypeptide chain.

4. Use of GFP fusions saves experimental time. For *in vitro* assays, proteins can be tested in crude extracts or immediately after column purification.

5. *In vivo* experiments on cells and tissues are possible. With chemical labeling, only cells that can be microinjected and only tissues in which one or a few cells contain the labeled protein can be investigated. With GFP fusions, these limitations do not apply.

6. Dickson *et al.* (1997) have shown that GFP fluorescence can be induced by light of the appropriate wavelength. As they discussed, this opens up possibilities for photoactivating fluorescence of GFP fusion proteins *in vitro* or in living cells, as well as potential applications in optical information storage

V. GFP *in Vitro* and *in Vivo* Assays

We have used the methods described in this chapter to image single-molecule movements of kinesin. Details of the assay and the preparation of fluorescent kinesin are given in a companion contribution to *Methods in Enzymology* (Pierce and Vale, 1998) and will not be recapitulated here, but we wish to give the reader a sense of the kinds of experiments that may be undertaken using this type of imaging system and GFP-labeled kinesin motors. A single molecule of conventional kinesin (Vale *et al.,* 1985) can walk over many tubulin subunits without detaching and diffusing away from the microtubule. Such continuous single-molecule movement is called "processivity," a term applied to polymerases to describe their long-range movements along a DNA polymer (Capson *et al.,* 1991). Elucidation of the mechanistic basis of kinesin processivity and determination of whether other motors in the kinesin superfamily are also processive are areas of active study (Case *et al.,* 1997).

Processive movement of conventional kinesin was first detected in conventional motility assays (Howard *et al.,* 1989; Block *et al.,* 1990) and relied upon the ability of this enzyme to adsorb to surfaces with very little loss of activity. Proof of single-molecule motility required painstaking statistical analysis. For both generality and convenience, we sought to develop a kinesin processivity assay in which single molecules were directly observed. Because the intensity of a fluorescent spot containing labeled kinesins gives a direct readout of the number of kinesins present, single-molecule imaging provides an attractive route to such an assay, which was realized in 1996 (Vale *et al.,* 1996). Fig. 8 shows the type of data generated by this assay, in which axonemes (a 9 + 2–microtubule array) are adsorbed to the surface of a quartz slide and single molecules of fluorescently labeled kinesin are observed binding to and moving along the axonemes. From the image data, the travel distance, association time, and spot intensity can be measured, allowing assessment of velocity, processivity (the mean distance traveled before dissociation), and the number of kinesins present in the moving

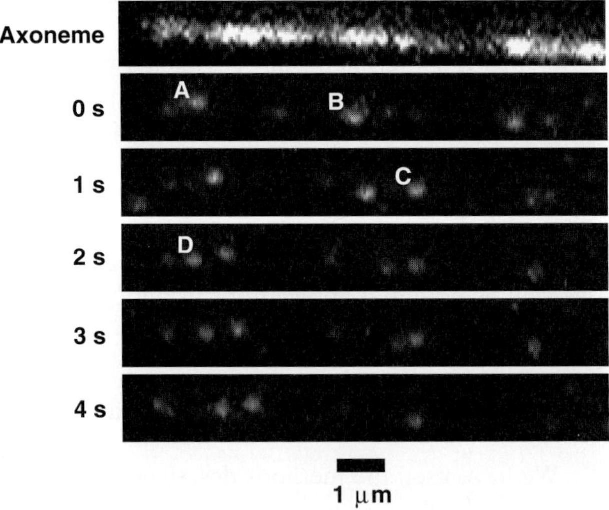

Fig. 8 Movement of kinesin–GFP on an axonemal microtubule. The topmost panel shows an image of a Cy5-labeled axoneme, recorded with a Cy5 filter set and 632-nm excitation. Subsequent panels show the same area imaged with a GFP filter set and 488-nm, 10-mW excitation. Each panel shows a single video frame acquired with a four-frame rolling average. The numbers on the left indicate time in seconds. The letters A, B, C, and D indicate kinesin–GFP molecules. Molecules A and B are bound to and moving along the axoneme over the entire 4-s interval displayed. Molecule B becomes dimmer between 1 and 2 s, probably due to photobleaching of one of the two GFP molecules present in the kinesin–GFP dimer. Molecule C binds to the axoneme or adjacent glass surface between 0 and 1 s, but does not move. Molecule D binds just behind molecule A between 1 and 2 s and maintains its position relative to molecule A. Reprinted from Pierce and Vale (1998).

spot. Ten minutes of recorded videotape are sufficient to score hundreds of movements.

Single-molecule GFP imaging could be applied to develop new *in vitro* assays in many other areas of biology. For example, it should be possible to develop single-molecule *in vitro* assays for DNA- or RNA-based motors such as helicases and polymerases. In some cases, the motion is slower than kinesin-driven movement, but slower processes can be captured with time-lapse data acquisition. These methods may prove useful in assaying the properties of protein complexes that are not motile. Particularly by using two or more fluorescent probes, it should be possible to measure the formation and dissociation of a variety of protein "machines" (e.g., signaling, transcription, and translation complexes). Moreover, by coating a surface with one component of a complex and monitoring the extent of surface binding of other (GFP-tagged) component(s) present at known concentrations in solution, precise affinity constants could be measured. Following substitution of the solution with buffer, the dissociation kinetics could then be measured by following the time course of loss of fluorescence spots from the surface, and association kinetics can be measured by the reverse experiment. Such measurements equate to a single-molecule version of surface plasmon resonance techniques, which are currently widely used to measure protein–protein interactions in bulk solution.

Lastly, we have preliminary evidence that it may be possible to image single GFP molecules in living cells. Images of the lamella of MDCK cells containing a GFP-E-cadherin construct (Y. T. Chen and W. J. Nelson; C. Adams and S. J. Smith, unpublished data) show fluorescent spots that have the intensities and photobleaching lifetimes expected for single GFP molecules. The contact regions between cells, where most of the E-cadherin is localized, are far brighter and show good structural detail. An advantage of imaging living cells on a highly sensitive TIR microscope is that photodamage is minimal. We have been able to image cells for tens of minutes without noticeable effects on cell viability. In contrast, the intense illumination of confocal microscopy, which penetrates the entire cell, can often produce damaging effects with much shorter illumination times. Obviously, a limitation of this approach is that it can only be used for GFP fusion proteins that are adjacent to the cell cortex. However, we envisage that this approach will be useful for providing new information on a variety of plasma-membrane-associated and cytoskeletal proteins.

Appendix I. Details of the TIR Microscope

Figure 2 shows the optical layout of the TIR microscope constructed in this laboratory. Illumination light is provided by an argon-ion (Ar^+) laser (Ion Laser Technology ILT5000 5490AWC) equipped with an intracavity Littrow prism for single-line operation. This laser provides a minimum of 40 mW of TEM0-0–mode, polarized light at 488 or 514 nm. The power can be adjusted between ~2 and

40 mW by adjustment of the tube current, and the light output is feedback controlled. A second helium–neon (HeNe) laser (Melles–Griot 05LHP927) provides 35 mW of polarized, TEM0-0 light at 632 nm without power adjustability or feedback control. It is important that laser sources provide polarized output in order that they may be effectively depolarized before illuminating the sample. Illumination with a polarized laser will result in photoselection (selective excitation of fluorophores depending on their orientation), a generally undesirable source of variation in the fluorescence intensity of isolated fluorophores or oriented sets of fluorophores. However, "unpolarized" laser sources are generally not actually unpolarized, but rather have an output polarization that varies unpredictably on all time scales. A polarized laser can be passed through a quarter-wave plate (oriented at a specific angle relative to the input polarization) to generate circularly polarized light, which may be regarded as depolarized for present purposes [see, however, Axelrod's discussion (Axelrod, 1989) of the polarization properties of the evanescent wave]. In addition to the polarization transformation discussed earlier, each laser is also passed through an interference filter (centered at 490, 514.5, and 632.8 nm, respectively, for the 488-, 514-, and 632-nm laser lines) to ensure the spectral purity of the light impinging on the sample. There are two reasons to further filter the already highly pure spectral output of the lasers: First, cavity fluorescence within the passband of the barrier filter used to detect fluorescence can increase the background, even if it is many orders of magnitude dimmer than the laser line itself (for instance, the 514-nm fluorescence of excited argon ions will pass through GFP barrier filters with high efficiency); second, short-wavelength cavity fluorescence can lead to photobleaching of samples with a quantum efficiency many orders of magnitude higher than that of visible light (K. Q. Lao, personal communication).

After circular polarization and filtering, the Ar^+ and HeNe beams are combined at a dichroic mirror, passed through an ND filter wheel to allow attenuation of the beams to less than the ~2 mW minimum output of the Ar^+ laser, and focused by a 100-mm lens into the sample. After this lens, a mirror mounted on a vertical translation stage serves to route the beam toward the sample at the correct angle for total internal reflection to occur; movement of the translation stage allows the beam to be centered over the objective lens without concomitant change in its angle of incidence. The lens is also mounted on a translation stage to allow for adjustment of the beam diameter at the sample. In practice the lens is not adjusted for the minimum spot size at the sample, as this would cause uneven illumination of the field of view imaged by the camera. The lens is translated away from the position that gives the smallest spot until the 1/e points are approximately twice as far apart as the image area is wide, yielding approximately even illumination. Under these conditions in our microscope, the area illuminated is ~30 μm × ~40 μm. Laser powers of 1–20 mW therefore yield power densities of ~80–1600 W/cm^2 in the quartz slide, and ~320–6400 W/cm^2 in the aqueous sample at the interface.

Fluorescence from the sample is collected by a Nikon PlanApo 100/1.4 objective lens. This lens is held in a standard Nikon nosepiece, which in turn is attached to a Nikon FX-A epifluorescence attachment. The dichroic mirror/barrier filter sets used to isolate the fluorescence bandpass are housed within this attachment and are selected by means of a thumbwheel. Because the objective is not designed for operation at infinite conjugate and the image light must be at infinite conjugate when passing through the dichroic mirror (angled at 45° to the optic axis), a negative telen lens (included with the nosepiece) acts to collimate the image light before it reaches the dichroic mirrors. After the original submission of this manuscript, we replaced this objective with an infinity-corrected Nikon Plan Apo 60×/1.40 objective and removed the telen lens. With either objective, the two lenses supplied in the FX-A epifluorescence attachment (which serve to collimate and then refocus the image light for noninfinity objectives when the nosepiece telen lens is not present) were removed.

The dichroic mirror/barrier filter sets in this instrument were custom made by Chroma, and their spectra are given in Fig. 6. The key features of these filter sets are that they provide at least six to eight orders of magnitude of attenuation of the laser wavelength used to excite the corresponding fluorophore (488 nm for GFP, 514 nm for Cy3, and 632 nm for Cy5) while passing as much of the fluorescence spectrum as possible. In this microscope, the dichroic mirrors do not fulfill their usual role of directing epifluorescence illumination into the objective. However, their presence is still important for two reasons: First, by reflecting >99% of the laser wavelength, they provide an additional two orders of magnitude of reduction of this background source, and second, because less laser light impinges on the barrier filter, background due to filter fluorescence is also reduced by the same factor.

After filtering, the image light is reflected by a visible-enhanced aluminum mirror on a turntable to allow the image to be directed to either a binocular head for viewing or towards the camera. In the latter case, a single planoconvex tube lens focuses the image light onto the active surface of the camera. The focal length of this lens is 561 mm (at 560 nm), and the ratio of this focal length to that of the objective (1.6 mm, obtained by dividing the standard tube lens focal length of 160 mm by the 100× magnification of the objective) gives the 350× magnification of the microscope. The magnification can be changed by either changing the objective or changing the focal length of the tube lens with concomitant movement of the camera to the new image plane.

To achieve the same magnification for all wavelengths across the visible spectrum, high-performance objectives are designed to be used with a particular tube lens located a particular distance from the shoulder of the objective. This design uses a nonstandard tube lens to achieve the desired magnification and is therefore not as well corrected as a standard microscope in this respect. However, the bandwidth of the fluorescence colors detected is relatively narrow and in practice the resolution is diffraction limited, with no observable differences in magnification or location of the image plane for fluorescence centered at ~510 nm (GFP) and ~667 nm (Cy5). This design employs as few optical elements as possible in the image path (two lenses, one dichroic mirror, one barrier filter, and one fold mirror) and therefore minimizes light losses. The fold mirror could in principle be eliminated for the path to the camera, but its presence greatly simplifies the mechanical layout of the microscope.

The entire apparatus is located on a vibration-isolated optical bench air table, and a dramatic loss of resolution is observed if the vibration isolation is compromised. Both the long optical path lengths and lack of a rigid, unitized body make this microscope highly vulnerable to building vibrations relative to a standard design.

The choice of camera is generally by far the most complex issue in component selection. If high frame rates and single-molecule sensitivity are desired, intensified video cameras are at present the best option. Many kinds of intensifier tubes are available. GEN III types are preferred based on sensitivity; a "selected SR UB GEN III+" is the best available intensifier for most visible applications. It has a quantum efficiency of 41% at the peak wavelength of 750 nm and a quantum efficiency of ~37% at 510 nm, near the center of the GFP fluorescence spectrum. However, GEN III intensifier tubes are not widely available, and GEN IV tubes may also offer adequate sensitivity and superior resolution. Either ICCD or ISIT cameras may be employed. ISIT cameras tend to give a slightly "quieter" image, but suffer from greater geometric distortions due to imperfections in the electron optics and lower effective time resolution due to the tendency of a transient bright object to fade slowly over several video frames rather than disappear abruptly from one frame to the next. ICCD designs differ in the manner in which light is coupled from the intermediate phosphor image to the CCD, in what type of CCD is used, and in the number of intensifier stages. Light may be relayed from the intermediate phosphor to the CCD by either a lens or a fiber-optic bundle. The fiber-optic bundle gives a higher effective NA than any lens design and is preferred. The CCD may be either a standard video chip or a scientific-grade sensor. The latter offers greater dynamic range and more pixels; however, depending on the gain setting of the camera, the dynamic range may be limited by intensifier-derived noise. The usefulness of designs incorporating two intensifier stages is sample dependent. In our experience, if other background sources have been carefully minimized and for samples with fluorophore concentrations in the range of 0.1 to 10 n*M*, the intrinsic background due to out-of-focus or rapidly diffusing fluorescence in the sample is significant and further amplification of the image by a second intensifier stage does not yield an image with a higher signal-to-noise ratio. The camera employed in our microscope is made by Stanford Photonics and incorporates a "selected SR UB GEN III+" intensifier tube, fiber-optic coupling of the phosphor to the CCD, and a standard 8-bit video sensor. The output of this camera is contrast enhanced using an Argus 20 image processor (Hamamatsu Photonics) before recording to sVHS videotape. Real-time rolling frame averaging is supported by the image processor, and data may optionally be averaged to increase the signal-to-noise ratio at the expense of temporal resolution.

Intensified cameras, particularly when operated at high gain, suffer from limited dynamic range. Due to intensifier noise, only about 6 bits or 64 gray levels of useful information are obtained, and it is not possible to quantitatively image bright and dim objects in the same field of view. However, the intensifier is able to amplify dim signals above the readout noise of the CCD. If longer (seconds) exposure times can be tolerated, cooled CCD cameras can also detect single molecules and can give images with far better dynamic range. Front-illuminated sensors suffer from low quantum efficiency and should be avoided if the ultimate in sensitivity is desired. Back-illuminated CCD sensors have quantum efficiencies of up to 80%. However, to prevent readout noise from obscuring faint signals, back-illuminated CCD cameras are usually supplied with relatively slow (100 kHz to 1 MHz) pixel clocks, which limits the frame rates that can be obtained (for example, a 1000- × 1000-pixel sensor read at 1 MHz can be read out in 1 s, so this is the fastest frame rate that can be accomplished, assuming no computer overhead time and zero integration time). An attractive option that offers a balance between frame rate and quantum efficiency is the interline transfer microlens array CCD. In this front-illuminated design, a readout register is located between each row of pixels, and light that would have been lost had it fallen on these readout registers is focused into the adjacent row of pixels by a microlens. This boosts the quantum efficiency to the 40 to 50% range, and the presence of the readout registers allows fast (10–20 MHz pixel clock rate) readout with low noise. In addition, like frame-transfer CCD sensors, charge integration in the pixels can occur simultaneously with readout. Although it is interesting, this technology is not mature, and the cameras currently available all fail to realize the potential of these sensors, although improvements are in progress.

Appendix II. Data Acquisition and Analysis

For video-rate systems such as ours and when hours of continuous data collection are anticipated, there is at present still no good digital alternative to sVHS videotape for data storage. The Argus 20 image processor is essentially a standalone frame grabber and allows for basic image quantitation on an individual captured frame, but for frame-by-frame analysis it is far more convenient to digitize limited portions of the image data. Given the speed limitations of available hard drives, it is not at present possible to capture and store to disk long (minutes) segments of 8-bit video data at 640- × 480-pixel resolution without compression. However, with a video capture board that supports on-board compression and a fast disk array, it is possible to capture and store minutes of images. Data such as the intensity records shown in Figs. 4 and 5 requires digitization of the video frames followed by frame-by-frame quantitation of the intensity present within a box of pixels centered on a fluorescence spot. Because analysis of 30 s of data requires quantitation of many such regions within each of 900 frames, there is no alternative to automation of this process. We have developed a set of virtual instruments in the LabVIEW programming environment (National Instruments), using image display and manipulation functions from the IMAQ VI library, that accomplish the necessary tasks. For instance, to define the center coordinates of fluorescent spots, one virtual instrument allows the user to view an image and approximately locate fluorescence spots with the computer mouse, after which the virtual instrument refines the spot positions by two cycles of searching for the brightest pixel within a mask centered on the initial location followed by two cycles of refinement based on calculation of the first moment of the intensity distribution. A second virtual instrument takes a set of refined spot locations as its input, quantitates the intensity within a mask centered at each location in a series of images, and writes out records of intensity versus time.

References

Axelrod, D. (1989). Total internal reflection fluorescence microscopy. *Meth. Cell Biol.* **30,** 245–270.

Betzig, E., and Chichester, R. J. (1993). Single molecules observed by near-field scanning optical microscopy. *Science* **262,** 1422–1428.

Block, S. M., Goldstein, L. S., and Schnapp, B. J. (1990). Bead movement by single kinesin molecules studied with optical tweezers. *Nature* **348,** 348–352.

Brejc, K., Sixma, T. K., Kitts, P. A., Kain, S. R., Tsien, R. Y., Ormo, M., and Remington, S. J. (1997). Structural basis for dual excitation and photoisomerization of the *Aequorea victoria* green fluorescent protein. *Proc. Natl. Acad. Sci. U. S. A.* **94,** 2306–2311.

Capson, T. L., Benkovic, S. J., and Nossal, N. G. (1991). Protein-DNA cross-linking demonstrates stepwise ATP-dependent assembly of T4 DNA polymerase and its accessory proteins on the primer-template. *Cell* **65,** 249–258.

Case, R. B., Pierce, D. W., Hom-Booher, N., Hart, C. L., and Vale, R. D. (1997). The directional preference of kinesin motors is specified by an element outside the motor catalytic domain. *Cell* **90,** 959–966.

Chattoraj, M., King, B. A., Bublitz, G. U., and Boxer, S. G. (1996). Ultra-fast excited state dynamics in green fluorescent protein: multiple states and proton transfer. *Proc. Natl. Acad. Sci. U. S. A.* **93,** 8362–8367.

Coppin, C., Finer, J., Spudich, J. S., and Vale, R. D. (1995). Measurement of the isometric force exerted by a single kinesin molecule. *Biophys. J.* **68,** 242S–244S.

Cormack, B. P., Valdivia, R. H., and Falkow, S. (1996). FACS-optimized mutants of the green fluorescent protein (GFP). *Gene* **173,** 33–38.

Dickson, R. M., Cubitt, A. B., Tsien, R. Y., and Moerner, W. E. (1997). On/off blinking and switching behaviour of single molecules of green fluorescent protein. *Nature* **388,** 355–358.

Enderle, T., Ha, T., Ogletree, D. F., Chemla, D. S., Magowan, C., and Weiss, S. (1997). Membrane specific mapping and colocalization of malarial and host skeletal proteins is the *Plasmodium falciparum* infected erythrocyte by dual-color near-field scanning optical microscopy. *Proc. Natl. Acad. Sci. U. S. A.* **94,** 520–525.

Finer, J. T., Simmons, R. M., and Spudich, J. A. (1994). Single myosin molecule mechanics: piconewton forces and nanometre steps. *Nature* **368,** 113–119.

Funatsu, T., Harada, Y., Tokunaga, M., Saito, K., and Yanagida, Y. (1995). Imaging of single fluorescent molecules and individual ATP turnovers by single myosin molecules in aqueous solution. *Nature* **374,** 555–559.

Ha, T., Enderle, T., Ogletree, D. F., Chemla, D. S., Selvin, P. R., and Weiss, S. (1996). Probing the interaction between two single molecules: fluorescence resonance energy transfer between a single donor and a single acceptor. *Proc. Natl. Acad. Sci. U. S. A.* **93,** 6264–6268.

Ha, T., Chemla, D. S., Enderle, T., and Weiss, S. (1997). Single molecule spectroscopy with automated positioning. *Appl. Phys. Lett.* **70,** 782–784.

Harada, Y., Sakurada, K., Aoki, T., Thomas, D. D., and Yanagida, T. (1990). Mechanochemical coupling in actomyosin energy transduction studied by *in vitro* movement assay. *J. Mol. Biol.* **216,** 49–68.

Heim, R., Prasher, D. C., and Tsien, R. Y. (1994). Wavelength mutations and posttranslational autoxidation of green fluorescent protein. *Proc. Natl. Acad. Sci. U. S. A.* **91,** 12501–12504.

Heim, R., Cubitt, A. B., and Tsien, R. Y. (1995). Improved green fluorescence. *Nature* **373,** 663–664.

Howard, J., Hudspeth, A. J., and Vale, R. D. (1989). Movement of microtubules by single kinesin molecules. *Nature* **342,** 154–158.

Ishijima, A., Harada, Y., Kojima, H., Funatsu, T., Higuchi, H., and Yanagida, T. (1994). Single-molecule analysis of the actomyosin motor using nanomanipulation. *Biochem. Biophys. Res. Commun.* **199,** 1057–1063.

Iwane, A. H., Funatsu, T., Harada, Y., Tokunaga, M., Ohara, O., Morimoto, S., and Yanagida, T. (1997). Single molecular assay of individual ATP turnover by a myosin-GFP fusion protein expressed *in vitro. FEBS Lett.* **407,** 235–238.

Lu, H. P., and Xie, X. S. (1997). Single-molecule spectral fluctuations at room temperature. *Nature* **385,** 143–146.

Macklin, J. J., Trautman, J. K., Harris, T. D., and Brus, L. E. (1996). Imaging and time-resolved spectroscopy of single molecules at an interface. *Science* **272,** 255–258.

Meyhofer, E., and Howard, J. (1995). The force generated by a single kinesin molecule against an elastic load. *Proc. Natl. Acad. Sci. U. S. A.* **92,** 574–578.

Moerner, W. E., and Kador, L. (1989). Optical detection and spectroscopy of single molecules in a solid. *Phys. Rev. Lett.* **62,** 2535–2538.

Neher, E., and Sakmann, B. (1976). Single-channel currents recorded from membrane of denervated frog muscle fibres. *Nature* **260,** 799–802.

Ormo, M., Cubitt, A. B., K., K., Gross, L. A., Tsien, R. Y., and Remington, S. J. (1996). Crystal structure of the *Aequorea victoria* green fluorescent protein. *Science* **273,** 1392–1395.

Pierce, D. W., Hom-Booher, N., and Vale, R. D. (1997). Imaging individual green fluorescent proteins. *Nature* **388,** 338.

Pierce, D. W., and Vale, R. D. (1998). Preparation and assay of fluorescent kinesin. *Meth. Enzymol.,* Molecular Motors and the Cytoskeleton, Part B (in press).

Prasher, D. C., Eckenrode, V. K., Ward, W. W., Prendergast, F. G., and Cormier, M. J. (1992). Primary structure of the *Aequorea victoria* green-fluorescent protein. *Gene* **111,** 229–233.

Svoboda, K., Schmidt, C. F., Schnapp, B. J., and Block, S. M. (1993). Direct observation of kinesin stepping by optical trapping interferometry. *Nature* **365,** 721–727.

Vale, R. D., Reese, T. S., and Sheetz, M. P. (1985). Identification of a novel force-generating protein, kinesin, involved in microtubule-based motility. *Cell* **42,** 39–50.

Vale, R. D., Funatsu, T., Pierce, D. W., Romberg, L., Harada, Y., and Yanagida, T. (1996). Direct observation of single kinesin molecules moving along microtubules. *Nature* **380,** 451–453.

CHAPTER 5

Targeting GFP to Organelles

Francesca De Giorgi, Zimran Ahmed, Carlo Bastianutto, Marisa Brini, Laurence Sophie Jouaville, Robert Marsault, Marta Murgia, Paolo Pinton, Tullio Pozzan, and Rosario Rizzuto

Department Biomedical Sciences and CNR
Center for the Study of Biomembranes
University of Padua
35121 Padua, Italy

I. Introduction

The possibility of directly identifying intracellular organelles in living cells has a major relevance in cell biology studies. Indeed, not only can key events, such as organelle distribution and dynamics, be monitored, in a variety of physiological phenomena (e.g., localized Ca^{2+} rises) the intracellular location can be unambiguously identified. In the past years, the imaging technology (traditional and two-photon confocal microscopy, deconvolution of wide-field images, etc.) has significantly improved, thus increasing the need for dyes capable of specifically labeling the various subcellular structures. GFP, the emerging tool in cell biology, to which this volume is devoted, appears ideally suited for this task. In fact, its

METHODS IN CELL BIOLOGY, VOL. 58
Copyright © 1999 by Academic Press. All rights of reproduction in any form reserved.
0091-679X/99 $25.00

fluorescence depends on an internal chromophore (thus does not require the addition of a cofactor), is not species specific (its simple recombinant expression yields a strong fluorescence signal in cell systems as diverse as bacteria, yeasts, plants, and mammalian cells), and is very resistant to photobleaching. Moreover, its sequence can be modified and appropriate targeting information can be added. In this chapter, we will show that, through suitable chimeras, GFP can be targeted to various intracellular locations (the cytosol, the nucleus, the mitochondria, the endoplasmic reticulum, the Golgi, the subplasmalemmal space), thus providing powerful *in vivo* markers of these compartments.

II. Construction and Expression of the Organelle-Targeted GFP Chimeras

The Cytosol. The starting point for the various GFP chimeras described in this article is the S65T mutant of GFP (Heim *et al.,* 1995). The GFP(S65T) cDNA, which will be also referred to in this chapter as cytGFP, has been modified at the 5′ end of the coding region to include the HA1 epitope tag and an appropriate cloning site (Fig. 1A); when expressed in mammalian cells, the recombinant protein shows mainly a cytosolic distribution with no nuclear exclusion (Fig. 2A). This modified cDNA, encoding a GFP moiety with the HA1

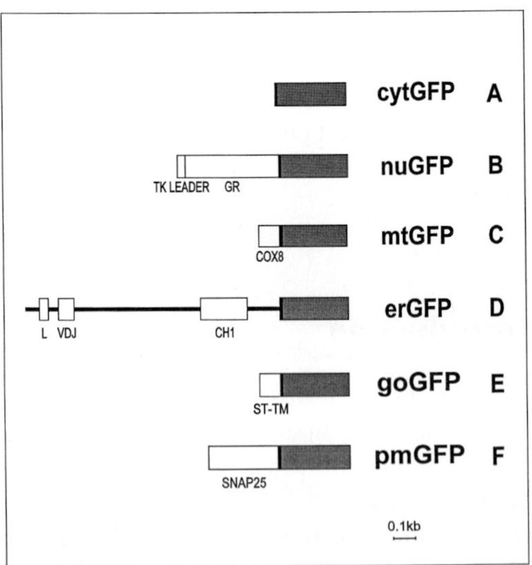

Fig. 1 Schematic representation of the GFP chimeras: cytGFP (A); nuGFP (B); mtGFP (C); erGFP (D) goGFP (E); pmGFP (F). The GFP moiety is indicated in gray; the HA1 epitope is indicated in black.

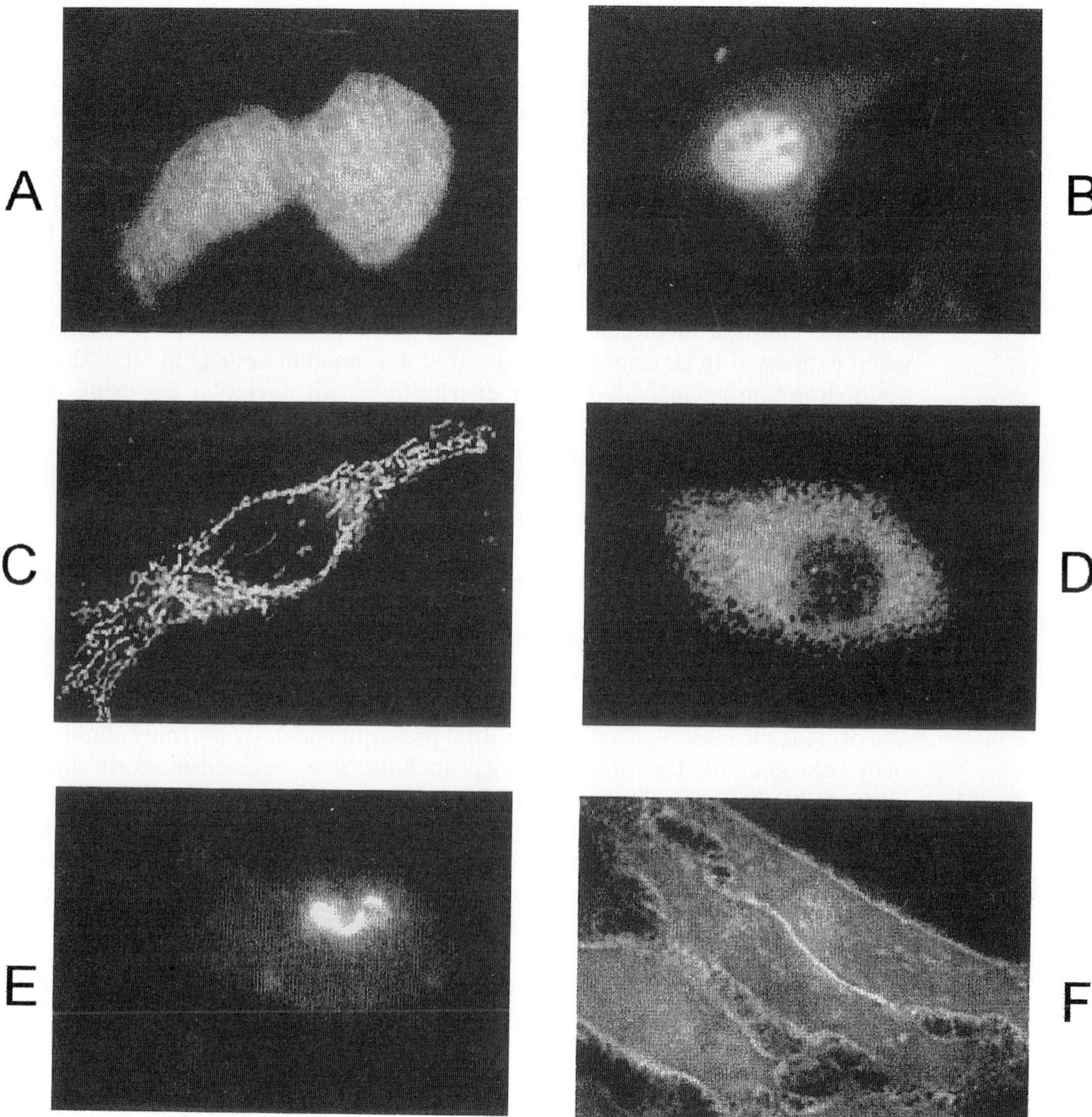

Fig. 2 HeLa cells expressing organelle-targeted GFP: cytGFP (A), nuGFP (B), mtGFP (C), erGFP (D), goGFP (E), pmGFP (F). Cells have been observed 36 h after transfection using the CCD camera. The nuGFP-expressing cell has been treated for 4 h with 10 μM dexamethasone.

epitope at the N terminus represents the initial building block of all targeted chimeras.

The Nucleus. The nuclear localization signal of nuclear-targeted GFP (nuGFP) was obtained from the glucocorticoid receptor (GR), a polypeptide that is translocated to the nucleus when, upon the conformational change induced by hormone binding, a typical nuclear localization signal (aa 497–795 of the GR sequence) is exposed (Picard and Yamamoto, 1987). A schematic map of nuclear-targeted GFP (nuGFP) is shown in Fig. 1B: nuGFP includes aa 407–794 of GR (i.e., both the NLS and the hormone-binding domain) and GFP, and thus can be expected to retain the sorting fate of the wild-type GR, as was previously shown for an aequorin chimera including the same GR domain (Brini *et al.,* 1993). Indeed, when expressed in dexamethasone-treated mammalian cells (Fig. 2B), the chimera shows an exclusively nuclear distribution with nucleolar exclusion. With wild-type GFP, the staining pattern was the same, but the intensity of fluorescence was markedly lower (Rizzuto *et al.,* 1996).

The Mitochondria. All mitochondrial proteins but the 13 encoded by the organellar genome are synthesized on cytoplasmic ribosomes and then imported into the organelle. In most cases, importation depends on the presence of a cleavable signal at the N terminus of a precursor protein (Hartl *et al.,* 1989). This signal (rich in basic and hydroxylated residues, and devoid of acidic ones), usually referred to as mitochondrial presequence, is removed after import by matrix proteases (Hendrick *et al.,* 1989). When added to a heterologous protein, a mitochondrial presequence is sufficient to drive its import into mitochondria. We previously employed a mitochondrial presequence derived from subunit VIII of cytochrome *c* oxidase to construct a mitochondrially targeted aequorin chimera (Rizzuto *et al.,* 1992). The same cDNA fragment (encoding the 25-aa-long presequence and 6 aa of the mature polypeptide) was fused in frame with HA1/GFP(S65T) to construct mitochondrially targeted GFP (mtGFP) (Rizzuto *et al.,* 1995) (Fig. 1C). When expressed in mammalian cells, the chimera shows a typical mitochondrial distribution (Fig. 2C), as is also verified by the colocalization with a mitochondrial resident protein (not shown).

The Endoplasmic Reticulum (ER). Resident ER proteins are usually correctly sorted because of the presence of two signals: an N-terminal hydrophobic leader sequence, and a "retention" signal, which prevents their escape into the secretory pathway. In most cases, the latter signal is a defined sequence (KDEL) located at the C terminus of the protein (Munro and Pelham, 1987). This tetrapeptide is recognized by a specific receptor, which mediates the retrieval into the ER of the protein. In other cases, however, the retention signal is not at the C terminus but is located in internal domains of the protein. One of these cases is the immunoglobulin heavy chain, which, in the absence of the light chain, is retained in the ER because of the binding of the CH1 domain to the resident ER protein BiP (Sitia and Meldolesi, 1992). This binding is displaced only by the light chain; in cells that don't express the latter polypeptide, the Ig heavy chain can thus be considered a bona fide ER protein. To include all the targeting information at one end of the protein, we decided to take advantage of this sorting mechanism.

The ER-targeted GFP chimera (erGFP) thus includes a portion of the μ heavy chain (leader, VDJ and CH1 domain) and GFP (Fig. 1D). When expressed in mammalian cells, the strong fluorescent labeling shows the typical reticular pattern of ER (Fig. 2D). As previously shown for a similar aequorin chimera (Montero *et al.,* 1995), this staining overlaps with that of calreticulin, an ER-resident Ca^{2+}-binding protein. In the case of erGFP, a dramatic difference was observed in the fluorescent signal of the chimera including wild-type GFP and the S65T mutant. Immunofluorescence revealed that the recombinant protein was properly sorted and expressed at comparable levels in the two cases. However, although with erGFPwt very few, weakly fluorescent cells could be identified, with erGFP(S65T) a strong labeling was observed in ~50% of the cells. The difference in fluorescence intensity, much more pronounced than with the other organelle-targeted chimeras, may depend, at least in part, on the different rate of chromophore formation (Heim *et al.,* 1995), which may be crucial in the intralumenal environment of the ER.

The Golgi. The Golgi retention of glycosylation enzymes resident in the Golgi has been shown to depend on their single transmembrane domain (TMD), in a process in which the length of the TMD appears to play a key role (Masibay *et al.,* 1993). Sialyltransferase (ST) has been employed as a model system for these studies, and, indeed, it has been shown that the TMD of ST causes the retention of heterologous proteins in the Golgi (Schwientek *et al.,* 1995). We thus constructed a cDNA chimera encoding the TMD of ST fused to GFP (Fig. 1E). When expressed in mammalian cells, the recombinant fluorescent protein appears largely confined to the Golgi (Fig. 2E), although in some cells a weak staining of the ER can also be observed.

The Plasma Membrane. Two different approaches could be employed for targeting GFP to the inner face of the plasma membrane: (*a*) fusing GFP to a cytosolic domain of an integral plasma membrane protein or (*b*) fusing to GFP the signal that recruits cytosolic proteins to this restricted space. Our past experience with aequorin chimeras showed that the latter approach is significantly more efficient. In fact, whereas a glutamate receptor/aequorin chimera was only partly sorted to the plasma membrane (Rizzuto, R., unpublished observations), a fusion protein including aequorin and SNAP25 (a polypeptide that is posttranslationally recruited to the plasma membrane upon palmitoylation of four internal cisteine residues), was fully sorted to the subplasmalemmal region (Marsault *et al.,* 1997). Figure 1F shows the schematic map of the plasma-membrane-targeted GFP (pmGFP). When pmGFP was expressed in mammalian cells, a clear membrane staining could be observed (Fig. 2F), which was also confirmed by the confocal analysis of the transfected cells.

III. Dynamic Monitoring of Organelle Structure with the Targeted GFPs

The GFP chimeras described in this chapter are currently utilized in our lab for monitoring the dynamic changes of organelle structure that occur in living cells. For this purpose, the cells are seeded onto round coverslips (diameter 24 mm) and transfected with the various GFP constructs, as described in the Section VI, "Protocols." The coverslip is fitted at the base of a thermostatted chamber, which is placed in the microscope stage. Two types of instruments are currently used for detecting GFP fluorescence: (*a*) a wide-field fluorescence imaging system, based on a Zeiss Axiovert inverted microscope and a Princeton Instruments back-illuminated camera, and (*b*) a Nikon RCM 8000 confocal microscope. In both cases, the microscope is equipped with motorized control of the Z stage, thus allowing the acquisition of three-dimensional images. Figure 3 shows the result of a typical experiment carried out (i.e., the dynamic monitoring of ER structure in Cos-7 cells, as obtained by confocal imaging of erGFP expressing cells). It is obvious that by confocal imaging, the dense ER structure of HeLa cells can be finely resolved, and the continuous rearrangement of organelle structure (the three images of the panel are taken 5 min apart) can be clearly appreciated.

IV. Expression in Primary Cultures

Figure 2 showed the expression of the GFP chimeras in the HeLa cell line using standard transfection procedures (see Section VI, "Protocols"). Similar results were obtained with a large variety of cell lines of different embryological origin: CHO (epithelial), L929 (fibroblast), N13 (glial), PC12 (chromaffin), and others (data not shown). Figure 4 (see color plate) shows that this approach can easily be extended to the study of primary cultures, which often require specific culturing procedures.

Figure 4A refers to the expression of the nuGFP chimera in a primary culture of skeletal myotubes. Under appropriate experimental conditions (see Section

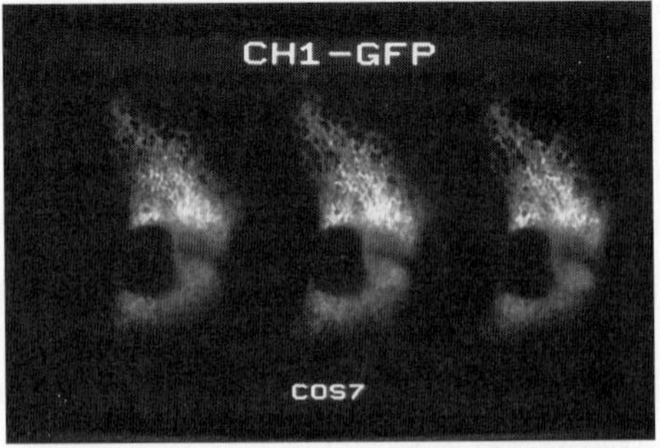

Fig. 3 Dynamic changes of the ER structure in a Cos7 cell in resting conditions: The three images correspond to the same cell expressing erGFP, observed on the confocal microscope at 5-min intervals.

VI, "Protocols"), the expression of a recombinant protein can be restricted, in the mixed culture, to the differentiated myotubes. Indeed, the fluorescent live image of a coverslip of cells reveals a strong, specific labeling of the nuclei of transfected myotubes. Figure 4B shows the expression of mtGFP in a primary culture of cortical neurons derived from rat embryos. Cells were transfected at the fifth or sixth day of culture, and the image was taken 36 h after the same transfection procedure employed for HeLa cells. The fluorescent labeling of the mitochondria both in the soma and in the fine processes can be clearly appreciated. Though for brevity no other example is provided, the technique also proved successful with the other chimeras in a variety of other cell lines and primary cultures (e.g., hepatocytes, hepatic ductal cells, etc.). Thus, although a transfection step is necessary, in our experience this does not represent a major limitation, and targeted GFPs may be regarded as a simple and powerful tool for labeling the various organelles of a living cell.

V. Visualizing GFP Chimeras with Different Spectral Properties

Although GFP itself rapidly emerged as an exciting new tool in cell biology, the successful isolation, in the past years, of various useful mutants has further expanded the possible applications. A first group of mutants essentially improves GFP light emission; among those, the S65T mutant, which, as the wild-type protein, can be excited with blue light and emits green light (peak ~510 nm), is in most applications (including the targeted chimeras described earlier) the GFP moiety of choice, thanks to the higher extinction coefficient and faster rate of chromophore formation. A second group of mutants also appears extremely interesting (i.e., those characterized by the possibility of being excited and/or emitting light of different wavelength) because they open the possibility of differentially labeling two or more structures of the same cell. The GFP mutants will be covered in detail by another chapter of this book, so we will not describe them in detail. We will rather provide a simple example of how the combined use of two GFP variants with different spectral properties allows us to unambiguously label two different intracellular organelles, namely the nucleus and the mitochondria. For this purpose, we employed the (Y66H,Y145F) mutant, which, when excited in the UV, emits blue light (peak ~445 nm) and thus can easily be distinguished from the green GFPs (wt or S65T). Figure 5A (see color plate) shows the fluorescence spectra of the two GFP moieties, thus revealing the major shift in the emission (~65 nm at the peak). Figure 5B (see color plate) shows the fluorescence image of a HeLa cell, transiently expressing nuGFPwt and mtGFP (Y66H,Y145F) upon illumination with UV light. The two GFP moieties can clearly be distinguished, thus allowing specific labeling of the two organelles.

VI. Protocols

A. Primary Cultures

1. Cultures of Rat Myotubes (Brini *et al.*, 1997)

Day 1

Remove posterior limb muscle from newborn rats (2–3 days).

Mechanically dissociate cells.

Incubate for 5 min with 0.125% trypsin in phosphate-buffered saline (PBS) and discard the harvest (containing mostly fibroblasts and endothelial cells).

Dissociate muscle cells by three successive 20-min treatments with 0.125% trypsin in PBS and collect the harvests, neutralizing trypsin with fetal calf serum (FCS).

Filter the cell suspension through a double gauze.

Collect cells by centrifugation (10 min at 1200 rpm in a Haereus Minifuge GL).

Resuspend cells in Dulbecco's modified Eagle medium (DMEM) supplemented by 10% FCS and 4.5 g/liter glucose.

Plate in 10-cm Petri dishes at a density of 10^6 cells/dish and incubate for 1 h at 37°C to decrease the number of fibroblasts.

Collect the nonadherent cells and seed at the density of 2×10^6 cells onto 24-mm coverslips coated with 2% gelatin.

Day 3

Change the medium.

Perform transfection as described here.

Day 4

Change the medium with DMEM supplemented with 10% horse serum to increase myoblast fusion.

Day 5

Add to the medium 5 μM 1β-D-arabinofuranosylcytosine to decrease the number of fibroblasts.

Day 6

Change the medium with a DMEM supplemented with 2% horse serum and keep under these conditions.

2. Cultures of Cortical Neurons

Day 1

Remove the uterus from a Wistar timed pregnant female rat (16–17 days of gestation).

Transfer the embryos in a 10-cm Petri dish with cold PBS supplemented with 6 mM glucose and 1% bovine serum albumin (BSA).

Remove the brains from the embryos and put them in the same buffer.

Using a dissection microscope and optical fibers, remove the meninges, extract the cortical regions, and incubate them in a 35-mm Petri dish containing the same cold buffer.

Transfer the cortex in a tube and incubate for 6–7 min with 4–5 ml of PBS supplemented with 0.33 mg/ml papain (prepare it 1 h before use and store at 37°C).

Change the enzyme solution with PBS and dissociate the cells by 5 or 6 passages through a narrowed bore of a fire-polished pasteur pipette.

Centrifuge cells for 5 min at 800 rpm in a cell centrifuge (Haereus Minifuge GL).

Resuspend in modified DMEM (Brewer and Cotman, 1989).

Plate the cells in 24-mm-diameter coverslips pretreated overnight with polylysine at the density of 2.5–3 $\times$ 10^6 cells/coverslip.

Change the medium when living cells are attached (normally after 4–5 h).

Day 2

After 18 h add 5 μM 1β-D-arabinofuranosylcytosine to the medium to block the glial cells grown.

Day 6–7

Perform transfection as described later.

B. Transfection (Calcium Phosphate Method)

Day 1

Precipitate 4–8 μg of plasmid DNA for each coverslip in 70% ethanol, incubate at -80°C for 30 min, centrifuge at 14,000 rpm in a Microcentrifuge Eppendorf, remove the ethanol, and let the pellet dry for 30–40 min.

Resuspend the DNA in 180 μl of TE (10 mM Tris, 1 mM EDTA, pH 8) and then add 20 μl of a 2.5 M $CaCl_2$ solution.

Add this solution dropwise under vortexing to a tube containing 200 μl of 2$\times$ HBS (280 mM NaCl, 50 mM HEPES, 1 mM Na_2HPO_4, pH 7.12 at 25°C).

Incubate for 30 min at room temperature.

Add the DNA precipitate to the coverslip dropwise.

Day 2

Change the medium.

C. Visualization of GFP Constructs

Mount a coverslip with transfected cells, after 36 h of transfection, on a thermostatted chamber containing saline buffer and observe with one of following systems.

1. Fluorescence Microscope

We use a Zeiss Axiovert inverted microscope with a mercury lamp as the light source.

For GFP S65T we use the following filter set:

excitation HQ480/40
dichroic Q480LP
emission HQ510LP

For BFP we use the following filter set:

excitation D390/20
dichroic 425DCLP
emission D460/50

For contemporary visualization of GFPwt and BFP:

excitation D360/40
dichroic 400DCLP
emission E420LP

All the filter sets are provided from Chroma Technology Corporation.

2. Videoimaging

The microscope previously described is equipped with the following devices to form a system for high-speed acquisition and processing of fluorescence images:

- A computer-controlled light shutter.
- A six-position filter wheel.
- A piezoeletric *z*-axis focus device.
- A back-illuminated 1000 × 800 CCD camera (Princeton Instruments)
- A computer equipped with a software for image acquisition, 2-D and 3-D visualization and analysis (Metamorph software).

3. Confocal Microscopy

We employ the Nikon RCM 8000 confocal microscope based on the Nikon Diaphot 300 inverted microscope, equipped with a 40× water immersion objective (NA = 1.1). The 365-nm band of an argon ion laser is used for excitation of wtGFP and BFP, and the emitted light, separated in two components by a dichroic mirror (475DXCR), is collected by two separate photomultipliers. For visualizing the GFPS65T fluorescence, the 488 band of a helium–neon laser is used and the emitted light (emission filter HQ525/50) is collected by one of the photomultipliers.

References

Brewer, G. J., and Cotman, C. W. (1989). Survival and growth of hippocampal neurons in defined medium at low density: advantages of a sandwich culture technique or low oxygen. *Brain Res.* **494,** 65–74.

Brini, M., Murgia, M., Pasti, L., Picard, D., Pozzan, T., and Rizzuto, R. (1993). Nuclear Ca^{2+} concentration measured with specifically targeted recombinant aequorin. *EMBO J.* **12,** 4813–4819.

Brini, M., De Giorgi, F., Murgia, M., Marsault, R., Massimino, M. L., Cantini, M., Rizzuto, R., and Pozzan, T. (1997). Subcellular analysis of Ca^{2+} homeostasis in primary cultures of skeletal myotubes. *Mol. Cell. Biol.* **8,** 129–143.

Hartl, F. U., Pfanner, N., Nicholson, D. W., and Neupert, W. (1989). Mitochondrial protein import. *Biochim. Biophys. Acta* **988,** 1–45.

Heim, R., Cubitt, A. B., and Tsien, R. Y. (1995). Improved green fluorescence. *Nature* **373,** 663–664.

Hendrick, J. P., Hodges, P. E., and Rosenberg, L. E. (1989). Survey of amino-terminal proteolytic cleavage site in mitochondrial precursor proteins: leader peptide cleaved by two matrix proteases share a three-amino-acid motif. *Proc. Natl. Acad. Sci. U. S. A.* **86,** 4056–4060.

Marsault, R., Murgia, M., Pozzan, T., and Rizzuto, R. (1997). Domains of high Ca^{2+} beneath the plasma membrane of living A7r5 cells. *EMBO J.* **16,** 1575–1581.

Masibay, A. S., Balaji, P. V., Boeggeman, E. E., and Qasba, P. K. (1993). Mutational analysis of the Golgi retention signal of bovine. *J. Biol. Chem.* **268,** 9908–9916.

Montero, M., Brini, M., Marsault, R., Alvarez, J., Sitia, R., Pozzan, T., and Rizzuto, R. (1995). Monitoring dynamic changes in free Ca^{2+} concentration in the endoplasmic reticulum of intact cells. *EMBO J.* **14,** 5467–5475.

Munro, S., and Pelham, H. R. B. (1987). A C-terminal signal prevents secretion of luminal ER proteins. *Cell* **48,** 899–907.

Picard, D., and Yamamoto, K. R. (1987). Two signals mediate hormone-dependent nuclear localization of the glucocorticoid receptor. *EMBO J.* **6,** 3333–3340.

Rizzuto, R., Simpson, A. W. M., Brini, M., and Pozzan, T. (1992). Rapid changes of mitochondrial Ca^{2+} revealed by specifically targeted recombinant aequorin. *Nature* **358,** 325–328.

Rizzuto, R., Brini, M., Pizzo, P., Murgia, M., and Pozzan, T. (1995). Chimeric green fluorescent protein: a new tool for visualizing subcellular organelles in living cells. *Curr. Biol.* **5,** 635–642.

Rizzuto, R., Brini, M., De Giorgi, F., Rossi, R., Heim, R., Tsien, R. Y., and Pozzan, T. (1996). Double labelling of subcellular structures with organelle-targeted GFP mutants *in vivo*. *Curr. Biol.* **6,** 183–188.

Schwientek, T., Lorenz, C., and Ernst, J. F. (1995). Golgi localization in yeast is mediated by the membrane anchor region of rat liver sialyltransferase. *J. Biol. Chem.* **270,** 5483–5489.

Sitia, R., and Meldolesi, J. (1992). Endoplasmic reticulum: a dynamic patchwork of specialized subregions. *Mol. Biol. Cell* **3,** 1067–1072.

CHAPTER 6

Cytoskeletal Dynamics in Yeast

Janet L. Carminati and Tim Stearns
Department of Biological Sciences
Stanford University
Stanford, California 94305

I. Introduction

All eukaryotic cells have at least two cytoskeletal filament systems: actin and microtubules. In different cells these filaments carry out different tasks, but one universal feature is that control over their polymerization and depolymerization is essential to their function. In some situations, the filaments are stabilized so that they provide a permanent framework, such as the actin filaments in muscle.

Copyright © 1999 by Academic Press. All rights of reproduction in any form reserved.
0091-679X/99 $25.00

In others, the filaments must constantly grow and shrink as part of their function, as for microtubules in the mitotic spindle. The dynamics of the cytoskeletal filaments in animal and plant cells have been extensively analyzed with fluorescent probes. Most commonly, the subunit of the filament system under study is purified and labeled with a fluorescent molecule. This labeled protein is then introduced back into cells, where it polymerizes along with the unlabeled subunits. This results in fluorescent filaments that can be studied over time. Such experiments have shown that for both actin and microtubules the length and number of filaments, as well as their location within the cell, can change rapidly.

Although the wealth of genetic information from yeast would be most useful for understanding how the cytoskeleton is controlled, it has not previously been possible to observe cytoskeletal dynamics directly in yeast cells. This is mostly due to the difficulty of introducing fluorescently labeled proteins into the small, cell-wall-bound yeast cell body. However, yeast cells can easily be transformed with exogenous DNA. Thus, by fusing the gene for GFP to the coding sequences for the major cytoskeletal proteins and associated proteins, it is possible to observe dynamics directly. There are interesting similarities and differences when comparing the results from yeast, animal, and plant cells. Finally, there are several advantages to studying these cytoskeletal processes in yeast. First, genetic and molecular techniques, used in combination with the complete genome sequence, allow GFP fusions to any protein to be rapidly constructed, expressed, and analyzed in cells. Second, cytoskeletal dynamics can be examined in the many mutants already known to affect the function of the cytoskeleton. Third, the small size and rapid growth of yeast cells allows observation of the entire cytoskeleton and examination of the changes that occur during the cell cycle.

II. Generating GFP Fusions

The remarkable ability of GFP to form fluorescent molecules in heterologous systems has led to its widespread use in the analysis of both gene expression and protein localization. In yeast cells, expression of GFP alone results in general cytoplasmic fluorescence; GFP is able to enter the nucleus, but is excluded from the vacuole. Cells remain viable, and photobleaching occurs only after several minutes of continued exposure (Stearns, 1995). For the purposes of observing cytoskeletal proteins, gene fusions with GFP must be constructed, and there are several points to consider in the design of a fusion. We will concentrate on the strategies and techniques that our lab has developed for generating and analyzing fusion proteins, although it is important to note that many variations are possible.

A. GFP Fusion Vectors

The goal in designing a GFP fusion protein is to make a fusion protein that is both functional and fluorescent. At the very least, the fluorescent protein should not negatively affect the function of the wild-type protein. Unfortunately, no definitive prediction can be made about whether a particular fusion construct will generate such a functional and fluorescent protein. Therefore, a reasonable approach is to create a variety of fusion constructs, to characterize the properties of the fusion proteins, and to use the "best" one for further analysis.

Two types of yeast vectors can be used to generate GFP fusions: Yeast integrating plasmids (YIps) allow the integration of the fusion protein into the genome, and yeast centromeric plasmids (YCps) allow the expression of the fusion protein from an autonomously replicating element. Centromeric plasmids contain a yeast origin of replication and centromeric sequences, allowing for their relatively stable segregation and maintenance at low copy numbers. Integrating plasmids do not contain a yeast replication origin and must be integrated into the genome for maintenance; integration is done using standard yeast molecular techniques (Stearns *et al.,* 1990). Both types of vectors are available for expression of GFP fusions in yeast, using the *URA3* gene as a selectable marker (Fig. 1). Overexpression of proteins sometimes causes novel phenotypes and altered localization properties, so it is advantageous in studying an uncharacterized protein to integrate the GFP fusion at the endogenous locus, such that the endogenous promoter drives the fusion protein. In this way, a fusion protein can be examined as it is being expressed by its own promoter in the genome. A second advantage of using integrating plasmids becomes apparent with very large genes, for which it is difficult to work with full-length versions on a centromeric plasmid. In such cases it is sufficient to clone a portion of the gene into the YIp vector, creating a GFP fusion to the full-length protein by an appropriate integration event at the chromosomal locus.

When higher levels of expression are desired, overexpression of GFP fusions from centromeric plasmids (YCps) with inducible promoters is favored. High expression levels might be necessary to visualize the localization of an uncharacterized protein and may be useful if overexpression does not result in deleterious phenotypes. High levels of expression can be attained using the galactose-inducible promoter *GAL1-10,* in which the fusion protein is induced by growing cells in galactose-containing media; all of our GFP vectors contain the *GAL1-10* promoter (Fig. 1). All vectors also contain termination sequences downstream of the GFP fusion to aid in its proper expression; the vectors described here contain 400 bp of the *ACT1* 3′ UTR. GFP vectors with different promoter/terminator sequences may be useful for specific needs and can be easily constructed. GFP fusion vectors containing the inducible *MET25* promoter and *CYC1* terminator sequences are also available, as is a gene replacement cassette allowing for expression levels of a promoter of interest to be determined (Niedenthal *et al.,* 1996).

GFP fusions can most easily be made to either the amino or carboxy terminus of proteins (Fig. 1). In most cases, it is not possible to predict a priori which fusion position will result in the most useful fusion protein, so it is best to try both. There are exceptions of course, for example, proteins in which localization

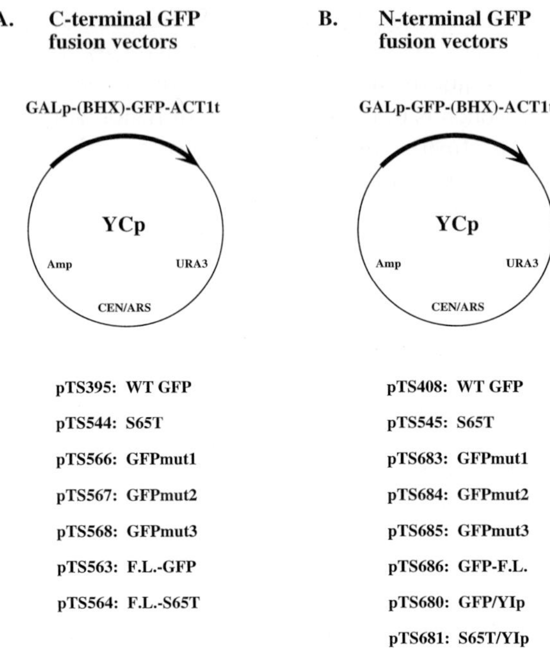

Fig. 1 A, Vectors for fusing GFP to the C terminus of a given protein. pTS395 was constructed by PCRing GFP from pGFP10.1 (Chalfie *et al.*, 1994) with the primers GFP.1 (GCTCTAGACTCGAGATGAGTAAAGGAGAAG) and GFP.2 (GCCGCTAGCGGCCGCTTATTTGTATAGTTCAT). The PCR fragment was cut with *Xba*I and *Nhe*I and inserted at the *Xba*I site of pTS210, a YCp50-based vector with the *GAL1-10* promoter (0.7 kb), a polylinker containing *Bam*HI-*Hind*III-*Xba*I (BHX) sites, and the *ACT1* terminator (0.4 kb) (Carminati and Stearns, 1997). The sequence of the polylinker is GGATCCGCAAGCTTCGTCTAGA. GFP can be fused to the C terminus of a protein by inserting a gene such that its coding sequence is in frame with the ATG of GFP in primer GFP.1. The plasmids listed below have GFP variants in place of wild-type GFP; S65T as described (Heim *et al.*, 1995), and GFPmut1-3 as described (Cormack *et al.*, 1996). pTS563 and pTS564 contain a flexible linker (F.L.) between the polylinker and GFP: GAGAPGAGA. B, Vectors for fusing GFP to the N terminus of a given protein. pTS408 was constructed by PCRing GFP from GFP10.1 with the primers GFP.3 (GCCAGATCTCAAAGATGAGTAAAG) and GFP.4 (CGGGGATCCTTTGTATAGTTCATCCAT). The PCR fragment was cut with *Bam*HI and *Bgl*II and inserted at the *Bam*HI site of pTS210. GFP can be fused to the N terminus of a protein by inserting a gene such that its coding sequence is in frame with the last codon of GFP (TTT in primer GFP.4). GFP variants are as described in A. In addition, two YIp plasmids were constructed from a YIp5 derivative; these can be digested and integrated into the genome. pTS680 and pTS681 are similar to pTS408 and pTS545, respectively, except for the absence of *ARS/CEN*.

information is known to reside in either the N or C terminus, in which case it would likely be best to make the fusion to the other end. Internal fusions are possible in theory, although there is little information on whether GFP can fold productively if both its N and C termini are attached to other protein sequences. It may be desirable to separate GFP and the target protein by a flexible linker sequence, with the notion that a linker without secondary structure would allow both proteins to fold and function without interfering with each other (Fig. 1). For example, a construct containing ten alanine residues between the actin gene *ACT1* and GFP resulted in lower cytoplasmic background than did constructs containing either no linker or a shorter linker of four residues (Doyle and Botstein, 1996).

B. Which Version of GFP?

GFP is a 238–amino acid polypeptide that forms a fluorescent protein in the absence of any specific cofactors. Chromophore formation occurs following the cyclization and oxidation of a tripeptide located at amino acid positions 65–67. Deletion analysis has shown that most of the protein is necessary for chromophore formation (Dopf and Horiagon, 1996), so fusions must be made to the entire GFP coding sequence. Wild-type GFP has a maximal excitation peak at 395 nm and a secondary peak at 470 nm. Excitation at 395 nm is often detrimental to cells and also causes rapid photobleaching of GFP. Excitation at approximately 470 nm results in better cell survival and less photobleaching. Mutant versions of GFP have been made in which the maximal excitation peak is shifted to 470 nm, resulting in brighter fluorescence at this wavelength; these are known as red-shifted mutants (Fig. 1). The most well-characterized of these is a mutation of serine to threonine at amino acid position 65 (S65T) (Heim *et al.,* 1995). GFP(S65T) shows faster folding, has six-fold increased fluorescence, and has been used extensively in yeast (Doyle and Botstein, 1996; Waddle *et al.,* 1996). However, under certain conditions, fusions using GFP(S65T) in yeast show much faster rates of photobleaching under 470-nm illumination when compared to the same fusions using wild-type GFP (Doyle and Botstein, 1996; I. Carminati, unpublished data). Several other versions of GFP have also been constructed; many contain a mutation at Ser-65 along with further mutations and are also useful in generating brighter fusions in yeast (Cormack *et al.,* 1996; Kahana and Silver, 1996) (Fig. 1).

More recently, blue-shifted GFP variants were made that have a single excitation peak at 380 nm and an emission peak in the blue at 445 nm (Heim *et al.,* 1994; Heim and Tsien, 1996). GFP variants that show emission spectra different from those of wild-type GFP will be useful in double-labeling experiments. By generating fusions with distinct GFP molecules, the localization of multiple fusion proteins could be tracked in the same cell *in vivo,* although this has yet to be put into common use in yeast.

C. Characterization of Fusion Proteins

Following the construction of GFP fusion protein constructs, standard techniques are used to generate yeast strains expressing them (Stearns *et al.,* 1990). In keeping with the goal of creating a functional and fluorescent fusion protein, it is useful to consider which strains to use for expression. It is often useful to transform the constructs into both a wild-type strain and a strain that is mutant for the gene under study so that the ability of the constructs to complement the mutant can be assayed. Diploid cells are often used for microscopy, as they are larger than haploid cells.

The fluorescent properties of a given construct are assessed simply by inducing expression of the fusion protein, if necessary, and examining the cells using a fluorescence microscope, as described later. In most cases it is desirable to use a liquid culture of exponentially growing cells, as this assures that most of the cells in the culture are in a similar physiological state. There is much anecdotal evidence to indicate that the growth state of a culture can affect fluorescence, although the parameters that are important have not been worked out. If a particular protein appears to be fluorescent but is only faintly visible, or is only visible in some cells in the population, it can be useful to examine cultures in different states of growth–for example, early log phase versus late log phase, or even stationary phase. Although it is not likely to be optimal, it is also possible to assay cells directly from agar plates, although the physiological state of cells in this case is not uniform. A common error when first looking at yeast cells expressing a fluorescent fusion protein is to mistake dead cells for the desired fluorescent cells. Dead cells are often brightly fluorescent under the conditions used to view GFP but can be distinguished from living cells by phase microscopy: dead cells have a crenelated or granular appearance in comparison to living cells.

It will often be necessary to adjust the expression level of fusion proteins that are being expressed from an inducible promoter, as the full expression level from these promoters is far greater than for most cytoskeletal proteins in yeast. For example, expression of *TUB4* from the *GAL* promoter results in a 300-fold increase over the normal level of Tub4p (Marschall *et al.*, 1996). Tub4p is normally localized to the spindle pole body, but at this high level of expression the protein was present at high levels in both the cytoplasm and the nucleus. The level of expression from the *GAL* promoter can be modulated in several ways. We have had the most experience with adding low levels of the repressing sugar glucose to a culture containing galactose as the main carbon source. For example, growing cells with the GAL–TUB4 construct described earlier in a mixture of 2% galactose and 0.5% glucose reduced expression levels to only a few-fold over those of wild type (Marschall *et al.*, 1996). Kahana *et al.* (1995) have reported obtaining similar results by growing cells initially in 2% raffinose and then inducing with galactose for 1 h, followed by a 6-h incubation in glucose. Modulation of fusion protein level is often important in achieving the best signal-to-noise ratio and in avoiding artifactual localization.

Assessing proper protein localization can be difficult when the protein in question is not a known component of the cytoskeleton but rather is being tested for association with the cytoskeleton by colocalization with cytoskeletal structures. Although the morphologies of the yeast actin and microtubule cytoskeletons are well characterized, there may be cases in which a protein will localize to only a subset of cytoskeletal structures, making colocalization with the filament system less obvious. In such cases, it is most useful to be able to fix the cells so that both GFP fluorescence and immunofluorescence can be used on the same cells. Although this should be straightforward, the fluorescence of some GFP fusion proteins is greatly reduced by formaldehyde fixation, making the experiment difficult. It has been observed that certain fusion proteins are completely nonfluorescent immediately after formaldehyde treatment, but regain their fluorescence after removal from formaldehyde and treatment for immunofluorescence (T. Stearns, unpublished data).

III. Imaging Considerations for Yeast Cells

Yeast cells are small, round, and relatively sensitive to the intense light used in fluorescent microscopy. Each of these properties makes for difficult imaging of fluorescent proteins and must be addressed. The small cell size requires that high magnification be used for imaging. This is typically achieved by using a high-quality 100× objective (typically 1.4 or 1.3 numerical aperture) and further magnifying the image for projection onto the camera. This can be achieved with an optovar type of lens or with an appropriate projection lens. The round shape of the yeast cell can be a problem, depending on the protein being imaged. With the objectives typically used for imaging, only a portion of the cell is in focus at any one focal plane. If it is necessary to follow the fluorescent protein through all focal planes, a microscope with an automated focus control must be used (discussed later). Photodamage is minimized by reducing the light intensity and exposure time as much as possible. For time-lapse analysis, an automatic shutter between the light source and microscope must be used so that the cells are only exposed to light during data collection.

Another consideration in imaging yeast cells is the method of immobilization of cells on the microscope slide. Yeast cells do not stick to untreated glass and will drift away unless care is taken to immobilize them. For relatively short time-lapse imaging (up to 30 min), an aliquot of cells in liquid culture (5 μl) can be spotted directly onto a glass slide and covered with a coverslip. To prevent drifting, the slide is blotted on absorbent tissue with light pressure. Care must taken to not to crush the cells, as they will lyse if too much pressure is applied. Coverslips can be sealed to slides with Valap, a 1:1:1 mixture of Vaseline, lanolin, and wax. Sealing the slides may help reduce photodamage during imaging, presumably because it blocks exchange with atmospheric oxygen and yeast cells use up the existing oxygen under the coverslip (Kahana *et al.,* 1995). For longer time-lapse imaging (several hours), cells can be mounted on low-melt agarose or 25% gelatin made with minimal media as described elsewhere (Doyle and Botstein, 1996; Shaw *et al.,* 1997a). Gelatin has the advantage that it more closely approximates the optical density of the cell wall, allowing for clearer imaging of internal structures by standard light microscopy techniques.

Procedure: Imaging Cells Containing GFP Fusion Proteins

1. Grow 3- to 5-ml cultures in minimal media containing 2% galactose; 0.5–1% glucose can be added to modulate expression. Grow cells overnight to log phase at 30°. Variability in the expression and fluorescence of fusion proteins is sometimes observed, so it is best to grow up several cultures at once.

2. Spot 5 μl onto a microscope slide and cover with a No. 1 coverslip (18 × 18 mm).

3. Turn the slide upside down onto two kimwipes and press gently until liquid appears at the edges of the coverslip. Blotting immobilizes the majority of cells; however, too much pressure will cause cells to lyse.

4. Image on a fluorescence microscope using a 100× objective of the highest numerical aperture available (example: 100×/1.4 Planapochromat from Carl Zeiss Inc., Thornwood, NY). Locate cells at this magnification. For time-lapse imaging, insert a 2.5× optovar if necessary with the microscope/camera system being used.

5. Set up time-lapse image acquisition using imaging software (MetaMorph; Universal Imaging, West Chester, PA). For short time periods (up to 30 min) take 100-ms exposures every 2–10 s.

IV. Time-Lapse Microscopy

A. Microscope Setup and Acquisition of Images

The fluorescence of microtubules generated with GFP–tubulin fusions is both variable and of fairly low intensity, so a microscope setup capable of detecting faint fluorescence is essential. Slow-scan CCD cameras can produce low-noise images suitable for imaging low levels of fluorescence. SIT cameras are also suitable for imaging yeast and have been used in studies of actin cytoskeletal dynamics (Doyle and Botstein, 1996; Waddle *et al.,* 1996). For imaging microtubule dynamics, a cooled CCD camera containing a TK512D chip with a readout noise of 8 e at 100 kHz readout was used in 16-bit mode (Princeton Research Instruments Inc., Princeton, NJ). The camera is cooled to −30°C and has a quantum efficiency of around 75% at 500 nm. The camera was attached via a basement port to an inverted fluorescence microscope (Axiovert; Carl Zeiss Inc.), and illumination was provided by a 100-watt Hg lamp in an Atto Arc housing for variable attenuation. A HiQ FITC cube #41001 (Chroma Technology Corp., Brattleboro, VT) was used with the following specifications: 480/80 exciter, 505 LP dichroic, and 535/50 emitter.

Because the yeast cell is a three-dimensional sphere roughly 5 μm in diameter, two techniques are available to image cytoskeletal dynamics within the cell. The first technique is to image only one focal plane; single microtubules can be followed during the time lapse by manually adjusting the fine focus such that a particular microtubule remains in focus. The second technique is to generate a time-lapse sequence consisting of consecutive Z-series images (Waddle *et al.,* 1996; Shaw *et al.,* 1997a). In this technique, 5–10 images are collected at 0.5- to 1-μm steps through the cell, and this is repeated every minute for several hours. When imaging cells over longer time periods, neutral density filters or an adjustable-intensity mercury lamp can be used to attenuate illumination in order to reduce photobleaching and toxicity. Using this method, microtubule and actin dynamics were examined during the course of a complete cell cycle (Waddle *et al.,* 1996; Shaw *et al.,* 1997a,b).

B. Analysis of Time-Lapse Data

The parameters of cytoskeletal dynamics, such as rates of movement and frequencies of events, are usually determined from the analysis of time-lapse data. As an example of how data is analyzed, we will describe calculations for determining rates of microtubule growth and shrinkage. First, time-lapse sequences are chosen in which a microtubule or microtubules remains in focus for long enough to determine the parameter of interest. Then, using a tracing tool (MetaMorph Imaging Software; Universal Imaging, West Chester, PA), the length of the microtubule in pixels is determined, and pixels are then converted into microns, using calibration information for the specific microscope setup used. Tracings of microtubules should be repeated in each frame and averaged; this reduces the error inherent in manual tracing. Measurements are repeated in successive frames for as long as a given microtubule can be tracked. Once all microtubules are measured for a given time-lapse stack of images, distances versus time can be plotted using a statistics software program (StatView; Abacus Concepts, Berkeley, CA). These plots are used to visually determine growing and shrinking phases for single microtubules. Regression analysis of each separate phase determines the slope of the line, and thus the rate of growth or shrinkage. Only regression analysis results having $r^2 > 0.7$ and $P \leq 0.5$ (95% confidence level) were considered statistically significant (Carminati and Stearns, 1997). Once growth and shrinkage rates are determined, the unpaired *t*-test determines whether growth and shrinkage values are significantly different from previously determined values.

C. Presentation of Time-Lapse Images

For generating figures from time-lapse images, the image stack file must first be separated into individual TIFF files. A graphics software program such as Adobe Photoshop (Adobe Systems Inc., San Jose, CA) can be used to create figures using TIFF files. Several methods exist for transferring the digital images generated by the CCD camera to analog videotape for presentation. One is to first transfer the images to an optical memory disk recorder (OMDR), at a rate of roughly 10 frames per second, then to transfer the analog images from the

OMDR to a video recorder and onto videotape. Alternatively, it is possible to outfit a personal computer so that it can output directly to an attached video recorder.

QuickTime movies for Internet Web pages can also be generated from time-lapse images (Table I); the time-lapse TIFF files must first be converted to a PICS-PICT sequence file using GraphicConverter shareware (Lemke Software, Peine, Germany). The brightness and sharpness of the PICS file can be altered using the "stacks" macro of NIH Image software, as can any cropping of the image that might be necessary. Finally, MoviePlayer software (QuickTime; Apple Computer Inc., Cupertino, CA) is used to convert the PICS file to a QuickTime movie. The compression (Motion JPEG A) and frame rate (usually ~10 frames/s) must be chosen before creating the movie. This JPEG file can now be inserted into a Web page and viewed using QuickTime.

V. Results: Cytoskeletal GFP Fusion Proteins

The microtubule and actin networks compose the two main cytoskeletal components of *Saccharomyces cerevisiae* (Botstein *et al.,* 1997; Winsor and Schiebel, 1997). In yeast, microtubules are required for only three processes, each involving movement of the nucleus or chromosomes. Nuclear microtubules form the mitotic and meiotic spindles, and are essential for the segregation of chromosomes in both processes. Cytoplasmic microtubules orient the nucleus and spindle, positioning them at the neck region prior to division and maintaining the proper orientation during spindle elongation. Cytoplasmic microtubules are also required for nuclear fusion during mating, acting to bring the two haploid nuclei together in the common zygote cytoplasm so that they can form a diploid nucleus. Both nuclear and cytoplasmic microtubules are nucleated from a microtubule organizing center, the spindle pole body (SPB), that is embedded into the nuclear envelope. During mitotic division, the SPB duplicates and nuclear microtubules from the separated SPBs form the mitotic spindle. Both SPBs also nucleate cytoplasmic microtubules, which are the equivalent of astral microtubules in animal cells.

In contrast to the roles of the microtubule network in yeast cell physiology, the function of the actin cytoskeleton is more complex and less well understood.

Table I.
Web Sites; Examples of Yeast Cytoskeletal Dynamics

Microtubules:	http://www.leland.stanford.edu/~stearns/carminati.html
	http://www.unc.edu/depts/biology/bloomlab/tubes1.mov
Actin:	http://www.genome.stanford.edu/group/botlab/images.html
	http://www.cooperlab.wustl.edu

Actin is important for polarized secretion and growth, cytokinesis, endocytosis, and movement of mitochondria and possibly other organelles. The actin cytoskeleton is composed of cytoplasmic actin cables and cortical actin patches associated with the plasma membrane. The distribution of both cables and patches is cell-cycle regulated, and polarized patch localization correlates spatially and temporally with polarized growth. Actin patches localize to sites of cell surface growth, mostly in the bud, whereas actin cables are associated with the cell periphery and are oriented toward the bud and the actin patches. At the end of the cell cycle, actin patches relocalize to the neck between the mother and bud cells, and the actin cables become randomly oriented in both the mother and bud.

A. GFP Fusions to Microtubule Proteins

1. α-Tubulin Fusions

To visualize the microtubule cytoskeleton in living yeast cells, fusions were constructed between GFP and the α-tubulin genes of yeast. Yeast contains two genes encoding α-tubulin, which differ in levels of expression: *TUB1,* the major essential gene, and *TUB3,* the minor nonessential gene. Overexpression of *TUB1* rescues a deletion in *TUB3,* and vice versa, suggesting that both genes are functionally equivalent. Fusions were made between GFP and both *TUB1* and *TUB3;* both N- and C-terminal fusions were made with *TUB1,* and an N-terminal fusion was made with *TUB3.* Although all fusions showed some level of microtubule fluorescence, the GFP–TUB3 fusion was the only construct to functionally rescue the benomyl supersensitivity of a *tub3* deletion strain (Carminati and Stearns, 1997). None of the constructs had deleterious effects on cell growth at normal temperature (30°C), so it is not clear why only one of the fusions is able to act functionally in the absence of the endogenous α-tubulin gene.

Using the GFP–TUB3 fusion to generate fluorescent nuclear and cytoplasmic microtubules, microtubules were shown to exhibit dynamic instability as in animal cells (Carminati and Stearns, 1997) (Fig. 2). Rates of microtubule growth and shrinkage were 0.5 and 1.4 μm/min, respectively, and catastrophe and rescue frequencies were 0.006 and 0.002 s^{-1}, respectively. A difference was seen in rates of microtubule shrinking when different stages of the cell cycle were examined; unbudded cells had a rate of 1.8 μm/min as compared with a rate of 0.8 μm/min in budded mitotic cells. This increase in microtubule dynamics during the early part of the cell cycle might be explained by the need for cytoplasmic microtubules to search for and locate the emerging bud tip; increased microtubule turnover may result in more efficient searching.

GFP–cytoskeletal fusion proteins can be introduced into mutant strains in order to characterize defects in living cells. The minus-end microtubule motor protein, dynein, has been shown to play a key role during the orientation of the mitotic spindle in yeast (Eshel *et al.,* 1993; Li *et al.,* 1993), and is necessary for microtubule-cortex interactions (Carminati and Stearns, 1997). In comparing microtubule dynamics of wild-type and dynein-deficient cells, a decrease was

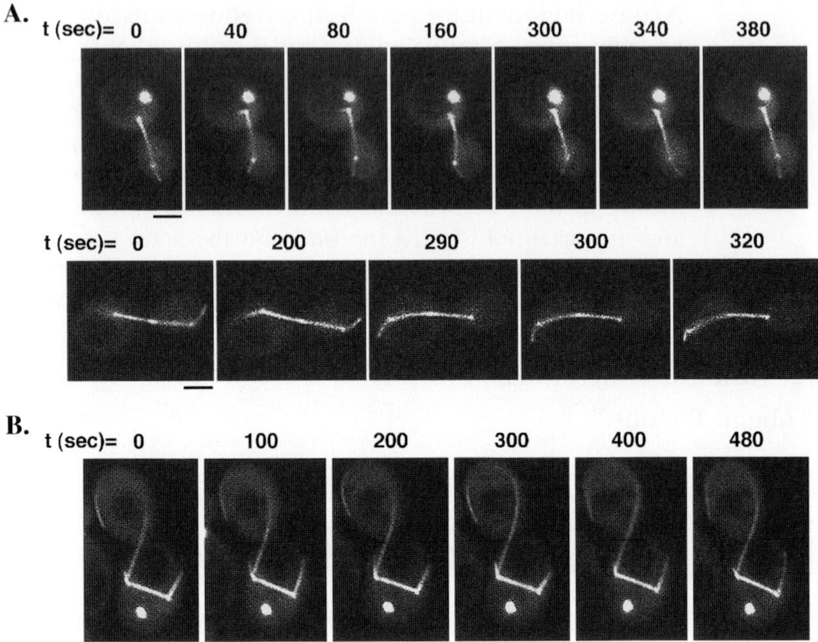

Fig. 2 Time-lapse images are shown of living yeast cells in which cytoplasmic microtubules, spindle microtubules, and SPBs are labeled following incorporation of GFP-TUB3. A, In wild-type cells containing long spindles, cytoplasmic microtubules interact with the cell cortex, resulting in proper spindle orientation during mitosis. In the top panel, a short cytoplasmic microtubule sweeps back and forth along the bud cortex; in the bottom panel, shrinking of a microtubule at the cortex results in spindle movement toward the direction of the cortical attachment. B. Microtubule and cortex interactions are aberrant in dynein-mutant cells, resulting in misaligned spindles during mitosis; note the location of the spindle entirely within the mother cell. Microtubule dynamics are also affected in the dynein mutant, resulting in longer cytoplasmic microtubules. (Reproduced from *The Journal of Cell Biology,* 1997, vol. 138, pp. 629–641, by copyright permission of The Rockefeller University press.)

seen in rates of growth and shrinkage, and catastrophe frequencies were reduced by half (Carminati and Stearns, 1997). More interestingly, characteristic interactions between cytoplasmic microtubule ends and the cell cortex were aberrant in the mutant cells during mitosis (Fig. 2). These cortical interactions result in associated spindle movement and the eventual positioning of the spindle at the bud neck prior to division. Many more questions can now be addressed as to how dynein and associated proteins become localized at the cell cortex prior to interaction with microtubules, and how this process is regulated during the cell cycle.

In other studies, both spindle microtubules and chromosomes were labeled with GFP fusion proteins using a technically impressive strategy (GFP–TUB1 and GFP–LacI/LacO) (Straight *et al.,* 1997). Several aspects of mitosis in living cells were characterized following Z-series imaging and deconvolution analysis. By observing chromosome dynamics in conjunction with spindle dynamics, it was shown that a metaphase plate is not present prior to chromosome separation in yeast and that anaphase A movement of chromosomes to spindle poles occurs during mitosis. Biphasic spindle elongation rates were similar to those observed in other studies (Kahana *et al.,* 1995; Yeh *et al.,* 1995).

2. Microtubule-Associated Proteins

Microtubule dynamic rates similar to those described earlier were also determined by examining microtubules following incorporation of GFP(S65T) fused to a truncation of the dynein heavy chain gene, *DYN1* (Shaw *et al.,* 1997b). Overexpression of DYN1–GFP(S65T) resulted in localization to cytoplasmic microtubules, and rates of polymerization and depolymerization were calculated following Z-series time-lapse imaging. Cytoplasmic microtubules were imaged during a complete cell cycle, and rates of growth and shrinkage were determined to range from 0.3 to 1.5 μm/min (Shaw *et al.,* 1997b). Whereas the DYN1–GFP(S65T) construct described earlier localizes only to cytoplasmic microtubules, overexpression of an integrated N-terminal fusion (GFP–DYN1) localizes to the SPB predominantly, yet is occasionally seen associated with both cytoplasmic and spindle microtubules (J. Carminati, unpublished data).

Other microtubule-associated proteins have been identified as well following localization of GFP fusions to microtubules. *BIM1* (binding to microtubules) was identified as a two-hybrid interactor with *TUB1,* and a deletion of *BIM1* results in phenotypes similar to known mutations affecting microtubules: temperature and cold sensitivity, benomyl supersensitivity, aberrant spindle morphology and nuclear migration, and defects in karyogamy (Schwartz *et al.,* 1997). Using a GFP(S65T)–BIM1 fusion driven by the *ACT1* promoter, it was shown that Bim1p localizes to both nuclear and cytoplasmic microtubules. Several lines of evidence, including the similarity of phenotypes resulting from the overexpression or loss of *BIM1,* suggest that Bim1p is a structural component of the microtubule cytoskeleton. Interestingly, the nearest homolog to *BIM1* is human EB1 protein, which was identified as a ligand to APC in a two-hybrid system (Su *et al.,* 1995). APC itself is a microtubule-binding protein, and the APC gene is responsible for a hereditary predisposition to a form of colon cancer (Groden *et al.,* 1991; Munemitsu *et al.,* 1994; Smith *et al.,* 1994).

B. Spindle Pole Body Components

1. γ-Tubulin

The γ-tubulin protein in yeast is encoded by *TUB4* and was localized to the SPB using a TUB4–GFP fusion protein; localization was not dependent on microtubules, showing that Tub4p is an intrinsic SPB component (Marschall *et al.,* 1996). A novel approach examining the dynamics of GFP–TUB4 showed that Tub4p is exchangeable at SPBs (Fig. 3). Using a *kar1-1* strain that is defective in nuclear fusion, cells were mated to a *KAR1* strain overexpressing GFP–TUB4 (Marschall *et al.,* 1996). Upon mating, GFP–TUB4 was initially localized only at the SPB of the *KAR1* nucleus. However, over a period of 30 min, the SPB of the *kar1-1* nucleus incorporated GFP–TUB4, showing that newly made γ-tubulin became incorporated into a preexisting SPB (Fig. 3).

2. Nuf2

The first *in vivo* study of a yeast cytoskeletal protein fused to GFP examined spindle dynamics by fusing GFP to the C terminus of a SPB component, Nuf2p (Kahana *et al.,* 1995). Overexpression of NUF2–GFP rescued the lethality associated with a *nuf2* conditional allele and resulted in SPB localization throughout the cell cycle that was similar to that of the endogenous Nuf2p. SPB separation during mitosis was examined and surprisingly revealed four distinct phases of spindle dynamics during anaphase. Following initial spindle formation in the mother cell, the small spindle eventually aligns at the bud neck along the mother–bud axis. A period of fast spindle elongation then occurs at a rate of 1.5 μm/min over 100–150 s; the spindle remains mainly in the mother cell at this time. The spindle is then translocated into the bud, and slow spindle elongation (0.7 μm/min) occurs over a 10- to 12-min period at the end of anaphase. Interestingly, the two different rates of spindle elongation indicate that distinct forces

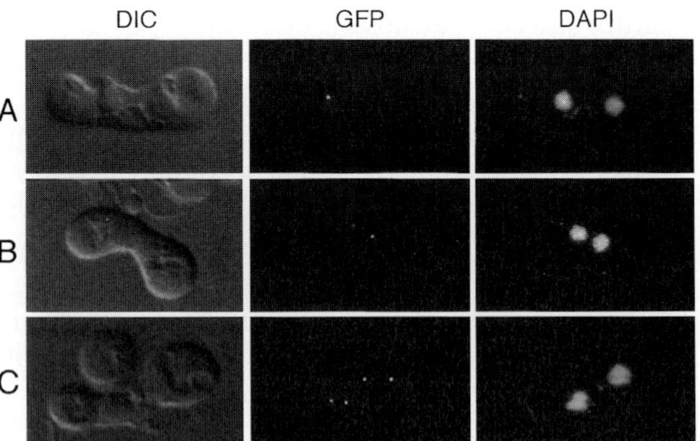

Fig. 3 Localization of TUB4–GFP following fusion of a *kar1-1* strain with a *KAR1* strain expressing TUB4–GFP. Cells were mated for 3 h and then fixed and stained with the DNA stain DAPI. The *kar1-1* mutation prevents fusion of nuclei in mating cells; thus, each zygote has two separate nuclei in a common cytoplasm. A, TUB4–GFP is localized to a single SPB in an unbudded zygote. B, The second SPB is faintly labeled with TUB4–GFP in a small-budded zygote. C, Both duplicated SPBs show TUB4–GFP localization in a large-budded zygote. (Reproduced from *The Journal of Cell Biology,* 1996, vol. 134, pp. 443–454, by copyright permission of The Rockefeller University Press.)

act on the spindle during anaphase (Kahana *et al.*, 1995). By tracking NUF2–GFP in cells mutant for several microtubule motor proteins, it was recently shown that *KIP3,* a kinesin-related gene, is required for the initial migration of the nucleus to the bud neck, whereas *DYN1,* the dynein heavy chain gene, is required for the later translocation of the nucleus through the bud neck (DeZwaan *et al.*, 1997). These and other motor proteins may contribute to the distinct forces acting on microtubules during mitosis.

C. GFP Fusions to Actin Cytoskeletal Proteins

1. Actin–GFP fusions

To visualize the actin cytoskeleton, GFP was fused to the C terminus of *ACT1,* the single essential gene in yeast encoding actin (Doyle and Botstein, 1996). Several constructs were also made containing alanine linkers between GFP and *ACT1,* and all were expressed on YCp plasmids. All fusion proteins became incorporated into cortical actin patches, had no deleterious effects on cell growth, yet did not act functionally to rescue a deletion of *ACT1.* The actin patches visualized by the fusion protein appeared similar to patches previously described by immunofluorescence in fixed cells, suggesting that the fusion protein became properly localized into the cytoskeleton. However, no actin cables were visualized by the ACT1–GFP fusion proteins, indicating either that the fusion protein is not assembled into all actin structures or that incorporation into actin cables is below the level of detection.

The visualization of actin patches in living cells led to the surprising finding that patches are very dynamic throughout the cell cycle (Doyle and Botstein, 1996; Waddle *et al.*, 1996) (Fig. 4). Generally, the movement of actin patches is confined to small distances; however, movement across larger distances occurs occasionally. It is thought that cytoplasmic actin cables are anchored in the cell membrane, forming cortical actin patches, and that secretory vesicles are delivered to the cell membrane via actin cables. The discovery that actin patches are dynamic suggests that membrane addition to the growing bud could be directed to different regions within the bud (Doyle and Botstein, 1996). Actin patches were also found to be dynamic in nature during cell mating and sporulation (Doyle and Botstein, 1996). Interestingly, actin patches were not dynamic once the mature ascus formed.

2. Actin-Associated Proteins

Several actin-associated proteins are associated with both actin cables and cortical patches, whereas others are associated with either cables or patches. *CAP2* encodes the β-subunit of capping protein and is localized only to actin patches (Amatruda and Cooper, 1992). In a similar study examining dynamic actin patches, GFP containing the S65T mutation was fused to the N terminus

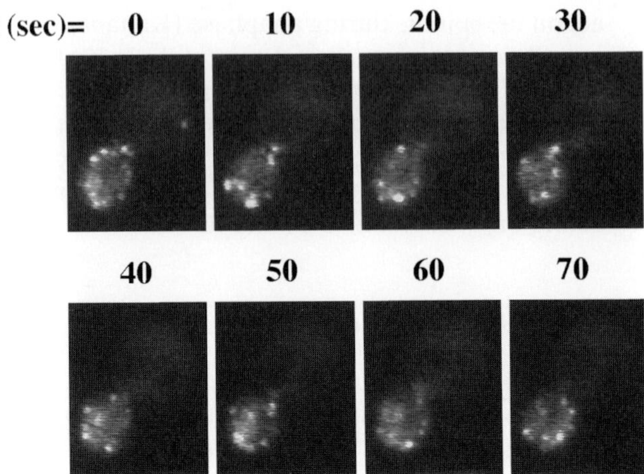

Fig. 4 Movement of cortical actin patches is shown in time-lapse images of *sac6::LEU2 abp1::LEU2* haploid cells containing a GFP(S65T)–SAC6 fusion protein. Patches are localized to the cell membrane of the daughter cell over the time course analyzed and are clearly dynamic, as can be seen from the change in distribution of patches over the 10-s intervals. Images were obtained using a SIT camera by T. Doyle, and methods are as described in Doyle and Botstein, 1996.

of *CAP2* and expressed on a YCp plasmid; this fusion protein rescued a synthetic lethality associated with a deletion of *CAP2* (Waddle *et al.,* 1996). To enhance the fluorescence of actin patches, the YCp plasmid was introduced into a diploid containing two integrated copies of GFP(S65T)–CAP2, replacing the endogenous loci.

This study also determined that actin patches moved rapidly along the cell periphery, independent of one another, and generally moved short distances with frequent changes in direction. Rates of patch movement were determined by examining time-lapse images, and a mean rate of 0.49 μm/s was calculated for dynamic patches (Waddle *et al.,* 1996). The clustering of patches that occurs during the cell cycle was also characterized; patches within clusters were also dynamic yet were restricted in movement so that a cluster remained together for a given time. Clusters of polarized patches were found to correlate with periods of bud growth, and patches became depolarized in the mother and bud when bud growth ended. To address whether movement of actin patches was powered by myosin motors, dynamics were analyzed in mutants of each of the five myosin heavy chain genes (*MYO1-5*). Patches remained dynamic, indicating that none of these genes were essential for movement. Similar results were found with the kinesin heavy chain gene, *SMY1,* and the tropomyosin genes, *TPM1* and *TPM2.* However, a quantitative analysis of rates of patch movement was not performed and might reveal subtle dynamic changes in these mutant backgrounds.

GFP has also been fused to several other actin-binding proteins; the S65T version of GFP was fused to the N terminus of both *ABP1* and *SAC6,* a fimbrin homolog (Doyle and Botstein, 1996). Both fusion proteins localize to actin patches, whereas GFP(S65T)–SAC6 localizes occasionally to actin cables; the dynamics of actin patch movement was similar to results using ACT1–GFP and GFP(S65T)–CAP2 (Fig. 4).

VI. Future

The results described here represent several different types of experiments that have been used to examine cytoskeletal dynamics. Using GFP fusions, many novel approaches remain available to continue the characterization of cytoskeletal dynamics in yeast. The development of GFP molecules with distinct spectral properties will allow double-labeling experiments *in vivo.* Several approaches are listed below that may prove valuable in future research.

A. Pharmaceutical Drug Studies

Using perfusion chambers, it should be possible to determine the immediate effects of drugs acting on the cytoskeleton. Drugs such as nocodazol and benomyl, which destabilize microtubules, and latrunculin A (Ayscough *et al.,* 1997), which destabilizes actin filaments, will be interesting to examine. The cytoskeletal response to drugs can be examined, as can the recovery of cells following removal of the drugs.

B. Photobleaching and Dynamics

Studies using GFP–tubulin fusions examined both the dynamic properties of cytoplasmic microtubules and aspects of spindle elongation during mitosis. However, the dynamics of individual spindle microtubules has not been determined due to the bundled nature of the mitotic spindle. Future studies aimed at examining spindle microtubule dynamics might be possible following photobleaching of the GFP fluorophore in a labeled spindle (Cole *et al.,* 1996). A second approach is to examine spindle microtubules in mutants resulting in monopolar spindles; these half spindles are often splayed apart and may allow single microtubules to be resolved and characterized.

C. Fluorescence Resonance Energy Transfer (FRET)

One of the most interesting new approaches using GFP fusion proteins is a method that allows the dynamic imaging of protein–protein interactions *in vivo.* Fluorescence resonance energy transfer (FRET) is the transfer of energy from a donor fluorophore to an acceptor fluorophore. When the emission spectrum of the donor overlaps with the excitation spectrum of the acceptor, the energy can be transferred if the molecules are within close proximity to each other. Using both blue-shifted and red-shifted GFP variants with distinct excitation and emission spectra, it has been shown that FRET occurs between fusion proteins that are physically close to one another (Heim and Tsien, 1996; Mitra *et al.,* 1996). The development of this technique as more GFP variants become available will be of enormous value in determining and monitoring *in vivo* protein associations.

Acknowledgments

The microscope setup described and used for studies of GFP–TUB3 is located in the laboratory of Dr. James Spudich (Dept. of Biochemistry, Stanford University, CA). We are grateful to Dr. James Sabry for help in dynamic imaging of GFP in yeast cells.

References

Amatruda, J. F., and Cooper, J. A. (1992). Purification, characterization, and immunofluorescence localization of *Saccharomyces cerevisiae* capping protein. *J. Cell Biol.* **117,** 1067–1076.

Ayscough, K. R., Stryker, J., Pokala, N., Sanders, M., Crews, P., and Drubin, D. G. (1997). High rates of actin filament turnover in budding yeast and roles for actin in establishment and maintenance of cell polarity revealed using the actin inhibitor Latrunculin-A. *J. Cell Biol.* **137,** 399–416.

Botstein, D., Amberg, D., Mulholland, J., Huffaker, T., Adams, A., Drubin, D., and Stearns, T. (1997). The yeast cytoskeleton. *In* "The Molecular and Cellular Biology of the Yeast *Saccharomyces*" (Pringle, J. R., Broach, J. R., and Jones, E. W., Eds.) pp. 1–90. Cold Spring Harbor Laboratory Press, Cold Spring Harbor, NY.

Carminati, J., and Stearns, T. (1997). Microtubules orient the mitotic spindle in yeast through dynein-dependent interactions with the cell cortex. *J. Cell Biol.* **138,** 629–641.

Chalfie, M., Tu, Y., Euskirchen, G., Ward, W. W., and Prasher, D. C. (1994). Green fluorescent protein as a marker for gene expression. *Science* **263,** 802–805.

Cole, N. B., Smith, C. L., Sciaky, N., Terasaki, M., Edidin, M., and Lippincott-Schwartz, J. (1996). Diffusional mobility of Golgi proteins in membranes of living cells. *Science* **273,** 797–801.

Cormack, B. P., Valdivia, R. H., and Falkow, S. (1996). FACS-optimized mutants of the green fluorescent protein (GFP). *Gene* **173,** 33–38.

DeZwaan, T. M., Ellingson, E., Pellman, D., and Roof, D. M. (1997). Kinesin-related *KIP3* of *Saccharomyces cerevisiae* is required for a distinct step in nuclear migration. *J. Cell Biol.* **138,** 1023–1040.

Dopf, J., and Horiagon, T. M. (1996). Deletion mapping of the *Aequorea victoria* green fluorescent protein. *Gene* **173,** 39–44.

Doyle, T., and Botstein, D. (1996). Movement of yeast cortical actin cytoskeleton visualized *in vivo. Proc. Natl. Acad. Sci. U. S. A.* **93,** 3886–3891.

Eshel, D., Urrestarazu, L. A., Vissers, S., Jauniaux, J. C., Van Vliet-Reedijk, J. C., Planta, R. J., and Gibbons, I. R. (1993). Cytoplasmic dynein is required for normal nuclear segregation in yeast. *Proc. Natl. Acad. Sci. U. S. A.* **90,** 11172–11176.

Groden, J., Thliveris, A., Samowitz, W., Carlson, M., Gelbert, L., Albertsen, H., Joslyn, G., Stevens, J., Spirio, L., Robertson, M., Sargeant, L., Krapcho, K., Wolff, E., Burt, R., Hughes, J. P., Warrington, J., McPherson, J., Wasmuth, J., Le Paslier, D., Abderrahim, H., Cohen, D., Leppert, M., and White, R. (1991). Identification and characterization of the familial adenomatous polyposis coli gene. *Cell* **66,** 589–600.

Heim, R., Prasher, D. C., and Tsien, R. Y. (1994). Wavelength mutations and posttranslational autoxidation of green fluorescent protein. *Proc. Natl. Acad. Sci. U. S. A.* **91,** 12501–12504.

Heim, R., Cubitt, A. B., and Tsien, R. Y. (1995). Improved green fluorescence. *Nature* **373,** 663–664.

Heim, R., and Tsien, R. Y. (1996). Engineering green fluorescent protein for improved brightness, longer wavelengths and fluorescence resonance energy transfer. *Curr. Biol.* **6,** 178–182.

Kahana, J. A., Schnapp, B. J., and Silver, P. A. (1995). Kinetics of spindle pole body separation in budding yeast. *Proc. Natl. Acad. Sci. U. S. A.* **92,** 9707–9711.

Kahana, J. A., and Silver, P. A. (1996). Use of the *A. victoria* green fluorescent protein to study protein dynamics in vivo. *In* "Current Protocols in Molecular Biology" (Ausubel, F. M., Brent, R., Kingston, R. E., Moore, D. E., Seidman, J. G., Smith, J. A., and Struhl, K., Eds.) vol. I, pp. 9.6.13–19.16.19. John Wiley & Sons, Inc., New York.

Li, Y.-Y., Yeh, E., Hays, T., and Bloom, K. (1993). Disruption of mitotic spindle orientation in a yeast dynein mutant. *Proc. Natl. Acad. Sci. U. S. A.* **90,** 10096–10100.

Marschall, L. G., Jeng, R. L., Mulholland, J., and Stearns, T. (1996). Analysis of Tub4p, a yeast γ-tubulin-like protein: implications for microtubule organizing center function. *J. Cell Biol.* **134,** 443–454.

Mitra, R. D., Silva, C. M., and Youvan, D. C. (1996). Fluorescence resonance energy transfer between blue-emitting and red-shifted excitation derivatives of the green fluorescent protein. *Gene* **173,** 13–17.

Munemitsu, S., Souza, B., Muller, O., Albert, I., Rubinfeld, B., and Polakis, P. (1994). The APC gene product associates with microtubules *in vivo* and promotes their assembly *in vitro. Cancer Res.* **54,** 3676–3681.

Niedenthal, R. K., Riles, L., Johnston, M., and Hegemann, J. H. (1996). Green fluorescent protein as a marker for gene expression and subcellular localization in budding yeast. *Yeast* **12,** 773–786.

Schwartz, K., Richards, K., and Botstein, D. (1997). *BIM1* encodes a microtubule-binding protein in yeast. *Mol. Biol. Cell* **8,** 2677–2691.

Shaw, S. L., Yeh, E., Bloom, K., and Salmon, E. D. (1997a). Imaging green fluorescent protein fusion proteins in *Saccharomyces cerevisiae. Curr. Biol.* **7,** 701–704.

Shaw, S. L., Yeh, E., Maddox, P., Salmon, E. D., and Bloom, K. (1997b). Astral microtubule dynamics in yeast: a microtubule-based searching mechanism for spindle orientation and nuclear migration into the bud. *J. Cell Biol.* **139,** 985–994.

Smith, K. J., Levy, D. B., Maupin, P., Pollard, T. D., Vogelstein, B., and Kinzler, K. W. (1994). Wild-type but not mutant APC associates with the microtubule cytoskeleton. *Cancer Res.* **54,** 3672–3675.

Stearns, T., Ma, H., and Botstein, D. (1990). Manipulating the yeast genome using plasmid vectors. *Methods Enzymol.* **185,** 280–296.

Stearns, T. (1995). The green revolution. *Curr. Biol.* **5,** 262–264.

Straight, A. F., Marshall, W. F., Sedat, J. W., and Murray, A. W. (1997). Mitosis in living budding yeast: anaphase A but no metaphase plate. *Science* **277,** 574–578.

Su, L. K., Burrell, M., Hill, D. E., Gyuris, J., Brent, R., Wiltshire, R., Trent, J., Vogelstein, B., and Kinzler, K. W. (1995). APC binds to the novel protein EB1. *Cancer Res.* **55,** 2972–2977.

Waddle, J. A., Karpova, T. S., Waterston, R. H., and Cooper, J. A. (1996). Movement of cortical actin patches in yeast. *J. Cell Biol.* **132,** 861–870.

Winsor, B., and Schiebel, E. (1997). An overview of the *Saccharomyces cerevisiae* microtubule and microfilament cytoskeleton. *Yeast* **13,** 399–434.

Yeh, E., Skibbens, R. V., Cheng, J. W., Salmon, E. D., and Bloom, K. (1995). Spindle dynamics and cell cycle regulation of dynein in the budding yeast *Saccharomyces cerevisiae. J. Cell Biol.* **130,** 687–700.

CHAPTER 7

Analysis of Nuclear Transport *in Vivo*

Paul Ferrigno and Pamela A. Silver
Department of Biological Chemistry and Molecular Pharmacology,
Harvard Medical School and Department of Cancer Biology,
Dana Farber Cancer Institute,
Boston, Massachusetts 02115

I. Introduction

A. Studying Dynamic Processes in Living Cells

Understanding the dynamics of intracellular processes is one of the most exciting and informative research areas today. The availability of the green fluorescent protein (GFP) from *Aequorea victoria* has revolutionized *in vivo* studies of dynamic processes. A simple combination of genetic engineering and cell biology has already allowed the determination of the intracellular location of a range of proteins, in many cases opening up new avenues of investigation following unexpected results. As part of our ongoing studies into the molecular mechanisms of protein targeting *in vivo,* we have been using GFP fusions to study nuclear protein transport. In this chapter, we will describe the approaches

METHODS IN CELL BIOLOGY, VOL. 58
Copyright © 1999 by Academic Press. All rights of reproduction in any form reserved.
0091-679X/99 $25.00

we and others have taken to address this problem. We will concentrate on studies in baker's yeast (*Saccharomyces cerevisiae*), because this model system has shown itself to be particularly amenable to *in vivo* studies.

B. Nuclear Transport

The import of proteins from the cytoplasm into the nucleus has been shown to be a rate-limiting step in many processes, such as T-cell activation, lending support to the idea that the acquisition of the nuclear membrane was a crucial step in the evolution of eukaryotic cells. Nuclear import is generally thought to comprise three distinct steps: recognition in the cytosol of the protein to be imported by the NLS receptor, targeting of this complex to the nuclear pore, and translocation through the pore. At the molecular level, many of the proteins involved in these events have been identified: the importin (or karyopherin) α/β-heterodimer recognizes proteins carrying a classical nuclear localization signal (NLS) in the cytoplasm, and the complex is targeted to the nuclear pore complex (NPC), where, in a process requiring the small GTPase Ran, translocation into the nucleus occurs (reviewed in Nigg, 1997). In contrast to nuclear import, nuclear export is much less well understood. RNA export is conceived to occur with the RNA in a multiprotein complex that may include heterogeneous nuclear ribonuclear proteins (hnRNPs) and components of the splicing machinery; the first targeting molecules involved in RNA export have recently been identified (Seedorf and Silver, 1997). Independent of RNA export, there is increasing evidence that proteins are actively exported from the nucleus (Fornerod *et al.*, 1997; Kutay *et al.*, 1997; Stade *et al.*, 1997). Finally, many proteins are known or proposed to shuttle between the nucleus and the cytoplasm (reviewed in Lee and Silver, 1997). The baker's yeast has been key to many of these advances, and later studies in vertebrate cells have served to confirm these findings (Nigg, 1997). This chapter will focus on working with proteins that are imported or that shuttle. It is clear that the nuclear export question remains to be addressed and that GFP fusions may well provide the key.

II. Experimental Approaches and Protocols

A. GFP Mutants Optimized for *in Vivo* Studies

Fluorescence of GFP requires correct folding of the protein, intramolecular oxidation to produce the chromophore, and excitation of the molecule with light of an appropriate wavelength (Cody *et al.*, 1993; Heim *et al.*, 1994; Inouye and Tsuji, 1994). Observation of GFP fusion proteins in living cells will thus be limited by the folding and stability of the fusion protein, and by the microscope set-up that is used. Although wild-type GFP can be detected using a standard fluorescein isothiocyanate (FITC) filter set, the excitation and emission spectra are not ideal. The original GFP isolate (Prasher *et al.,* 1992), referred to as wild-type GFP, had a maximal excitation at 395 nm and a maximum peak of emission at 509 nm. Random mutagenesis of the GFP gene resulted in the isolation of a mutant (S65T) for which the excitation peak was shifted to 490 nm (Heim *et al.,* 1995; Kahana and Silver, 1996), that was more stable at 37°C, and whose folding was approximately twice as rapid as that of wild-type GFP; a secondary mutation (V163A) in our mutant was found to further increase the rate of folding of GFP *in vivo* (Kahana and Silver, 1996). We find that our S65T V163A mutant GFP appears at least 10 times brighter than wild-type GFP using standard fluorescence filter sets (see Section II.B).

B. Microscopy

The limiting factor in GFP detection will usually be the microscope setup. It is imperative that the GFP be excited at the optimal wavelength, with minimal interference, and that the emitted light be transmitted as efficiently as possible to the observer or camera. This section describes the system in use in our lab. A full discussion is given in Kahana and Silver (1996).

1. Excitation and Emission Filter Sets and Objectives

a. Excitation Source

We find that a 75-W xenon source is optimal in terms of high excitation and low bleaching, although a 100-W mercury lamp also gives good results. Laser scanning confocal microscopes can also be used but generally lead to increased photobleaching because increased exposure times are required.

b. Filter Sets

Standard FITC longpass filters (excitation from 450 to 490 nm; emission 520 nm) are optimal for use with the commonly used S65T mutants of GFP, as described in the previous section. Bandpass filter sets with wider excitation and emission ranges that bracket these values may lead to increased background fluorescence from the specimen. Commercially available GFP-specific filters use lower wavelengths for excitation, which can lead to photobleaching of the GFP signal.

c. Objectives

We use lenses with the highest numerical aperture available (currently 1.4 for the 60× and 100× objectives necessary for working with yeast cells). Use of a lens with a numerical aperture of 1.3 results in the loss of more than a third of the signal. Unfortunately, the higher the magnification of the lens, the less light is transmitted. Thus, for weak signals or for time-lapse microscopy, we use the lowest magnification that allows visualization of the required cellular structures.

2. Expression of the GFP Fusion Protein

To obtain the utmost information from a localization experiment, the fusion protein is optimally expressed from the gene's endogenous promoter so that physiological levels are obtained. Assuming that the GFP fusion can functionally substitute for the wild-type gene, information derived from these studies can be extrapolated with a high degree of confidence to the wild-type protein. In many cases, however, proteins are expressed at levels too low to allow the detection of a GFP signal or are turned over so rapidly that the GFP moiety does not have time to fold into its active, fluorescent, state. In these cases, it will become appropriate to express the fusion from a constitutively highly expressing promoter, such as *HIS3* [used by Straight *et al.,* (1997) to express Tub1–GFP] or *ADH1.* Alternatively, it may be convenient to express the GFP fusion from an inducible (e.g., *GAL1* or *GAL10*) or a repressible (*MET*) promoter. Because expression levels from the *GAL* promoters can be very high, it will be important to attempt to titrate the fusion protein's expression levels by decreasing the concentration of galactose used for induction over a range from 0.001% to 2%, or by varying the concentration of glucose (which represses expression from the *GAL* promoters) in the medium.

One major limiting factor in the detection of GFP in living cells can be the background autofluorescence of the cells. In particular, the red pigment that accumulates in *ade2* cells fluoresces strongly in both the FITC/GFP and the rhodamine/Texas Red channels, and can completely mask a weak or tightly localized signal. For this reason, we strongly recommend working in cells that are at least $ADE2^+$; if possible they should be wild type for all the genes involved in adenine biosynthesis, as even *ade3* or *ade8* cells have a higher than wild-type level of autofluoresence that appears yellow, but that the camera will be unable to distinguish from genuine green fluoresence in the FITC/GFP channel. *ade2* cells can usually be restored to *ADE2* by transformation with the 4.2-kb BamHI fragment of this gene described by Aparicio *et al.* (1991).

3. Colocalization of GFP Fusion Proteins with Organelles

The enormous advantage of GFP over conventional microscopic probes is that it allows the observation of proteins in living cells. Thus, to maximize the potential of GFP, it is useful to use vital dyes in colocalization experiments. Where possible, this is the method of choice for two reasons: avoidance of fixation artifacts, and preservation of the GFP fluorescence. We have found that fixation of cells with formaldehyde appears to cause some GFP fusions to precipitate out of solution, forming bright spots throughout the cell (fixation artifact), whereas other fusions are unable to renature following treatment with fixing agents, leading to loss of signal. It is possible to visualize the nucleus in cells using brief treatment with methanol or ethanol before resuspending the cells in a solution containing DAPI (4′, 6-diamidino-2-phenylindole); however, these treatments kill the cells, and may adversely affect the GFP signal. An alternative is given here. Mitochondria can be visualized *in vivo* using rhodamine 123 (Skowronek *et al.,* 1990) or rhodamine B hexyl ester (Terasaki and Reese, 1992; Molecular Probes, Inc., Oregon). Vacuoles can be visualized in living cells with *N*-(3-triethyammoniumpropyl)-4-(*p*-diethylaminophenylhexatrienyl) pyridinium dibromide (FM 4-64; Molecular Probes, Inc, Oregon), which also emits red fluoresence (Vida and Emr, 1995).

a. Protocol I: In Living Cells

There follows a protocol for the localization of GFP fusion proteins with respect to DAPI staining of the yeast nucleus. For other organelles, appropriate markers include the organelle-specific dyes mentioned earlier (vacuole, mitochondrion), or costaining with antibodies to proteins known to localize to each organelle (e.g., Sec63p or BiP/KAR2 as an ER marker; see next protocol). To visualize the yeast nucleus in living cells, simply add DAPI to 1μg/ml to the growth medium 1 h prior to microscopic observation of the cells. For optimal results, we grow the cells in synthetic complete medium; complete media such as yeast extract peptone dextrose (YEPD) have a high background of fluorescence. If medium fluorescence is a problem, the cells can be washed into water; however, this decreases their longevity on the microscope slide so that observations have to be made within 10 min. After this time, fragile cells begin to lyse and it becomes obvious that a fresh sample is required. Finally, some workers prefer to add membrane permeabilizing agents such as dimethyl sulfoxide (DMSO) (0.5%) to the cells, to aid the uptake of DAPI.

b. Protocol II: Colocalization of GFP Fusion Proteins with Other Proteins in Fixed Cells

Frequently, the *in vitro* characterization of a protein reveals interactions with proteins whose relevance needs to be confirmed. An initial approach is to demonstrate that the two proteins are present in the same subcellular domain. For example, in nuclear transport, potential mediators of transpore trafficking may be found in proximity to nuclear pores. Standard protocols for immunofluorescence in yeast have been discussed (Adams and Pringle, 1984; Rose *et al.,* 1990; Pringle *et al.,* 1991). The following protocol is used for the colocalization of GFP fusion proteins with nuclear pore proteins recognized by the anti-nucleoporin antibody MAb 414 [initially raised against rat nuclear pore proteins (Davis and Blobel, 1986), this monoclonal antibody also recognizes a range of yeast proteins (Davis and Fink, 1990)]. Despite the inherent advantages of GFP, observation of subcellular structures in fixed cells remains the most widely recognized method of determining a protein's localization.

1. *Grow 5 ml of cells to mid-log phase (~5 × 10^6 to 10^7 cells/ml).* If using a reporter protein under the control of a *GAL* promoter, the cells should be grown overnight in raffinose before expression of the reporter is induced by the addition of galactose. Optimal galactose induction requires the presence of less than 0.01% glucose. Induction times will vary with each reporter: long times may give high

levels of expression, which should be checked first by Western analysis but can also lead to cellular toxicity. In addition, high levels of overexpression can lead to the accumulation of the reporter in regions of the cell that do not reflect a physiological localization.

2. *Fix the cells by the method of choice.* For most proteins, including the nuclear pore proteins recognized by MAb 414 (Davis and Fink, 1990), the addition of 1/10 volume of 37% formaldehyde (of the highest possible quality) is sufficient. (Davis and Fink, 1990) fixed for 2 h on ice, but room temperature fixation for 60 min also works. The cells can be left in formaldehyde for as little as 5 min on ice or as long as 90 min at 30°C. This will depend on the protein's location in the cell: soluble proteins generally require much longer fixation times. Varying the fixation conditions can further distinguish soluble pools (which require long fixation times to be seen) of a protein from those associated with cellular structures such as nuclear pore complexes, which can be visualized with a shorter fixation, where the soluble pool is washed away during the subsequent steps of the protocol. Alternatively, for other epitopes, the cells may need to be treated with ice-cold methanol for 6 min, followed by ice-cold acetone for 30 s. The acetone should be rapidly evaporated. *Note:* methanol/acetone treatment is most conveniently achieved after cells have been placed on poly-L-lysine–coated slides (step 7), with the methanol and the acetone in Coplin jars into which the slides may be gently slid. Some protocols combine formaldehyde fixation with the methanol/acetone treatment. For some applications, formaldehyde-fixed cells may be permeabilized by treatment with methanol rather than detergent. We find that GFP fused to Nuf2 is sensitive to methanol/acetone fixation but is resistant to formaldehyde, whereas a fusion of Cdc23 to GFP is sensitive to formaldehyde but retains the ability to fluoresce after methanol/acetone treatment. We therefore recommend that both methods be tested initially for each new application.

3. *Harvest the cells* by centrifugation at 800 $\times$ g.

4. *Wash the cells* twice in 5 ml of 0.1 *M* potassium phosphate, pH 6.5, and once in 5 ml of P solution (1 *M* sorbitol in 0.1 *M* potassium phosphate, pH 6.5). The aim of this step is to eliminate traces of formaldehyde that may adversely affect antibodies.

5. *Resuspend the cells* in 0.5 ml of P solution. Add dithiothreital (DTT) to 25 m*M*. Incubate at room temperature for 10 min.

6. *Add 15 μl of 10 μg/ml "100,000 T" zymolyase.* Incubate at room temperature for 5 min to half an hour. An alternative is to use oxalyticase to digest the cells. Both of these reagents can be stored at −20°C in potassium phosphate buffer, 25% glycerol, pH 7, for prolonged periods of time, possibly years. The length of digestion varies enormously with the experimental conditions (e.g., growth in galactose can affect the composition of the cell wall. Unfortunately, this is also the most critical step in the protocol: one wishes to digest the cells sufficiently to allow the antibodies access to the cell interior without destroying cellular

structure(s). The extent of digestion can be monitored using a phase microscope. As a rule of thumb, if 50% of the cells lyse (turn dark) in 0.1% sodium dodecyl sulfate (SDS), then they are sufficiently digested. Observing a parallel aliquot of digesting cells in water may allow a finer distinction to be made between intact cells (bright appearance in phase microscopy), partially digested, light gray cells, and dark, overdigested cells.

7. *While the cells are digesting, prepare the microscope slides.* The most convenient are Teflon-coated slides (Cel-Line Associates, New Jersey), in which the Teflon coating defines wells into which cells can be placed. These wells help to prevent sample cross-contamination. Into each well, pipette 1 drop (20–30 μl) of 0.3% poly-L-lysine (high-molecular-weight polymer). Allow to sit for 5 minutes at room temperature, then aspirate off, wash rapidly with distilled water and allow to air dry. These aspiration and wash steps, like those described in all the following steps, are best performed with a vacuum line with a yellow pipette tip to remove the solution from each well in one hand, and a P200 micropipette containing the solution to be added in the other hand. This arrangement allows a rapid replacement of the solutions in each well, which is particularly important for preventing the cells from drying out and losing structure. Touch the vacuum line tip just outside the well to aspirate off, and gently pipette the fresh solution in from the opposing side of the well.

8. *Pipette 1 drop (20–30 μl) of cells into each well.* Allow them to attach to the poly-L-lysine for 15 min.

9. If the cells have not been treated with methanol (see step 2), *permeabilize the cells by* treating them with 0.5% NP-40 in P solution. Leave the detergent on for 5 min.

10. *Remove the detergent and replace with antibody blocking buffer* (e.g., 5% heat inactivated fetal calf serum, 0.3% Triton X-100 in 0.1 *M* Tris–Cl, 0.15 *M* NaCl, pH 9.0 or 1% fat-free dried milk reconstituted in this buffer). Block for 15 min to 1 h at room temperature. To prevent cells drying out during this and the other prolonged incubations that follow, the cells should be kept in a moist, sealed box. A box that does not allow light in will be best, to preserve any antibody-conjugated fluorophores. Due to the reduced surface tension of this buffer (caused by detergent and protein), liquid is more likely to spill from the wells whenever antibody-blocking buffer solutions are used. This has two disadvantages: samples or antibody solutions can become cross-contaminated, making interpretation difficult; and the wells are more likely to dry out, due to the increase in surface area of evaporation. Therefore, extreme care should be taken to avoid jolting the slides at these times.

11. *Add appropriately diluted antibody in antibody-blocking buffer to each well.* Purified MAb 414 (Babco, Richmond, CA, catalog no. MMS-120P-500) is used at 1:5000. Depending on your antibody, incubate for 1 h to overnight. *Remember to take precautions to prevent your cells from drying out.*

12. *Wash the cells once with Antibody Wash Buffer 1* (AWB1; 0.1 *M* Tris–Cl, 0.15 *M* NaCl, pH 9.0) or 10 min, then again for 30 min.

13. *Wash the cells once with AWB 2* (0.1 *M* Tris–Cl, 0.1 *M* NaCl, 50 m*M* $MgCl_2$, pH 9.0) for 10 min, then again for 30 min.

14. *Add fluorophore-conjugated antibodies and incubate for 1 h at room temperature.* For optimal signal, preserve from light! We obtain our conjugated secondary antibodies from Jackson, and use them at 1 : 1000 in antibody-blocking buffer.

15. *Repeat steps 12 and 13.*

16. *Stain DNA with 1 μg DAPI/ml in AWB 2 for 5 min at room temperature.* Keep a stock 1 mg/ml solution at −20°C, and make a fresh dilution for each experiment.

17. *Rinse twice for 5 min each with AWB 2.* Aspirate the supernatant, and allow the slide to air dry. This may not be necessary but prevents dilution of the antifade solution.

18. *Mount the cell in antifade. Antifade is toxic: wear gloves!* To make antifade, dissolve 100 mg of *p*-phenylenediamine in 10 ml PBS. pH to 8 with 0.5 *M* sodium carbonate (pH 9.0). Bring to 100 ml with glycerol. Store at −20°C and protect from light. Add 1 drop of antifade to each well on the slide. Carefully drop a coverslip (24 × 40 to 24 × 60 mm, depending on the size of your microscope slide) on the samples. Fold the slide into an absorbent paper towel, and gently squeeze out the excess antifade from between the slide and the coverslip. It is important not to allow the coverslip to slip over the slide, as this may cause sample mixing between wells.

19. *Seal the slide with clear nail polish.* Use as little as necessary to completely seal the slide: although we have not experienced this, it has been reported that nail polish can affect GFP.

C. Fusion Proteins That Have Been Used in the Study of Nuclear Transport

1. Protein Fusions That Have Been Used to Study Nuclear Import

Since the first use of GFP to study a dynamic process in yeast (Kahana *et al.*, 1995), many studies have used GFP fusions to localize proteins involved in nuclear transport or whose location in the nucleus has implications for their function; some of the more recent studies are summarized in Table I. Broadly speaking, these studies fall into two categories: those intended to study a dynamic process, such as nuclear pore assembly or the nuclear transport of a particular protein; and those that aim simply to localize a protein implicated in nuclear function. Protocols involving each of these are given next.

2. Specific Examples: NPL3

a. Protocol III: In Vivo Assay for Nuclear Import

A convenient reporter protein for nuclear protein localization *in vivo* was reported by Corbett *et al.* (1995). A fusion of GFP to a small portion of the

Table I
A Sample of Applications of GFP in Studies of the Yeast Nucleus

Application	Reference
Studies of Nuclear Pore Dynamics	
GFP fusions of Nup49 and Nup133 to study nuclear pore assembly	Belgareh and Doye, 1997
GFP–Nup49; in heterokaryons, NPCs diffuse through the NE	Bucci and Wente, 1997
Studies of Subnuclear Structures	
First application of GFP to dynamic studies: SPB dynamics *in vivo*	Kahana *et al.,* 1995
Calmodulin–GFP fusions localize to SPB and sites of cell growth	Moser *et al.,* 1997
S. pombe GFP fusion library to identify nuclear-targeted proteins	Sawin and Nurse, 1996
Integrated *E. coli* Lac operator sends GFP–Lac repressor to the centrosome	Straight *et al.,* 1996
GFP–Tubulin and GFP–centrosome dynamics through mitosis	Straight *et al.,* 1997
Studies of Nuclear Protein Targeting	
Regulated nucleocytoplasmic shuttling of transcription factor, Mig1–GFP	De Vit *et al.,* 1997
First report of GFP fusion in yeast; localizes Np13 to the nucleus	Flach *et al.,* 1994
Novel assay for protein shuttling between nucleus and cytoplasm	Lee *et al.,* 1996
Localization of Pse1 and Sxm1, RNA export factors, to the nuclear rim	Seedorf and Silver, 1997
Mex67, genetically identified RNA export factor, localized to the NPC	Segref *et al.,* 1997
A fusion of SV40 T NLS to GFP is specifically targeted to the nucleus	Shulga *et al.,* 1996
Novel assay for nuclear protein/RNA export, using NLS-GFP-GFP-NES	Stade *et al.,* 1997
Nuclear localization of a putative Histone H1 homolog in *S. cerevisiae*	Ushinsky *et al.,* 1997
New Strategies for Creating GFP Fusions	
PCR-based method for GFP-tagging genes in the yeast genome	Wach *et al.,* 1997

yeast-RNA-binding protein Npl3p that was sufficient to target the protein to the nucleus (Flach *et al.,* 1994) was placed under control of the GAL promoter. Cells are grown to mid-log phase in selective medium supplemented with 2% raffinose, which neither induces nor represses expression from the GAL promoter. Expression of the reporter protein is induced by the addition of galactose (2% final) to the medium, and the cells are allowed to grow for another 4–5 h to allow folding and fluorescence of the expressed GFP fusion. If further treatment of the cells is required, glucose can be added to the medium to prevent further expression of the reporter. To test whether the "classical" nuclear transport pathway is involved, this experiment can be performed in a temperature sensitive strain defective in a nuclear transport component, such as *srp1-31* (Loeb *et al.,* 1995) or *rna1-1* (Corbett *et al.,* 1995).

b. Protocol IV: Assay for Shuttling Between the Nucleus and the Cytosol

There is an increasing awareness of the importance of protein export from the nucleus. Investigation of this process has been hampered by the fact that assays for protein export require that the substrate first enter the nucleus. This assay was devised to address this limitation with respect to RNA export, which requires previous protein import into the nucleus, but should also be useful for studies of nuclear protein export; a related assay, specifically designed to look at nuclear protein export, has recently been reported (Stade *et al.,* 1997). Our assay uses a fusion of GFP to an RNA-binding protein, Npl3p, previously shown to shuttle (Flach *et al.,* 1994; Lee *et al.,* 1996). At steady state, the bulk of Npl3p is nuclear in wild-type cells, but mutants have been identified that mislocalize to the cytoplasm (Lee *et al.,* 1996). The assay uses a yeast strain harboring a mutant form of a nuclear pore complex protein, Nup49p, to investigate the shuttling of Npl3-GFP between the nucleus and the cytoplasm. At the restrictive temperature of 36°C, *nup49-313* cells are defective in protein import, but not in RNA export (Schlenstedt *et al.,* 1993; Doye *et al.,* 1994). Thus, shuttling proteins that leave the nucleus in a complex with RNA are able to do so but will be trapped in the cytoplasm by the block in protein import.

1. *Grow 5 ml of nup49-313 cells carrying the desired expression vector* [e.g., pPS 811, where the expression of the Npl3-GFP fusion is controlled by a galactose-inducible promoter, (Lee *et al.,* 1996)] overnight in the appropriate selection medium supplemented with 2% glucose.

2. *Dilute the overnight culture* at least 200-fold into dropout medium containing 2% raffinose as the sole carbon source, and grow to mid-log phase (usually requires an overnight incubation). Induction of gene expression from the *GAL1* or the *GAL10* promoter is exquisitely sensitive to glucose. Replacing glucose with raffinose, which neither represses nor induces the GAL promoters, ensures that any residual glucose is metabolized. If the number of cells in the glucose culture is limiting, as can be the case with slow-growing mutant strains, the whole overnight culture can be recovered by centrifugation at $800 \times g$ for 5 min, washed twice with distilled water, and resuspended in the desired volume of raffinose-containing growth medium.

3. *Induce the expression of the GFP fusion reporter protein* by the addition of galactose to 2%. As a negative control, add 2% glucose to a parallel aliquot of the raffinose culture or use cells carrying a galactose-inducible version of the reporter protein lacking GFP. The amount of expression will increase with time, and in some strains can be varied by varying the final concentration of galactose over a 0.003 to 2% (final) range. Each reporter will behave differently. Optimal induction periods can be conveniently determined by Western analysis of samples taken from an induced culture over a time course of ~6 hours, with an anti-GFP antibody; this has the added advantage of confirming that the GFP signal detected in the microscope derives from the fusion protein, not from free GFP that may be generated by faulty translation or proteolytic clipping of the reporter fusion.

4. *Repress the expression of the fusion protein* by the addition of glucose to 2%. Confirm by microscopy that the GFP reporter is nuclear. Glucose repression of reporter gene expression ensures that any protein accumulating in the cyto-

plasm derives from preexisting protein you will have seen in the nucleus, rather than from newly synthesized and hence cytoplasmic protein. For this reason, it is good practice to wait for about 30 min after the addition of glucose before proceeding with the experiment, to allow any newly synthesised protein to fold, and the GFP to fluoresce, so that all the reporter can be reliably stated to be nuclear at the beginning of the experiment.

5. *Incubate an aliquot of the repressed culture at the restrictive temperature* (36°C for *nup49-313*) and a parallel aliquot at the permissive temperature (e.g., 25°C). Over time, microscopically examine the cells for the GFP signal. If the reporter protein is able to exit the nucleus and is stable in the cytoplasm, it should accumulate in these cells at 36°C, but not at 25°C. Note that not all cells will mislocalize the reporter. Lee *et al.* (1996) report a strong effect after 5 h at 36°C in only 30–35% of the *nup49-313* cells, with a weaker effect in 30–35% of the remaining cells.

D. Making New Fusions

1. Available GFP Vectors

The most convenient way of making fusion proteins is a PCR-based method. A GFP-containing vector with a polylinker either upstream (for C-terminal fusions) or downstream (for N-terminal fusions) of the GFP open reading frame (ORF) allows the insertion of the ORF of interest in frame with GFP. This is achieved by producing the ORF of interest in a PCR reaction where the primers add appropriate restriction sites to either end of the ORF, so that the PCR product can be digested and ligated into the GFP vector. For an example of a series of vectors allowing C-terminal GFP fusions, see Kahana and Silver (1996). Many GFP fusions have now been published (Table I), and, rather than making your own GFP vector, it may be quicker to obtain a ready-made expression vector and insert your ORF in place of the one inserted by others.

2. Considerations in Designing an *in Vivo* Localization Experiment

When considering the backbone to be used in a GFP fusion expression experiment, it can be useful to know a little about the protein beforehand. Is the protein toxic when overexpressed? Is it stable? Is the addition of GFP at the N or the C terminus likely to destablize the protein or adversely affect its function? Ideally, for a localization experiment, one would like the fusion protein to functionally replace the endogenous protein, confirming that the fusion is localized in the place where its function is required; overexpressed proteins can titrate out their normal binding sites and mislocalize to nonspecific and irrelevant locations. However, detection of low-abundance proteins may not be possible due to low signal, and proteins of short half-life may turn over too rapidly for the GFP cyclization reaction to occur so that the GFP signal never gets generated. Under these circumstances, it may be more appropriate to use a constitutive promoter (e.g., the *HIS3* promoter), a high-level constitutive promoter (e.g., *ADH)*, or, for very low-level or toxic proteins, a high-level inducible promoter (e.g., *GAL1* or *GAL10*).

a. Protocol VI: Screening for Mislocalization of GFP Fusion Proteins

We have recently reported a visual screen for temperature-sensitive mutations that mislocalize a nuclear reporter protein (Npl3p) to the cytoplasm (Corbett and Silver, 1996). This screen identified a number of genes, some of which remain to be characterized, and two essential nuclear transport factors, *PRP20* and *NTF2* (Corbett and Silver, 1996). The following protocol is an adaptation of this screen using a GFP fusion rather than laborious indirect immunofluorescence. Such a screen may be useful in dissecting the mechanism of nuclear uptake of proteins whose localization appears not to depend upon the classical nuclear transport pathway. Corbett and Silver (1996) made use of a library of temperature-sensitive (ts) mutants (Amberg *et al.,* 1992) and screened for the mislocalization of an endogenous protein.

1. *The first step in a GFP-based screen requires the generation of a yeast strain,* in which the gene encoding the protein of interest has been replaced with a GFP-tagged version by homologous recombination. A convenient method for achieving this is to clone GFP in frame, either at the N terminus (after the initiating ATG) or at the C terminus, before the STOP codon, of the ORF of interest. A series of vectors carrying yeast-selectable markers but lacking yeast origins of replication has been described (Sikorski and Hieter, 1989). For homologous integration to occur, it is essential that GFP be flanked by sequences belonging to the target gene or the GFP-encoding sequences will be lost upon recombination. To integrate the GFP fusion, linearize the plasmid carrying the fusion construct by cutting with a restriction enzyme at a unique site within the ORF of your gene and transform a diploid strain with 0.2–0.5 μg of digested plasmid (Rose *et al.,* 1990); we find this to give the most reliable integration at the target site in the yeast genome. Sporulate the diploid, and select one spore of each mating type that correctly expresses the GFP fusion. Correct integration is best confirmed by Southern hybridization; PCR-based techniques can give rise to spurious results. Expression of the intact fusion protein should be confirmed by Western blotting with an anti-GFP antibody (Clontech).

2. Perform the mutagenesis following the standard yeast techniques given in Lawrence (1991) and Rose *et al.* (1990).

E. Making Movies

1. Practical Considerations

There are two limiting factors to be taken into consideration in the making of a time-lapse series of micrographs: cell viability and photobleaching. Prolonged incubations on the microscope slide require that the cell be adequately supplied with water and nutrients; this is achieved by placing the cells on an agar pad containing SC medium and 2% glucose (or other carbon source), sealed with lanolin and liquid paraffin. Cell viability will also be adversely affected by the irradiation required for detection of the GFP signal, and care must be taken both to minimize this exposure and to control for its effects on the outcome of the experiment. These precautions will also go some way to reducing the problem of photobleaching, defined as the irreversible loss of fluorescence induced by irradiation damage. Typically, one has to try several different combinations of exposure times and frequencies before hitting on the optimal conditions for each GFP fusion.

Exposure times can be reduced by the use of an electronically controlled fluorescence shutter (e.g., Uniblitz no. D122), which can be driven by appropriate computer software (e.g., Metamorph, from Universal Imaging, Inc., MA) to open for the exact time interval required to capture the image. This also has the advantage that one does not have to remember to open and close the shutter before and after each shot. Similarly, the use of an electronic remote focus accessory (e.g., Nikon no. 79588) allows rapid and fine adjustments to the focus without the danger of causing microscope or camera vibration. Finally, it will clearly be important to choose the most sensitive image acquisition device available. We use either a Photometrics (cooled) or a Princeton Instruments CCD (charge-coupled-device) camera; both of these are equipped with a KAF1400 chip, for maximal resolution.

2. Protocol VII: Using Metamorph to Make a Movie

The following protocol is derived from Kahana *et al.* (1995) and Kahana and Silver (1997). Although it deals specifically with the Metamorph software we use in the lab, most image-acquisition packages are able to perform similar functions. The most important consideration is that the software be compatible with the camera.

1. *Grow cells to mid-log phase in liquid culture.*

2. *Prepare the agar pad.* Mix 100 ml of synthetic complete medium containing 2% glucose with 1 g of agarose, and microwave until the agar is dissolved. Pipette 1 ml onto a clean microscope slide. Typically, five slides should be prepared at once, as there is a high failure rate. This allows the use of a 5-ml glass pipette, which can be prewarmed briefly over a Bunsen burner to avoid solidifying of the agar within the pipette. *Gently* drop a second microscope slide onto the molten agar without causing the liquid to overflow from the lower slide and without the creation of air bubbles. Some bubbles almost always arise; as long as they do not affect the thickness of the pad and do not occupy the center of the slide, these may be permissible. Allow to solidify for 5–10 min on the bench.

3. *Carefully remove the top slide.* This should leave a 1-mm-thick agar pad over the surface of the slide.

4. *Pipette 3 μl of yeast culture into the center of the slide. Gently* drop a 22- × 22-mm coverslip onto the cells. Using a clean razor blade or scalpel, trim away the excess agar from around the cover slip, leaving a square in the center of the slide. Seal the cover slip with molten VALAP (5 g lanolin, 5 g paraffin chips, 5 g petroleum jelly mixed together and melted in a boiling water bath; this wax can be cooled and reheated several times over). The wax is best applied with a Q-tip, but care is needed.

5. *Place the slide on the microscope.* Wherever possible, use bright (not fluorescent) light to focus on the cells, to avoid unnecessary exposure to fluorescent light.

6. *Set up the Metamorph software to acquire from the digital camera.* First, define the acquisition settings; the exposure time (typically 100 to 500 ms) will have been determined previously as the minimum required to obtain an identifiable image; cells suffer side effects at around 15 s total exposure. The other major variable is the area of the slide to be imaged: large areas will include more cells and hence more information, but will also create more background noise and hence will result in longer exposure times. Our compromise is to image small numbers of cells (as low as one) and to repeat the experiment many times over. This further allows us to identify subnuclear structures in single cells that would be out of focus in the majority of a field of cells.

For an optimal signal-to-noise ratio, we use the "Autoscale 16-bit image" function of Metamorph, which redefines the wide range of brightness values obtained by each well of the charge-coupled camera into 256 shades of gray (the brightest camera well is assigned a value of 256, the darkest = 0). This maximizes the contrast of the displayed image, making weak GFP signals much easier to identify and interpret, and allows minimal fluorescence exposure times. If little spatial resolution is required, it may also be useful to "bin" the image; this allows the information that is normally contained in a number (e.g., 4 or 9) of neighboring pixels on the screen to be combined, giving rise to an even brighter signal.

7. *Set up the "Acquire Timelapse" menu* to capture images at the desired interval (e.g., 10 s; if the exposure is 100 ms, and allowing a maximum cellular exposure of 15 s; this would permit 150 frames in a movie that would be 25 min long).

8. *To confirm that the cells have not been damaged by the experiment, always check after 1–2 h growth on the slide that the cells have continued to bud and divide.*

Acknowledgments

Jason Kahana set up GFP and the CCD camera in this lab, and has been inspirational to many in the field. Jen Hood provided insightful comments on the manuscript. Anita Corbett, Jason Kahana, Deanna Koepp, Margaret Lee, and Darren Wong stand out among the interactive members of the Silver lab: thanks to all. Barry Alpert has provided unflagging technical support. P.F. thanks Bruce

Futcher for 10 interactive days that helped to shape this Chapter. This work was supported by a Long Term Fellowship from the Human Frontiers in Science Organization to P.F., and by grants from NIH and the Claudia Adams Barr Investigator Program to P.A.S.

References

Adams, A. E. M., and Pringle, J. (1984). Relationship of actin and tubulin distribution to bud growth in wildtype and morphogenetic-mutant *Saccharomyces cerevisiae. J. Cell Biol.* **98,** 934–945.

Amberg, D. C., Goldstein, A. L., and Cole, C. N. (1992). Isolation and characterization of *RAT1:* an essential gene of *Saccharomyces cerevisiae* required for the efficient nucleocytoplasmic trafficking of mRNA. *Genes Dev.* **6,** 1173–1189.

Aparicio, O. M., Billington, B. L., and Gottschling, D. E. (1991). Modifiers of position effect are shared between telomeric and silent mating-type loci in *S. cerevisiae. Cell* **66,** 1279–1287.

Belgareh, N., and Doye, V. (1997). Dynamics of nuclear pore distribution in nucleoporin mutant yeast cells. *J. Cell Biol.* **136,** 747–759.

Bucci, M., and Wente, S. R. (1997). *In vivo* dynamics of nuclear pore complexes in yeast. *J. Cell Biol.* **136,** 1185–1199.

Cody, C. W., Prasher, D. C., Westler, W. M., Prendergast, F. G., and Ward, W. W. (1993). Chemical structure of the hexapeptide chromophore of the *Aequorea* green-fluorescent protein. *Biochemistry* **32,** 1212–1218.

Corbett, A. H., Koepp, D. M., Schlenstedt, G., Lee, M. S., Hopper, A. K., and Silver, P. A. (1995). Rnalp, a Ran/TC4 GTPase activating protein, is required for nuclear import. *J. Cell Biol.* **130,** 1017–1026.

Corbett, A. H., and Silver, P. A. (1996). The *NTF2* gene encodes an essential, highly conserved protein that functions in nuclear transport *in vivo. J. Biol Chem.* **271,** 18477–18484.

Davis, L. I., and Blobel, G. (1986). Identification and characterization of a nuclear pore complex protein. *Cell* **45,** 699–709.

Davis, L. I., and Fink, G. R. (1990). The *NUP1* gene encodes an essential component of the yeast nuclear pore complex. *Cell* **61,** 965–978.

De Vit, M. J., Waddle, J. A., and Johnston, M. (1997). Regulated nuclear translocation of the Mig1 glucose repressor. *Mol. Biol. Cell* **8,** 1603–1618.

Doye, V., Wepf, R., and Hurt, E. C. (1994). A novel nuclear pore protein Nup133p with distinct roles in poly(A)$^+$ RNA transport and nuclear pore distribution. *EMBO J.* **13,** 6062–6075.

Flach, J., Bossie, M., Vogel, J., Corbett, A., Jinks, T., Willins, D. A., and Silver, P. A. (1994). A yeast RNA-binding protein shuttles between the nucleus and the cytoplasm. *Mol. Cell Biol.* **14,** 8399–8407.

Fornerod, M., Ohno, M., Yoshida, M., and Mattaj, I. W. (1997). CRM1 is an export receptor for leucine-rich nuclear export signals. *Cell* **90,** 1051–1060.

Heim, R., Prasher, D. C., and Tsien, R. Y. (1994). Wavelength mutations and posttranslational autoxidation of green fluorescent protein. *Proc. Natl. Acad. Sci. U. S. A.* **91,** 12501–12504.

Heim, R., Cubitt, A. B., and Tsien, R. Y. (1995). Improved green fluorescence [letter]. *Nature* **373,** 663–664.

Inouye, S., and Tsuji, F. I. (1994). Evidence for redox forms of the *Aequorea* green fluorescent protein. *FEBS Lett.* **351,** 211–214.

Kahana, J. A., Schnapp, B. J., and Silver, P. A. (1995). Kinetics of spindle pole body separation in budding yeast. *Proc. Natl. Acad. Sci. U. S. A.* **92,** 9707–9711.

Kahana, J. A., and Silver, P. A. (1996). Use of *A. victoria* green fluorescent protein to study protein dynamics *in vivo. In* "Current Protocols in Molecular Biology" (Ausubel, F. M., Brent, R., Kingston, R. E., Moore, D. E., Seidman, J. G., Smith, J. A., and Struhl, K., Eds.) vol. 1, pp. 9.6.13–9.6.19. John Wiley and Sons, Inc., New York.

Kahana, J. A., and Silver, P. A. (1997). The uses of green fluorescent proteins in yeasts. *In* "GFP: Green Fluorescent Protein Strategies and Applications" (Chalfie, M., and Kain, S., Eds.). John Wiley and Sons, Inc., New York.

Kutay, U., Bischoff, F. R., Kostka, S., Kraft, R., and Gorlich, D. (1997). Export of Importin a from the nucleus is mediated by a specific nuclear transport factor. *Cell* **90,** 1061–1071.

Lawrence, C. W. (1991). Classical mutagenesis techniques. *Methods Enzymol.* **194,** 273–281.

Lee, M. S., Henry, M., and Silver, P. A. (1996). A protein that shuttles between the nucleus and the cytoplasm is an important mediator of RNA export. *Genes Dev.* **10,** 1233–1246.

Lee, M. S., and Silver, P. A. (1997). RNA movement between the nucleus and the cytoplasm. *Curr. Opin. Genet. Dev.* **7,** 212–219.

Loeb, J. D., Schlenstedt, G., Pellman, D., Kornitzer, D., Silver, P. A., and Fink, G. R. (1995). The yeast nuclear import receptor is required for mitosis. *Proc. Natl. Acad. Sci. U. S. A.* **92,** 7647–7651.

Moser, M. J., Flory, M. R., and Davis, T. N. (1997). Calmodulin localizes to the spindle pole body of *Schizosaccharomyces pombe* and performs an essential function in chromosome segregation. *J. Cell Sci.* **110,** 1805–1812.

Nigg, E. A. (1997). Nucleocytoplasmic transport: signals, mechanisms and regulation. *Science* **386,** 779–787.

Prasher, D. C., Eckenrode, V. K., Ward, W. W., Prendergast, F. G., and Cormier, M. J. (1992). Primary structure of the *Aequorea victoria* green-fluorescent protein. *Gene* **111,** 229–233.

Pringle, J. R., Adams, A. E. M., Drubin, D. G., and Haarer, B. K. (1991). Immunofluorescence methods for yeast. *In* "Guide to Yeast Genetics and Molecular Biology" (Guthrie, C., and Fink, G. R., Eds.) vol. 194, pp. 565–602. Academic Press, Inc., San Diego, CA.

Rose, M. D., Winston, F., and Hieter, P. (1990). "Methods in Yeast Genetics: A Laboratory course manual." Cold Spring Harbor Laboratory Press, Cold Spring Harbor, New York.

Sawin, K. E., and Nurse, P. (1996). Identification of fission yeast nuclear markers using random polypeptide fusions with green fluorescent protein. *Proc. Natl. Acad. Sci. U. S. A.* **93,** 15146–15151.

Schlenstedt, G., Hurt, E., Doye, V., and Silver, P. A. (1993). Reconstitution of nuclear protein transport with semi-intact yeast cells. *J. Cell Biol.* **123**(4), 785–798.

Seedorf, M., and Silver, P. A. (1997). Importin/karyopherin protein family members required for mRNA export from the nucleus. *Proc. Natl. Acad. Sci. U. S. A.* **94**(16), 8590–8595.

Segref, A., Sharma, K., Doye, V., Hellwig, A., Huber, J., Luhrmann, R., and Hurt, E. (1997). Mex67p, a novel factor for nuclear mRNA export, binds to both poly(A)$^+$ RNA and nuclear pores. *Embo J.* **16,** 3256–3271.

Shulga, N., Roberts, P., Gu, Z., Spitz, L., Tabb, M. M., Nomura, M., and Goldfarb, D. S. (1996). *In vivo* nuclear transport kinetics in *Saccharomyces cerevisiae:* a role for heat shock protein 70 during targeting and translocation. *J. Cell Biol.* **135,** 329–339.

Sikorski, R. S., and Hieter, P. (1989). A system of shuttle vectors and yeast host strains designed for efficient manipulation of DNA in *Saccharomyces cerevisiae. Genetics* **122,** 19–27.

Skowronek, P., Krummeck, G., Haferkamp, O., and Rodel, G. (1990). Flow cytometry as a tool to discriminate respiratory-competent and respiratory-deficient yeast cells. *Curr. Genet.* **18,** 265.

Stade, K., Ford, C. S., Guthrie, C., and Weis, K. (1997). Exportin 1 (Crm1p) is an essential nuclear export factor. *Cell* **90,** 1041–1050.

Straight, A. F., Belmont, A. S., Robinett, C. C., and Murray, A. W. (1996). GFP tagging of budding yeast chromosomes reveals that protein–protein interactions can mediate sister chromatid cohesion. *Curr. Biol.* **6,** 1599–1608.

Straight, A. F., Marshall, W. F., Sedat, J. W., and Murray, A. W. (1997). Mitosis in living budding yeast: anaphase A but no metaphase plate. *Science* **277**(5325), 574–578.

Terasaki, M., and Reese, T. S. (1992). Characterization of endoplasmic reticulum by colocalization of BIP and dicarbocyanine dyes. *J. Cell Sci.* **101,** 315.

Ushinsky, S. C., Bussey, H., Ahmed, A. A., Wang, Y., Friesen, J., Williams, B. A., and Storms, R. K. (1997). Histone H1 in *Saccharomyces cerevisiae. Yeast* **13,** 151–161.

Vida, T. A., and Emr, S. D. (1995). A new vital stain for visualizing vacuolar membrane dynamics and endocytosis in yeast. *J. Cell Biol.* **128,** 779–792.

Wach, A., Brachat, A., Alberti-Segui, C., Rebischung, C., and Philippsen, P. (1997). Heterologous HIS3 marker and GFP reporter modules for PCR-targeting in *Saccharomyces cerevisiae. Yeast* **13,** 1065–1075.

CHAPTER 8

GFP Fusion Proteins as Probes for Cytology in Fission Yeast

Kenneth E. Sawin
Imperial Cancer Research Fund
London WC2A 3PX, United Kingdom

I. Introduction

Since the introduction of green fluorescent protein (GFP) (Prasher *et al.,* 1992) as a marker of gene expression *in vivo* (Chalfie *et al.,* 1994) and as a cytological marker fused to protein-coding sequences (Wang and Hazelrigg, 1994), scores of investigators have used GFP fusion proteins to monitor protein expression and localization within cells of all types. Here we focus on expression of GFP fusion proteins in the fission yeast *Schizosaccharomyces pombe,* with specific applications to studying intracellular architecture. Fission yeast has been used widely as a model organism for the control of the eukaryotic cell cycle (MacNeill and Nurse, 1997) as well as many of the events of cell division (Su and Yanagida, 1997), but only more recently for detailed molecular–genetic investigations into cell structure and cytoskeletal organization. As a consequence, the cytology of

METHODS IN CELL BIOLOGY, VOL. 58
Copyright © 1999 by Academic Press. All rights of reproduction in any form reserved.
0091-679X/99 $25.00

S. pombe has remained relatively underdeveloped, and in this area the use of GFP fusion proteins can advance our understanding considerably. There are primarily two areas in which GFP fusion proteins may be particularly useful in fission yeast. First, new types of experiments, such as determining protein localization in cells without the potential artifacts of fixation or filming protein localization and dynamics in living yeast cells, have now become possible. In addition, using GFP fusion proteins rather than more conventional methods (such as immunofluorescence) for seeing proteins inside cells may make specific techniques, such as visual screens for protein localization, much more simple and rapid in execution. In this chapter we discuss many practical considerations for expressing GFP fusion proteins in fission yeast, with an emphasis on guidelines and examples rather than specific rules or recipes. Related topics are also discussed in an excellent recent paper from the laboratory of M. Yanagida (Nabeshima *et al.,* 1997).

II. Expressing GFP Fusion Proteins

A. Plasmids

Expressed GFP fusion proteins generally need to be visible to the naked eye under the microscope, although if sensitive imaging equipment is available, expression only just at the limit of detection by eye may be sufficient (Moser *et al.,* 1997; T. Davis, personal communication). Producing a visible protein may often require higher than normal expression levels and can be achieved by using heterologous promoters, multicopy plasmids (especially if endogenous regulation of expression is desired), or both. In addition, GFP mutants can now be used in place of the wild-type GFP.

1. Mutant GFPs

The development of brighter, mutant GFPs such as GFPS65T (Heim *et al.,* 1995) and GFPmut2 (Cormack *et al.,* 1996) represents a major technical advance. By FACS analysis these mutants show an approximate 6- to 10-fold fluorescence increase *in vivo* in fission yeast as compared to wild-type GFP, with S65T and GFPmut2 producing nearly equal intensities (Figs. 1 and 2). In some cases, a single-copy GFP fusion protein under its own promoter can be detected without difficulty using the S65T mutant (Moser *et al.,* 1997; Saitoh *et al.,* 1997; J. Baehler, personal communication), and when such a construct complements a deletion mutant, this is ideal (see Table I). There is probably no reason not to use these mutants, unless one is interested in specialized techniques such as enhanced fluorescence by microbeaming (discussed later).

2. Plasmid Copy Number

Because the *S. cerevisiae LEU2* gene complements leu1–32 mutations in *S. pombe* only weakly, plasmids containing *LEU2* as a selectable marker force an

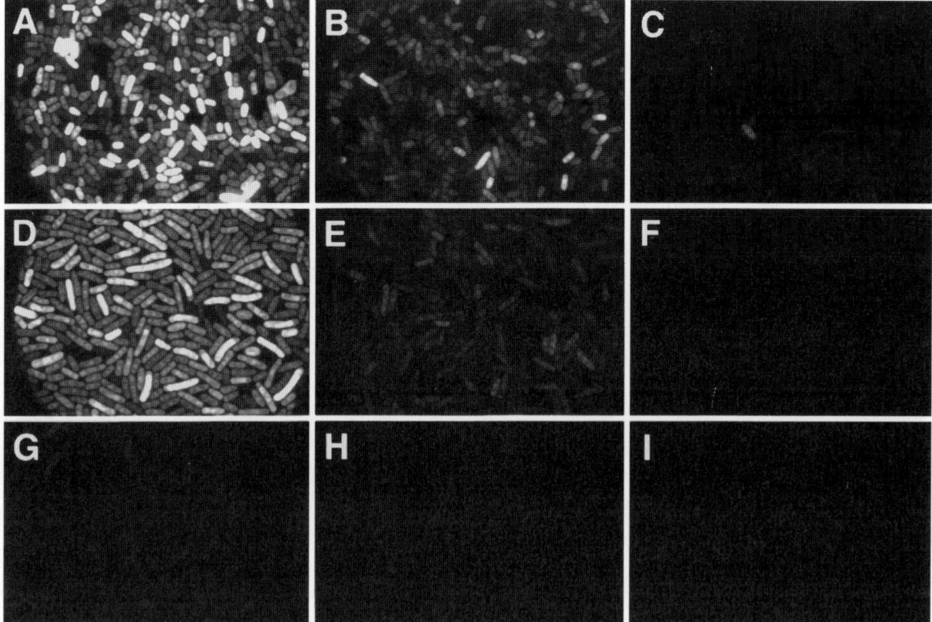

Fig. 1 Relative intensity of different GFPs under different promoters. All are on multicopy *LEU2* plasmids, and all exposure times are identical. A, *nmt1*–GFPmut2; B, *nmt41*–GFPmut2; C, *nmt81*–GFPmut2; D, *nmt1*–GFPS65T; E, *nmt41*–GFPS65T; F, *nmt81*–GFPS65T; G, *nmt1*–GFP (wild type); H, *nmt41*–GFP (wild type); I, *nmt81*–GFP (wild type). Note that some constructs appear faint only because longer exposure times would saturate the images in A and D; under the microscope, all constructs can be seen with the naked eye except the *nmt81*–GFP (wild type) in I.

increase in plasmid copy number (estimated to be 3 to 10 copies per cell), and thus also total expressed protein levels (Nabeshima *et al.,* 1995). It is important to recognize that the copy number can vary widely from cell to cell (see Figs. 1–3), with up to 50% of cells having lost the plasmid and en route to nutrient starvation. This variability can often present an experimental advantage, in that one quickly gets a sense of what may occur with a fusion protein at several different expression levels, at what levels localization may become artifactual, and at what levels expression may be deleterious to cells, and as such, multicopy plasmids may be preferred for preliminary work as well as for screening large numbers of fusion proteins (Sawin and Nurse, 1996). The plasmid copy number could also be modulated by using plasmids conferring G418 resistance and varying the concentration of G418 (Giga-Hama *et al.,* 1994), although this has not yet been applied specifically to GFP.

Integrating plasmids can go into the genome in several places (e.g., pREP6X, which carries the *sup3–5* nonsense supressor tRNA as a selectable marker; J. Hayles, personal communication), at specific loci (e.g., pJK148, which tends to

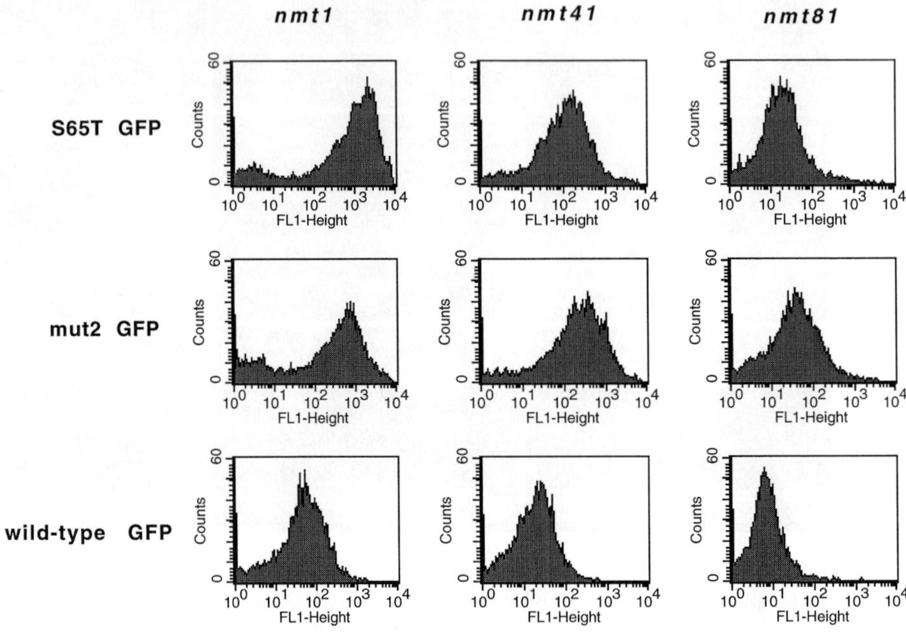

Fig. 2 FACS analysis of the strains shown in Fig. 1, with promoter on (A) and promoter off (B). Small differences seen between the S65T and mut2 GFP mutants are likely due to variation in culture conditions.

integrate at the *leu1* locus (Keeney and Boeke, 1994), or at the locus of the gene of interest itself. In all cases this presents the advantage of having similar levels of expression in all cells, as compared to multicopy plasmids, and integrated plasmids might therefore be preferred for detailed studies of a specific protein, such as filming protein dynamics, quantitating fluorescence localization in different contexts, or identifying mutants defective in localization of a given fusion protein. It is worth noting here that if integration can occur at several different sites, expression may vary among different colonies depending on the site used, and it is worthwhile to screen several colonies to find desired levels of expression.

3. Promoters

For genes whose endogenous expression is insufficient for easy visualization, how high should expression be? The safest answer is "as low as possible," because overexpression can lead to mislocalization of fusion proteins and/or abnormal cell

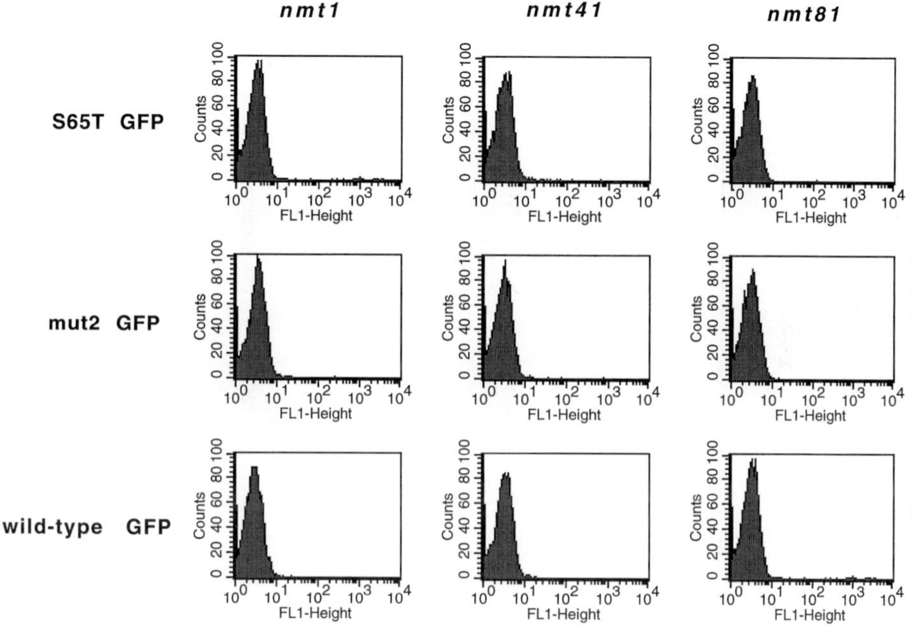

Fig. 2 *(continued)*

physiology. Many current expression vectors contain the thiamine-repressible *nmtl* promoter (Maundrell, 1993), which in the absence of thiamine is one of the strongest promoters characterized in *S. pombe* (Forsburg, 1993), and most GFP fusion proteins described to date have used the *nmtl* promoter or its weakened derivatives *nmt41* and *nmt81* (Basi *et al.,* 1993), either with LEU2 plasmids or as integrated constructs (Table I). By FACS analysis, *nmt41*-driven expression of GFP is about sixfold lower than expression from *nmtl,* and *nmt81*-driven expression approximately another 10-fold lower beyond that (Figs. 1 and 2). The use of other, heterologous promoters such as *adh,* SV40, or CMV (Forsburg, 1993) to drive GFP fusion protein expression has not been described to date. In our work, GFP and GFP fusion proteins expressed from multicopy LEU plasmids under the control of the *nmtl* and *nmt41* promoters are very easy to see with a normal fluorescence microscope. We cannot see *nmt81*–GFP alone above background when expressed from a *LEU2* plasmid unless a mutant GFP is used.

Fluorescence from an expressed GFP-fusion protein can be lower than that of GFP alone (see Fig. 3), because fusion proteins may not accumulate to as high levels as GFP alone, due to increased protein turnover and/or abnormalities caused by expression of the fusion gene. At the same time, however, many fusion

Table I
GFP-Fusion Proteins and Their Behavior *in vivo*

Gene	Promoter	Plasmid	Expression OFF	Expression ON	Localization	Reference
cam1$^+$	endogenous	integrant	N/A	rescues *cam1*Δ	spindle poles body, growing ends of cells	Moser *et al.*, 1997
cor1$^+$	*nmt1*	*LEU2* (multicopy)	rescues *cor1*Δ (no visible fluorescence)	rescues *cor1*Δ	nucleus	K.E.S. unpublished data
dis1$^+$	endogenous	*LEU2*	N/A	rescues *dis1*Δ	microtubules	Nabeshima *et al.*, 1995
mal2$^+$	*nmt1*	*LEU2*	rescues *mal2–1*^{ts} (no visible fluorescence)	rescues *mal2–1*^{ts}	nucleus	Fleig *et al.*, 1996
mei2$^+$	*nmt81*	*LEU2*	fails to rescue *mei2*Δ meiosis defect	rescues *mei2*Δ meiosis defect	nuclear spot	Watanabe *et al.*, 1997; Y. Watanabe, personal communication
mis6$^+$	endogenous	*LEU2*	N/A	complete rescue of *mis6–302*	centromeres	Saitoh *et al.*, 1997
	endogenous	integrant	N/A	rescues *mis6–302*^{ts} phenotype	centromeres	Saitoh *et al.*, 1997
myo2$^+$	*nmt81*	*LEU2*	?	rescues *myo2*Δ lethality	cytokinetic ring	Kitayama *et al.*, 1997
ral3$^+$ (*scd2*$^+$)	*nmt41*	*LEU2*	rescues *ral3*Δ sterility	rescues *ral3*Δ sterility; slight abnormal morphology even in wild-type cells	growing ends of cells	K.E.S., unpublished data
ssm4$^+$	*nmt1*	*LEU2*	?	fails to rescue *ssm4*Δ	microtubules	Yamashita *et al.*, 1997
sts5$^+$	*nmt1*	*LEU2*	rescues *sts5* mutant	toxic, fails to form colonies	cytoplasmic spots	Toda *et al.*, 1996
tea1$^+$	*nmt41*	*LEU2*	?	abnormal cell morphology	microtubules and cytoplasmic aggregates	J. Mata, personal communication

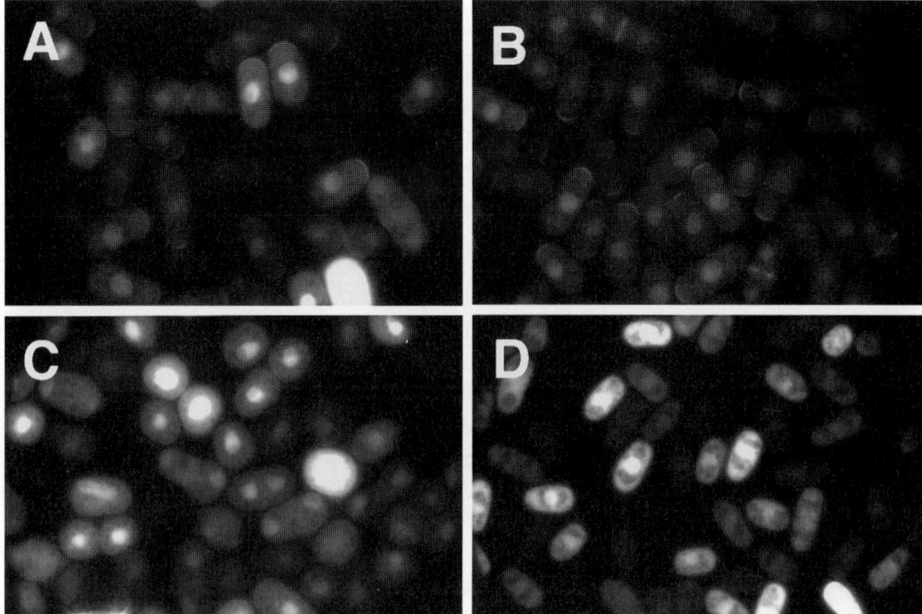

Fig. 3 Localization and intensity of a Ral3p–GFP fusion protein, relative to GFP alone. A, Ral3p–GFP on LEU2 plasmid under *nmt41* promoter, 8-s exposure. B, Ral3p–GFP integrant under *nmt1* promoter, 8-s exposure. C, Ral3p–GFP on LEU2 plasmid under *nmt1* promoter, 2-s exposure, D, GFP alone on LEU2 plasmid under *nmt41* promoter, 2-s exposure. All constructs use GFPmut2. The cortical localization is of interest, while the nuclear localization is most likely an artifact of overexpression (see faint cells in A and B). *Note:* (*a*) the *nmt1*-promoter integrant in B gives roughly the same intensity as the *nmt41*-promoter multicopy plasmid in A, but with less variation from cell to cell; (*b*) after correcting for exposure time, the *nmt1*-promoter multicopy plasmid in C gives considerably more intensity than the *nmt41*-promoter multicopy plasmid in A, but cell morphology is profoundly altered and cortical fluorescence can barely be seen above background; and (*c*) GFP alone on an *nmt41*-promoter multicopy plasmid in D gives as much fluorescence intensity as Ral3p–GFP on an *nmt1*-promoter multicopy plasmid in C, in spite of the 10-fold lower promoter strength of *nmt41*.

proteins may be considerably easier to see than GFP alone if they are present at very high local concentrations. For example, expression of a spindle pole-localized Cut7p–GFP can be seen when integrated under *nmt81* control (D. Drummond and I. Hagan, personal communication). We have easily detected expressed fusion proteins with a more disperse distribution when integrated as GFPmut2 fusion genes under the *nmt1* promoter (Fig. 3), and would expect that for many fusion proteins, integrants under *nmt41* control should also be visible. For a given protein, what can be seen easily will clearly depend intimately on its pattern of localization; fortunately it is usually not difficult to exchange among *nmt* promoters to obtain appropriate expression levels.

Over the long term, especially with multicopy plasmids, *nmt1*-driven expression may be too strong for cells to grow normally in the absence of thiamine, and therefore to determine "normal" protein localization one may need to monitor cells carefully to find the earliest possible times of expression, before things go awry. In living cells this can represent a very narrow window, but in some cases this window can be expanded by growing cells in limiting amounts of thiamine (Toda *et al.,* 1996; D. Drummond and I. Hagan, personal communication). In 25–50 n*M* thiamine, cells ultimately reach high levels of expression, but with slower kinetics, allowing more freedom to evaluate localization patterns *in vivo.*

4. Available Expression Vectors

pSGA, pSGB, pSGC, pMGA, pMGB, pMGC, pWGA, pWGB, and pWGC from our laboratory are LEU2 plasmids with *nmt1* ("S" series), *nmt41* ("M" series), and *nmt81* ("W" series) promoters, which allow cloning into a number of sites 3′ to GFP (Sawin and Nurse, 1996). "B" and "C" versions have 1 and 2 base insertions, respectively, just 5′ to the SalI site (see Fig. 1 in Sawin and Nurse, 1996). These were originally constructed with wild-type GFP but have been remade using GFPmut2. GFP fusion genes can be cut out easily from these vectors by Xho/BamHI digestion and subcloned into the integrating vector pREP6X (J. Hayles, personal communication). A more versatile set of vectors, pGFT1, pGFT41, and pGFT81, have been constructed by Y. Watanabe in the laboratory of M. Yamamoto (Kitayama *et al.,* 1997; Yamashita *et al.,* 1997; Y. Watanabe, personal communication). These are also LEU2-based and use *nmt1, nmt41,* and *nmt81* promoters to control expression of genes that can be cloned either 5′ or 3′ to the S65T GFP coding sequence.

B. Behavior of Fusion Proteins

In fission yeast, proteins have been successfully fused to GFP at both the C terminus and the N terminus, as well as in the middle of three-way fusions. As with any protein-tagging method, fusion proteins may not always function like their wild-type counterparts, either as a consequence of increased expression levels or because of the fusion to GFP. In extreme cases, overexpression may be toxic (Toda *et al.,* 1996). With at least one fusion protein we have worked with, Ral3p-GFP (Fukui and Yamamoto, 1988; Chang *et al.,* 1994), high levels of protein are obviously deleterious, albeit nonlethal. Lower levels are more tolerated but can still cause slight abnormal cell morphology, and even lower levels of visible expression produce the most clear localization and fewest abnormalities (Fig. 3). In screening GFP fusion libraries we have found that plasmids producing an apparently normal localization at lower levels of expression can, at higher levels, spill out into neighboring regions of the cell or form large clumps, which may be analogous to inclusion bodies. In any case, even if a given GFP fusion protein fails to behave perfectly normally or suppress a deletion phenotype completely (although, as shown in Table I, some do quite successfully), as long as some normal functions are retained, it can serve as a useful tool. For example, although expression of GFP–Tea1p produces an artifactual localization to cytoplasmic microtubules (J. Mata, personal communication) this initial evidence ultimately led to the finding of an association of endogenous Tea1p with microtubule ends (Mata and Nurse, 1997). Similarly, although GFP–Ssm4p does not rescue an *ssm4* deletion, it does rescue an *sme2* deletion (which was the basis for the original isolation of the *ssm4*$^{+}$ cDNA) and shows an interesting localization to microtubules (Yamashita *et al.*, 1997). Overall, however, it is important to emphasize that for critical experiments one should be extremely careful to avoid overinterpretation of results based solely on the localization of GFP fusion proteins, even under what might appear to be benign conditions.

III. Applications of GFP Fusion Proteins

A. Screening Living Cells

In our laboratory we are interested in identifying protein components of cellular architecture in *S. pombe* by direct visual fluorescent screening of cells transformed with libraries of random GFP fusion proteins (Sawin and Nurse, 1996). Although this work is still in progress, the screening methods used are both simple and fast, allowing screening of perhaps up to 1000 colonies per day, and could easily be adapted to additional purposes such as isolating mutants defective in the localization of a specific GFP fusion protein of interest. Colonies expressing fusion proteins are spotted in a 16 × 24 grid pattern (Fig. 4) using 384-well plates (Nunc No. 242757; plates are also available from other manufacturers) and a 384-

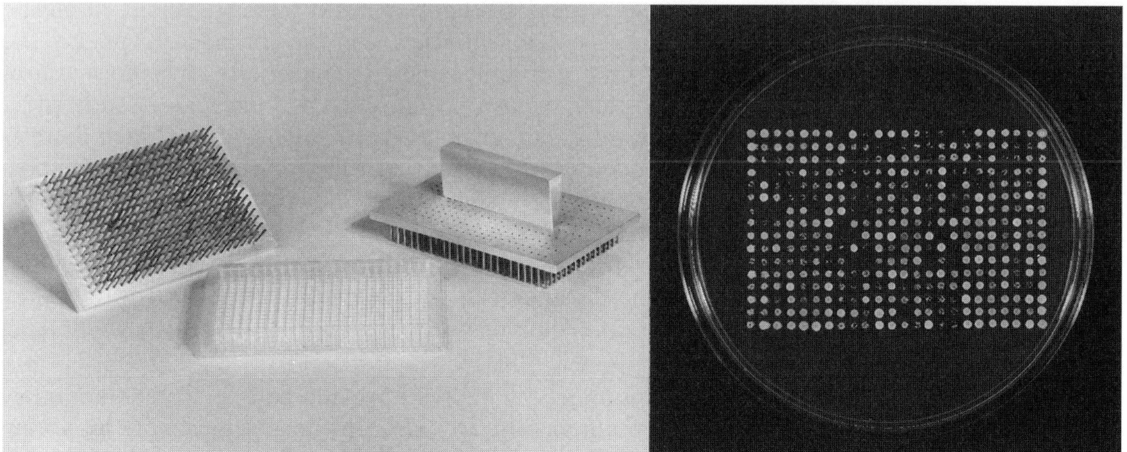

Fig. 4 A 384-well plate and spotter used for producing grids of fission yeast colonies, and a colony grid.

pin spotting device with hollow pins, which was adapted from high-throughput genomics systems and can be flame-sterilized (designed by Mr. J. Kitau, ICRF; disposable plastic 384-pin spotters can also be used). After 1–2 days of growth, the spots are replica-plated in small groups to untreated microscope slides using filter paper and observed directly under a conventional fluorescence microscope. The advantages of screening a grid pattern vs. a random array of colonies are that (*a*) each colony has a distinct "address" and thus is instantly located on a master plate, and (*b*) because colonies are spaced apart by a uniform distance, no time is wasted searching for the next colony under the microscope.

1. Fill wells of a 384-well plate with 50–100 μl of medium with a multichannel pipetter. Pick colonies from plates with toothpicks and place in wells manually, leaving toothpicks in wells to keep track of which ones have been filled. This is the most tedious part of the method, and it usually takes at least an hour to fill a plate completely. When finished, remove toothpicks from wells, avoiding cross-contamination.
2. Sterilize spotter with ethanol, touching to a paper towel to remove excess alcohol before flaming. (This is *extremely* important with a hollow pin spotter, which can suck up a considerable amount of liquid by capillary action.) Place the spotter in a multiwell plate, remove it, and place it on 150-mm agar-medium plates to be used for screening (e.g., minus thiamine, to induce expression). The plates should be premarked so that the orientation of spotting is obvious; the spots will be about 2 mm in diameter. Repeat (without flaming) to a fresh plate, to create a master plate (e.g., plus thiamine, to repress expression). After drying, incubate the plates at an appropriate temperature until screening. Dense growth is not required for making replicas. It is useful to prepare several replicas with which to experiment.
3. Before screening, cut pieces of Whatman 3MM paper to about 20 mm $\times$ 40 mm (sterile technique is not essential). Using flat, broad forceps, lay the filter paper over a 4 $\times$ 8 grid of colonies and let it sit briefly until moist, about 10–30 s. Remove it and place it on a standard 1- $\times$ 3-inch microscope slide. Press down on the paper with another microscope slide (a 2- $\times$ 3-inch double-width slide is easier). Add 1 μl of medium at a few places between the spots and cover with a 25- $\times$ 40-mm coverslip. The medium assures that the cells do not dry out and provides a continuity of refractive index for better optics. Cells from neighboring spots should not blend significantly with each other for about 30 min, during which time the slide can be screened. If 32 spots are too many to screen in this time, the procedure can be modified to groups of 16 spots, using the same method and a 22- $\times$ 22-mm coverslip.
4. Scan the spot-replicas under the fluorescence microscope. Colonies of interest can be noted either by marking a premade grid sheet or simply by noting their grid position. Although some parts of a spot-replica can be too dense for viewing, other areas should give a clear view of localization. When this is complete, the next replica is made and the process repeated. A clear evaluation

of an interesting localization usually takes 1–2 min, and obvious "negatives" significantly less time.

B. Photographing and Filming Cells

Detailed methods for photographing and filming cells are beyond the scope of this chapter (see Nabeshima *et al.,* 1997, for more information); however, some hints may be useful with respect to preparing and mounting cells. For still photography it is often sufficient to immobilize living cells by using a small volume of liquid culture on a slide (e.g., 2.5–3.5 μl underneath a 22- × 22-mm coverslip). This also serves to flatten the cells, which can further improve images. In some cases slides or coverslips can be pretreated with poly-L-lysine (Sigma No. P-1524; 10 mg/ml in water with 0.1% azide, store at 4°C; coat glass with 2–3 μl just before use, wiping dry with a tissue or Q-tip after a few seconds), although we do not prefer this for living cells because it seems to cause vacuolation. In addition, in our hands poly-L-lysine seems to interfere with counterstaining of cell wall with Calcofluor. If long-term filming is desired (i.e., over several hours), the cells should be mounted under conditions that favor growth. Traditionally fission yeast have been filmed by mounting them on a pad of medium agar or agarose, under a coverslip; more recently a simple method of embedding cells in low-melt agarose has been used (Saitoh *et al.,* 1997). In our own limited experience, mounting cells on top of an agarose pad has worked better for cell growth over several generations, and it also puts all the cells in the same focal plane. Other methods for mounting cells can be found in Nabeshima *et al.* (1997).

Making an Agarose Pad for Filming

1. Melt agarose (final 2%) in a few milliliters of minimal medium in a microwave oven (yeast extract medium contributes to autofluorescence but could be used in some instances). For short-term viewing, sterile technique may not be essential.

2. Using a 2- × 3-inch clean microscope slide as a base, apply Scotch "Magic" tape (3M) at either end to act as spacers, and place the slide on a hotplate to warm. Without prewarming, the agarose will set too quickly and the pad will not have a uniform thickness.

3. Place 15–50 μl of melted agarose medium on the warm slide and cover it with a 1- × 3-inch slide, applying a small amount of pressure to flatten the pad to the thickness of the tape (approx. 50 μm). Remove the pad from the hotplate and allow it to set on the bench. In some cases one may want to let the agarose set for up to 30 min.

4. Carefully remove the lower base slide, keeping the upper slide and the pad together (if this is difficult, try pretreating the base slide with a siliconizing agent, such as Sigmacote, before preparing the pad). Apply cells to pad in 2–3 μl of medium; it may be necessary to remove excess liquid with a tissue. Cells can be preconcentrated by centrifugation, bearing in mind how many cells will be on

the pad relative to nutrient amounts, which could be limiting. One to ten thousand cells per pad is probably a safe range.

5. If desired, seal the chamber with melted paraffin wax to avoid drying out, applying the wax to the coverslip edges with a small artist's paintbrush. Other methods of sealing, such as VALAP, a 1 : 1 : 1 mixture of vaseline, lanolin, and paraffin, can also be used.

When observing faint signals it is critical to have an objective that transmits as much light as possible. The best and brightest objectives tend to be the most expensive, but because fission yeast are relatively small, less-expensive objectives that do not give a flat field of view are usually adequate for filming a small number of cells in the center of a field. We use a nonplan Zeiss Fluar 100×/1.4 objective (*not* Neofluar), which is approximately as bright as its flat-field Planapochromat 100×/1.4 counterpart, at about one third of the cost. Other manufacturers are likely to offer similar objectives.

C. Photoactivation of GFP

One potentially very useful application of GFP in conjunction with filming is the the study of protein dynamics *in vivo.* GFP inside living tissue culture cells can be photobleached under the microscope by local microbeam irradiation (Cole *et al.,* 1996), and the fluorescence of wild-type GFP (but not the brighter mutants) can be enhanced by illumination with 365-nm light (Yokoe and Meyer, 1996). By following the position and intensity of a photobleached or enhanced marks under the microscope over time, in what is essentially a pulse-chase experiment in real time, one can learn a great deal about the dynamics components and underlying mechanisms involved in a particular process. Recently we have found that under certain conditions GFP can be photoactivated to a red fluorescent form by illumination with blue light (Sawin and Nurse, 1997; Elowitz *et al.,* 1997). The mechanisms responsible for this conversion are unknown, but for practical purposes, enough is known to do interesting experiments. We have not yet applied this method to biological questions, but photoactivation of GFP should greatly improve the potential to study protein dynamics in fission yeast.

With both GFP alone and GFP fusion proteins, photoactivation is effective only when cells are isolated from the atmosphere, either taken from the bottom of a centrifuged pellet or under a coverslip. Cells taken from a pellet and then seeded lightly on an agar pad should be photoactivated without much difficulty, and a brief exposure (1–5 s) to blue light (460–500 nm from a fluorescein excitation filter set and an HBO100 mercury arc lamp) is sufficient to convert GFP into a red fluorescent protein with reasonable efficiency (Fig. 5, see color plate). The fate of the photoactivated population of red fluorescent protein can then be followed by videomicroscopy (see Nabeshima *et al.,* 1997) for further details). For any serious work, local microbeam irradiation is necessary; this can be achieved either with a second mercury arc lamp introduced into the microscope epi-illumination path (Mitchison *et al.,* 1998) or by using a laser, in which case a confocal microscope might be adapted to the purpose.

One important practical consideration is that one may want to observe changes in the position and/or intensity of a photoactivated subpopulation of a protein relative to the global distribution of the same protein. With only a single GFP fusion protein this may be difficult, as wide-field illumination of the cells with blue light would result in photoactivation of the entire field of view. It may be possible to get around this problem by coexpressing a blue fluorescent protein (BFP) (Heim *et al.,* 1995) fusion of the same type, in conjunction with a GFP fusion protein, and carefully controlling the different wavelengths used for photoactivation and observation. Whether such a method would actually work depends on further evaluation of the action spectrum for photoactivation and whether BFP can be observed without photoactivation of GFP.

D. Viewing Fixed Cells

Often one may want to observe GFP fusion proteins in fixed rather than living cells. Two different fixation methods tend to be used for *S. pombe*: dehydration in cold (−20°C) methanol, and cross-linking with paraformaldehyde (3–4%), with or without glutaraldehyde (0.1–1%). The two methods are known to preserve at least some intracellular structures differentially; for example, with respect to microtubules, methanol fixation favors cytoplasmic microtubules but preserves nuclear and spindle structures only weakly, whereas aldehydes give significantly higher background and poor preservation of the cytoplasmic microtubules but are preferred for immunofluorescence of spindle microtubules (Hagan and Hyams, 1988). By the same token, staining of filamentous actin with rhodamine–phalloidin does not work with methanol fixation and is best with formaldehyde alone (Marks and Hyams, 1985). In general, methanol fixation appears to preserve GFP fluorescence better than aldehydes do, but in cases when aldehyde fixation would be preferred (e.g., if counterstaining with rhodamine–phalloidin) it is certainly worth trying, especially if the signal is strong. In our hands glutaraldehyde often destroys GFP fluorescence, but in other cases it is tolerated (D. Drummond and I. Hagan, personal communication), so it is best to try a wide variety of fixation conditions. With some fusion proteins no method may prove ultimately satisfactory; attempts in our laboratory to fix cells expressing Ral3p–GFP have always produced significantly lower signals than with living cells, and sometimes even nonphysiological relocalization of the fusion protein (K. E. S., unpublished).

1. Methanol Fixation

Starting with cells between OD 0.2 and 0.5, concentrate the cells either by centrifugation or filtration (Moreno *et al.,* 1991). For cell pellets, resuspend in 1–2 culture volumes of methanol at −20°C. For filtered cells, place the filter on the inside wall of a 50-ml conical tube (Falcon) and wash with cold methanol with a pipette. Return the cells to −20°C and allow them to fix for at least 10 min. Cells can usually be left overnight in methanol at −20°C. To rehydrate the cells, centrifuge and resuspend them in a small volume of PEM (100 m*M* NaPIPES, pH 6.8, 1 m*M* $MgCl_2$, 1 m*M* EGTA) containing 0.05% azide.

2. Formaldehyde Fixation

The simplest method of formaldehyde fixation involves adding formaldehyde [either low-grade formalin or methanol-free EM grade, e.g., from Polysciences (U.S.) or TAAB (U.K.)] directly to cultures, and continuing incubation for 30–90 min, followed by washing in PEM or phosphate-buffered saline (PBS). In several cases we have found that long fixation times can completely obliterate GFP fluorescence. An alternative fixation method would be to centrifuge cells as for methanol fixation and then resuspend them in PEM containing 3–4% formaldehyde (EM grade may be preferred), fixing for 10 min. Centrifuge the cells and resuspend them in PEM; repeat this two additional times, and resuspend the cells in PEM with 0.05% azide. In some cases, however, this second method may be suboptimal for "true" preservation of structure (Pringle *et al.,* 1991).

Fixed cells can be mounted on slides with or without polylysine, as described earlier. When absolute immobility of the cells is essential, they can be centrifuged onto polylysine-coated 13-mm round coverslips, using Corex tubes modified to this purpose (Evans *et al.,* 1985; Sawin and Nurse, 1996). Cell suspensions in PEM are layered above a PEM/20% glycerol cushion or added directly over the coverslip and centrifuged for 5–10 min at 3000 rpm in a tabletop centrifuge. After a few rinses in PEM, one is left with a monolayer of well-stuck cells. In cases in which numbers are limiting, cells can be first centrifuged in a microcentrifuge and then 1–3 μl of the pellet laid onto a polylysine-coated coverslip and spun dry in Corex tubes, before rinsing with PEM.

3. Immunoelectron Microscopy

In addition to looking at the localization of GFP fusion proteins in fixed cells at the light level, we have also recently followed them under the electron microscope with considerable success by immunoelectron microscopy, using aldehyde-fixed cells, cryosectioning, and an affinity-purified polyclonal antibody raised against GFP (K. E. S., N. Hajibagheri, and P. Nurse, unpublished observations). Thus, for many proteins it should be possible to use the same tag at both the light and ultrastructural level. What is not yet clear is the sensitivity of this technique; thus far we have used it only with multicopy plasmids under the *nmt1* promoter. However, the very low background obtained thus far suggests that lower levels of expression will probably produce excellent images as well.

Acknowledgments

I thank J. Bähler, T. Davis, D. Drummond, J. Hayles, J. Mata, P. Nurse, S. Saitoh, and M. Yanagida for helpful discussions and comments. This work has been supported by the Imperial

Cancer Research Fund, the Helen Hay Whitney Foundation, and the Leukemia Society of America; during this time I have been a fellow of the Helen Hay Whitney Foundation and more recently a Special Fellow of the Leukemia Society of America.

Note Added in Proof

Since the submission of this chapter, the work of Drummond and Hagan described in the text has been published (Drummond, D., and Hagan, I. (1998). Mutations of the bimC box of *cut7*$^+$ indicate divergence of regulation within the bimC family of kinesin related proteins. *J. Cell Sci.* **111,** 853–865). In addition, a simple and powerful PCR-based approach to expressing tagged proteins and creating gene deletions in fission yeast has recently been developed, providing a rapid means to expressing GFP fusion proteins *in vivo* using both endogenous and exogenous promoters (Bähler, J., and Pringle, J. R. (1998). Pom1p, a fission yeast protein kinase that provides positional information for both polarized growth and cytokinesis. *Genes Dev.* **12,** 1356–1370; and Bähler, J., Wu, J.-Q., Longtine, M., Shah, N. G., McKenzie, A., Steever, A. B., Wach, A., Philippsen, P., and Pringle, J. R. (1998). Heterologous modules for efficient and versatile PCR-based gene targeting in *Schizosaccharomyces pombe. Yeast* **14,** (in press)).

References

Basi, G., Schmid, E., and Maundrell, K. (1993). TATA box mutations in the *Schizosaccharomyces pombe nmt1* promoter affect transcription efficiency but not the transcription start point or thiamine repressibility. *Gene* **123,** 131–136.

Chalfie, M., Tu, Y., Euskirchen, G., Ward, W. W., and Prasher, D. C. (1994). Green fluorescent protein as a marker for gene expression. *Science* **263,** 802–805.

Chang, E. C., Barr, M., Wang, Y., Jung, V., Xu, H.-P., and Wigler, M. H. (1994). Cooperative interaction of *S. pombe* proteins required for mating and morphogenesis. *Cell* **79,** 131–141.

Cole, N. B., Smith, C. L., Sciaky, N., Terasaki, M., Edidin, M., and Lippincott-Schwartz, J. (1996). Diffusional mobility of golgi proteins in membranes of living cells. *Science* **273,** 797–801.

Cormack, B., Valdivia, R. H., and Falkow, S. (1996). FACS-optimized mutants of the green fluorescent protein. *Gene* **173,** 33–38.

Elowitz, M. B., Surette, M. G., Wolf, P.-E., Stock, J., and Leibler, S. (1997). Photoactivation turns green fluorescent protein red. *Curr. Biol.* **7,** 809–812.

Evans, L., Mitchison, T., and Kirschner, M. (1985). Influence of the centrosome on the structure of nucleated microtubules. *J. Cell Biol.* **100,** 1185–1191.

Fleig, U., Sen-Gupta, M., and Hegemann, J. H. (1996). Fission yeast *mal2*$^+$ is required for chromosome segregation. *Mol. Cell Biol.* **16,** 6169–6177.

Forsburg, S. L. (1993). Comparison of *Schizosaccharomyces pombe* expression systems. *Nucleic Acids Res.* **21,** 2955–2956.

Fukui, Y., and Yamamoto, M. (1988). Isolation and characterization of Schizosaccharomyces pombe mutants phenotypically similar to ras1$^-$. *Mol. Gen. Genet.* **215,** 26–31.

Giga-Hama, Y., Tohda, H., Okada, H., Owada, M. K., Okayama, H., and Kumagai, H. (1994). High-level expression of human lipocortin I in the fission yeast *Schizosaccharomyces pombe* using a novel expression vector. *Biotechnology* **12,** 400–404.

Hagan, I. M., and Hyams, J. S. (1988). The use of cell division cycle mutants to investigate the control of microtubule distribution in the fission yeast *Schizosaccharomyces pombe J. Cell Sci.* **89,** 343–357.

Heim, R., Cubitt, A. B., and Tsien, R. Y. (1995). Improved green fluorescence. *Nature* **373,** 663–664.

Keeney, J. B., and Boeke, J. D. (1994). Efficient targeted integration at *leu1-32* and *ura4-294* in *Schizosaccharomyces pombe. Genetics* **136,** 849–856.

Kitayama, C., Sugimoto, A., and Yamamoto, M. (1997). Type II myosin heavy chain encoded by the *myo2* gene composes the contractile ring during cytokinesis in *Schizosaccharomyces pombe. J. Cell Biol.* **137,** 1309–1319.

MacNeill, S. A., and Nurse, P. (1997). Cell cycle control in fission yeast. *In*: "The Molecular and Cellular Biology of the Yeast Saccharomyces. Cell Cycle and Cell Biology" (Pringle, J. R., Broach, J. R., and Jones, E. W., Eds.) pp. 697–763. Cold Spring Harbor Laboratory Press, Cold Spring Harbor, NY.

Marks, J., and Hyams, J. S. (1985). Localisation of F-actin through the cell-division cycle of *Schizosaccharomyces pombe. Eur. J. Cell Biol.* **39,** 27–32.

Mata, J., and Nurse, P. (1997). Tea1 and the microtubular cytoskeleton are important for generating global spatial order within the fission yeast cell. *Cell* **89,** 939–949.

Maundrell, K. (1993). Thiamine-repressible expression vectors pREP and pRIP for fission yeast. *Gene* **123,** 127–130.

Mitchison, T. J., Sawin, K. E., Theriot, J. A., Gee, K., and Mallavarapu, A. (1998). Caged fluorescent probes. *Methods Enzymol.* **291,** 63–78.

Moreno, S., Klar, A., and Nurse, P. (1991). Molecular genetic analysis of fission yeast Schizosaccharomyces pombe. *Methods Enzymol.* **194,** 795–823.

Moser, M. J., Flory, M. R., and Davis, T. N. (1997). Calmodulin localizes to the spindle pole body of *Schizosaccharomyces pombe* and performs an essential function in chromosome segregation. *J. Cell Sci.* **10,** 1805–1812.

Nabeshima, K., Kurooka, H., Takeuchi, M., Kinoshita, K., Nakaseko, Y., and Yanagida, M. (1995). p93dis1, which is required for sister chromatid separation, is a novel microtubule and spindle-body-associating protein phosphorylated at the Cdc2 target sites. *Genes. Dev.* **9,** 1572–1585.

Nabeshima, K., Saitoh, S., and Yanagida, M. (1997). Use of green fluorescent protein for intracellular protein localization in living fission yeast cells. *Methods Enzymol.* **283,** 459–471.

Prasher, D. C., Eckenrode, V. K., Ward, W. W., Prendergast, F. G., and Cormier, M. J. (1992). Primary structure of the *Aequorea victoria* green fluorescent protein. *Gene* **111,** 229–233.

Pringle, J. R., Adams, A. E. M., Drubin, D. G., and Haarer, B. K. (1991). Immunofluorescence methods for yeast. *Methods Enzymol.* **194,** 565–602.

Saitoh, S., Takahashi, K., and Yanagida, M. (1997). Mis6, a fission yeast inner centromere protein, acts during G1/S and forms specialized chromatin required for equal segregation. *Cell* **90,** 131–143.

Sawin, K. E., and Nurse, P. (1996). Identification of fission yeast nuclear markers using random polypeptide fusions with green fluorescent protein. *Proc. Natl. Acad. Sci. U. S. A.* **94,** 15146–15151.

Sawin, K. E., and Nurse, P. (1997). Photoactivation of green fluorescent protein. *Curr. Biol.* **7,** R606–R607.

Su, S. S. Y., and Yanagida, M. (1997) Mitosis and cytokinesis in the fission yeast, Schizosaccharomyces pombe. *In* "The Molecular and Cellular Biology of the Yeast Saccharomyces. Cell Cycle and Cell Biology" (Pringle, J. R., Broach, J. R., and Jones, E. W., eds.) pp. 765–825. Cold Spring Harbor Laboratory Press, Cold Spring Harbor, NY.

Toda, T., Niwa, H., Nemoto, T., Dhut, S., Eddison, M., Matsusaka, T., Yanagida, M., and Hirata, D. (1996). The fission yeast *sts5*$^{+}$ gene is required for maintenance of growth polarity and functionally interacts with protein kinase C and an osmosensing MAP-kinase pathway. *J. Cell Sci.* **109,** 2331–2342.

Wang, S., and Hazelrigg, T. (1994). Implications for *bcd* mRNA localization from spatial distribution of *exu* protein on *Drosophila* oogenesis. *Nature* **369,** 400–403.

Watanabe, Y., Shinozaki-Yabaua, S., Chikashige, Y., Hiraoka, Y., and Yamamoto, M. (1997). Phosphorylation of RNA-binding protein controls cell cycle switch from mitotic to meiotic in fission yeast. *Nature* **386,** 187–190.

Yamashita, A., Watanabe, Y., and Yamamoto, M. (1997). Microtubule-associated coiled-coil protein Ssm4 is involved in the meiotic development in fission yeast. *Genes Cells* **2,** 155–166.

Yokoe, H., and Meyer, T. (1996). Spatial dynamics of GFP-tagged proteins investigated by local fluorescence enhancement. *Nat. Biotech.* **14,** 1252–1256.

CHAPTER 9

GFP Variants for Multispectral Imaging of Living Cells

James Haseloff
MRC Laboratory of Molecular Biology
Cambridge CB2 2QH, England

I. Introduction

Developing multicellular tissues or organs generally demonstrate a capacity for self-organization. For example, wounded tissues can generally respond in a robust and coordinated fashion to allow repair, and local induction events can initiate prolonged and coordinated developmental processes (such as in limb bud formation). These types of developmental plasticity and functional autonomy are particularly evident in plant tissues. The basic features of a plant's body plan are established during embryogenesis; however, its final form results from the continued growth of meristems and the formation of organs throughout its life, often in a modular and indeterminate fashion. Plant cells are constrained by rigid cell walls and are generally nonmotile, so there is the clear possibility that cell fates within a meristem are determined by lineage. However, evidence from plant chimera and wounding studies have demonstrated a more important role for cell–cell interactions during fate determination (reviewed in Steeves and Sussex, 1989). It is likely that positional information during plant development

Copyright © 1999 by Academic Press. All rights of reproduction in any form reserved.
0091-679X/99 $25.00

is obtained via cell–cell contact and that the coordination and fate of cells within a developing meristem may be determined by a network of local cellular interactions. It is also likely that intercellular communication plays a coordinating role in the development of all multicellular organisms. We are using the *Arabidopsis thaliana* root meristem as a model system for investigating intercellular interactions. The root meristem possesses indeterminate growth and has a simple and transparent architecture. *Arabidopsis* is amenable to genetic manipulation, and one can routinely generate transgenic lines for work with the intact organism. To dissect local cell–cell interactions it is crucial that we be able to (*a*) clearly image individual cells inside living meristems and (*b*) have the means to perturb them.

Therefore, we have adapted the jellyfish green fluorescent protein (GFP) for use as a directly visible gene marker in *Arabidopsis* and have developed genetic and optical techniques that allow us to visualize and manipulate cells within living plants.

II. Green Fluorescent Protein Markers

Marker genes have proven extremely useful for reporting gene expression in transformed cells, and the β-glucuronidase (GUS) gene (Jefferson *et al.,* 1987) has been used extensively in transgenic plants. Transformed tissues or patterns of gene expression can be identified histochemically, but this is generally a destructive test and is not suitable for assaying primary transformants, for following the time course of gene expression in living plants, or as a means of rapidly screening segregating populations of seedlings. The GFP from the cnidarian jellyfish *Aequorea victoria* shares none of these problems, and there has been much interest in using the protein as a genetic marker in transgenic *Arabidopsis thaliana.*

Aequorea victoria are brightly luminescent, with glowing points around the margin of the jellyfish umbrella. Light arises from yellow tissue masses that each consist of about 6000–7000 photogenic cells. The cytoplasm of these cells is densely packed with fine granules that contain the components necessary for bioluminescence (Davenport and Nichol 1955; Morin and Hastings 1971). In other bioluminescent coelenterates these have been characterized as 0.2-micron-diameter particles enclosed by a unit membrane, and have been termed lumisomes (Anderson and Cormier 1973). The components required for bioluminescence include a Ca^{2+}-activated photoprotein, aequorin, that emits blue-green light, and the accessory protein, GFP, which accepts energy from aequorin and reemits it as green light (Morise *et al.,* 1974). GFP is an extremely stable protein of 238 amino acids (Prasher *et al.,* 1992). The fluorescent properties of the protein are unaffected by prolonged treatment with 6 *M* guanidine HCl, 8 *M* urea, or 1% sodium dodecyl sulfate (SDS), and two-day treatment with various proteases such as trypsin, chymotrypsin, papain, subtilisin, thermolysin, and pancreatin at concentration of up to 1 mg/ml fail to alter the intensity of GFP fluorescence (Bokman and Ward 1981). GFP is stable in neutral buffers up to 65°C, and displays a broad range of pH stability from 5.5 to 12. The protein is intensely fluorescent, with a quantum efficiency of approximately 80% and a molar extinction coefficient of 2.2×10^4 cm^{-1} M^{-1} (Morise *et al.,* 1974) (after correction for the known molecular weight). GFP fluoresces maximally when excited at 400 nm, with a lesser peak at 475 nm, and fluorescence emission peaks at 509 nm.

The intrinsic fluorescence of the protein is due to a unique covalently attached chromophore that is formed posttranslationally within the protein upon cyclization and oxidation of residues 65–67, Ser-Tyr-Gly (Prasher *et al.,* 1992; Cody *et al.,* 1993; Heim *et al.,* 1994). Several genomic and cDNA clones of *gfp* have been obtained from a population of *A. victoria* (Prasher *et al.,* 1992). The *gfp* gene contains at least three introns, and the coding sequence derived from one of the cDNA clones, pGFP10.1 has been used for protein expression, first in *Escherichia coli,* then in *Caenorhabditis elegans* (Chalfie *et al.,* 1994; Heim *et al.,* 1994; Inouye and Tsuji, 1994), and *Drosophila melanogaster* (Wang and Hazelrigg 1994). Fluorescent protein has now been produced in a number of heterologous cell types, and there appears to be little requirement for specific additional factors for posttranslational modification of the protein, which may be autocatalytic or require ubiquitous factors.

The bright intrinsic green fluorescence of GFP allows it to be directly visualized in transformed cells, and we wished to use the protein as a simple marker for transformation and misexpression studies. However, we found that the wild-type GFP cDNA was not expressed properly in *Arabidopsis,* and we have needed to extensively modify the gene (Haseloff and Amos, 1995; Siemering *et al.,* 1996; Haseloff *et al.,* 1997).

1. *Removal of a cryptic intron.* We discovered that the GFP mRNA sequence is efficiently mis-spliced in transgenic *Arabidopsis* plants, resulting in the removal of 84 nucleotides from within the coding sequence, between residues 380 to 463. We removed the cryptic intron by mutagenesis, allowing proper expression (Haseloff *et al.,* 1997).

2. *Subcellular localization of GFP.* We found that GFP accumulates within the nucleoplasm of plant cells, as it does in other organisms, and that it was difficult to regenerate plants from very brightly transformed tissues. This apparent mild toxicity may be due to the generation of fluorescence-related free radicals during growth under light, which might then lead to DNA damage. We have therefore targeted GFP to different subcellular compartments in transgenic plants and assayed for improved regeneration and fluorescence. One of our constructions, which is excluded from nuclei and is retained within the endoplasmic reticulum in *Arabidopsis,* consistently produces bright and healthy transformants.

3. *Thermotolerant GFP mutants.* We have shown that wild-type GFP is thermosensitive and fluoresces poorly at temperatures above 25°C. We have subjected

our modified GFP coding sequence to PCR-based mutagenesis and have isolated a thermotolerant mutant with improved fluorescence. The mutant contains two altered amino acids (V163A, S175G) that greatly improve folding of the apoprotein at elevated temperatures (Siemering *et al.,* 1996).

4. *Altered spectral properties.* The fluorescence excitation of wild-type GFP peaks at wavelengths of 400 and 475 nm, with the 400-nm peak predominating. We have recombined a published mutant of GFP (I167T; Heim *et al.,* 1994) with our improved mutant and have produced a variant that has dual excitation peaks of almost equal amplitude (Siemering *et al.,* 1996). This allows the efficient use of techniques that require either UV or blue light excitation of the protein, for example when screening GFP-expressing plants with a UV lamp or when using blue laser-light-excited confocal microscopy.

All of these alterations have been incorporated into a single modified form of the gene (*mgfp5-ER*), which we now routinely use for monitoring gene expression and marking cells in live transgenic plants. These improved mutant genes have also proven useful for studies in animal cells (Zernicka-Goetz *et al.,* 1996; Zernicka-Goetz *et al.,* 1997).

Other groups have been struggling with similar problems in other systems, and a number of functionally similar *gfp* variants are now available. For example, *gfp* variants with "humanized" or other optimized codon usage have been produced for better expression. These also show improved levels of expression in plants (Chiu *et al.,* 1996; Haas *et al.,* 1996; Pang *et al.,* 1996). Although these genes were expected to provide better translation efficiency, it is likely that these alterations also confer some degree of immunity from aberrant RNA processing in plants. A number of workers have obtained GFP mutants that show brighter fluorescence in heterologous cell types, and it is likely that the improved properties result from better folding of the proteins. For example, the V163A mutation has been generated independently by at least three different groups (Crameri *et al.,* 1996; Heim and Tsien 1996; Kohler *et al.,* 1997), and this residue may play a pivotal role in folding of the protein. In addition, Cormack *et al.* (1996) introduced large numbers of random amino acid substitutions into the 20 residues flanking the chromophore of GFP. They used fluorescence-activated cell sorting to select variants that fluoresced 20- to 35-fold more intensely than wild type. They also showed that the folding of these mutant proteins was more efficient during expression in bacteria. One of these variants, GFPmut1 (Cormack *et al.,* 1996), contains two amino acid differences, F64L and S65T, located within the central α-helix of the protein, adjacent to the chromophore. The V163A and S175G mutations that we have isolated are positioned on the outer surface of the protein (Ormo *et al.,* 1996; Yang *et al.,* 1996), and recombination of these two sets of mutations results in an exclusively blue-light-excited GFP with markedly improved fluorescence properties (Zernicka-Goetz *et al.,* 1996, 1997). The beneficial effect of both sets of mutations on protein folding and their apparent additive effect suggest that they may play separate roles in the folding or maturation process.

III. Imaging of Living Cells

GFP expression and localization can be visualized directly, without a prolonged and lethal staining procedure. The expression of GFP within an organism produces an intrinsic fluorescence that can used to "paint" particular cells or cellular processes. The fluorescent properties of the protein allow (*a*) simple screening for gene expression in living plants by inspection with a UV lamp, and (*b*) high-resolution imaging of subcellular events in living cells using fluorescence microscopy.

GFP can possess alternate protonated and anionic forms of its chromophore, which allow excitation with long-wavelength UV (395 nm) or blue light (475 nm), respectively (Cody *et al.,* 1993; Heim *et al.,* 1994). The relative amplitude of the two excitation peaks is predominantly determined by the amino acid sequence and protein environment around the chromophore, rather than the solvent, and mutant forms of GFP have characteristic excitation spectra. Most available forms of GFP have been optimized for excitation at either of these wavelengths; for example GFP variants containing the S65T mutation (Heim *et al.,* 1995) are widely used and can be excited by blue, but not UV, light. In contrast, we have chosen to use GFP derivatives (mGFP5) with more versatile spectral properties for our initial work in *Arabidopsis.* The mGFP5 protein is equally well excited with either long-wavelength UV or blue light. This allows the use of ultraviolet excitation for simple inspection of transformed material, because the illuminating wavelength is poorly detected by the human eye. In addition, transformed plant tissue or seedlings growing in sterile culture can be simply scored for GFP expression using fluorescence microscopy, and we have used the marker to simplify an enhancer-trap screen. Efficient blue-light excitation of GFP also allows work with commonly available microscope filter sets and laser sources, such as those used during confocal microscopy (Fig. 1, see color plate).

Confocal imaging allows precise visualization of fluorescent signals within a narrow plane of focus, with exclusion of out-of-focus blur, and the technique permits the reconstruction of three-dimensional structures from serial optical sections. Intact plant tissue proves a difficult subject for fluorescence microscopy because it consists of deep layers of highly refractile cell walls and aqueous cytosol, and contains various autofluorescent and light scattering components. There are two approaches to the difficulties imposed by these conditions. (*a*) to fix and to clear the tissue with a high-refractive-index mounting medium, or (*b*) to directly image living tissue using suitably corrected microscope optics. In our experience, it has proved difficult to effectively clear *Arabidopsis* wholemounts without causing artifacts or losing GFP fluorescence, and there are considerable advantages to working with living tissues, so we have pursued the second approach. *Arabidopsis* seedlings can simply be mounted in water for microscopy and examined using a long-working-distance water-immersion objective to minimize the effects of spherical aberration when focusing deep into an aqueous sample (Haseloff and Amos, 1995). Even with the use of such a specialized objective (Nikon 60× planapochromat, numerical aperture 1.2, working distance 220 μm), image quality degrades rapidly for optical sections deeper than 60–80 μm within the tissue. However, the small size of *Arabidopsis* seedlings allows very useful imaging despite this limitation and, for example, median longitudinal optical sections can be obtained from intact roots (Fig. 2A, see color plate).

Protocol: Mounting and Observing GFP-expressing *Arabidopsis* Seedlings

A. Growth of *Arabidopsis* in sterile culture.

1. Twenty to one hundred transgenic *Arabidopsis* seeds were placed in a 1.5-ml microfuge tube and washed for about 1 min with 1 ml of ethanol.

2. The seeds were then incubated with 1 ml of a surface-sterilizing solution containing 1% (w/v) sodium hypochlorite and 0.1% (v/v) NP40 detergent for 15 min at room temperature.

3. The seeds were then washed three times with 1 ml of sterile water and transferred by pipette to agar plates containing GM medium (Valvekens *et al.,* 1988):

1× Murashige and Skoog basal medium with Gamborgs B5 vitamins (Sigma)

1% sucrose

0.5 g/liter 2- (*N*-morpholino) ethanesulfonic acid (MES)

0.8% agar (adjusted to pH 5.7 with 1 *M* KOH)

Twenty-five mg/liter kanamycin was added if antibiotic selection of transgenic seedlings was necessary.

These procedures were performed in a laminar flow hood.

Alternatively, for extended time-lapse imaging of roots, sterile seeds were sown in coverslip-based vessels (Nunc) that comprised 4 wells, each containing about 400 μl of low-gelling temperature agarose with GM medium. The roots of these plants grow down through the medium and then along the surface of the coverslip. The roots are then ideally positioned for high-resolution microscopic imaging through the base of the vessel.

4. Sealed plates or vessels were incubated for 1–3 days in the dark at 4°C, and then transferred to an artifically lit growth room at 23°C for germination.

5. *Arabidopsis* seedlings germinate after 3 days and can be used for microscopy for several weeks. Root and shoot tissues can be directly scored for GFP expression using a inverted fluorescence microscope (Leitz DM-IL) fitted with filter sets suitable for UV (Leitz-D; excitation filter 355–425 nm, dichroic mirror 455 nm, longpass emission filter 460 nm) and blue (Leitz-I3; excitation filter 450–490 nm, dichroic mirror 510 nm, longpass emission filter 520 nm) light excitation of GFP. Roots, which grow along the base of the Petri dish can be observed directly by epifluorescence microscopy through the clear plastic base.

Shoot tissues were directly observed in inverted dishes by using one or two 7-mm threaded extension tubes with a 4× objective (EF 4/0.12), which gave greater working distances. Epifluorescence images were captured in Adobe Photoshop using a Sony DXC-930P 3-chip CCD video camera and F100-MPU integrating frame store, connected to a NuVista+ video digitizer in an Apple Macintosh computer.

B. Confocal Imaging

GFP-expressing *Arabidopsis* seedlings were removed from the agar media and simply mounted in water under glass coverslips for microscopy. Growing roots could also be directly viewed through coverslip-based vessels. The specimens were examined using a BioRad MRC-600 laser-scanning confocal microscope equipped with a 25-mW krypton–argon or argon-ion laser and filter sets suitable for the detection of fluorescein and Texas Red dyes (BioRad filter blocks K1/K2 with the krypton–argon-ion laser, and A1/A2 with the argon-ion laser). We routinely use a Nikon 60× PlanApo NA 1.2 water-immersion objective to minimize loss of signal through spherical aberration at long working distances. For the collection of time-lapse images, the laser light source was attenuated by 99% using a neutral density filter, the confocal aperture was stopped down, and single scans were collected at 2-s intervals. The large data files were transferred to an Apple Macintosh computer, and the programs PicMerge, authored by Eric Sheldon, and 4DTurnaround, authored by Charles Thomas, were used with Adobe Photoshop and Premiere to produce QuickTime movies for display and analysis. More examples of captured images and movies are available on a Web site, which can be accessed at http://www.mrc-1mb.cam.ac.uk.

Direct visualization of GFP fluorescence in living tissues is not prone to fixation or staining artifacts, and can provide images of exceptional clarity. Moreover, the activities of living cells, such as cytoplasmic streaming, are clearly evident during microscopy. Ordinarily, movement within a sample is a nuisance, placing constraints on the use of sometimes lengthy techniques for noise reduction during confocal microscopy, such as frame averaging. However, it is also possible to monitor dynamic events by time-lapse confocal microscopy, and this combination of a vital fluorescent reporter with high-resolution optical techniques shows much promise for use in cell biological and physiological experiments. We have also found that autofluorescent chloroplasts, normally present in the upper parts of the plant, and certain red fluorescent dyes can provide useful counterfluors for GFP. For example, propidium iodide can be applied to live seedlings in water, to specifically label root cell walls and allow accurate identification of GFP expressing cells.

Protocol: Fluorescent Counterstaining

A. Labeling Root Meristem Cell Walls with Propidium Iodide

Arabidopsis seedlings were grown in sterile culture, removed from agar media, and placed in a well of a microtiter dish with 1 ml of staining solution for 10–

20 min at room temperature. An aqueous 10 μg/ml solution of propidium iodide (Sigma) was used to stain the cell walls of the *Arabidopsis* root meristem. The seedlings were then mounted in water under a coverslip for direct microscopic observation. Propidium iodide is red fluorescent and can be detected using a filter set suitable for Texas Red fluorescence, with little spillover between the GFP (fluorescein) and propidium iodide channels. The cationic dye does not readily cross intact membranes, yet it penetrates throughout the meristem and binds to cell walls, forming an outline of the living cells. The dye is excluded by the Casparian strip present in older parts of the root and does not penetrate shoot tissue well, and thus it is best suited for use in the root meristem. An example is shown in Fig. 2A.

B. Labeling Plasma Membranes with FM 1-43

The cationic styrylpyridinium dye FM 1-143 (Molecular Probes Inc.) provides a useful stain for the plasma membrane in root and shoot tissue of *Arabidopsis.* It is particularly useful for specifically labeling the plasma membrane of shoot epidermal cells, and we have been using this to characterize GFP expression patterns in *Arabidopsis* cotyledons and leaves. The seedlings were removed from sterile culture and placed in 1 ml of 1 μg/ml FM 1-43 in water for 10 min at room temperature. They were mounted in water under a coverslip for direct microscopic observation. FM 1-43 emits a broad orange fluorescence, and the signal can be detected in both the red and green emission channels. We generally use 488-nm laser light to excite both GFP and FM 1-43, and to collect the emissions of both fluors in the same channel. This is possible because of the very localized distribution of Fm 1-43 in shoot epidermal cells. An example is shown in Fig. 2B (see color plate).

IV. Marking Different Cell Types in *Arabidopsis*

It is now possible to genetically mark cells within a living organism using GFP and to visualize these cells directly during development. To provide such markers for work in *Arabidopsis* and to allow genetic manipulation of cells during meristem development, we have adapted a scheme for targeted gene expression used in *Drosophila* (Brand and Perrimon, 1993). Brand and Perrimon used a P-element-based "enhancer-trap" strategy to generate *Drosophila* lines that express different patterns of the yeast transcription activator, GAL4. A chosen target gene could then be placed under the control of GAL4 upstream activation sequences (UASs), transformed, and maintained silently in the absence of GAL4. Genetic crosses between this single line and any of the library of GAL4-containing lines could specifically activate the target gene in a particular tissue or cell type. The phenotypic consequences of misexpression, including those lethal to the organism, could then be conveniently studied.

We found that GAL4 is not expressed in *Arabidopsis* due to a high A/T content, which can interfere with mRNA processing in plants. Elevated A/U content plays a major role in intron recognition during plant pre-mRNA splicing, and we have found that this can poison expression of heterologous A/T-rich genes such as *GAL4* and *GFP.* It was necessary to alter the codon usage of the gene and to use a derivative, *GAL4-VP16,* to ensure efficient expression in *Arabidopsis.* We have randomly inserted the modified *GAL4-VP16* gene into the *Arabidopsis* genome, using *Agrobacterium tumefaciens*–mediated transformation. Expression of the *GAL4-VP16* gene is dependent upon the presence of adjacent genomic enhancer sequences, so different patterns of expression are generated. The inserted DNA also contains a GAL4-responsive *mGFP5* gene (Fig. 3A, see color plate), so patterns of *GAL4-VP16* gene expression are immediately detectable, with each GAL4-expressing cell marked by green fluorescence.

In vivo detection of GFP can be used as a simple genetic screening procedure for plants growing in normal culture, and we have used the marker to greatly speed and improve a large enhancer-trap screen. Because our particular interest lies in the cells of the *Arabidopsis* root tip, we have modified the plant transformation protocol to include an auxin induction of roots from regenerating shootlets. More than 7500 transformants were then generated, planted in grid patterns in sterile culture dishes, and directly screened for GAL4-mediated GFP expression within roots. Several hundred lines with interesting patterns of root expression were choosen, documented, transferred to soil, and grown to seed, to both amplify and self-hybridize the lines. As a result, we have a collection of 250 *Arabidopsis* lines with distinct and stable patterns of GAL4-VP16 and mGFP5 expression in the root (Fig. 3B, see color plate). These are being made available through the *Arabidopsis* stock center, and we have created a graphical database of the expression patterns to allow easy computer and Web access (http://www.mrc-lmb.cam.ac.uk). These lines provide a valuable set of markers in which particular cell types are tagged and can be visualized with unprecedented ease and clarity in living plants. The physical arrangements of cells and common fields of gene expression are highlighted within the root tip. More important, GAL4-VP16 expression within these same lines will allow precise targeted gene misexpression. A chosen target gene can be cloned under the control of GAL4 UASs, transformed, and maintained silently in the absence of GAL4. Genetic crossing of this single line with any of the library of GAL4-containing lines allows specific activation of the target gene in particular tissue and cell types. The phenotypic consequences of misexpression, including those lethal to the organism, can be conveniently studied. We have targeted the expression of toxic and regulatory proteins to particular cells of the root meristem.

V. Spectrally Distinct Fluorescent Proteins for Multichannel Confocal Microscopy

Subcellular structures can be decorated *in vivo* with GFP-tagged proteins and observed microscopically. These benefits would be greatly extended if we could use additional and distinct fluorescent protein markers in our experiments. This would immediately enable one to examine in detail the dynamic behavior of a GFP-tagged protein or cell with respect to another fluorescent structure.

Several spectral variants of GFP have already been described (Chapter 2 of this book; Heim *et al.,* 1994; Ormo *et al.,* 1996). We have been anxious to obtain one or more new fluorescent protein markers that can be used with GFP and that can be used for dual-channel confocal imaging. A major constraint in choosing a second fluorescent marker is that its excitation should be compatible with commonly used (i.e., cheap) laser sources. Thus, we have experimented with variants of GFP that have yellow-shifted emission (YFP, excitation maximum 514 nm, emission maximum 527 nm) and cyan-shifted emission (CFP, excitation maximum 440 nm, emission maximum 485 nm).

To produce GFP variants that are expressed well in plants, we have modified the genes so that they also contain altered codon usage and mutations that confer improved folding properties, and this has resulted in the production of mYFP (S65G, S72A, V163A, I167T, S175G, T203Y) and mCFP (Y66W, V163A, S175G). The mCFP and mYFP variants are comparable in brightness to our best GFPs but possess quite distinct spectral properties (Fig. 4, see color plate). Although the emission spectra of mCFP, mGFP5, and mYFP overlap to a large degree, laser scanning confocal microscopy allows the use of monochromatic light for selective excitation of the proteins. For example, common multiline argon-ion lasers emit light mainly at discrete wavelengths of 458, 477, 488, and 514 nm, and we have adapted suitable laser line excitation and emission filters for the detection of the GFP variants. An example is provided later in which we have used this technique to visualize the subcellular localization of particular homeodomain proteins in transgenic *Arabidopsis* plants.

The GAL4 system can be used to view the dynamics of protein localization in living plants. We have cloned the cDNA for a homeodomain protein, KNAT3 (Serikawa *et al.,* 1996), from the C24 ecotype of *Arabidopsis* (this is the ecotype that we use for GAL4-directed misexpression), and have generated transgenic plants containing the gene under the control of GAL4-responsive or constitutive promoters. We have fused mYFP to the KNAT3 protein. GAL4-directed expression of the fusion protein in intact plants has shown that subcellular localization of fusion proteins is precisely regulated. The proteins initially accumulate within the cytoplasm of meristematic cells but are progressively concentrated within nuclei as the cells age (Fig. 5, see color plate). KNAT3 is a putative transcription factor that would not be active if it were excluded from the nucleus, and there appears to be clear posttranslational regulation of the protein's activity. Interestingly, transcription of the KNAT3 promoter seems limited to those cells in which the mYFP-protein fusion is nuclear localized (Serikawa *et al.,* 1997). We have now mutagenized the KNAT3 sequence and are screening for mutant forms that escape this regulation, and that should provide dominant gain of function phenotypes.

Protocol: Double Labeling with GFP Variants

1. Plants expressing mCFP, mGFP5, and/or mYFP proteins were grown in sterile culture and mounted in water for microscopy.

2. A Biorad MRC-600 microscope was equipped with an 80-mW argon ion laser and a motorized excitation filter wheel containing narrow bandpass filters to select laser lines at 458 nm for excitation of mCFP, 477 nm for excitation of mGFP5 (mGFP5), and 514 nm for excitation of mYFP. A multiline argon-ion laser of higher power (>50 mW) is generally needed to provide 458-nm illumination of useful intensity. We have used mCFP and mYFP, and mGFP5 and mYFP together for double-labeling experiments. The proteins were sequentially excited using the appropriate laser lines, and the signals were collected through specialized emission filter blocks: mCFP/mYFP = 495-nm longpass dichroic mirror, 485 ± 30 nm, 540- ± 30-nm bandpass filters, or mGFP5/mYFP = 527 nm longpass dichroic mirror. 500-nm longpass and 540- ± 30-nm bandpass filters (Omega Optical). The use of selective monochromatic excitation allows useful discrimination between mGFP5 and mYFP, which have overlapping fluorescent spectra. The greater spectral differences between mCFP and mYFP result in clean discrimination of the fluorescent signals. Sequentially collected images were merged and pseudocolored using Adobe Photoshop.

VI. Summary

Unlike enzyme markers, green fluorescent protein can be visualized at high resolution in living cells using confocal microscopy. The images are not prone to fixation or staining artifacts, and can be of exceptional clarity. Moreover, the activities of living cells, such as cytoplasmic streaming, are clearly evident during microscopy. Ordinarily, movement within a sample is a nuisance, placing constraints on the use of sometimes lengthy techniques for noise reduction during confocal microscopy, such as frame averaging. However, it is possible to monitor dynamic events by time-lapse confocal microscopy, and this combination of a vital fluorescent reporter with high-resolution optical techniques shows much promise for use in cell biological and physiological experiments.

Genetic systems such as that of *Arabidopsis* provide a large resource of potentially informative mutants, and there has been much recent improvement in techniques for determining the molecular basis of a particular phenotype. The use of fluorescent proteins will provide further tools for examining the biology of mutant cells. The precision with which particular cellular structures can be decorated with GFP and the ease with which subcellular traffic can be monitored indicate that this approach will be very useful for cell biological and physiological observations, particularly for detailed examination of plant mutant phenotypes.

References

Anderson, J. M., and Cormier, M. J. (1973). Lumisomes, the cellular site of bioluminescence in coelenterates. *J. Biol. Chem.* **248,** 2937–2943.

Bokman, S. H., and Ward, W. W. (1981). Renaturation of *Aequorea* green fluorescent protein. *Biochem. Biophys. Res. Commun.* **101,** 1372–1380.

Brand, A. H., and Perrimon, N. (1993). Targeted gene expression as a means of altering cell fates and generating dominant phenotypes. *Development* **118,** 401–415.

Chalfie, M., Tu, Y., Euskirchen, G., Ward, W. W., and Prasher, D. C. (1994). Green fluorescent protein as a marker for gene-expression. *Science* **263,** 802–805.

Chiu, W. L., Niwa, Y., Zeng, W., Hirano, T., Kobayashi, H., and Sheen, J. (1996). Engineered GFP as a vital reporter in plants. *Curr. Biol.* **6,** 325–330.

Cody, C. W., Prasher, D. C., Westler, W. M., Prendergast, F. G., and Ward, W. W. (1993). Chemical structure of the hexapeptide chromophore of the *Aequorea* green fluorescent protein. *Biochemistry* **32,** 1212–1218.

Cormack, B. P., Valdivia, R. H., and Falkow, S. (1996). FACS-optimized mutants of the green fluorescent protein (GFP). *Gene* **173,** 33–38.

Crameri, A., Whitehorn, E. A., Tate, E., and Stemmer, W. P. C. (1996). Improved green fluorescent protein by molecular evolution using DNA shuffling. *Nat. Biotech.* **14,** 315–319.

Davenport, D., and Nichol, J. A. C. (1955). Luminescence in hydromedusae. *Proc. R. Soc. Ser. B.* **144,** 399–411.

Haas, J., Park, E. C., and Seed, B. (1996). Codon usage limitation in the expression of HIV-1 envelope glycoprotein. *Curr. Biol.* **6,** 315–324.

Haseloff, J., and Amos, B. (1995). GFP in plants. *Trends Gene.* **11,** 328–329.

Haseloff, J., Siemering, K. R., Prasher, D. C., and Hodge, S. (1997). Removal of a cryptic intron and subcellular localization of green fluorescent protein are required to mark transgenic *Arabidopsis* plants brightly. *Proc. Natl. Acad. Sci. U. S. A.* **94,** 2122–2127.

Heim, R., Prasher, D. C., and Tsien, R. Y. (1994). Wavelength mutations and posttranslational autoxidation of green fluorescent protein. *Proc. Nat. Acad. Sci. U. S. A.* **91,** 12501–12504.

Heim, R., Cubitt, A. B., and Tsien, R. Y. (1995). Improved green fluorescence. *Nature* **373,** 663–664.

Heim, R., and Tsien, R. Y. (1996). Engineering green fluorescent protein for improved brightness, longer wavelengths and fluorescence resonance energy-transfer. *Curr. Biol.* **6,** 178–182.

Inouye, S., and Tsuji, F. I. (1994). *Aequorea* green fluorescent protein—expression of the gene and fluorescence characteristics of the recombinant protein. *FEBS Lett.* **341,** 277–280.

Jefferson, R. A., Kavanagh, T. A., and Bevan, M. W. (1987). GUS fusions: β-glucuronidase as a sensitive and versatile gene fusion marker in higher plants. *EMBO J.* **6,** 3901–3907.

Kohler, R. H., Zipfel, W. R., Webb, W. W., and Hanson, M. R. (1997). The green fluorescent protein as a marker to visualize plant mitochondria *in vivo. Plant. J.* **11,** 613–621.

Morin, J. G., and Hastings, J. W. (1971). Energy transfer in a bioluminescent system. *J. Cell. Physiol.* **77,** 313–318.

Morise, H., Shimomura, O., Johnson, F. H., and Winant, J. (1974). Intermolecular energy transfer in the bioluminescent system of *Aequorea. Biochemistry* **13,** 2656–2662.

Ormo, M., Cubitt, A. B., Kallio, K., Gross, L. A., Tsien, R. Y., and Remington, S. J. (1996). Crystal structure of the *Aequoria victoria* green fluorescent protein. *Science* **273,** 1392–1395.

Pang, S. Z., DeBoer, D. L., Wan, Y., Ye, G., Layton, J. G., *et al.* (1996). An improved green fluorescent protein gene as a vital marker in plants. *Plant Physiol.* **112,** 893–900.

Prasher, D. C., Eckenrode, V. K., Ward, W. W., Prendergast, F. G., and Cormier, M. J. (1992). Primary structure of the *Aequorea victoria* green fluorescent protein. *Gene* **111,** 229–233.

Serikawa, K. A., Martinez-Laborda, A., and Zambryski, P. (1996). Three *knotted1*-like genes in *Arabidopsis. Plant Mol. Biol.* **32,** 673–683.

Serikawa, K. A., Martinez-Laborda, A., Kim, H. S., and Zambryski, P. C. (1997). Localisation of expression of KNAT3, a class 2 *knottedl*-like gene. *Plant J.* **11,** 853–861.

Siemering, K. R., Golbik, R., Sever, R., and Haseloff, J. (1996). Mutations that supress the thermosensitivity of green fluorescent protein. *Curr. Biol.* **6,** 1653–1663.

Valvekens, D., Van Montagu, M., and Van Lijsebettens, M. (1988). *Agrobacterium tumefaciens*-mediated transformation of *Arabidopsis thaliana* root explants by using kanamycin selection. *Proc. Nat. Acad. Sci. U. S. A.* **85,** 5536–5540.

Wang, S. X., and Hazelrigg, T. (1994). Implications for *bcd* mRNA localization from spatial distribution of *exu* protein in *Drosophila* oogenesis. *Nature* **369,** 400–403.

Yang, F., Moss, L. G., and Phillips, G. N. J. (1996). The molecular structure of green fluorescent protein. *Nat. Biotech.* **14,** 1246–1251.

Zernicka-Goetz, M., Pines, J., Ryan, K., Siemering, K. R., Haseloff, J., *et al.* (1996). An indelible lineage marker for *Xenopus* using a mutated green fluorescent protein. *Development* **122,** 3719–3724.

Zernicka-Goetz, M., Pines, J., Siemering, K. R., Haseloff, J., and Evans, M. J. (1997). Following cell fate in the living mouse embryo. *Development* **124,** 1133–1137.

CHAPTER 10

GFP Fusions to a Microtubule Motor Protein to Visualize Meiotic and Mitotic Spindle Dynamics in *Drosophila*

Sharyn A. Endow
Department of Microbiology
Duke University Medical Center
Durham, North Carolina 27710

I. Introduction

During cell division, the chromosomes become attached to the spindle fibers and oriented on the spindle before segregating toward opposite poles. Rapid poleward and away-from-the-pole chromosome movements occur during most of meiosis and mitosis, mediated by microtubule and spindle dynamics, and culminate with the segregation of homologous chromosomes in meiosis I and sister chromatids in meiosis II and mitosis.

Copyright © 1999 by Academic Press. All rights of reproduction in any form reserved.
0091-679X/99 $25.00

Despite intense study for more than a century, the basis of the movements of the spindle and chromosomes during cell division is still not clearly established. This is especially true of the meiotic divisions of oocytes, which have been less widely studied due to the relative difficulty in obtaining oocytes compared to mitotic cells and the difficulty in visualizing the meiotic apparatus, which is often obscured by yolky cytoplasm. A major breakthrough in recent years has been the discovery that microtubule motors are intimately involved in all aspects of chromosome and spindle movement during meiosis and mitosis (Sawin and Endow, 1993). Microtubule motors use the energy from ATP hydrolysis to bind to and move on microtubules, generating force for spindle assembly and function as well as attachment of chromosomes to spindle fibers and probably also movement of chromosomes along microtubules. One of the microtubule motors now known to be required for spindle assembly and maintenance during meiosis in *Drosophila* oocytes is Ncd (Matthies *et al.,* 1996; Endow and Komma, 1997). Ncd also functions during mitosis in early embryos to maintain attachment of centrosomes to spindle poles and chromosomes to spindle fibers, preventing loss of chromosomes at the metaphase/anaphase transition (Endow and Komma, 1996).

To better understand the role of Ncd in the spindle and to permit the analysis of spindle dynamics during meiosis and mitosis, we have made fusions of *ncd* to *gfp* and transformed the gene fusions into *Drosophila* (Endow and Komma, 1996, 1997). The Ncd–GFP fusion proteins, regulated by the native *ncd* promoter, rescue an *ncd* null mutant for chromosome segregation and embryo viability, providing evidence that the fusion protein can replace the function of wild-type Ncd. Green fluorescent spindles are observed in *ncd–gfp* oocytes and embryos, demonstrating that the fusion protein is expressed and localizes to meiotic and mitotic spindles. Both wild-type and mutant forms of Ncd were made and used to obtain information about spindle assembly and function by following spindle dynamics in live wild-type and loss-of-function oocytes and embryos.

The use of *ncd–gfp* to visualize spindles is minimally perturbing to the cells and represents an advance over previous methods that have been used to label spindle microtubules. These have involved the injection into live cells of fluorescently labeled tubulin, which is incorporated into the spindles. Analysis of the transgenic *ncd–gfp* oocytes and embryos is technically simpler than injection of labeled tubulin and avoids the problem of unintentional activation of the oocytes due to injection. It also does not require the incorporation of exogenous tubulin into spindle fibers or result in high local concentrations of foreign tubulin in the cells, which might cause abnormalities in division.

II. Labeling Strategies

A. Overall Strategy for Labeling Cytoskeletal Structures

The overall strategy for labeling the spindle is to fuse a microtubule-associated protein in-frame to GFP and to use the fusion protein to label the spindle microtubules. An alternate strategy, of labeling the spindle microtubules directly with tubulin–GFP, has worked less well in yeast and is less attractive for higher eukaryotes because of the multiplicity of tubulin isoforms, which are expressed in a tissue- and developmentally specific manner. This means that structures such as spindles, which comprise different tubulin isoforms, might be better labeled in some tissues than in others, and better labeled at certain times during development than at others. This is also true of microtubule motors, although the motors that are highly expressed in given cells, such as Ncd in oocytes, are likely to uniformly label the spindle fibers in these cells and therefore to be optimal for use in analyzing spindle dynamics in these cells. In addition, the use of genes for cytoskeletal proteins such as tubulin in constructing GFP fusions may be limited by structural considerations—perturbations in overall structure due to the presence of the GFP moiety could disrupt the cytoskeletal structures under study. In yeast, a tubulin–GFP fusion protein that has been analyzed failed to complement the corresponding tubulin null mutant (Marshall *et al.,* 1996). This is presumably due to the requirement to preserve the overall structure of the tubulin protein to permit interactions with other tubulin proteins during the assembly of microtubules.

The strategy of fusing GFP to proteins *associated* with given cytoskeletal structures, rather than to proteins that form an integral part of the structures, could become a method of choice for labeling other cytoskeletal components. For example, a cytoplasmic myosin fused to GFP has been used to label actin fibers associated with the contractile disk in *Dictyostelium* (Moores *et al.,* 1996) and GFP fused to a microtubule-associated protein, MAP4, has been used to label cytoplasmic microtubules (Olson *et al.,* 1995).

B. General Strategy for Labeling Ncd

The general strategy that we have used for labeling the Ncd motor protein is to fuse GFP, either wild-type or mutant, in-frame to the C terminus of the full-length motor protein (Fig. 1). The motor domain of Ncd, with the conserved

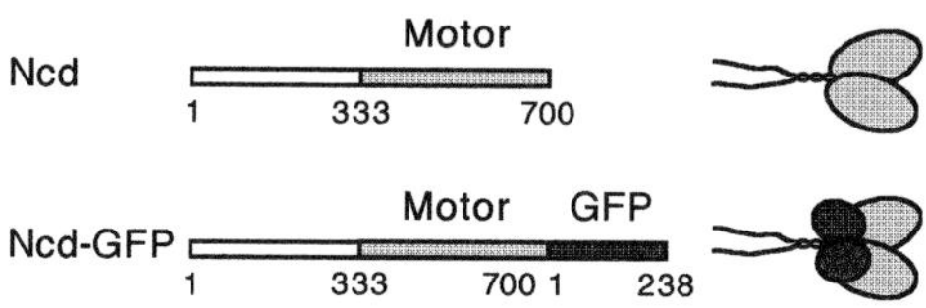

Fig. 1 The Ncd–GFP fusion protein. Ncd–GFP consists of wild-type GFP fused in-frame to the C terminus of full-length Ncd, adjacent to the motor domain. The motor domain contains ATP- and microtubule-binding sites that are required for motor function. Amino acid 699 of Ncd was changed from aspartic acid to leucine during the construction of the gene fusion. The Ncd and Ncd–GFP proteins are shown schematically, based on biochemical analysis of truncated Ncd proteins (Chandra *et al.,* 1993) and structural analysis of the Ncd motor domain (Sablin *et al.,* 1996) and GFP (Ormo *et al.,* 1966; Yang *et al.,* 1996).

ATP- and MT-binding sites, is at the C terminus of the protein. The fusion therefore places the GFP adjacent to the motor domain, as depicted schematically in Fig. 1. An insertion of GFP into the tail–stalk junction of Ncd did not rescue an *ncd* null mutant and did not result in GFP fluorescence.

C. Choice of GFP

In addition to wild-type GFP from *Aequorea victoria* and *Renilla reniformis,* many mutant forms of GFP are now available for use in making fusion proteins. The mutant GFPs have been selected for brighter green fluorescence or fluorescence emission at different wavelengths, sometimes resulting in different colors of fluorescence. A comprehensive table of available forms of GFP can be found in Endow and Piston (1998).

One property of GFP fluorescence that may limit the choice of GFP is the ability of wild-type *A. victoria* GFP to undergo photoisomerization, resulting in fluorescence photoactivation. This results in a brighter fluorescence with increased exposure to excitation light and is thought to be caused by rotation of the chromophore around a bond (isomerization) in the GFP. Photoisomerization does not occur in several of the mutant forms of GFP that have been examined, including S65T and F64L, S65T (*EGFPmut1*) GFP. For photobleaching experiments, such as those described in Chapter 16, photoactivation is undesirable, so wild-type *A. victoria* GFP should not be used for plasmid construction for these studies. The ability to photoactivate GFP could potentially be useful in studying the dynamics of cellular structures, much as the uncaging of fluorochromes has been used (Mitchison, 1989), but this application remains to be exploited. Experiments demonstrating the photoactivation of wild-type GFP are described in Section IV.C.

In general, brighter GFP fluorescence is desirable to reduce imaging time and photodamage to the live cells induced by the excitation light. This makes the brighter mutant forms of GFP, such as S65T and F64L, S65T (*EGFPmut1*) the GFPs of choice for most studies. Brighter fluorescence can also be achieved by increasing the copy number of the *gfp* fusion gene in the cells under study or by optimizing the collection of the fluorescence emission with the use of custom GFP filter sets. Filters sets designed to increase the GFP signal are described in Section III.C.

III. Imaging GFP

A. Specimen Preparation

Specimens for imaging GFP, in principle, can be either fixed or live. For Ncd–GFP in *Drosophila* oocytes and early embryos, however, the GFP fluorescence is destroyed or greatly diminished by all fixatives that have been tried, including brief fixation in formaldehyde, which has been reported to preserve Exu–GFP fluorescence in egg chambers of *Drosophila* (Wang and Hazelrigg, 1994).

By contrast, the GFP fluorescence is very bright in live *ncd–gfp* oocytes and embryos. The major considerations in preparing live eggs for imaging are to reduce autofluorescence due to the chorion, vitelline membrane, and yolk granules, and to avoid damage to the meiotic/mitotic apparatus due to anoxia. The chorion can be easily removed by pulling on the dorsal appendages of oocytes with fine-tipped forceps or by rolling embryos gently on a piece of double-stick tape. The egg is then placed in a drop of light halocarbon oil (Halocarbon Oil 27, Sigma Chemical Co.) on a slide, and a coverslip fragment is mounted over it, supported on either side by two layers of double-stick tape. The halocarbon oil can be briefly bubbled with O_2 to ensure adequate oxygen. Oxygenation of the halocarbon oil should be omitted when nonactivated oocytes are being analyzed to avoid unintentional activation of the oocytes. An open slide without a coverslip (for use with an inverted microscope) or an oxygen-permeable membrane instead of a slide has been used by some workers to ensure adequate oxygen exchange for imaging live egg chambers or embryos of *Drosophila.*

Removal of the chorion should not be performed when nonactivated oocytes are being imaged, as this can cause activation of the oocytes (Endow and Komma, 1997). The use of a brighter GFP or increasing the *ncd–gfp* copy number can permit the collection of adequate to good images through the chorion. The use of a bandpass emission filter instead of a longpass filter can also greatly improve image quality by reducing the amount of chorion and vitelline membrane autofluorescence and scattered light that is collected (Endow and Piston, 1998).

Mature oocytes of *Drosophila* can be routinely activated by brief immersion in aqueous solution [e.g., *Drosophila* PBS (Robb, 1969)] during preparation (Endow and Komma, 1997). Ovaries are dissected directly into undiluted *Drosophila* PBS and, after 3–5 min, mature oocytes are teased from the ovarioles and transferred to a drop of light halocarbon oil on a slide. The chorion is partially removed from the anterior of the oocyte by pulling on the dorsal appendages with fine-tipped forceps, and the oocyte is positioned dorsal side up and covered by a coverslip fragment mounted onto two layers of double-stick tape placed on either side.

Spindles in activated oocytes should be visualized immediately after specimen preparation to avoid missing events that occur soon after activation. The time required for completion of the meiosis I and II divisions has been estimated to be ~5–10 min (Riparbelli and Callaini, 1996; Endow and Komma, 1997), so little time can be spent searching for the spindle or optimizing the images following activation of the oocytes.

For early embryos, the spindles can be best visualized following their migration to the cortex during interphase of mitotic cycle 9. Earlier divisions can also be imaged, but the spindles in many cases are positioned perpendicular to the imaging plane, rather than parallel, so a cross section of the spindle is obtained rather than a pole-to-pole view. The early divisions of the *Drosophila* embryo occur very rapidly, with a division time of 8–10 min (Foe *et al.,* 1993). It is therefore necessary to begin imaging immediately after identifying an embryo at a suitable stage to avoid missing events that occur during the mitotic cycle.

B. Time-Lapse Imaging

A detailed protocol for time-lapse confocal imaging of mitotic spindles in early *ncd–gfp* embryos has been reported in Endow and Piston (1998). Imaging of meiotic spindles is carried out in an identical manner (Endow and Komma, 1997). Briefly, live oocytes or embryos are imaged using the 488-nm line of a krypton–argon laser, a custom GFP filter block (discussed later), and a Bio-Rad MRC 600 confocal scanning unit mounted on a Zeiss Axiophot microscope. A 63×/1.4 NA Planapochromat oil-immersion objective is used to collect the images. Images are collected into stacks of 60 at 15 to 25 s intervals using the time-lapse feature of COMOS software (Bio-Rad) and three or five Kalman-averaged slow scans per image. Animation of time-lapse sequences, recording to videotape, and conversion of image stacks to QuickTime movies are described in detail elsewhere (Endow and Piston, 1998).

C. GFP Filters

For initial visualization of GFP, FITC filter sets can be used to determine whether cells are actively fluorescing with GFP. It is useful to keep in mind, however, that FITC filter sets are not optimized for GFP fluorescence activation and emission. Wild-type GFP has excitation peaks at 395 and 475 nm, and an emission peak at 508 nm with a shoulder at 540 nm, and S65T GFP has an excitation peak at 488 nm and emission peak at 511 nm. For extensive studies involving GFP imaging, it is advantageous to purchase one of the custom GFP filter sets now available. These can improve GFP fluorescence by optimizing the excitation wavelength, the fluorescence emission that is collected, or both. The filter set that has been found to be most generally useful for imaging GFP is one that is optimized for both excitation and emission of wild-type and S65T GFP (Endow and Komma, 1997). Comparable filter sets have been designed for fluorescence and confocal microscopy (Table I), and tested. The custom confocal GFP filter set increases the signal collected over the Bio-Rad BHS and GR2 FITC filter blocks by ~1.3× and ~1.8×, respectively (Endow and Komma, 1997). The use of a bandpass rather than a longpass emission filter can greatly improve image quality by reducing the collection of signal due to specimen autofluorescence (Endow and Piston, 1998), as noted earlier. The GFP filter sets described here are available from Chroma Technology Corp., and the custom confocal GFP block is now being marketed by Bio-Rad.

Table I
Filter Sets for FITC and GFP

Filter set	Excitation filter	Dichroic filter	Emission filter
Fluorescence Microscopy			
Leica K3 cube	BP 470-490	RKP 510	LP 520
Leica L4 cube	BP 470-490	RKP 510	BP 515-560
Zeiss Set 09	BP 450-490	FT 510	LP 520
Zeiss Set 10	BP 450-490	FT 510	BP 515-565
Chroma #41017, (Endow GFP BP Set)	HQ 470/40	Q 495LP	HQ 525/50
Chroma #41018 (Endow GFP LP set)	HQ 470/40	Q 495LP	HQ 500LP
Confocal Microsopy			
Bio-Rad BHS	488/10 (filter wheel)	510LP	OG 515LP
Bio-Rad GR2/T3	488/10 (filter wheel)	T3 (trichroic)	522/32BP
Endow Confocal GFP	488/10×	Q 498LP	HQ 518/40BP

IV. Applications of Ncd–GFP Imaging

A. Protein Localization

The subcellular localization of proteins labeled by fusion to GFP can be determined by the patterns of GFP fluorescence. In the case of Ncd, extensive antibody labeling experiments had been carried out previously, both during meiosis in oocytes (Hatsumi and Endow, 1992) and mitosis in early embryos (Endow *et al.,* 1994). Through these studies, the distribution of Ncd was found to closely parallel the distribution of tubulin. By carrying out time-lapse imaging of live *ncd–gfp* embryos, it has been possible to confirm the localization of Ncd throughout the mitotic cell cycle.

An unexpected observation was that Ncd–GFP in metaphase is present in brightly fluorescent pole-to-pole fibers that cross the chromosomes at the metaphase plate, unlike tubulin, which is diminished in the plate region (Endow and Komma, 1996). This distribution of Ncd was also observed in anaphase (Fig. 2, see color plate) but was not observed in fixed embryos stained with anti-Ncd antibodies; instead, the Ncd localization was similar to that of tubulin. Fixed *ncd–gfp* embryos that were stained with anti-Ncd antibodies, however, showed the same distribution of Ncd–GFP as fixed wild-type embryos, indicating that the fibers of Ncd–GFP observed in live embryos are labile to fixation for antibody staining. These observations imply that the true distribution of Ncd during metaphase and anaphase includes pole-to-pole fibers, as was observed in the live *ncd–gfp* embryos but not in the fixed, antibody-stained embryos. These fibers are likely to be important for Ncd mitotic function in maintaining chromosome attachment to the spindle fibers at the metaphase-to-anaphase transition. The ability to visualize protein localization in live cells using fusions with GFP should prove valuable for analysis of other proteins that are present in subcellular structures that are labile to fixation.

B. Mutant Analysis

A mutant Ncd–GFP fusion protein has also been used to study the effects of loss of Ncd function on chromosome distribution. The first *ncd* mutant that has been analyzed by analysis of live eggs is *ncd*^{2}*–gfp* (Endow and Komma, 1996, 1997), which behaves like a complete loss-of-function mutant, although the mutant protein is expressed and localizes to spindle microtubules. Time-lapse imaging of the mutant *ncd*^{2}*-gfp* oocytes and embryos showed the same abnormal spindle structures that have been observed using antibody staining, but has been crucial in revealing how the spindle abnormalities arise—this could be inferred from analysis of fixed oocytes and embryos but not determined with certainty. The origin of the abnormal spindles is critical to determining the functions that are defective in the mutant Ncd motor protein. Analysis of live activated oocytes further showed that the spindle microtubules are destabilized in the mutant oocytes and undergo depolymerization during the meiotic divisions. This was not observed in previous analysis of antibody-stained oocytes, but could be easily observed in fixed oocytes once it was looked for.

The Ncd–GFP fusion protein can also be used as a spindle marker to analyze other mutants that affect chromosome movement or spindle function during meiosis or mitosis by crossing the mutants into the *ncd–gfp* stock. Many meiotic and mitotic mutants are now known in *Drosophila* and can be used to gain insights into the roles of the mutant proteins during cell division. Interactions between these mutants and *ncd* mutants can be analyzed using *ncd*^{2}*–gfp* and other *ncd–gfp* mutant transgenes that are currently being constructed and transferred into *Drosophila.*

C. Photoactivation

As noted earlier, wild-type GFP fluoresces more brightly after an initial exposure either to ultraviolet or blue excitation light. Although this property has not yet been fully exploited, the ability to undergo photoactivation upon exposure to 488-nm light can be demonstrated using confocal microscopy and cellularized cycle 14 *ncd–gfp* embryos in which the Ncd–GFP is localized to the cytoplasm.

The extended length of interphase of mitotic cycle 14 (Foe *et al.,* 1993) allows adequate time for experimental manipulations using the confocal microscope. These were performed by focusing on the embryo surface under white light under 2× zoom magnification, then scanning the embryo continuously for 1 min with the 488-nm line of the krypton–argon laser using a 10% transmission neutral density filter (Neutral Density 1), the Bio-Rad BHS filter block, and manual gain and black settings. The zoom setting was then reduced to 1×, and the embryo was rescanned at low magnification using eight Kalman-averaged slow scans to obtain a final image. The embryos imaged in this way showed a clear enhancement of fluorescence in the prescanned region (Fig. 3). The final image was analyzed using NIH Image to measure pixel density in the prescanned region compared to the nonprescanned region. Analysis of prescanned and nonprescanned cells (4–10 cells each) in three independent embryos resulted in an estimate of ~2.1× increased fluorescence in the prescanned region (range = 1.7×–2.5× for individual embryos). Although this method of quantitation is relatively crude, it demonstrates that exposure to blue light can photoactivate wild-type GFP.

The ability of wild-type GFP to undergo photoactivation has potential applications with respect to Ncd–GFP. For example, it could be used to monitor spindle

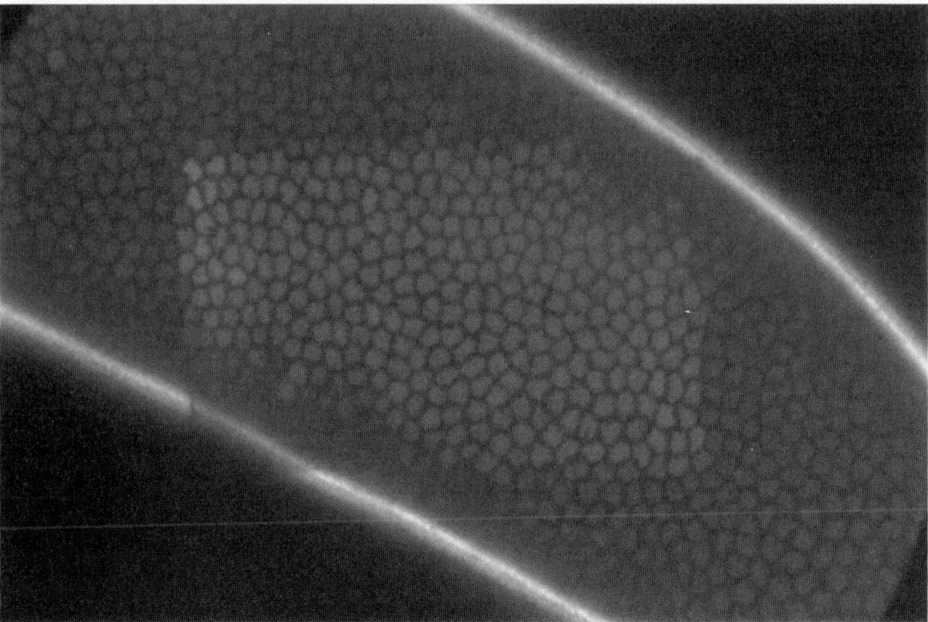

Fig. 3 Early embryo demonstrating photoactivation of Ncd–GFP. The Ncd–GFP is localized to the cytoplasm of the embryo, which is in interphase of mitotic cycle 14. After focusing on the embryo surface under white light, the embryo was scanned at 2× zoom magnification for 1 min with the 488-nm line of the laser, then rescanned at reduced (1× zoom) magnification. The brighter rectangular region is the prescanned region and shows ~2.5× brighter fluorescence than the nonprescanned region. The bright autofluorescence of the vitelline membrane is readily apparent when a longpass emission filter is used to collect the images, but can be reduced by use of a bandpass emission filter. The membrane autofluorescence is photobleached in the prescanned region, in contrast to the photoactivated Ned–GFP fluorescence.

dynamics in much the same way as caged fluorescent tubulin has been used to monitor dynamics of spindle microtubules (Mitchison, 1989).

V. Perspectives

Although the use of GFP has been relatively recent, it has now been shown to be effective in many of the applications that have previously been used for fluorescently labeled proteins. These include subcellular localization of proteins and *in vivo* labeling of cytoskeletal structures, as described in this chapter, single-molecule tracking (Pierce *et al.,* 1997), and photobleaching experiments (Cole *et al.,* 1996). The use of GFP in these applications is advantageous over previous methods in the specificity and efficiency of labeling, and in specific targeting to given cellular structures or organelles.

The use of GFP in the analysis of spindle dynamics during meiosis and mitosis represents an advance over methods that have been used previously to label spindle fibers *in vivo,* and may eventually replace these methods. Analysis of meiosis in oocytes has been refractory to conventional methods due to problems with unintentional activation or lysis of oocytes. Here the use of GFP is especially advantageous, and promises to revolutionize methods currently used for analysis.

Finally, the potential exists for applications of GFP that go beyond conventional fluorescence methods. These include fluorescence resonance energy transfer (FRET), which may become a standard method for colocalization of proteins within the cell. The development of this and other innovative methods will help to realize the potential of GFP while providing important information about basic cellular processes.

Acknowledgments

This work was supported by grants from the National Institutes of Health (GM 46225) and the American Cancer Society (CB-47).

References

Chandra, R., Salmon, E. D., Erickson, H. P., Lockhart, A., and Endow, S. A. (1993). Structural and functional domains of the *Drosophila* ncd microtubule motor protein. *J. Biol. Chem.* **268,** 9005–9013.

Cole, N. B., Smith, C. L., Sciaky, N., Terasaki, M., Edidin, M., and Lippincott-Schwartz, J. (1996). Diffusional mobility of Golgi proteins in membranes of living cells. *Science* **273,** 797–801.

Endow, S. A., Chandra, R., Komma, D. J., Yamamoto, A. H., and Salmon, E. D. (1994). Mutants of the *Drosophila* ncd microtubule motor protein cause centrosomal and spindle pole defects in mitosis. *J. Cell Sci.* **107,** 859–867.

Endow, S. A., and Komma, D. J. (1996). Centrosome and spindle function of the *Drosophila* Ncd microtubule motor visualized in live embryos using Ncd–GFP fusion proteins. *J. Cell Sci.* **109,** 2429–2442.

Endow, S. A., and Komma, D. J. (1997). Spindle dynamics during meiosis in *Drosophila* oocytes. *J. Cell Biol.* **137,** 1321–1336.

Endow, S. A., and Piston, D. W. (1998). Methods and protocols. *In* "GFP: green fluorescent protein strategies and applications" (M. Chalfie and S. Kain, Eds.). John Wiley & Sons, Inc., New York.

Foe, V. E., Odell, G. M., and Edgar, B. A. (1993). Mitosis and morphogenesis in the *Drosophila* embryo: point and counterpoint. *In* "The Development of *Drosophila melanogaster*" (M. Bate and A. M. Arias, Eds.), pp. 149–300. Cold Spring Harbor Laboratory Press, Cold Spring Harbor, NY.

Hatsumi, M., and Endow, S. A. (1992). The *Drosophila* ncd microtubule motor protein is spindle-associated in meiotic and mitotic cells. *J. Cell Sci.* **103,** 1013–1020.

Marshall, L. G., Jeng, R. L., Mulholland, J., and Stearns, T. (1996). Analysis of Tub4p, a yeast γ-tubulin-like protein: implications for microtubule-organizing center function. *J. Cell Biol.* **134,** 443–454.

Matthies, H. J. G., McDonald, H. B., Goldstein, L. S. B., and Theurkauf, W. E. (1996). Anastral meiotic spindle morphogenesis: role of the Non-Claret Disjunctional kinesin-like protein. *J. Cell Biol.* **134,** 455–464.

Mitchison, T. J. (1989). Polewards microtubule flux in the mitotic spindle: evidence from photoactivation of fluorescence. *J. Cell Biol.* **109,** 637–652.

Moores, S. L., Sabry, J. H., and Spudich, J. A. (1996). Myosin dynamics in live *Dictyostelium* cells. *Proc. Natl. Acad. Sci. U. S. A.* **93,** 443–446.

Olson, K. R., McIntosh, J. R., and Olmstead, J. B. (1995). Analysis of MAP 4 function in living cells using green fluorescent protein (GFP) chimeras. *J. Cell Biol.* **130,** 639–650.

Ormo, M., Cubitt, A. B., Kallio, K., Gross, L. A., Tsien, R. Y., and Remington, S. J. (1966). Crystal structure of the *Aequorea victoria* green fluorescent protein. *Science* **273,** 1392–1395.

Pierce, D. W., Hom-Booher, N., and Vale, R. D. (1997). Imaging individual green fluorescent proteins. *Nature* **388,** 338.

Riparbelli, M. G., and Callaini, G. (1996). Meiotic spindle organization in fertilized *Drosophila* oocyte: presence of centrosomal components in the meiotic apparatus. *J. Cell Sci.* **109,** 911–918.

Robb, J. A. (1969). Maintenance of imaginal discs of Drosophila melanogaster in chemically defined media. *J. Cell Biol.* **41,** 876–885.

Sablin, E. P., Kull, F. J., Cooke, R., Vale, R. D., and Fletterick, R. J. (1996). Crystal structure of the motor domain of the kinesin-related motor ncd. *Nature* **380,** 555–559.

Sawin, K. E., and Endow, S. A. (1993). Meiosis, mitosis and microtubule motors. *BioEssays* **15,** 399–407.

Wang, S., and Hazelrigg, T. (1994). Implications for *bcd* mRNA localization from spatial distribution of *exu* protein in *Drosophila* oogenesis. Nature **369,** 400–403.

Yang, F., Moss, L. G., and Phillips, G. N., Jr. (1996). The molecular structure of green fluorescent protein. *Nat. BioTech.* **14,** 1246–1251.

CHAPTER 11

GFP as a Cell and Developmental Marker in the *Drosophila* Nervous System

Andrea Brand
Wellcome/CRC Institute and Department of Genetics
Cambridge University
Cambridge CB2 1QR, United Kingdom

I. Introduction

The ideal way to follow the development of individual cells or tissues is in living organisms, in which every developmental stage can be observed in a single intact animal. An ideal vital marker could be used in all cell types and would reveal the morphology of the cell, to enable the identification of different cell types without compromising cell viability. Green fluorescent protein (GFP) fulfills all of these requirements. It is a naturally fluorescent protein that has been shown

Copyright © 1999 by Academic Press. All rights of reproduction in any form reserved.
0091-679X/99 $25.00

to function in a wide variety of transgenic animals, requires no subtrate, and is nontoxic (Prasher *et al.,* 1992; Chalfie *et al.,* 1994; Cubitt *et al.,* 1995; Prasher, 1995; Gerdes and Kaether, 1996). GFP is a small protein (27 kDa) that freely enters the nucleus, fills the cytoplasm of the cell, and is able to diffuse into small cytoplasmic extensions. The excitation and emission spectra of GFP are similar to those of FITC, such that the protein can be visualized with most conventional epifluorescence and confocal filter sets, GFP does not bleach readily and is therefore ideal for long-time-course analyses.

A. Nervous System Development

We are studying the development of the embryonic nervous system, a tissue in which a vital marker is particularly advantageous. During neurogenesis, thousands of neurons are born, each of which migrates to a characteristic position within the brain or nerve cord from where it extends an axon toward, and synapses with, a specific target cell. These dynamic properties are well suited to real-time analysis, from the birth of neurons and glial cells through cellular migration, axon extension and synaptogenesis, and finally muscle contraction. Furthermore, the two cell types found in the nervous system, neurons and glia, are indistinguishable by nuclear markers but are easily identified by their characteristic cellular morphology.

In *Drosophila,* cell labeling in the nervous system has relied almost exclusively on invasive techniques such as dye microinjection or on antibody staining of fixed tissues. The advantage of using dyes, such as lucifer yellow and DiI, is that they fill the entire cell and label axonal or glial extensions. Because labeling is carried out on living embryos, individual cells can be followed throughout development in a single animal. However, dye microinjection is time-consuming and requires great dexterity, and for this reason, immunohistochemistry has been more widely used. Antibody staining also has its limitations. Samples must first be fixed, which can adversely affect the morphology of the sample and induce artifacts. If antibodies against nuclear proteins are used, cell shapes are not revealed, and it can be difficult to identify individual cells. Furthermore, the cuticles of older embryos, larvae, and adult flies are impermeable to antibodies. It is necessary to analyze a large number of animals of different ages to construct a comprehensive time course of development. This can be particularly difficult in certain mutant backgrounds, in which some of the developmental landmarks may be distorted or missing altogether. Furthermore, it is easier to interpret expression patterns in a whole mounts than to reconstruct a three-dimensional picture from a series of tissue sections.

Since the introduction of GFP as a live marker, it has become not only feasible, but also relatively easy, to study neurogenesis in living animals. In a single embryo, after a simple dechorionation step, it is possible to visualize the segregation of neural precursors, the subsequent mitotic divisions that produce first ganglion mother cells and then neurons or glia, neuronal and glial cell migration, and axon outgrowth (Brand, 1995; A.H.B., C. M. Davidson, E. L. Dormand, U. P. John, J. A. Kaltschmidt, and A. J. Schuldt, unpublished observations). In pupae, it should be feasible to observe the rewiring of the nervous system as larval tissues are histolyzed and adult structures take their place. In the embryonic, larval, and adult brains, GFP or GFP fusion proteins are being used to identify and to trace individual cells.

It is now possible to label specific cells or tissues, and to follow their fates in both wild-type animals and in a variety of mutant backgrounds. GFP can be used as a lineage tracer, either by inducing constitutive GFP expression in clones using the Flp/FRT recombination system (Struhl and Basler, 1993) or by double labeling GFP-positive cells with DiI (Bossing and Technau, 1994). In genetic screens, mutant animals can be identified on the basis of altered GFP expression patterns and then recovered as individuals to establish a stock.

B. Targeted Expression—The GAL4 System

To target expression of GFP *in vivo,* we use the GAL4 system (Brand and Perrimon, 1993; Brand *et al.,* 1994; Brand and Dormand, 1995), which allows the selective activation of any cloned gene in a wide variety of tissue- and cell-specific patterns. The GAL4 system is designed to generate lines that express a transcriptional activator, rather than a target gene, in numerous expression patterns. First the yeast transcriptional activator, GAL4, is expressed in specific cell or tissue types. It is then possible to activate a synthetic gene that contains GAL4-binding sites in its promoter in those cells in which GAL4 is expressed.

Two approaches can be used to generate different patterns of GAL4 expression. The first is to drive *GAL4* transcription using characterized promoters (Fischer *et al.,* 1988; Brand and Perrimon, 1993). The second is based on "enhancer detection" (O'Kane and Gehring, 1987; Bellen *et al.,* 1989; Bier *et al.,* 1989; Wilson *et al.,* 1989). The *GAL4* gene is randomly integrated in the *Drosophila* genome, bringing it under the control of an array of different genomic enhancers that direct expression in a wide range of patterns in embryos, larvae, and adults (Brand and Perrimon, 1993). This eliminates the need to link numerous different promoters to the *GAL4* gene and allows expression in novel patterns from enhancers that have not yet been described. In addition, the enhancer-detection GAL4 vector can be mobilized to new genomic sites simply by P-transposition (Robertson *et al.,* 1988). In this way, a single transformant can be used to generate a large number of transgenics, each exhibiting a different GAL4 expression pattern.

GAL4-responsive target genes can be created using the vector pUAST (Brand and Perrimon, 1993), into which sequences can be subcloned behind a tandem array of five optimized GAL4-binding sites (hereafter referred to as the UAS, for upstream activation sequence), and upstream of the SV40 transcriptional terminator. It is possible, then to (*a*) subclone any sequence behind GAL4-

binding sites and (*b*) activate that target gene only within cells in which GAL4 is expressed.

In *Drosophila,* and more recently in *Arabidopsis* (see Chapter 9), it is possible to disrupt development specifically by targeted expression of a gene using the GAL4 system. If the GAL4-expressing cells are simultaneously labeled with GFP, any alteration in cell fate, pattern of cell division, axon extension, or target recognition can be conveniently monitored. We have constructed GAL4-responsive *GFP* transgenes with which it is possible to label almost any cell type *in vivo* (Brand, 1995; A.H.B., C. M. Davidson, E. L. Dormand, J. A. Kaltschmidt, C. B. Phelps, and A. J. Schuldt, unpublished observations).

II. Targeted Expression of GFP in *Drosophila*

GFP is produced as an apoprotein that is converted to an actively fluorescent form after cyclization and autoxidation of a chromophore consisting of just three amino acids, serine, tyrosine, and glycine (Heim *et al.,* 1994). The requirement for autoxidation leads to a delay between expression of GFP and the appearance of the actively fluorescent molecule (Heim *et al.,* 1994). Consistent with this observation, we observe fluorescence of wild-type GFP in *Drosophila* embryos at stage 12, approximately 3 h after β-galactosidase expression from a similar transgene.

Although it is possible to visualize cells labeled with wild-type GFP, the advent of mutants that are brighter, undergo fluorophore formation more rapidly, and absorb at wavelengths approaching 488 nm [a wavelength produced by most laser scanning confocal microscopes (Heim *et al.,* 1994; Delagrave *et al.,* 1995; Heim *et al.,* 1995; Cormack *et al.,* 1996; Siemering *et al.,* 1996)] has greatly facilitated *in vivo* labeling.

III. Lines for Expression of GFP

A. Selecting a GAL4 Driver Line

1. GAL4 Databases

GAL4-expressing lines that drive expression in diverse patterns in embryos, larvae, and adults have been generated. Descriptions of many of these lines are available on FlyBase, and a number of the lines are available from the *Drosophila* stock center in Bloomington, Indiana.

B. Generating a UAS-Target Gene

1. The UAS Vector

The gene of interest is cloned into the polylinker of the vector pUAST (UAS terminator) (Brand and Perrimon, 1993; Brand *et al.,* 1994) downstream of the five optimized GAL4 binding sites (the "ScaI site") and a synthetic TATA box, and upstream of the SV40 transcriptional terminator. In our experience, we have found it best to minimize the amount of 5′ leader sequence included when subcloning the target gene into pUAST, although there are reports that in some cases the 5′ untranslated sequence may increase translational efficiency. The SV40 sequence downstream of the target gene may stabilize the mRNA. UAS transgenes are subject to position effects, such that insertions at different genomic sites are expressed at varying levels. We therefore test and maintain between five and six independent insertion lines. A phenotypic series can then be generated by crossing a single GAL4 line to several UAS lines, each of which expresses the same target gene but at a different level.

2. UAS–GFP Reporters

Wild-type GFP suffers from a long lag between transcription and the first detectable signs of fluorescence, due to the requirement for autoxidation (Heim *et al.,* 1994). For example, we have observed wild-type GFP fluorescence approximately 3 h after β-galactosidase expression from a similiar transgene. However, mutant GFPs have been isolated that have a greatly reduced time of onset (Heim *et al.,* 1994, 1995; Cormack *et al.,* 1996; Siemering *et al.,* 1996).

- *UAS–GFP* encoding wild-type GFP has been generated (Brand, 1995; Yeh *et al.,* 1995). GFP is a small protein (27 kDa) that seems to concentrate in the nucleus but is excluded from the nucleolus. GFP diffuses throughout the cytoplasm. Whereas a single GFP-expressing cell is easily resolved when surrounded by nonexpressing cells, it can be difficult to distinguish individual cells when they are part of a GFP-positive cell cluster (Fig. 1).
- *UAS–GFP S65T* (B. Dickson, unpublished observations) carries a single amino acid substitution, serine to threonine at position 65, that shifts the major excitation peak from 395 to 475 nm (Heim *et al.,* 1995). This variant has been reported to fluoresce approximately six times more brightly than wild-type GFP (Heim *et al.,* 1995).
- *UAS–GFP S65T/I167T* (C. M. Davidson and A.H.B., unpublished observations) carries a second mutation, isoleucine to threonine at position 167 (Heim *et al.,* 1994), which also shifts the major GFP excitation peak to 475 nm. The combination of these two mutations yields a brighter GFP variant than the S65T mutation alone.
- *UAS–mGFP6* (J. Haseloff, C. M. Davidson, and A.H.B., unpublished observations) expresses a variant of mGFP5 (Siemering *et al.,* 1996) that was developed for use in *Arabidopsis* but has also proved useful for labeling *Xenopus* and mouse cells (Zernicka-Goetz *et al.,* 1996, 1997). mGFP6 carries mutations that improve the maturation and spectral properties, and enhance the solubility and fluorescence of GFP: F64L, S65T (Heim *et al.,* 1995; Cormack *et al.,* 1996) and V163A, I167T, S175G (Heim *et al.,* 1994; Siemering

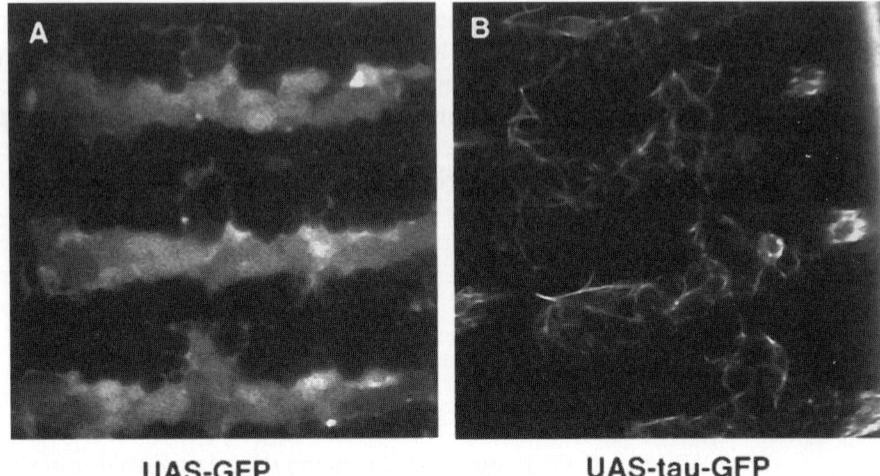

Fig. 1 GFP labeling in the embryonic CNS of living embryos. (a) *Engrailed–GAL4* drives expression of *UAS–gfp* in epidermal stripes and a subset of neurons. The proximity of the cell bodies makes it difficult to identify individual neurons. (b) *Engrailed–GAL4* drives expression of *UAS–tau–gfp* expression in the same cells as in (a). tau–GFP is excluded from nuclei and individual cells can be easily distinguished.

et al., 1996). Codon usage changes remove a cryptic plant intron and optimize expression (J. Haseloff *et al.,* unpublished observations). In our hands, mGFP6 is more fluorescent than GFP S65T/I167T.

3. Fusion Proteins

We have taken advantage of the fact that GFP remains active as a fusion protein (Wang and Hazelrigg, 1994) and have linked GFP to the microtubule binding protein tau (Butner and Kirschner, 1991; Callahan and Thomas, 1994). Using laser scanning confocal microscopy, we are able to resolve the cytoskeleton within individual cells of a living larva (Fig. 3). Although wild-type GFP is a small protein (~27 kDa; Prasher *et al.,* 1992) that can readily diffuse into axons, we find that tau–GFP is a better marker for labeling individual neurons (Figs. 1 and 4; discussed later). We do find, however, that GFP fluoresces more brightly than do the GFP fusion proteins. GFP has also been fused to nuclearly localized β-galactosidase to generate an *in vivo* nuclear marker.

- *UAS–nuclear GFP–lacZ.* Shiga *et al.* (1996) fused GFP to a NLS in an attempt to generate a nuclearly localized GFP. However, fluorescence was detectable in polytene, but not diploid, tissues. GFP fusions to a nuclearly localized form of β-galactosidase produced a marker that fluoresces brightly in the nuclei of diploid embryonic and larval tissues.

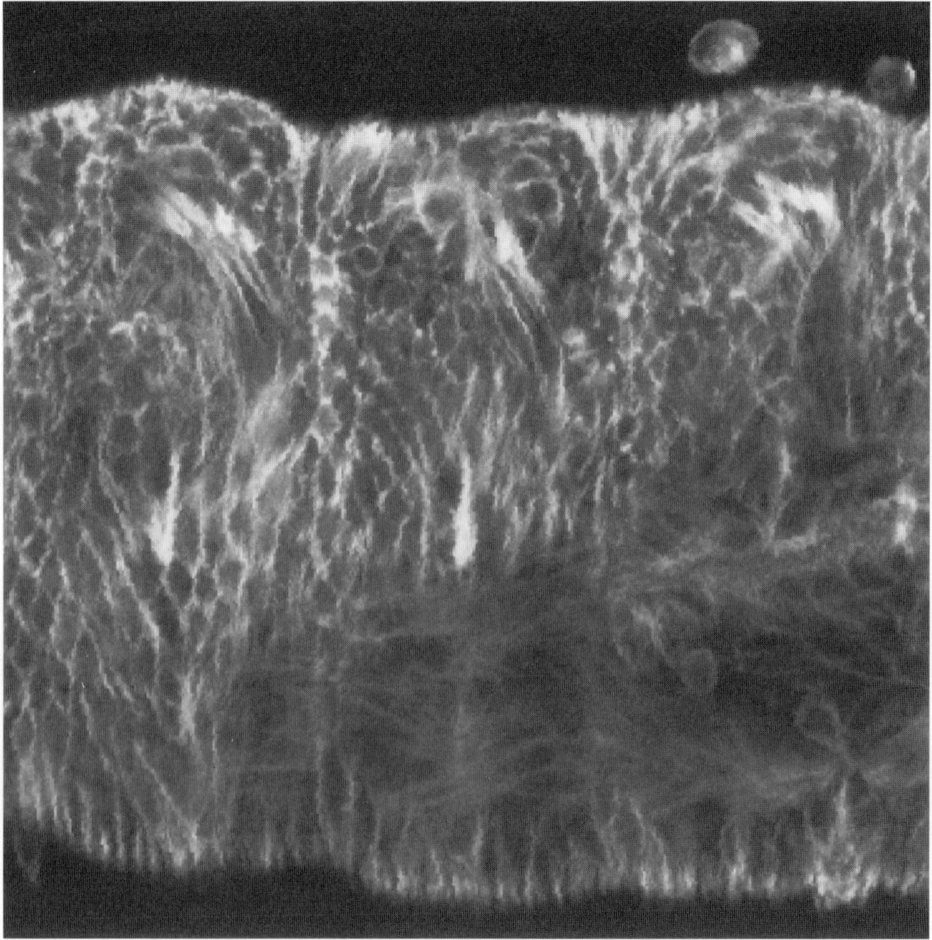

Fig. 2 Tau–GFP labels epidermal cells and muscles in a living embryo.

- *UAS–tau–GFP* (Brand, 1995). Tau is a microtubule-binding protein that highlights the cytoskeleton within every cell (Butner and Kirschner, 1991; Callahan and Thomas, 1994). Tau is normally found in neuronal lineages, where it binds and stabilizes microtubules; however, it can also bind microtubules in nonneuronal cell types (Callahan and Thomas, 1994). Fusing tau to GFP produces a reporter whose expression pattern can be followed in real time while also revealing the cytoskeletal architecture (Fig. 3). Tau-GFP is excluded from the nucleus and labels the microtubule network, thereby revealing the morphology of the cell and facilitating its identification. Tau reporter fusions have proved extremely useful in labeling neurons because they label not only the cell body but also the axonal projection (Callahan and Thomas, 1994; Brand, 1995; Hidalgo and Brand, 1997). This is particularly valuable when the axonal projection pattern of a neuron must be known in order to verify its identity.

When expressed at high levels, tau–β-galactosidase can interfere with mitosis, causing phenotypes that are most readily seen in rapidly dividing tissues. For example, when *UAS–tau–lacZ* is driven in the imaginal disks by crossing to *engrailed–GAL4,* it leads to pupal lethality: pupae dissected from their pupal cases completely lack thoracic structures (A.H.B., unpublished observations).

Expression of tau–GFP causes much less severe phenotypes than tau–β-galactosidase. Crosses to *engrailed–GAL4* yield valuable adults with posterior wing defects. Tau–β-galactosidase may cause more severe defects because β-galactosidase can tetramerize, perhaps enhancing the cross-linking and stabilization of microtubules.

- *UAS–tau–GFP S65T/I167T* (Dormand and Brand, 1998) carries both the S65T and I167T mutations (see III.B.2) (Heim *et al.,* 1994, 1995). The combi-

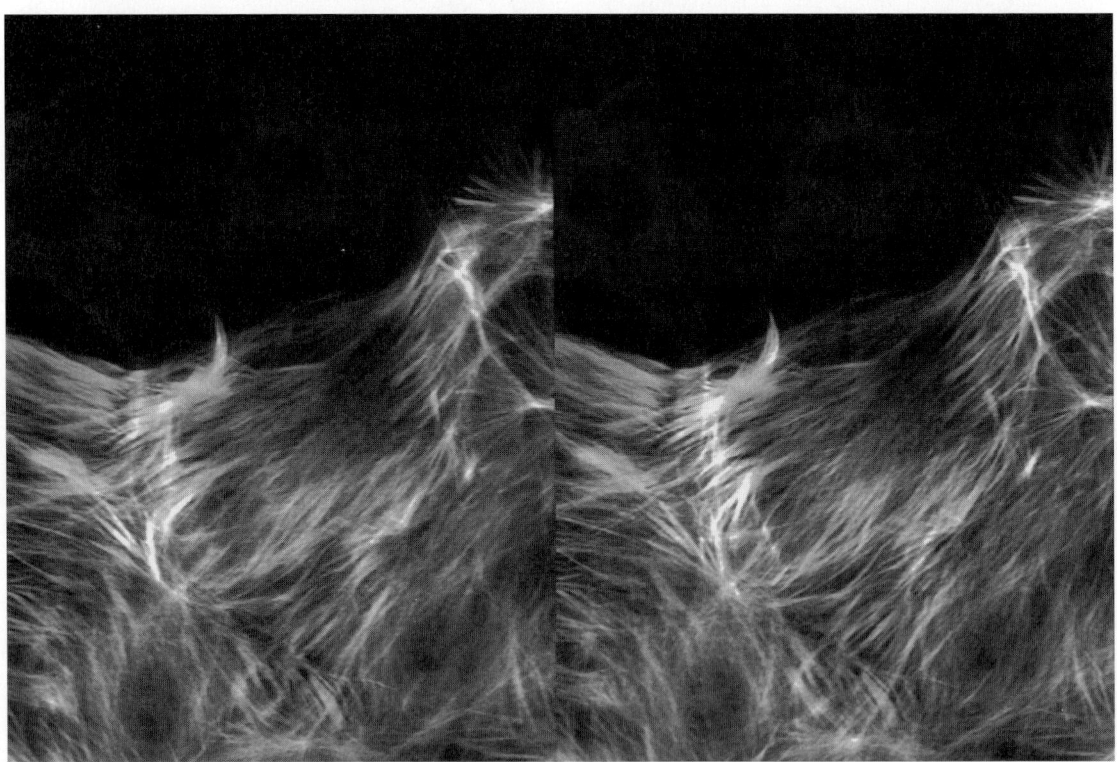

Fig. 3 Tau–GFP binds to microtubules, highlighting the cytoskeleton within each epidermal cell of a living larva. The image is composed of a stereo pair of confocal micrographs.

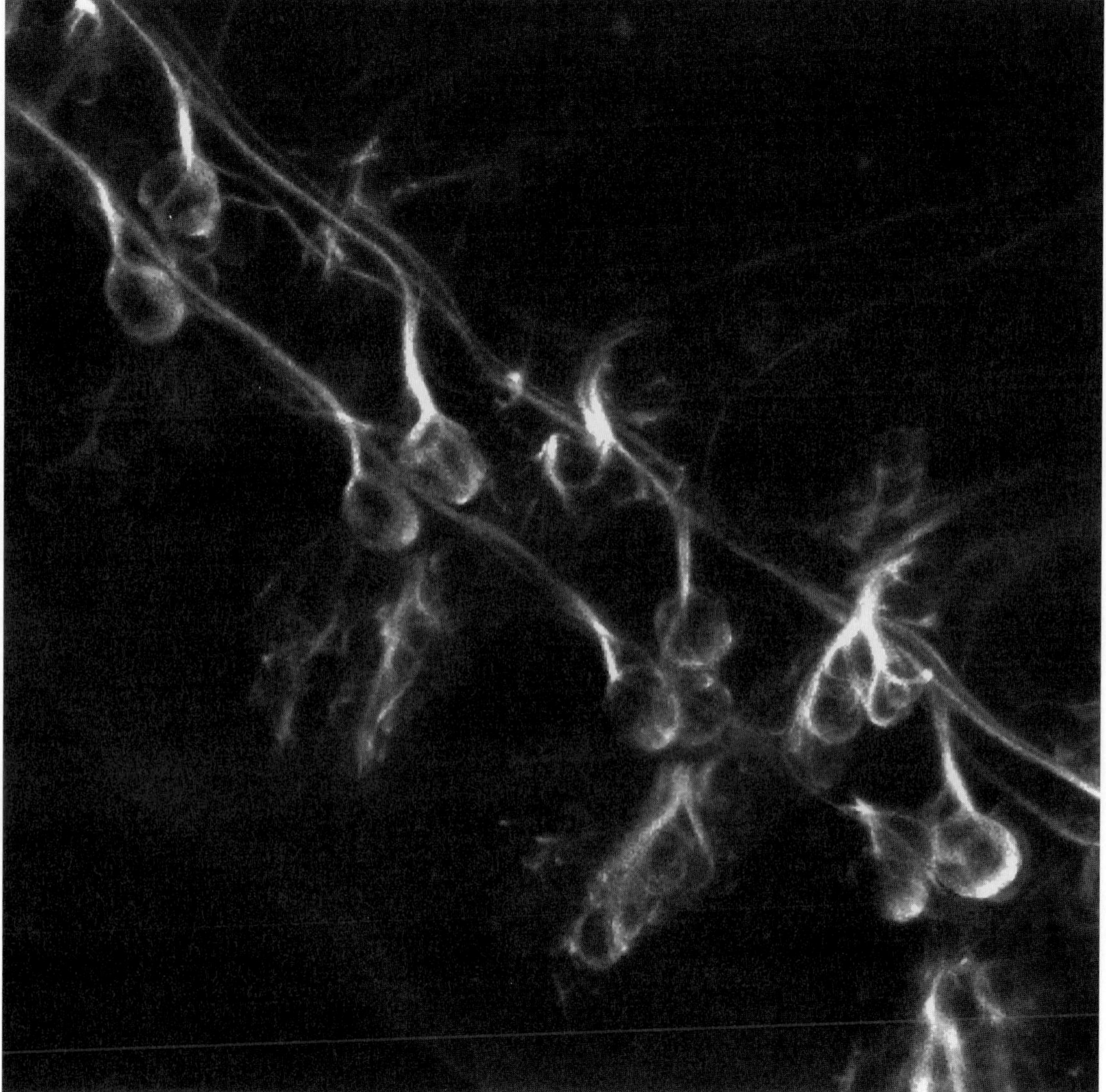

Fig. 4 Targeted expression of tau–GFP in motor neurons and interneurons enables individual neurons to be traced as they extend their axons in the central nervous system of living embryos.

nation of these two mutations yields a GFP variant that is much brighter than wild-type GFP.

- *UAS–tau–mGFP6* (J. Haseloff, C. M. Davidson, and A.H.B., unpublished) expresses a tau fusion with mGFP6 (see III.B.2.). Tau–mGFP6 is generally more fluorescent than tau–GFP S65T/I167T.

C. Maternal and Early Embryonic Expression

In contrast with enhancer-detection/*lacZ* screens in which one third of the insertion lines express β-galactosidase in the female germ line (Fasano *et al.*, 1988), we have seen no GAL4-mediated misexpression in the germ line. To express GFP in ovaries or early embryos, the GFP coding sequence is fused directly to promoters that are active in the female germ line.

- *exu–GFP* (Wang and Hazelrigg, 1994). GFP fused to either the N or C terminus of *Exuperantia* is driven from the *exu* promoter. In ovaries, exu–GFP is concentrated in particles at the ring canals, which connect the nurse cells and the oocyte.
- *ncd–GFP* (Endow and Komma, 1996; Endow and Komma, 1997). Ncd–GFP is expressed from the *ncd* promoter. Ncd is a minus-end directed kinesin motor protein, and ncd–GFP labels meiotic spindles in oocytes and mitotic spindles in the early embryo (see Chapter 10).
- *PUbnlsGFP* (Davis *et al.*, 1995). Nuclearly-localized GFP is expressed from the polyubiquitin promoter.
- *PUbGFP* (Davis *et al.*, 1995). GFP is expressed from the polyubiquitin promoter.
- *α4 tubulin–tau–GFP* (Micklem *et al.*, 1997; Schuldt *et al.*, 1998). Tau–GFP is expressed in the oocyte, where it binds to microtubules, and in early embryos, where it binds to microtubules and labels mitotic spindles.
- *α4 tubulin–GFP–stau* (Schuldt *et al.*, 1998). GFP–staufen is expressed in the oocyte and early embryo. Staufen is a double-stranded RNA binding protein, and GFP–staufen appears to bind to *prospero* mRNA injected into early embryos.
- *hsp70–moesin* (Edwards *et al.*, 1997). A GFP–moesin fusion protein is expressed ubiquitously from the *hsp70* promoter after heat shock. GFP–moesin labels cortical actin and membrane processes.

Much of the power of the GAL4 system stems from the range of GAL4 lines available and the great diversity of ectopic expression patterns that these lines afford. However, the temporal diversity of ectopic expression patterns has been curtailed by a block to GAL4-mediated expression prior to stage 6 of embryogenesis (about three hours of development) (Brand *et al.*, 1994). The *GAL4* expression vectors include the 5′ and 3′ UTRs from the *hsp70* gene, which may inhibit translation during oogenesis. The temporal block might also result from the requirement for a transcriptional coactivator that is not expressed until later stages of embryogenesis. D. St. Johnston (personal communication) has achieved GAL4-mediated expression at earlier stages of embryogenesis by modifying both the GAL4 UTRs and the GAL4 protein. With these modifications, the onset of GAL4-mediated expression has been advanced to stage 4.

IV. Visualizing GFP Expression

The following protocols are for examining GFP fluorescence in either live or fixed whole-mount embryos. We have also stained embryos with rabbit anti-GFP antibodies (Clontech) with excellent results, although we have had less success using monoclonal anti-GFP antibodies (Clontech).

A. Materials

1. Live samples

   - Home-made sieves. Cut the top off a 15-ml polypropylene Falcon tube, about 4 cm from the screw-on cap. Cut a wide hole in the cap. Place fine gauze over the end of the Falcon tube and hold it in place by screwing on the cap.
   - 18- × 18-mm coverslips (Menzel–Glaser)
   - 22- × 40-mm coverslips (Menzel–Glaser)
   - 50% Clorox
   - Glue (Double-sided Scotch tape glue dissolved in heptane)
   - Voltalef oil, 10S
   - Halocarbon oil, 50% Halocarbon 27 and 50% Halocarbon 700 (Sigma)
   - Air-permeable Teflon membrane mounted on perspex frame
   - Parafilm

2. Fixed samples

   - PBT [PBS, 0.1% Triton X-100 (Sigma)]
   - 50% Clorox
   - 4% Formaldehyde (BDH) in PBT (make up fresh)
   - Heptane (Sigma–Aldrich)
   - Methanol (Fisher Scientific International)
   - Vectashield (Vector Labs)

B. Live Samples

Collect the embryos in a vial or on an apple juice agar plate. To ensure that most of the embryos are at the appropriate stage, it is helpful to do short collections (2–4 h) and then age the embryos until the appropriate stage of development. Wash the eggs into a sieve, placed on a plastic tray, with water and a paintbrush. Dechorionate the embryos in 50% Clorox for 3 min, then wash them thoroughly with water. If you plan to use an upright microscope, transfer the embryos with a paintbrush into a drop of Voltalef or Halocarbon oil placed in the middle of an air-permeable teflon membrane stretched over a perspex frame (designed by E. Wiechaus). Place an 18- × 18-mm coverslip on either side of the embryos, about 2–3 cm apart, to prevent the embryos from being squashed, and cover them with a 22- × 40-mm coverslip. The embryos will develop normally through embryogenesis in most cases.

We do not seal the coverslips, as it is often useful to roll the embryos to obtain a dorsal, ventral, or lateral view. In addition, the solvents in many nail varnishes have been reported to inhibit GFP fluorescence (Chalfie *et al.,* 1994).

When using an inverted microscope, the embryos may be mounted simply on a coverslip. With a razor blade, cut a square hole in the middle of a small piece of parafilm and place this "frame" around the embryos to prevent the oil running off.

C. Fixed Samples

GFP remains fluorescent after fixation and, although real-time analysis is sacrificed, GFP expression can be correlated with endogenous gene expression (Fig. 5, see also color plate). GFP fluorescence is preserved, if somewhat diminished, after formaldehyde fixation. Transfer the dechorionated embryos to an Eppendorf tube and half fill it with heptane. Remove the residual PBT and top off with 4% formaldehyde in PBT. The embryos should float at the interface between the formaldehyde and heptane. Fix for 30 min with gentle rolling. Replace the formaldehyde with methanol and remove the vitelline membranes by shaking vigorously for 30 s. Wash in methanol for 10 s, then replace with PBT to rehydrate the embryos. Do not leave the embryos in methanol longer than necessary, because this has been reported to lead to a rapid and irreversible loss of GFP fluorescence (Ward *et al.,* 1980). Proceed with antibody staining as usual (for example, see Patel, 1994). Mount the embryos in Vectashield. Use two 18- × 18-mm coverslips flanking the embryos as supports, then cover with a 22- × 40-mm coverslip. We do not seal the coverslips (discussed earlier).

In our hands, GFP fluorescence does not survive *in situ* hybridization protocols, which include an overnight incubation in formamide.

D. Imaging GFP

1. Epifluorescence

GFP can be detected by epifluorescence (using most FITC filter sets or a Chroma GFP filter set) or by laser scanning confocal microscopy. We observe GFP fluorescence through the embryonic vitelline membrane, the larval cuticle, and the early, unpigmented pupal case. When examining living whole-mount preparations by conventional epifluorescence microscopy (as opposed to laser scanning confocal microscopy) autofluorescence can be a problem. In embryos, yolk autofluorescence can obscure the signal from GFP, and in larvae reflection

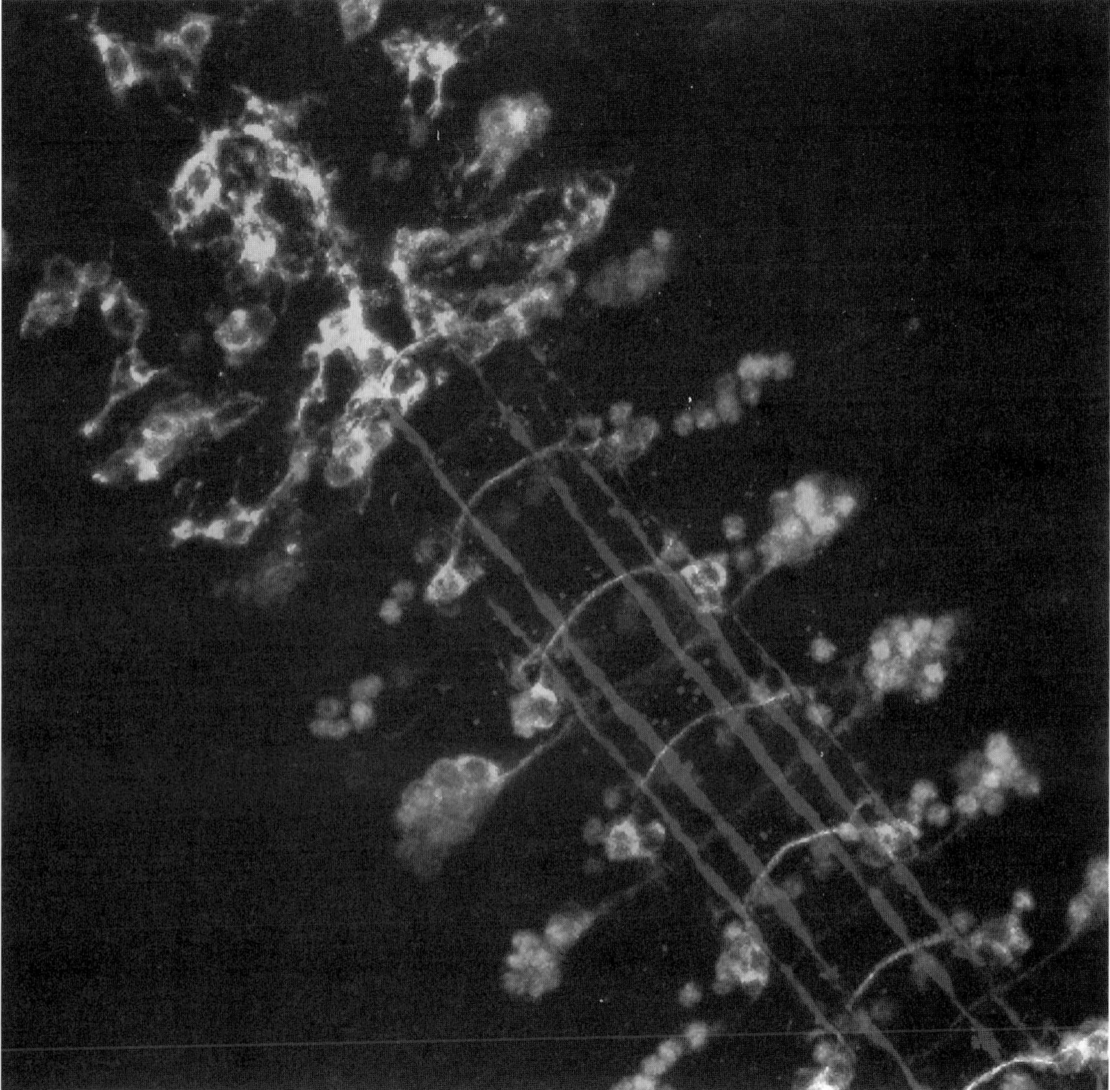

Fig. 5 Targeted expression of tau–GFP labels individual neurons as they extend their axons in the central nervous system of living embryos. Tau–GFP continues to fluoresce after fixation, which allows immunolabeling of other proteins, such as Even-skipped and Fasciclin II (see also color plate).

from the cuticle can present difficulties. In both cases background autofluorescence is indistinguishable from GFP fluorescence using a fluorescein filter set (Zeiss filter set 10: excitation 450–490 nm, emission 515–565 nm). A DAPI filter set (Zeiss filter set 2: excitation 365 nm, emission 420 nm and above) can be used to discriminate between the two. Using the DAPI set (or Zeiss filter set 18: excitation 390–420 nm, emission 450 nm and above) the background fluorescence is blue and GFP fluorescence is green. Ideally, filter sets that are tailor-made for GFP are used. Chroma Technology Corp. supplies bandpass [HQ470/40 excitation (i.e., excitation wavelengths between 450 and 490 nm), Q495LP dichroic, HQ525/50BP emission (i.e., emission wavelengths between 500 and 550 nm)] and longpass filter sets [HQ470/40 excitation (i.e., excitation wavelengths between 450 and 490 nm), Q495LP dichroic, HQ500LP emission (i.e., emission wavelengths above 500 nm)]. We have also had success with the bandpass and longpass filters supplied with our Nikon microscope: B2A (excitation, 450–490 nm; 505-nm dichroic; 520-nm longpass) and B2EC (excitation, 465–495 nm; 505-nm dichroic; 515 to 555-nm bandpass). In both cases, the bandpass filters allow GFP fluorescence to be distinguished from background fluorescence.

2. Confocal Microscopy

Confocal microscopy, which generates optical sections of a fluorescently labeled sample, holds great advantages for imaging cells labeled with GFP. For example, when imaging the CNS, which lies on the ventral side of the embryo, the background fluorescence emanating from the more dorsally placed yolk can be completely eliminated by optical sectioning. The resolution of confocal images is much greater than can be obtained by conventional epifluorescence techniques. By collecting a stack of optical sections (or Z-series) the sample can be reconstructed in three dimensions, then rotated or tilted to view cells or axons that would otherwise be obscured. Most confocal software enables time-lapse imaging in one focal plane and 4-D imaging, where Z-series are collected over time. These series of images can then be viewed as movies.

We collect our images using a BioRad MRC1024 confocal scan head and krypton–argon mixed gas laser, on a Nikon BX60 upright microscope. Wild-type GFP, GFP S65T, GFP S65T/I167T, and mGFP6 are excited by the 488-nm laser line. We use a laser power between 1 and 30%, and the standard setup for imaging FITC, which uses a 522/32 emission filter (i.e., wavelengths between 506 and 538 nm are transmitted). The Chroma HQ500LP emission filter, which transmits light of wavelengths greater than 500 nm, gives a brighter GFP signal.

We routinely use GFP labeling in conjunction with other vital dyes, such as DiI, for cell lineage tracing, and acridine orange, to detect dying cells. Filter settings for both dyes can be found as part of the basic set up on the MRC1024.

Embryos can be viewed with a 60× planapochromat, 1.4-NA oil-immersion objective. To reduce spherical aberration when focusing deep into an aqueous sample (Haseloff and Amos, 1995), embryos are mounted in water or another aqueous mounting medium and viewed with a long-working-distance, coverslip-corrected, water-immersion objective (for example, a Nikon 60× planapochromat, 1.2 NA, working distance 220 μm). When imaging more than 60 to 80 μm deep in the tissue, even under the conditions described here, image quality degrades rapidly. To reduce noise and sharpen the image, we Kalman average between 2 and 15 frames whenever possible (see IV.D.2.a.).

a. Movement

During embryogenesis, many cells and tissues undergo significant morphogenetic movements, such as ventral furrow formation, germ band extension and retraction, dorsal closure, and nerve cord condensation. Once functional neural circuits have been established, muscle contraction takes place and, later in development, larvae tend to crawl out of the field of view and, eventually, off the microscope slide. For this reason, we will confine our discussion to embryonic development.

When movement is rapid, it is not possible to Kalman average several scans, as this blurs rather than sharpens the image. Similarly, it may not be possible to project a Z-series, as each optical section will not be aligned with its neighbors. For example, when imaging the synchronous nuclear divisions in precellular embryos, we scan a single focal plane continuously on a slow setting. When observing macrophage movements, we collect an image every 15–20 s on a slow setting and Kalman average 2–3 frames.

As mentioned in IV.B., we do not seal our coverslips. For this reason, we collect Z-series from the top of the sample to the bottom, thereby forcing the coverslip against the embryo. If the focus motor is set in the opposite direction, it lifts the coverslip and the embryo with it. It is possible to glue the embryos, in an appropriate orientation, to the coverslip using a nontoxic glue such as Scotch tape dissolved in heptane, although this may reduce image quality.

b. Bleaching

GFP is an ideal marker for time-lapse studies because of its resistance to photobleaching. When examining intact embryos in their vitelline membranes, we can collect images in a single focal plane, every 15 s, over a 2-h period without a major loss of signal. When collecting a 4-D series, it has been possible to collect a Z-stack of optical sections every 5 min over 5 h. To limit bleaching and photodamage to the specimen we try to keep the number of scans for Kalman averaging and the number of sections per Z-series to a minimum. In living samples, cells often continue to express GFP during the period of the time course, thereby replenishing the fluorescent protein.

We do almost all of our imaging on whole-mount embryos within their vitelline membranes. In our experience, animals that have been dissected prior to viewing (for example as "flat preps") tend to bleach more readily.

E. Making Movies

The BioRad Laser Sharp confocal software allows confocal images to be converted to PICT or TIFF format. When converting large Z- or time-series, it is faster to use Confocal Assistant to generate TIFF files. These can then be exported to a Macintosh computer and assembled using Adobe Photoshop or Illustrator. To make movies that can be transferred to videotape, we assemble Z- or time-series and in Adobe Premiere.

Acknowledgments

Many thanks to Emma Dormand, Chris Phelps, and Jim Haseloff for help with this chapter.

References

Bellen, H. J., O'Kane, C. J., Wilson, C., Grossniklaus, U., Pearson, R. K., and Gehring, W. J. (1989). P-element-mediated enhancer detection: a versatile method to study development in Drosophila. *Genes & Development* **3,** 1288–1300.

Bier, E., Vaessin, H., Shepherd, S., Lee, K., McCall, K., Barbel, Ackerman, L., Carretto, R., Uemura, T., and Grell, E. (1989). Searching for pattern and mutation in the Drosophila genome with a P-lacZ vector. *Genes & Deveopment* **3,** 1273–1287.

Bossing, T., and Technau, G. M. (1994). The fate of the CNS midline progenitors in *Drosophila* is revealed by a new method for single cell labelling. *Development* **120,** 1895–1906.

Brand, A. H., and Perrimon, N. (1993). Targeted gene expression as a means of altering cell fates and generating dominant phenotypes. *Development* **118,** 401–415.

Brand, A. H., Manoukian, A. S., and Perrimon, N. (1994). Ectopic expression in *Drosophila. In* "*Drosophila melanogaster*: Practical Uses in Cell and Molecular Biology" (Goldstein, L. S. B., and Fyrberg, E. A., Eds.) pp. 635–654. Academic Press, San Diego.

Brand, A. (1995). GFP in *Drosophila. Trends Genet.* **11,** 324–325.

Brand, A. H., and Dormand, E. L. (1995). The GAL4 system as a tool for unravelling the mysteries of the nervous system. *Curr. Opin. Neurobiol.* **5,** 572–578.

Butner, K. A., and Kirschner, M. W. (1991). Tau protein binds to microtubules through a flexible array of distributed weak sites. *J. Cell Biol.* **115,** 717–730.

Callahan, C. A., and Thomas, J. B. (1994). Tau–β-galactosidase, an axon-targeted fusion protein. *Proc. Natl. Acad. Sci. U. S. A.* **91,** 5972–5976.

Chalfie, M., Tu, Y., Euskirchen, G., Ward, W. W., and Prasher, D. C. (1994). Green fluorescent protein as a marker for gene expression. *Science* **263,** 802–805.

Cormack, B. P., Valdivia, R. H., and Falkow, S. (1996). Facs optimized mutants of the green fluorescent protein (GFP). *Gene* **173,** 33–38.

Cubitt, A. B., Heim, R., Adams, S. R., Boyd, A. E., Gross, L. A., and Tsien, R. Y. (1995). Understanding, improving and using green fluorescent proteins. *Trends Biochem. Sci.* **20,** 448–455.

Davis, I., Girdham, C. H., and O'Farrell, P. H. (1995). A nuclear GFP that marks nuclei in living *Drosophila* embryos: maternal supply overcomes a delay in the appearance of zygotic fluorescence. *Dev. Biol.* **170,** 726–729.

Dormand, E. L., and Brand, A. H. (1998). Runt determines cell fates in the *Drosophila* embryonic CNS. *Development* **125,** 1659–1667.

Edwards, K. A., Demsky, M., Montague, R. A., Weymouth, N., and Kiehart, D. P. (1997). GFP-moesin illuminates actin cytoskeleton dynamics in living tissue and demonstrates cell shape changes during morphogenesis in *Drosophila. Dev. Biol.* **191,** 103–117.

Endow, S. A., and Komma, D. J. (1996). Centrosome and spindle function of the Drosophila Ncd microtubule motor visualised in live embryos using Ncd–GFP fusion genes. *J. Cell Sci.* **109,** 2429–2442.

Endow, S. A., and Komma, D. J. (1997). Spindle dynamics during meiosis in *Drosophila* oocytes. *J. Cell Biol.* **137,** 1321–1336.

Fasano, L., Core, N., and Kerridge, S. (1988). Expression of a reporter gene resembles that of its neighbour: an insertion in the *hairy* gene of *Drosophila. Roux's Arch. Dev. Biol.* **197,** 507–512.

Fischer, J. A., Giniger, E., Maniatis, T., and Ptashne, M. (1988). Gal4 activates transcription in *Drosophila. Nature* **332,** 853–856.

Gerdes, H., and Kaether, C. (1996). Green fluorescent protein: applications in cell biology. *FEBS Lett.* **389,** 44–47.

Haseloff, J., and Amos, B. (1995). GFP in plants. *Trends Genet.* **11,** 328–329.

Heim, R., Prasher, D. C., and Tsien, R. Y. (1994). Wavelength mutations and post-translational autoxidation of green fluorescent protein. *Proc. Natl. Acad. Sci. U. S. A.* **9,** 12501–12504.

Heim, R., Cubitt, A. B., and Tsien, R. Y. (1995). Improved green fluorescence. *Nature* **373,** 663–664.

Hidalgo, A., and Brand, A. H. (1997). Targeted neuronal ablation: the role of pioneer neurons in guidance and fasciculation in the CNS of *Drosophila. Development* **124,** 3253–3262.

Micklem, D. R., Dasgupta, R., Elliott, H., Gergely, F., Davidson, C., Brand, A., Gonzalez-Reyes, A., and St. Johnston, D. (1997). *mago nashi* is required for the polarisation of the oocyte and the formation of perpendicular axes in *Drosophila. Curr. Biol.* **7,** 468–478.

O'Kane, C. J., and Gehring, W. J. (1987). Detection *in situ* of genomic regulatory elements in Drosophila. *Proceedings of the National Academy of Sciences of the United States of America* **84,** 9123–9127.

Patel, N. H. (1994). Imaging neuronal subsets and other cell types in whole-mount *Drosophila* embryos and larvae using antibody probes. *In* "*Drosophila melanogaster*: Practical Uses in Cell and Molecular Biology" (Goldstein, L. S. B., and Fyrberg, E. A., Eds.) pp. 446–485. Academic Press, San Diego.

Prasher, D. C., Eckenrode, V. K., Ward, W. W., Prendergast, F. G., and Cormier, M. J. (1992). Primary structure of the *Aequorea victoria* green fluorescent protein. *Gene* **111,** 229–233.

Prasher, D. C. (1995). Using GFP to see the light. *Trends Genet.* **11,** 320–323.

Robertson, H. M., Preston, C. R., Phillis, R. W., Johnson-Schlitz, D., Benze, W. R., and Engels, W. R. (1988). A stable source of P-element transposase in *Drosophila melanogaster. Genetics* **118,** 461–470.

Schuldt, A. J., Adams, J. H. J., Davidson, C. M., Micklem, D. R., Haseloff, J., St. Johnston, D., and Brand, A. H. (1998). Miranda mediates asymmetric protein and RNA localization in the developing nervous system. *Genes Dev.* (in press).

Shiga, Y., Tanakamatakatsu, M., and Hayashi, S. (1996). A nuclear GFP beta-galactosidase fusion protein as a marker for morphogenesis in living *Drosophila. Dev. Growth Differ.* **38,** 99–106.

Siemering, K. R., Golbik, R., Sever, R., and Haseloff, J. (1996). Mutations that suppress the thermosensitivity of green fluorescent protein. *Curr. Biol.* **6,** 1653–1663.

Struhl, G., and Basler, K. (1993). Organizing activity of wingless protein in *Drosophila. Cell* **72,** 527–540.

Wang, S., and Hazelrigg, T. (1994). Implications for *Bcd* messenger RNA localization from spatial distribution of exu protein in *Drosophila* oogenesis. *Nature* **369,** 400–403.

Ward, W. W., Cody, C. W., Hart, R. C., and Cormier, M. J. (1980). Spectrophotomeric identity of the energy transfer chromophores in *Renilla* and *Aequorea* green fluorescent proteins. *Photochem. Photobiol.* **31,** 611–615.

Wilson, C., Pearson, R. K., Bellen, H. J., O'Kane, C. J., Grossniklaus, U., and Gehring, W. J. (1989). P-element-mediated enhancer detection: an efficient method for isolating and characterizing developmentally regulated genes in Drosophila. *Genes & Development* **3,** 1301–1313.

Yeh, E., Gustafson, K., and Boulianne, G. L. (1995). Green fluorescent protein as a vital marker and reporter of gene-expression in *Drosophila. Proc. Natl. Acad. Sci. U. S. A.* **92,** 7036–7040.

Zernicka-Goetz, M., Pines, J., Ryan, K., Siemering, K. R., Haseloff, J., Evans, M. J., and Gurdon, J. B. (1996). An indelible lineage marker for *Xenopus* using a mutated green fluorescent protein. *Development* **122,** 3719–3724.

Zernicka-Goetz, M., Pines, J., Siemering, K. R., Haseloff, J., and Evans, M. J. (1997). Following cell fate in the living mouse embryo. *Development* **124,** 1133–1137.

CHAPTER 12

Using Time-Lapse Confocal Microscopy for Analysis of Centromere Dynamics in Human Cells

Kevin F. Sullivan and Richard D. Shelby

Department of Cell Biology
The Scripps Research Institute MB21
La Jolla, California 92037

I. Introduction

The genome is a highly dynamic entity in growing cells, active in gene expression, DNA replication, and chromosome segregation as the cell cycle proceeds. Although cytological observation of chromosome behavior in living mitotic cells has a long and productive history, the direct visualization of individual chromosomes terminates in telophase because chromosome decondensation obliterates the visible boundaries between chromosomes. However, the nucleus is not simply

METHODS IN CELL BIOLOGY, VOL. 58
Copyright © 1999 by Academic Press. All rights of reproduction in any form reserved.
0091-679X/99 $25.00

a "soup" of chromatin. Rather, individual chromosomes maintain their identity as discrete domains during interphase, and evidence points to the idea that the genes adopt nonrandom spatial organization within chromosomal territories (Kurz *et al.,* 1996). Similarly, the distribution of centromere DNA sequences (Manuelidis, 1984) and chromosomes within the nucleus (LaSalle and Lalande, 1996) is nonrandom and may be related to nuclear function (Janevski *et al.,* 1995). Knowledge of how chromosomes adopt their organization within the nucleus and the dynamic events associated with different nuclear functions remains quite limited, however, due largely to technical problems associated with visualizing nuclear events. The introduction of GFP-fusion-protein technology has provided new approaches for visualization of chromosomes and chromosomal loci *in vivo.* In Chapter 13, Belmont *et al.* describe a method for labeling and visualizing discrete DNA sequences inserted into mammalian chromosomes. The present chapter describes the development and application of a GFP labeling strategy that allows visualization of the centromeres of living human chromosomes. Our initial work with this system has focused primarily on centromere dynamics in mitotic cells (Shelby *et al.,* 1996), but centromere labeling also provides a unique window into genomic organization throughout the cell cycle.

This chapter describes the strategy we have used to label centromere DNA sequences *in vivo,* confocal microscopy methods used to collect time-lapse images of labeled cells, and analytical methods used to extract quantitative morphometric data and motility analyses from time-lapse images. It also details some of the methods we use to generate images and movies for display.

II. GFP Fusion Proteins

CENP-B is a sequence-specific DNA binding protein that recognizes a 17-bp sequence element (CENP-B box) in centromeric satellite DNA of human and other mammalian chromosomes (Masumoto *et al.,* 1989; Kipling *et al.,* 1995). The DNA binding activity of CENP-B is specified by an N-terminal domain of about 150 amino acids (Yoda *et al.,* 1992). To take advantage of the repetitive nature of centromeric satellite DNA, we constructed a fluorescent derivative of CENP-B by fusing its DNA binding domain, residues 1–157, to the N terminus of GFP (Shelby *et al.,* 1996). As a control, a parallel construct was prepared using a CENP-B derivative deleted for residues 4–16 and lacking DNA-binding activity. Both fusion genes were cloned into an expression vector driven by the highly active CMV immediate early promoter (pCDNA-3, Invitrogen, Carlsbad, CA). When transfected into human cells, CENP-B–GFP was localized in discrete nuclear foci that were shown by immunofluorescence to correspond to centromeres (Fig. 1, see color plate). CENP-B–GFP(Δ4–16) was efficiently transported into the nucleus, but was distributed throughout the nucleoplasm and nucleoli. Preliminary experiments demonstrated that HeLa cells with CENP-B–GFP–labeled nuclei executed anaphase with normal kinetics. Subsequently, stably transformed cell lines have been isolated that show normal rates of proliferation and no difficulties executing mitosis. Thus, CENP-B–GFP targets to centromeres on the basis of its DNA-binding specificity and incorporation of the probe into centromeres has no detectable effect on mitotic chromosome segregation.

The control experiments described are important to consider when designing and using GFP fusion proteins. Independent verification that the fusion protein localizes correctly is particularly important when working with proteins whose endogenous localization is not known, but it also rules out promiscuous or nonspecific localization at ectopic sites. In some organisms it is possible to use genetic methods to determine whether a GFP derivative is fully functional (see Chapters 6 and 10). Without genetics, comparison of the behavior of transfected cells with wild-type cells is a crucial step in development of a useful GFP-based probe.

The fluorescent signal of CENP-B–GFP is extremely robust and tolerant of a variety of fixation conditions. We have used up to 4% formaldehyde or −20°C MeOH routinely for immunofluorescence applications. An illustration of the stability of GFP for a variety of biological applications is shown in Fig. 2 (see color plate), in which CENP-B–GFP has been colocalized with BrdU after metabolic labeling of DNA during S phase. In this experiment cells were fixed at pH 2.0 in 70% ethanol, followed by DNAse treatment (Boehringer-Mannheim, Indianapolis, IN). As compared with signal obtained under other fixation conditions, incubation at pH 2.0 had little or no effect on the fluorescent signal.

CENP-B–GFP does not localize specifically to centromeres in nonhuman cells, even those that have CENP-B box sequences in their centromeres, for reasons we do not fully understand. Expression of CENP-B–GFP in cultured mouse cells, even after reconstructing the protein using the mouse CENP-B DNA-binding domain, results in uniform nuclear fluorescence in live cells (data not shown). We suspect that this is related to the number of CENP-B boxes at each centromere and the efficiency of competition of CENP-B–GFP with endogenous protein. To date, use of other centromere proteins as labeling reagents has been complicated by phenotypic effects of constructs developed to date for CENP-E–GFP derivatives (Schaar *et al.,* 1997) and failure to observe mitotic cells following transfection with CENP-C–GFP (John Tomkiel, personal communication).

III. Microscopy

We have used confocal microscopy almost exclusively in the analysis of centromere dynamics in mitotic cells. This has been essential for observing the behavior of individual centromeres in human cells because as cells round up during mitosis and the metaphase plate develops, the alignment of centromeres along the optical (z-) axis results in substantial blur, obscuring individual centromeres. The optical sectioning capability of confocal microscopy yields isolated focal planes *ca.*

0.5 μm deep, allowing good resolution of individual centromeres. We collect a series of images at a single focal plane, creating a three-dimensional x–y–t time-lapse image allowing detailed analysis of centromere dynamics. An important limitation of this method is that only movement in the x–y plane is recorded, since motion greater than about 0.3 to 0.5 μm in the z-axis results in loss of objects out of the visible plane. Nevertheless, because most centromere movement in mitosis is perpendicular to the optical axis, it is possible to collect data for motility and morphometric analysis with good efficiency.

A. Cell Culture on the Microscope

For live observations we use a Dvorak–Stotler chamber (Lucas-Highland Co., Chantilly VA), a closed perfusable chamber system using 24.5-mm no. 1½ coverslips as the optical elements. A stage adaptor is necessary to mount the chamber; most microscope manufacturers make adjustable culture dish holders for inverted microscope stages. Temperature control is provided by an air-curtain incubator obtained from the same company. Cells are grown up on acid-washed coverslips in 35-mm tissue culture dishes or, more typically, 6-well plates. Conditions for transfecting cells are detailed in other chapters in this volume and also in Shelby *et al.* (1996). We have empirically observed that it is necessary to wait at least 24 h following transient transfection to see labeled cells in mitosis. We believe that this is because DNA does not enter the nucleus until mitosis, when the nuclear envelope is broken down. Analysis of transfection with soluble GFP vectors after very short-term exposure to DNA (3–5 h) reveals that almost all fluorescent cells occur in pairs, often still linked by a midbody (Sullivan, 1998); thus, cells expressing the transfected gene product are not observed until the second mitosis after transfection.

In preparation for time-lapse microscopy, the microscope stage and objective turret must be equilibrated to 37°C. The incubator is placed in a convenient spot next to the microscope with the fan pointed toward the stage, 6–12 in. from the microscope, and run for 60–90 min. Using this simple system, incubator settings are determined empirically because there is no feedback between the heater and the microscope stage. Adjust the heater output so that it feels warm on your hand at a distance of 6 in. from the fan head. After an hour, determine the temperature at the objective. A thermocouple thermometer with a small wire probe is ideal and can be placed directly on the body of the objective lens. Alternatively, a standard laboratory thermometer can be used and is placed on the stage with the bulb directly over the objective. Adjust the heater output to produce a 36–37°C reading at the objective. This may take about 30–60 min the first time. Insufficient preheating will result in drift of the focal plane as the microscope approaches thermal equilibrium during observations. Once set, the precise position of the incubator is crucial, and it should not be moved. If it is necessary to move it, the footprint should be carefully marked on the table to facilitate subsequent setup. There is a small degree of vibration produced by the fan, but we damp this by placing the incubator on a rubber mouse pad.

Prior to mounting cells in the observation chamber, an aliquot of medium is equilibrated for pH and temperature. Pipette 10–20 ml of complete medium into a tissue culture flask or dish and place this in the incubator for ~30 min or until warm. Supplement with 25 m*M* HEPES, pH 7.4. For the human cell lines we have used, this medium can be used for at least 4 h without changing. The chamber does have perfusion ports available, however. If perfusion is to be performed, the perfusion medium must be pre-equilibrated and maintained at 37°C during the experiment. We accomplish this by connecting a loaded syringe to the perfusion port by a length of short flexible tubing and resting the syringe on the microscope stage. Care must be taken not to move the chamber while perfusing, which we do by gently flushing ~1 ml of medium through the chamber over the course of about 1–2 min. Excessive flow rates may dislodge cells or crack the coverslips. Alternative stage heating and culture systems are available, and these guidelines should be adapted according to the equipment available. In addition, Chapter 13 provides a detailed discussion of long-term (>6-h) mammalian cell culture on the microscope.

A note on monitoring cell viability is in order. We have focused on analysis of mitotic cells and have found that the metaphase–anaphase transition is an exquisitely sensitive process that is easily arrested. During a collection session, we occasionally assay the overall health of the cells by scanning the coverslip to determine whether cells are actively executing anaphase. Because anaphase only lasts ~10–15 min in most cells, this provides a fairly timely report on the viability of cells on the stage. Our experience is that coverslips remain viable for up to 6 h under good conditions.

B. Optics and Image Collection

The confocal microscope systems we have used are the BioRad MRC series 600 and 1024 systems built on Zeiss Axiovert microscopes. The discussions in this chapter refer to specific features of the COMOS or LaserSharp software systems used for microscope control on the MRC instruments, but we have tried to make them general enough to apply to other microscope systems. For single-cell observations choose an objective lens with the highest numerical aperture available, for example, a 63× 1.4-NA Zeiss Apochromat or Neofluar objective lens. Using the BioRad MRC600 confocal scanning system, a zoom factor of about 2.5 provides a pixel size of about 0.1 μm for a full 768- × 512-pixel image. This is sufficient to record the full resolution of the objective lens. Zoom factors higher than this increase the dwell time of the laser without providing higher resolution. For observation of larger fields, a 40× 1.3-NA Neofluar lens is suitable. The S65T spectral mutant of GFP is ideally suited for excitation with the 488-nm line of the krypton–argon laser (Heim and Tsien, 1996). The standard fluorescein emission filter set supplied with the MRC600 and MRC1024 allows sensitive detection of the GFP signal.

It is important to adjust the confocal image collection parameters so that the image is optimally recorded for subsequent analysis. The parameters are laser power, slit width, gain, and black level. The first thing to determine is the laser power necessary to collect a good image. It is necessary to consider the trade-offs between spatial resolution, temporal resolution and duration, and cell photo-stability. The ideal is to irradiate the sample as little as possible for each frame. Accordingly, we usually work at maximal gain with signal enhancement to boost photomultiplier tube output. The slit width is kept below 50% full and ideally below 25% full using the vertically mounted sliders on the MRC-600. On the MRC-1024, we keep the slit width below 2.5.

To calibrate the image, use a calibrating LUT (look-up table) that displays pixel levels below 5 and above 250 as a color (e.g., red). The LUT maps the signal intensity from the photomultiplier to an output intensity for the computer display. The calibrating LUT allows easy determination of the high and low extremes of the image. Start with a laser power of 10% and scan an image. Adjust the gain so that only a very small number of pixels are at >250. If a large portion of the image is of saturating intensity, reduce the laser power; if the signal is weak, increase the laser power to 30%. We find that 30% laser power is the practical maximum due to unacceptable rates of photobleaching and phototoxicity at full laser power. Adjust the black level to reduce noise between cells to below 5, but do not impinge on signal-containing elements of the image. This is somewhat a subjective method, but it is effective for most applications involving morphometry and motility analysis. It is not suitable for quantitation of fluorescence signal intensity; for a more rigorous method for setting collection parameters see Chapter 3.

Once the image collection parameters are set, an image collection mode must be decided. Digital signal averaging using the Kalman filter improves image quality at a cost of increased scans and scan times. For most of our work, we have used the fast-scan mode and average 2–4 images using the Kalman algorithm. This provides an image of better quality than a single normal scan with about the same dose of radiation. We have been able to collect up to 200 images using this approach without significant photobleaching. If necessary, a transmitted light image can be collected simultaneously with each frame. In this case, we use simple reflectance-mode imaging rather than DIC, because the polarizing filters required for DIC imaging necessitate high-laser-power illumination.

To collect a time series, access the series collection dialog from the collection menu. On the MRC600, inactivate the "collect Z-series" box and enter the time series collection parameters: total number of images and the interval between images. The product of these values is the total time interval for the time-lapse collection. So, for example, a 15-min series collected at 10 frames/min would require 90 images. The time required to collect and store an image is ultimately the limiting factor in determining image collection frequency. On the MRC600, four fast scans of the full 768- × 512-pixel frame require about 2.5 s of scan time and another 3–4 s to save the image to disk, for a maximal collection rate of 8–10 frames/min. This time can be reduced by using a smaller image collection box. For single cells, an image of 256 × 256 pixels at 0.2 μm/pixel provides adequate resolution and reduces the collection time and size of the images. Using small fields, collection rates of 20–30 frames/min can be achieved. The key factors to consider are the resolution and field size required for the experiment and the speed of the events to be observed. During image collection the stage motor is engaged to assist holding the stage against drift. Occasionally, the focal plane drifts slowly during the experiment or the cell moves in the focal plane. The focus can be reset by hand, if necessary, but we try to avoid this. The best protection against focal plane drift is full thermal equilibration of the microscope.

We have observed that the metaphase–anaphase transition is quite sensitive to photodamage. Collecting images at a rate of 10/min using 10% laser power on the MRC-600, we do not observe cells executing anaphase after about 5–7 min of observation; if they do not enter anaphase within this time, they remain in a prometaphase-like state exhibiting oscillatory movement. Cells at metaphase that execute anaphase within 5–7 min of laser irradiation show normal kinetics of anaphase chromosome movement, cytokinesis, and telophase. Similarly, cells exposed to lower levels of irradiation by imaging at a rate of 2–3 images/min execute mitosis normally. Our interpretation of this is that a photosensitive process can influence checkpoint control, but that the basic processes of motility and chromosome decondensation are less susceptible to photodamage.

C. Data Storage and Transport

For the best system performance during image acquisition, files should be saved to the local hard disk. Although most confocal microscopes have some image-processing and analysis tools, most image processing and analysis is done offline on a laboratory PC, so the first step is to export image files to the laboratory. The best option for transporting files is across a local Ethernet connection. If this is not available, we have found that Zip disks (Iomega Corp., Roy, UT) provide a cheap, widely used removable disk system for distributing large files to different laboratories. Transferring and storing images can be facilitated by using file compression software, such as PKZip (Pkware, Brown Deer, WI) or a backup utility, to decrease the disk space required for file storage. Typically, complex images (gray scale) compress by a factor of 2 and simple images, containing just a few fluorescent objects on a black background, can compress by a factor of 10 or more. PKZip provides a lossless compression method that can be helpful in long-term storage of confocal image archives.

It is easy to underestimate the amount of disk storage required for time-lapse imaging projects at the outset. A file containing 60 full-frame images from the MRC600 will be almost 25 MB in size, uncompressed. Data storage systems are rapidly evolving, but it is worth putting some consideration into the means of long-term data storage before accumulating gigabytes of data. We use magneto-optical drives (DynaMO, Fujitsu Computer Products, Palo Alto, CA; Apex, Pinnacle Micro, Irvine, CA) for archiving image files. These systems provide optimal storage stability and respectable performance, albeit at high initial hardware cost. Magneto-optical storage media are relatively inexpensive at ~\$0.02–0.03/MB. Removable magnetic disk systems, such as the Jaz drive (Iomega) or the Sy-Jet, are less expensive at the hardware level, but require more expensive media (~\$0.10/MB) and lack the stability of magneto-optical media. Recordable CD is an attractive recent alternative, with intermediate purchase cost, a cheap long-lived medium, and a platform available on any PC. Finally, tape storage systems in the 2- to 4-GB capacity range have become quite affordable and probably represent the most economical method for data storage. For working with time-lapse data, it is recommended that at least 2 GB of hard disk space be available on the system used for analysis. This allows ample room to store and manipulate several files that may be associated with one project.

IV. Analysis

A. Observing Dynamics: Making and Playing Movies

After image collection, data are analyzed and displayed using a set of several software tools for the PC. BioRad Confocal Assistant 4.02 is a freeware program provided by BioRad that duplicates most of image processing and display functions of the COMOS and LaserSharp software on the microscope itself (available by anonymous FTP from ftp.genetics.bio-rad.com public\confocal\cas\setup402.exe). Adobe Photoshop and Adobe Premiere are used for assembling images and computer video, respectively (Adobe Systems Inc., Palo Alto, CA). We use Image Pro Plus (Media Cybernetics, Gaithersburg, MD) for quantitative image analysis as described later. NIH Image is a free image analysis package available from the NIH (http://rsb.info.nih.gov/NIH-Image/Default.html). A PC version of NIH image has recently been developed that incorporates many features of the Macintosh version.

Images from the MRC microscope systems are recorded in a proprietary file format known as the BioRad PIC format (the exact format specifications are available with the microscope documentation). This is a simple 8-bit digital image format in which a short header (76 bytes) at the beginning of the file specifies image parameters such as height (h) and width (w) in pixels. The image is stored as an array of pixel intensity values at one byte/pixel, with each set of w bytes corresponding to one row of the image, which is h rows in height. If it is a serial image, the first byte of the next image follows directly after the last byte of the previous image. The individual frames are recorded successively in the file, hence the term "stack" of images. Additional data, specifying the microscope settings, time, date and so on, are appended after the last image byte. This information can be useful for recovering experimental data from the image using an ASCII text editor or viewer.

The simplest way to view a time-lapse PIC image for laboratory use is with Confocal Assistant, which allows viewing the animated series in a continuous loop or as a rocking forward–reverse repetition. I find that it is useful to have a the Confocal Assistant software open in a window while analyzing a stack of images to assist in identifying specific objects in a complex field. Confocal Assistant also has a utility that will convert a PIC file into Microsoft Video format (AVI) for viewing with other programs. However, for more options in assembling digital video, we export the PIC file as a stack of .TIF images using the batch conversion utility of Confocal Assistant. This feature exports each frame as a .TIF image in a numbered series, which can then be used with any graphics software package.

Digital video files in the form of Quicktime movies have become the lingua franca for distributing time lapse and other dynamic microscopy data to different laboratories via the Internet. Adobe Premiere is a simple and powerful desktop video editor that can be used to create Quicktime movies using the stack of serially numbered .TIF images as input. NIH image also provides the ability to make Quicktime movies on Macintosh platforms. It is often useful to "preprocess" the .TIF images, which can be easily done using the batch Action script feature of Photoshop 4.0. To be practical, it is almost always necessary to use software compression to generate quicktime files of manageable size. Image quality and file size are directly proportional to each other, so it is important to plan a digital movie to show the essential features of a time-lapse sequence in the most economical way. One of the most important methods is to crop the image, showing only the essential elements. This can be done within Premiere, or by using the batch Action scripting feature of Adobe Photoshop 4.0. After assembling an image sequence, it is advisable to experiment with compression algorithms and compression ratio, to achieve maximize resolution while minimizing file size. Although digital image files are suitable for electronic distribution of time lapse data, it is usually necessary to output data to video tape for presentation. A simple method for making video tapes, if equipment is available, is to write the stack of .TIF images in sequential order to a laserdisc or OMDR device and then play this directly to a video tape recorder. In this case the original images can be used to maintain the highest resolution possible. For an excellent discussion of the production of digital movies, see Waterman-Storer *et al.* (1998).

B. Quantitative Analysis–Morphometrics and Motion Analysis

Quantitative data regarding dynamic processes can be extracted from confocal images using simple image-processing techniques. Any image processing necessary for analysis should be applied to the entire time-lapse sequence uniformly. This can be done prior to exporting the data using the tools available in Confocal Assistant. Alternatively, batch processing can be performed with Photoshop or the analytical software. I generally avoid image processing prior to quantitative analysis and limit it to light smoothing or sharpening algorithms that may aid object identification. We have used Image Pro Plus (Media Cybernetics, Silver Springs, MD) as a good, versatile image analysis platform. Two experiments will be described later to illustrate methods we have used in our projects.

C. Motion Analysis

Software methods for automated object tracking in time-lapse images based on cross-correlation analysis have been reported (Gelles *et al.,* 1988; Skibbens *et al.,* 1993). However, in our analysis of centromere dynamics we have found that automated object identification is often unsuitable for tracking centromeres within crowded fields. We have worked out semiautomated procedures using Image Pro Plus to collect position data from images followed by analysis in a spreadsheet (Excel, Microsoft, Redmond, WA). The first step in this analysis is creation of a stack of .TIF files from the original time-lapse data, as described earlier. I generally create individual folders for each stack of .TIF files associated with a .PIC file. Preview the series by animation to identify specific objects to be analyzed. For example, in a sequence showing centromeres executing anaphase, previewing is used to identify those centromeres that are clearly resolved throughout the time interval to be examined. I find it helpful to run this animation in a window as the analysis proceeds.

Centromeres are identified using the count/measure function of Image Pro Plus, which identifies discrete objects using a thresholding routine (Fig. 3A, see color plate). Prior to counting, it is necessary to specify what quantitative parameters are to be recorded from each object. For motion analysis, the position of each centromere is the critical measurement, determined as the x- and y-coordinates of the centroid of each particle. It is also possible to specify the scale of the image in pixels per micron, although this can easily be done subsequent to data collection. By running the count/measure routine, a threshold value for signal is automatically determined from the distribution of pixel intensity within the image. A perimeter is drawn automatically, essentially a single slice of a contour map, and each separate area enclosed within a common perimeter is considered to represent one object and is assigned an ID number. This works well for punctate signal sources such as centromeres. Generally, it is sufficient to let the software automatically set the threshold intensity, but it can be set manually if necessary. The automated object identification is not perfect, however. Once objects are "counted," it is necessary to identify and split compound objects that are erroneously grouped together. This can be done using a watershed splitting method, essentially a refined algorithm allowing the software to identify object boundaries, but there are generally a few objects that must be split by hand, using the mouse to draw the boundary between them. Once all the objects are identified, the measurement data can be transferred to Excel using a dynamic data exchange method or by manually copying and pasting data into the spreadsheet. A caveat of this method is that the object ID number is not generally the same from one frame to the next, requiring the user to record the ID for each centromere from frame to frame. An alternative method is to collect data from a few sets of centromeres at a time, using the manual tagging measurement feature, which allows one to use the mouse to define centromere position directly by clicking on the screen. Using this method, it is helpful to enlarge the image severalfold to aid in the estimation centroid of each particle. For both methods, it is straightforward to generate macros that automate data transfer into the spreadsheet.

Once the data from each centromere is collected and collated in the spreadsheet, it is a straightforward matter to plot and analyze the positional data. Figure 3B (see color plate) illustrates the raw positional data for a sister centromere pair during a 15-min period spanning metaphase and most of anaphase A. To aid interpretation of these type of data, we examine the relative motion of centromeres, either between sister centromere pairs or with respect to fixed points in the field. For anaphase motion, we identify a pair of fixed points approximating the position of the spindle poles as the vertex of a right triangle whose base is the band of centromeres at the end of anaphase. Position data is converted to distance from the "poles," d, using the Pythagorean theorem $d = [(x_{\mathrm{CEN}} - x_{\mathrm{POLE}})^2 + (y_{\mathrm{CEN}} - y_{\mathrm{POLE}})^2]^{1/2}$. Similarly, analysis of elastic distortions of the centromere involves determining the center-to-center distance between pairs of sister centromeres during chromosome oscillations (Fig. 3C, see color plate). For analysis of centromere movements in interphase nuclei, we have simply plotted the raw position of centromeres in successive frames, determining distance from their origin as a function of time. One important complication of analyses of absolute position data is that cells can move by translation and rotation, sometimes over significant distances, during the time period of interest. We are currently developing methods to characterize global movements and develop correction methods for centromere motility analysis. Alternatively, measuring centromere-to-centromere distances, as in the elastic analysis cited earlier, can reveal relative movement independent of global movements of the cell (see also Marshall *et al.,* 1997).

D. Morphometric Analysis

We have used morphometry to analyze changes in centromere shape during the cell cycle and mitosis using similar analytical methods. For these experiments using fixed cells, an optical section series was collected from each cell under analysis. After examination of the image stack, each centromere was measured in the optical section (image frame) that showed the largest profile using the automated object identification method described earlier. In this case, the measurement parameters selected were length of the long and short axes of each centromere, and after each frame was counted, the data were transferred to a spreadsheet. Rather than perform multiple measurements on each cell to ensure the accuracy of the measurements, data was collected from several cells at different stages of the cell cycle. Comparison of the mean and standard deviation of centromere long-axis length in different cells at the same stage of the cell cycle showed that this method is highly accurate (Shelby *et al.,* 1996). Although it is feasible to perform such morphometric analyses on three-dimensional reconstructions of optical section data (Dernburg *et al.,* 1996; Marshall *et al.,* 1997), software for such analysis is not directly available for most confocal microscopes. In addition, the method described does not rely directly on quantitation in the poorly resolved z-axis.

V. Summary

Using these methods we have shown that the alpha-satellite domain of the human centromere behaves as an elastic element, stretching in response to spindle forces applied during prometaphase and metaphase (Shelby *et al.,* 1996). These data complement previous observations of centromere stretching during mitosis (e.g., Skibbens *et al.,* 1993) by demonstrating a specific molecular compartment within the centromere, the satellite heterochromatin domain, that supports this strain. Centromere stretching reports on the net force applied across the centromere during mitosis and the availability of a fluorescence-based assay system in human cells provides a robust assay system to complement the elegant DIC-based methods that have been perfected using marsupial and newt cell cultures (Cassimeris *et al.,* 1990; Skibbens *et al.,* 1993, 1995; Rieder *et al.,* 1994). Current applications of this method are directed toward examining the relationship between centromere tension and microtubule dynamics using pharmacological approaches and the behavior of kinetochore-associated regulatory proteins, such as Mad2 and Bub1 (Li and Benezra, 1996; Chen *et al.,* 1996; Taylor and McKeon, 1997), as a function of centromere distortion.

In addition, GFP-labeled centromeres can be observed during interphase, providing a novel window into chromosome organization within the nucleus. Our observations show that centromeres distribute into the newly forming nucleus at telophase by what is apparently a uniform isometric expansion, with little evidence for directed motion of individual centromeres contributing to the formation of the G1 nucleus. During interphase, centromeres show very little movement in general, behaving as though embedded in a rigid matrix. Sustained movements of individual centromeres or groups of centromeres are occasionally observed, however, suggesting that chromosome position is subject to change during interphase (Shelby *et al.,* 1996). These experiments complement those described by Belmont and colleagues, who have developed a method to mark specific chromosomal sites with GFP for analysis *in vivo* (see Chapter 13; Robinett *et al.,* 1996; Straight *et al.,* 1996). These new GFP-based techniques for direct observation of defined DNA sequence domains *in vivo* carry the logic of *in situ* hybridization analysis into living cells and, coupled with new methods for observing global chromatin architecture as well as functional nuclear protein domains (Huang *et al.,* 1997; Misteli *et al.,* 1997), promise significant progress toward understanding the dynamic organization of the genome within the living nucleus.

Appendix: Handling Confocal Images on the Laboratory Computer

Photoshop Techniques. Image presentation for slides and publication is an essential feature of any microscopic technique. This appendix will discuss methods developed using Adobe Photoshop to assemble images for display or further analysis. (*Note:* these methods depend on the features of Photoshop 4.0, particularly in the paste feature which generates a new layer for each pasted element. In Photoshop 3.0, use the Paste Layer command instead of Paste.)

Importing Images into Photoshop. The best portable format for images is TIFF (tagged image file format), because it can be read by essentially any graphics program and encodes the image with a lossless algorithm. Many microscopes produce .TIF images as their primary output. The MRC series microscopes utilize a proprietary format with the extension .PIC. (This is not to be confused with the Macintosh .PIC format.) The Confocal Assistant utility provides routines to export .PIC files into .TIF format, and the analytical component of the MRC-1024 LaserSharp software also allows export to a variety of image formats. In addition, Bio-Rad provides an excellent set of Photoshop plug-ins that facilitate importing .PIC files into Photoshop. However, it is possible to import these images directly into Photoshop without prior export using the Photoshop Open as Raw method.

1. Select File . . . Open as, to bring up the open file dialogue.
2. Select RAW from the list of formats.
3. Select a .PIC file and open it.
4. A dialog box opens with three sections: dimensions, channels, and header.
5. Enter the size of the image frames in pixel width and height. Usually 768×512, 1024×1024, 512×512, etc.
6. In Channels count, enter the number of frames in the image. Photoshop can only open 24 channels at a time. If your image contains more than this, you can modify the header to skip to the desired place in the stack.
7. .PIC files are noninterleaved images, so make sure interleaved is not selected. Select 8-bit depth unless your file has been saved in words rather than bytes, in which case select 16 bits.
8. The header size for a Bio-Rad .PIC file is 76 bytes. If you are skipping ahead to images past the first 24, increase the header by the appropriate

size. For example, to load images 30–54 in a 768 × 512 stack, include the first 29 frames in the header: $76 + 29(h \times w) = 11{,}403{,}340$.

9. The program reports that the specified image is smaller than the file. This is because of the ancillary data appended at the end of the file after the image data. Open the file anyway.
10. When finished viewing or copying frames, close the file but *do not* save because this will overwrite the original data with a truncated dataset.

The images are loaded into separate channels and can be viewed using the channels window in Photoshop. Channels 1–3 correspond to the red, green, and blue channels respectively and are displayed by default. To view other channels simply select them, viewing them one at a time. An interesting feature that is occasionally useful for identifying or visualizing movement is revealed using this method. Spots that don't move are equivalent in the R, G, and B channels and so appear white. Spots that move, however, create a rainbow effect because the spots separate each other and overlap only partially or not at all.

Colorizing Fluorescence Images. Often, the different channels that make up a fluorescence image will be saved as individual 8-bit TIFF files. If so, it may be necessary to change them to color images for display or for assembling 2- or 3-color composite images. This is done easily by editing in the Photoshop channels window, as outlined next.

Single color images: Converting gray to color.

1. Open an 8-bit TIFF file.
2. Change the image mode to RGB.
3. Open the channels window.
4. Select a channel that will be absent from the color image (e.g., to display GFP as green, select either the red or blue channels).
5. Using the menu Edit, Fill . . . , fill the channel with black.
6. Select the other nondisplaying color in the channels window and fill it with black.
7. Select RGB in the channels window to display the color image.

Multi-color images: The method for mixing individual 8-bit grayscale images to produce a 24-bit RGB image is similar to the single-color method, but instead of the channels being filled with black, images are pasted into them. In the example here, three separate 8-bit images are placed into the three color channels of an RGB image. When mixing two colors, simply fill the unoccupied channel with black.

1. Open all three grayscale images in Photoshop.
2. Select the image corresponding to the red signal and change the image mode to RGB.
3. On the channels window, select the green channel of the RGB image.

4. Select the image corresponding to the green signal and copy it to the clipboard (Select-all, Edit-copy or Ctl-A, Ctl-C).
5. Select the RGB image and paste the copy. Because the green channel is selected, it pastes only into the green channel (Edit, paste or Ctl-V).
6. Repeat steps 4 and 5 for the image that is to appear blue in the RGB composite, using the blue channel.
7. Select RGB to display the 3-color image.

Grayscale. Often there is very little contrast in a phase or DIC image collected from the confocal. We also find that the field is unevenly illuminated. This method outlined here is a useful way of improving the quality of a continuous tone micrograph for display purposes. The first operation is equivalent to a histogram equalization or contrast stretch that remaps the tightly distributed pixel intensity values of a low-contrast image out to the 256 intensity levels available in an 8-bit image. The second step uses the Gaussian blur filter to correct uneven background illumination in the image. A background image is generated by running the Gaussian blur at a large radius, generating an image that contains the average gray levels of the image without any image detail. By subtracting this background image from the equalized image, variations in the illumination intensity over the field are corrected. The remaining manipulations, smoothing, sharpening, and a final equalization, improve the visual quality of the image. Note that the pixel values described correspond to a 1024 × 1024-pixel image. They should be adjusted to compensate for differences in the size of the image and features within the image.

1. Open the file containing the grayscale image in Photoshop.
2. Use Image, adjust, Autolevels to equalize the image.
3. Copy and paste to new layer = Layer 1 (Select-all, Edit-copy, Edit-paste or Ctl-A, Ctl-C, Ctl-V).
4. Filter, Gaussian blur, 15–35 pixel radius. Blur until no actual cellular details are visible.
5. From the Image menu, select calculations. For the parameters enter: Source 1 = background, black channel; Source 2 = Layer 1, black channel. Blend by subtraction with opacity at 100% (default), Scale of 1 (default) and an offset of usually 64–128 levels. The offset function essentially places the mean density of the resulting image to a level near 50% gray. Send the result to a new file, which will be appear as Untitled-1 (unless there are other untitled images open).
6. Select and copy the background corrected image in Untitled-1. Paste it into the main image to create a new Layer 2.

The subsequent steps are optional, depending on the quality of the resulting image.

7. Filter, Gaussian blur, 0.3–0.6 pixels. This smoothes the image with a result similar to the built in smoothing algorithms of the confocal software.
8. Filter, Unsharp mask, 100%, 2- to 4-pixel radius, 0–5-level threshold.
9. Image, adjust, Autolevels to equalize the final image.

Merging Fluorescence onto a Continuous Tone Image.

1. Open continuous tone image and change image mode to RGB.

2. Open the color fluorescence image. If the fluorescence image is still grayscale, convert it to color as described earlier.

3. Copy the fluorescence image and paste it into the continuous tone image.

4. Open the layers window. The fluorescence image should be Layer 1 and currently selected. If not, select it by clicking on the corresponding layer.

5. The mixing mode for Layer 1 onto the background image is changed to "color". This can be done using the drop-down box in the upper left of the layers window or through the layer menu under layer options.

6. Mixing the two images is a matter of reaching a workable compromise between the details of the two images. It often helps to have the continuous tone image darker than one would normally use in order to enhance the contrast and prevent washing out the color(s). This can be done using the brightness adjustment for the background layer. It may be necessary to reduce the intensity of the color layer in order to reveal details from the continuous tone image. This can be done using the transparency slider at the upper right of the layers window, with the color layer highlighted.

Using the Action Feature of Photoshop 4.0 to Automate Tasks. Often a manipulation or series of manipulations needs to be performed on each image in a time-lapse series. This can be done using tools provided in LaserSharp or Confocal Assistant software prior to exporting to a stack of .TIF images. However, it is occasionally necessary to manipulate the entire image stack for image processing, such as intensity adjustments or histogram equalization, image cropping, or other modifications. Two examples are given here to illustrate the uses of Photoshop actions for digital microscopy.

Colorizing a Stack. Although images representing a single fluorochrome are adequately represented in their native grayscale, it is attractive to display GFP probes as green images. The technique used is essentially that described above for single images, but the actions are recorded as for a macro and then applied to a stack.

1. Create two directories off the main directory holding the image files, named action and test. Copy (do not move) the set of files into the action directory and also copy a few files into the test directory. The original files in the main directory are the archive set and should not be changed.

2. Open a file from the test directory in Photoshop.

3. Activate the Actions palette from the Windows menu.

4. Create a New Action from the Actions palette menu or the "new" icon at the bottom of the palette. Give it a name. Click Record and note that recording mode is indicated by a red circle at the bottom of the Actions palette.

5. Perform the actions required for colorizing a grayscale image
   A. Image . . . mode . . . RGB.
   B. Open the channels palette and click Red to activate the red channel (alternatively Ctl-1).
   C. Edit . . . fill . . . black
   D. On the channels palette, click Blue to activate the blue channel (alternatively Ctl-3).
   E. Edit . . . fill . . . black
   F. On the channels palette, click RGB to view the color image and verify the procedure.
   G. If done, save the image with File . . . save.
   H. Close with File . . . close.
   I. Turn off Action recording by clicking the black box at the bottom of the Actions palette.

6. Open another file in the test directory and run the Action by selecting it and clicking the white arrow at the bottom of the Actions palette. Because the Action closes the file, reopen it to verify that the procedure took place correctly.

7. If necessary, correct the action by identifying the faulty procedure(s).

8. To execute the Action on the stack of files, select the Action and then select Batch from the palette menu (black arrowhead at top right of palette).

9. Specify the folder containing files to be modified. Batch operates on every image file in the folder, so be sure it contains only the files to be modified.

10. If not already done, specify the Action to be executed and the destination as Save and Close.

11. Click OK to start. Each file is successively opened, modified, saved, and closed. The action subdirectory now contains the modified image stack.

Cropping a stack.

1. Set up action and test directories as discussed earlier.
2. To make sure that the crop to be performed is appropriate to the entire stack, open the first, middle, and last images in the set.
3. Select and copy one image: Ctl-A, Ctl-C.
4. Select File . . . new to open a new file. Photoshop automatically defaults to the size of the image just placed on the clipboard. Tell it OK. This will be the template image for the crop selection.
5. Go back to an image file and use the marquee tool to define an area containing all the elements to appear in the cropped image.

6. Click within the defined area and hold down the Shift key while dragging to the second of the test images. Shift-drag places the selection outline in precisely the same place as it was on the original image.
7. Repeat for the third image. Is the defined area sufficient? Too big? Correct with this process until a suitable region is defined.
8. Shift-drag the selection outline into the template image.
9. From the Select menu, save selection.
10. Specify the destination (template) document and a new channel.
11. Go back to an image file and remove any selection outline on the image (Select . . . none)
12. From the Select menu, choose load selection. Make sure the template is the source document and the appropriate channel is specified (usually the default values)
13. Selection outline appears on your image file.
14. From the Image menu, select crop.
15. The system is ready to record the crop Action. Close all but one image file and the template. Make sure nothing is selected in the image file and that it is the active file.
16. Start a new action as discussed earlier.
    A. Select . . . load selection.
    B. Image . . . crop.
    C. File . . . save.
    D. File . . . close.
17. Stop recording action. It is always a good idea to test the Action on a few files in the test directory before unleashing it on the whole stack. When the Action is working properly, execute it on the images in the action directory as described earlier.

These two examples should illustrate the key features of Action use and recording; they are by no means a definitive demonstration of the Action techniques. There may be more efficient methods for performing these procedures, and it is important to experiment. This is essentially a simple scripting or macro language now available in Photoshop that will be an important tool for scientific image manipulations.

Acknowledgments

We thank Roger Tsien and Roger Heim for generously sharing GFP mutants and Martin Friedlander for access to the Bio-Rad MRC-1024 microscope. Work in the authors' laboratory is supported by a grant from the National Institutes of General Medical Sciences, GM39068.

References

Cassimeris, L., Rieder, C. L., Rupp, G., and Salmon, E. D. (1990). Stability of microtubule attachment to metaphase kinetochores in PtK1 cells. *J. Cell Sci.* **96,** 9–15.

Chen, R. H., Waters, J. C., Salmon, E. D., and Murray, A. W. (1996). Association of spindle assembly checkpoint component XMAD2 with unattached kinetochores. *Science* **274,** 242–246.

Dernburg, A. F., Broman, K. W., Fung, J. C., Marshall, W. F., Philips, J., Agard, D. A., and Sedat, J. W. (1996). Perturbation of nuclear architecture by long-distance chromosome interactions. *Cell* **85,** 745–759.

Gelles, J., Schnapp, B. J., and Sheetz, M. P. (1988). Tracking kinesin-driven movements with nanometre-scale precision. *Nature* **331,** 450–453.

Heim, R., and Tsien, R. Y. (1996). Engineering green fluorescent protein for improved brightness, longer wavelengths and fluorescence resonance energy transfer. *Curr. Biol.* **6,** 178–182.

Huang, S., Deerinck, T. J., Ellisman, M. H., and Spector, D. L. (1997). The dynamic organization of the perinucleolar compartment in the cell nucleus. *J. Cell Biol.* **137,** 965–974.

Janevski, J., Park, P. C., and De Boni, U. (1995). Organization of centromeric domains in hepatocyte nuclei: rearrangement associated with *de novo* activation of the vitellogenin gene family in *Xenopus laevis. Exp. Cell Res.* **217,** 227–239.

Kipling, D., Mitchell, A. R., Masumoto, H., Wilson, H. E., Nicol, L., and Cooke, H. J. (1995). CENP-B binds a novel centromeric sequence in the Asian mouse *Mus caroli. Mol. Cell Biol.* **15,** 4009–4020.

Kurz, A., Lampel, S., Nickolenko, J. E., Bradl, J., Benner, A., Zirbel, R. M., Cremer, T., and Lichter, P. (1996). Active and inactive genes localize preferentially in the periphery of chromosome territories. *J. Cell Biol.* **135,** 1195–1205.

LaSalle, J. M., and Lalande, M. (1996). Homologous association of oppositely imprinted chromosomal domains. *Science* **272,** 725–728.

Li, Y., and Benezra, R. (1996). Identification of a human mitotic checkpoint: hsMAD2. *Science* **274,** 246–248.

Manuelidis, L. (1984). Different central nervous system cell types display distinct and nonrandom arrangements of satellite DNA sequences. *Proc. Natl. Acad. Sci. U. S. A.* **81,** 3123–3127.

Marshall, W. F., Straight, A., Marko, J. F., Swedlow, J., Dernburg, A., Belmont, A., Murray, A. W., Agard, D. A., and Sedat, J. W. (1997). Interphase chromosomes undergo constrained diffusional motion in living cells. *Curr. Biol.* **7,** 930–939.

Masumoto, H., Masukata, H., Muro, Y., Nozaki, N., and Okazaki, T. (1989). A human centromere antigen (CENP-B) interacts with a short specific sequence in alphoid DNA, a human centromeric satellite. *J. Cell Biol.* **109,** 1963–1973.

Misteli, T., Caceres, J. F., and Spector, D. L. (1997). The dynamics of a pre-mRNA splicing factor in living cells. *Nature* **387,** 523–527.

Rieder, C. L., Schultz, A., Cole, R., and Sluder, G. (1994). Anaphase onset in vertebrate somatic cells is controlled by a checkpoint that monitors sister kinetochore attachment to the spindle. *J. Cell Biol.* **127,** 1301–1310.

Robinett, C. C., Straight, A., Li, G., Willhelm, C., Sudlow, G., Murray, A., and Belmont, A. S. (1996). *In vivo* localization of DNA sequences and visualization of large-scale chromatin organization using lac operator/repressor recognition. *J. Cell Biol.* **135,** 1685–1700.

Schaar, B. T., Chan, G. K., Maddox, P., Salmon, E. D., and Yen, T. J. (1997). CENP-E function at kinetochores is essential for chromosome alignment. *J. Cell Biol.* **139,** 1373–1382.

Shelby, R. D., Hahn, K. M., and Sullivan, K. F. (1996). Dynamic elastic behavior of alpha-satellite DNA domains visualized *in situ* in living human cells. *J. Cell Biol.* **135,** 545–557.

Skibbens, R. V., Skeen, V. P., and Salmon, E. D. (1993). Directional instability of kinetochore motility during chromosome congression and segregation in mitotic newt lung cells: a push-pull mechanism. *J. Cell Biol.* **122,** 859–875.

Skibbens, R. V., Rieder, C. L., and Salmon, E. D. (1995). Kinetochore motility after severing between sister centromeres using laser microsurgery: evidence that kinetochore directional instability and position is regulated by tension. *J. Cell Sci.* **108,** 2537–2548.

Straight, A. F., Belmont, A. S., Robinett, C. C., and Murray, A. W. (1996). GFP tagging of budding yeast chromosomes reveals that protein-protein interactions can mediate sister chromatid cohesion. *Curr. Biol.* **6,** 1599–1608.

Sullivan, K. F. (1998). Enlightening mitosis: construction and expression of GFP fusion proteins. *In* "Mitosis" (Rieder, C. L., Ed.). Academic Press, San Diego.

Taylor, S. S., and McKeon, F. (1997). Kinetochore localization of murine Bub1 is required for normal mitotic timing and checkpoint response to spindle damage. *Cell* **89,** 727–735.

Waterman-Storer, C. M., Shaw, S. L., and Salmon, E. D. (1998). Production and presentation of digital movies. *Trends Cell Biol.* **7,** 503–506.

Yoda, K., Kitagawa, K., Masumoto, H., Muro, Y., and Okazaki, T. (1992). A human centromere protein, CENP-B, has a DNA binding domain containing four potential alpha helices at the NH_2 terminus, which is separable from dimerizing activity. *J. Cell Biol.* **119,** 1413–1427.

CHAPTER 13

Visualization of Large-Scale Chromatin Structure and Dynamics Using the *lac* Operator/*lac* Repressor Reporter System

Andrew S. Belmont, Gang Li, Gail Sudlow, and Carmen Robinett

Department of Cell and Structural Biology
University of Illinois, Urbana-Champaign
Urbana, Illinois 61801

I. Introduction

In this chapter we review our initial efforts to develop an approach that allows *in situ* localization and direct *in vivo* visualization of specific chromosome regions through protein/DNA recognition. As our model system we have used a 256-copy direct repeat of the *lac* operator combined with a *lac* repressor–NLS (nuclear localization signal) or GFP–*lac* repressor–NLS fusion protein for detection. This methodology is already proving useful in visualizing chromosome segregation in

METHODS IN CELL BIOLOGY, VOL. 58
Copyright © 1999 by Academic Press. All rights of reproduction in any form reserved.
0091-679X/99 $25.00

organisms that lack obvious, cytologically visible chromosomes (Minshull *et al.*, 1996; Straight *et al.*, 1996; Straight *et al.*, 1997; Webb *et al.*, 1997). Here, however, we focus on our early efforts to exploit this system to facilitate direct *in vivo* visualization of large-scale chromatin dynamics within mammalian interphase nuclei (Robinett *et al.*, 1996). By combining transfection of the 256-copy direct repeat of the *lac* operator with gene amplification to obtain large, amplified chromosome regions, or HSRs, tens of Mbp in size, containing the *lac* operator repeats, we have been able to visualize large-scale chromatin fibers *in vivo*.

II. Overview of Methodology

Investigation of nuclear architecture and interphase chromosome structure has been severely hindered by a number of serious technical problems reviewed in more detail elsewhere (Belmont, 1997a,b). Briefly, these difficulties include problems related to preservation of *in vivo* structure, selective staining of specific proteins or DNA sequences within densely packed chromosomal and nuclear structures, and high-resolution imaging and three-dimensional visualization of complex nuclear structures that extend over large nuclear regions. Unfortunately, methods that address any one of these problems often exacerbate the others; this is especially true for the inherent conflict between optimizing selective staining of nuclear components and preserving ultrastructure.

In situ hybridization methods have been pivotal in allowing exploration of chromosome and gene arrangements within interphase nuclei. However, the requirement of DNA denaturation for *in situ* hybridization is incompatible with direct *in vivo* observations. Moreover, the harsh conditions required for DNA denaturation are antagonistic to ultrastructural preservation. This has led to a large gap between primarily light microscopy studies, using *in situ* hybridization to examine the locations of specific chromosomal regions or loci, and ultrastructural analysis of higher order, "generic" chromatin structure, primarily by electron microscopy.

In an attempt to bridge these two structural approaches, we decided to explore the use of protein-DNA recognition as a tag for localizing specific chromosome regions to provide an alternative to *in situ* hybridization. We chose the *lac* operator/repressor system for our initial test case for several reasons. The *lac* repressor had been demonstrated as having the capability of binding *in vitro* to an operator sequence packaged within nucleosomes (Chao *et al.*, 1980) with only a small decrease in binding affinity. Later *in vivo* experiments confirmed the specific binding of *lac* repressor to *lac* operator within eukaryotic cells (Brown *et al.*, 1987; Hu and Davidson, 1987). Important for the ultimate signal sensitivity was the high ratio between *lac* repressor binding to specific versus nonspecific sequences, estimated as $\sim 10^9$ and near the maximum for DNA-binding proteins (Rawn, 1989). Finally, the *lac* repressor has been extensively characterized over the last two decades, so a number of mutants in both the *lac* repressor and *lac* operator sequence are available that could be used for later optimization of the system.

Our initial application of this approach (Robinett *et al.,* 1996) to the study of interphase chromosome dynamics within mammalian cells has two components as outlined in Fig. 1. First is the construction of a large *lac* operator repeat, the introduction of this repeat into a mammalian expression vector, and the selection of stable transformants of this vector in CHO cells. Detection of the *lac* operator repeat in these cells can be accomplished by three methods: (*a*) Staining of fixed cells with purified *lac* repressor followed by immunostaining, (*b*) *in vivo* expression of a *lac* repressor–NLS fusion protein followed by immunostaining, and (*c*) *in vivo* expression of a GFP–*lac* repressor–LS fusion protein and direct observation.

Previous work using *in situ* hybridization has demonstrated a high DNA compaction ratio within interphase nuclei ranging from hundreds to thousands to one (Lawrence *et al.,* 1990). This means that both single-copy vector insertions and multiple-copy vector insertions up to several hundred kilobases in size will appear in the light microscope as a shapeless, diffraction-limited spot. To be able to visualize large-scale chromatin dynamics within interphase nuclei, we followed the generation of stable transformants containing the *lac* operator repeats with a second stage of selection using gene amplification to generate amplified chromosome regions thousands of kilobases in size containing the *lac* operator repeats. This has enabled us to visualize chromosome segments the size of bands or even small chromosome arms. Gene amplification was accomplished using an expression vector containing DHFR (dihydrofolate reductase) and transfecting this vector into CHO DG44 cells that contain a double deletion for the endogenous DHFR locus (Urlaub *et al.,* 1986). Stepwise selection with methotrexate, an inhibitor of DHFR, then allows selection for cells that have undergone gene amplification (Delidakis *et al.,* 1989).

Amplification of the endogenous DHFR locus typically results in tens to hundreds of copies of a several-hundred- to thousand-kilobase region surrounding the selectable marker. This results in new chromosome bands or even larger segments, which typically at later stages of amplification stain homogeneously using mitotic chromosome-banding methods; therefore, these regions have been called HSRs, or homogeneously staining regions. As described elsewhere (Robinett *et al.,* 1996), we have isolated cell clones containing HSRs with very different patterns of interphase chromosome condensation. Detailed analysis of the A03 clone containing a late-replicating, heterochromatic HSR has shown that this HSR undergoes a complex but reproducible choreography of condensation and decondensation during the interphase cell cycle (Li *et al.,* 1998). The entire HSR behaves relatively uniformly with respect to timing of replication and degree of chromosome condensation, facilitating our analysis of large-scale chromatin structure. Next we provide more detailed information describing the experimental methods used in this overall approach.

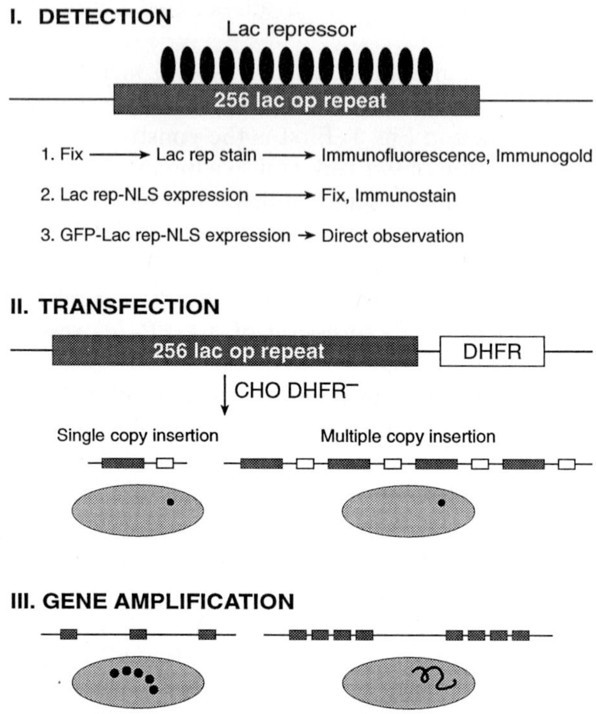

Fig. 1 Methodology for *in vivo* visualization of large-scale chromatin organization. (I) Detection of specific chromosomal sites using *lac* operator/*lac* repressor recognition. A 256-copy direct repeat of the *lac* operator sequence was constructed. *Lac* repressor binding to this repeat can be detected in three ways: (*a*) staining of fixed cells with purified *lac* repressor followed by immunostaining; (*b*) expression *in vivo* of a *lac* repressor–NLS fusion protein followed by fixation and immunostaining; (*c*) expression *in vivo* of a green fluorescent protein–*lac* repressor–NLS fusion protein and direct observation. Immunostaining can be done using immunofluorescence for light microscopy or immunogold staining for electron microscopy. (II) Transfection of *lac* operator repeats into CHO cells. The *lac* operator repeat was linked to an expression vector for dihydrofolate reductase (DHFR) and used for transfection into CHO cells with a double DHFR deletion. Stable transformants containing either multiple-vector copies or single-vector copies were selected. (III) Gene amplification was used to create large chromosomal regions that could be visualized by *lac* repressor binding. Methotrexate selection was used to select for cells undergoing amplification. Typical amplification units are rough repeats of DNA segments hundreds to thousands of kilobases in size, containing the DHFR selectable marker. Amplified chromosome regions after *lac* repressor staining appear as an array of spots or contiguous fibers depending on the relative sizes of the vector repeats and flanking coamplified genomic DNA in the amplification units.

III. Construction of the *lac* Operator Repeat

Previous work had shown that inverted repeats of the *lac* operator were unstable when propagated in *Escherichia coli* (Sasmor and Betz, 1990). Cloning high-copy repeats of the *lac* operator therefore required forming direct repeats. One way to force direct repeats using simple ligation is by using a nonpalindromic restriction site that yields overhanging ends after restriction. Ava I has been used in this way to construct multiple repeats of a DNA segment (Hartley and Gregori, 1981). In our hands a slower but more consistent method used a directional cloning scheme that doubled the number of direct repeats with each cloning cycle. This requires the use of three restriction enzymes, of which two have compatible sticky ends as outlined in Fig. 2. A fringe benefit of this approach is that one obtains a set of repeat sizes in powers of 2.

A key concern at the beginning of this project was whether a suitably large direct repeat would be sufficiently stable in *E. coli* to allow cloning. The wild-type *lac* operator deviates in only 7 out of 35 positions from a perfect palindromic sequence. Although a symmetric, perfect palindromic sequence binds *lac* repressor better than the wild-type *lac* operator, we deliberately used the wild-type sequence in our initial work due to these concerns. A second question relates to the optimal spacing between *lac* operator repeats. Earlier experiments using linearized DNA had suggested that no more than one *lac* repressor tetramer could bind per 4 to 8 direct *lac* operator repeats (Sasmor and Betz, 1990), implying that spacing the *lac* operator repeats closer than a critical distance might lead to an increased instability without any increase in total repressor binding at saturation. Also unknown was what effect having large numbers of contiguous *lac* repressor proteins would have on DNA function, including DNA replication and chromatin packing. For this reason we initially made two related constructs, designed for cloning either a direct repeat of the *lac* operator alone or a direct repeat of a single *lac* operator ligated to a spacer DNA sequence.

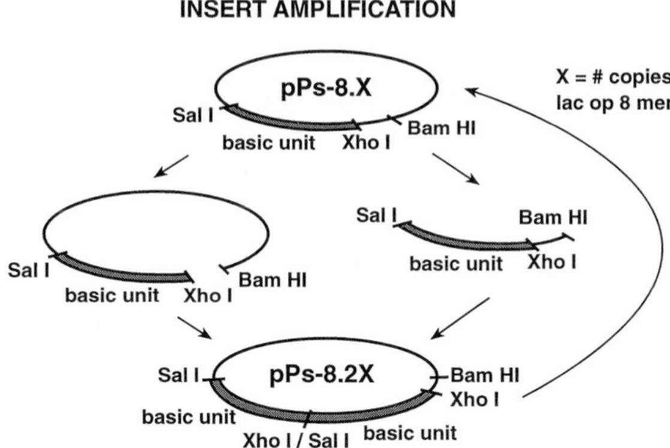

Fig. 2 Directional cloning scheme for building *lac* operator direct repeats. The use of two enzymes with compatible sticky ends allowed the doubling of direct repeat number of a basic unit between the Sal I and Xho I sites with each cloning cycle. We began with a basic repeating unit already consisting of a *lac* operator 8mer (see text); 5 cycles led to a 256-copy direct repeat.

For cloning the direct repeat of the *lac* operator, we modified the pUC 18 plasmid by making small changes in the polylinker and deleting EcoR I and Nar I restriction sites in the plasmid (Robinett *et al.,* 1996). The polylinker then was modified to HindIII-SphI-PstI-SalI-XbaI-Nar1-EcoR1-XhoI-BamH1-XmaI-KpnI-SacI, creating the vector pBS. An 8mer direct repeat of the *lac* operator (Sasmor and Betz, 1990) containing EcoR I ends was cloned into the EcoR I site of this polylinker. Five cycles of the directional cloning strategy outlined in Fig. 2 produced the pBS-8.32 vector containing 256 copies of the *lac* operator (10.1 kb). The pSV2-DHFR plasmid was modified by inserting an adapter, containing KpnI-SalI-SphI restriction sites, at the EcoR I site; the adapter was designed with overhanging ends which eliminated the EcoR I site. The SalI-KpnI fragment from pBS-8.32 containing the 256-copy direct *lac* operator repeat was then inserted at this adapter, creating the pSV2-DHFR-8.32 vector with the insert SphI-SalI-[XbaI-NarI-EcoRI-8mer-EcoRI-(SalI/XhoI)]$_{31x}$-XbaI-NarI-EcoRI -8mer-EcoRI-XhoI-BamHI-XmaI-KpnI.

A similar scheme was used to clone 32 direct repeats of a single *lac* operator and spacer combination. In this case a *lac* operator oligonucleotide with Nar I and Sal I ends was ligated into pPS to create the pPS OP vector. Random pieces of Taq I-digested human DNA were inserted at the Nar I site, destroying the Nar I site and adding a spacer DNA fragment. Again, 5 cycles of directional cloning as outlined in Fig. 2 created 32 copies of the single *lac* operator and spacer DNA. Two random, roughly 200-bp spacers, were carried through the operator amplification. Direct repeats with one of these spacers were very unstable and it was not able to obtain 32 direct repeats. However, with the second spacer 32 direct repeats were obtained, creating the pBSOP-6.32 vector. These direct repeats were excised with a SalI-KpnI digest and inserted into the modified pSV2–DHFR vector, as described earlier, creating pSV2–DHFR-6.32.

IV. Manipulation of the *lac* Operator Repeats

Although it was possible to clone up to 256 direct repeats of the *lac* operator sequence, the stability of the repeat was a serious problem at the higher repeat numbers (8.16, 8.32). This stability was also very plasmid dependent; for instance, the pSV2–DHFR-8.32 plasmid showed greater stability of repeat size than the pBS-8.32. Other examples were encountered as well that appeared to correlate at least to some extent with plasmid copy number. This instability was addressed through the replacement of single large-scale preps with many minipreps. Each miniprep was screened using a restriction digest to select those with full-size *lac* operator repeats, and the DNA from these full-size minipreps was pooled. One strategy that increased DNA yield was to inoculate 10–100 ml of broth using 1–2 ml of overnight culture from a miniprep that screening showed was positive for a full-length repeat.

The instability of the operator repeat was increased after transformation, with electroporation greatly increasing the instability. In fact, in our hands we were unable to clone the *lac* operator repeat into *E. coli* using electroporation. With pSV2–DHFR-8.32 if we used bacteria from a miniprep that gave a full-length insert to restreak a fresh plate, more than half of the colonies would give rise to minipreps showing full-length repeats. After retransformation using DNA from the original miniprep, this number would drop to as low as 1–2/10 using DH5alpha-competent cells. Therefore for routine preparation of additional DNA, bacteria should be restreaked from frozen stocks preferentially over using DNA for a new transformation.

An example of a gel showing the screening of a number of minipreps is shown in Fig. 3. These numbers are better than our results using the original pBS-8.32 vector, which may be related to the lower copy, pBR322 origin used with pSV2–DHFR.

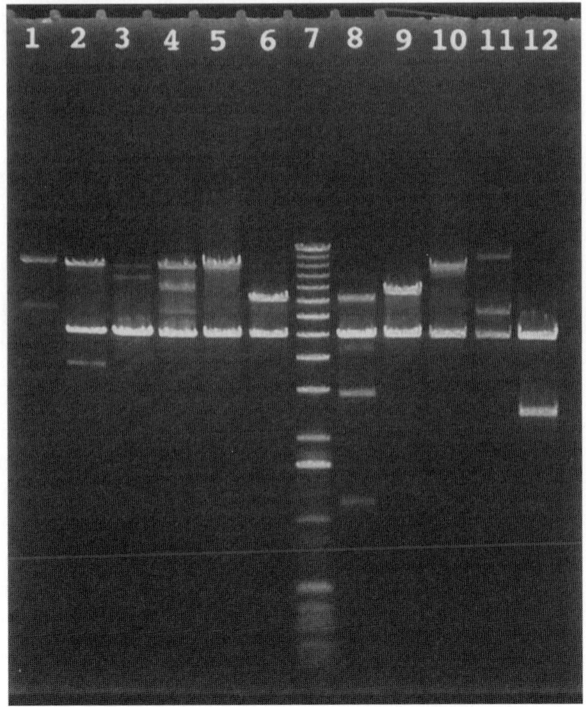

Fig. 3 Screening minipreps for full-size *lac* operator repeat inserts. pSV2–DHFR-8.32 DNA containing the 256-copy operator repeats was used to transform DH5alpha cells. Individual colonies were used to inoculate overnight cultures. DNA isolated from minipreps was cut with Sal I and Kpn I, which should produce a 10.1-kb operator repeat fragment and a 5-kb vector backbone. Two of 11 colonies showed the expected bands, corresponding to lanes 5 and 10. The others show varying degrees of recombination. Lane 7 is a 1-kb DNA standard ladder (from Gibco BRL).

Our best results have been achieved using STBL2-competent cells (Life Technologies, Gibco BRL), which were selected for their ability to propagate plasmids containing direct repeats. To achieve the advertised stability, it is important to grow the cells at 30°C as described in the directions for their use. Transformations are done with 17- × 100-mm polypropylene tubes (Falcon 2059) to optimize the heat shock step, which is 25 s at 42°C. The cells go on ice for 2 min, and SOC medium is added. After a 90-min incubation at 30°C the cells are plated out and grown overnight, again at 30°C. The colonies will be small in size. STBL2-competent cells that have been transformed with plasmids should be grown at 30°C overnight in TB. These overnight cell cultures can be frozen with glycerol at −80°C. Minipreps are done from 3-ml overnights. We usually start with at least ten overnight cultures. The QIAprep Spin Miniprep kit (Qiagen) works well, and the elution is done in 100 μl of 10 m*M* Tris–HCl, pH 8.5, to get a greater return rather than a higher concentration. pSV2–DHFR-8.32 is a low-copy-number plasmid, so we digest about 8 μl with Sal I and Bam HI. This digestion clips out the *lac* operator repeat, leaving the backbone intact. In the absence of recombination this digestion should give rise to two bands at ~10 and 5 kb.

V. Repressor–NLS and GFP–Repressor–NLS Constructs

Initial detection of the operator array in CHO cells was through either staining of fixed samples with purified *lac* repressor followed by immunostaining or *in vivo* expression of a *lac* repressor–NLS fusion protein followed by immunostaining. Previous measurements of the binding constants and IPTG inducibility of several different *lac* repressor–NLS fusion proteins revealed that there was a significant variation dependent on the location of the NLS fusion and the linker used (Fieck *et al.,* 1992). For *in vivo* expression we used the p3′SS plasmid available from Stratagene, which codes for a *lac* repressor fused at its COOH terminus via a short amino acid linker to the SV40 NLS. In the design of this vector a number of different linker and fusion sites were tested for their effect on *lac* repressor binding and IPTG inducibility (Fieck *et al.,* 1992). The fusion chosen for the p3′SS vector was demonstrated to preserve normal *lac* repressor binding and IPTG inducibility.

Using this construct we obtained stable transformants expressing the *lac* repressor–NLS fusion protein in CHO cells. Initial experiments revealed that stable expression of the *lac* repressor–NLS fusion protein led to greatly increased instability of amplified chromosome regions containing the *lac* operator repeats. Specifically, there appeared to be a significant increase in observations of chromosome bridging and nondisjunction during anaphase. Also transformation of a cell line, A03, which without *lac* repressor expression had a single HSR of stable size and chromosome number, resulted in stable transformants that were highly unstable in the number of HSRs per cell and their size. This effect was rationalized as being a result of the bivalent binding of the tetrameric wild-type *lac* repressor protein. A similar phenomenon was seen in yeast, where expression of the tetrameric form of repressor prevented disjunction of daughter chromatids containing the *lac* operator repeat inserted near the centromere when microtubules were depolymerized (Straight *et al.,* 1996) in cells with a mitotic checkpoint mutant.

We addressed this problem by making a 5–amino acid deletion at the COOH end of *lac* repressor that had been shown to eliminate tetramer formation without affecting dimerization (Chen and Matthews, 1992). Expression of the dimer form of the *lac* repressor greatly reduced effects on HSR stability, enabling the isolation of A03 clones expressing the dimer repressor that were relatively stable in HSR size, number, and location (Robinett *et al.,* 1996). However, the percentage of cells with a single HSR was reduced from ~98% to ~90%, indicating that there still might be a long-term, residual effect on chromosome stability with expression of the repressor. Systems allowing inducible expression of the *lac* repressor are now being developed. In yeast, use of a similar 11–amino acid COOH deletion to yield a dimer form of the *lac* repressor was observed to solve the problem of chromosome nondisjunction (Straight *et al.,* 1996).

To construct a *lac* repressor fusion with GFP, a COOH terminus fusion with the *lac* repressor–NLS was attempted first, which added the GFP after the NLS. Initial screening suggested that this was negative; in retrospect, the lack of signal may very well have been due to poor expression of wild-type GFP expressed at 37°C (Siemering *et al.,* 1996). However, a fusion of GFP with the NH_2 terminus of the repressor was successful in yeast (Straight *et al.,* 1996) at lower temperatures, and this same fusion was cut and pasted into our mammalian expression vector, p3′SS, which preserved the original repressor–NLS fusion (Robinett *et al.,* 1996). This construct produced cells with sufficient GFP–repressor expression to visualize chromosome HSRs containing the *lac* operator repeat, but only in a very small percentage of cells; typically in a transient transfection, only several cells on an entire coverslip showed good expression levels, as compared to several percent of cells that expressed the *lac* repressor–NLS construct. Antibody staining and Western blotting revealed that this was due to low protein expression levels rather than problems with forming the GFP fluorophore.

An "improved" GFP (iGFP) consisting of the S65T mutation (Helm *et al.,* 1995) was introduced into the fusion construct by site-directed mutagenesis. This increased the percentage of positive cells, particularly during transient expression. However, the percentage of strongly expressing stable transformants was still very low and the fluorescence brightness was very dependent on the growth conditions of the cell culture. A cell clone that would show bright fluorescence on one day might show barely detectable fluorescence one or two days later. This variation appeared to depend on the number of days after passage and/or change of media. In yeast it was discovered that a similar GFP–repressor–NLS fusion also showed poor fluorescence when the culture conditions were raised to 37°C (A. Straight and A. Murray, personal communication). Growing stable transformants of CHO cells at 32°C produced a dramatic improvement of :GFP fluorescence. Mixed populations of stable transformants might show 20–50% of cells with HSRs showing obvious fluorescence as compared to a few percent or less at 37°C. This behavior pointed to a known problem regarding the poor expression of GFP at higher temperatures (Siemering *et al.,* 1996).

More recently we have introduced the human-codon-optimized version of GFP from Clontech (EGFP) into our fusion protein constructs. This version includes two point mutations, S65T and F64L. Transient expression of this construct works well at 37°C, and fluorescence intensity of stable transformants does not show any obvious temperature dependence. Also, stable clones can be isolated that show uniform fluorescence characteristics independent of culture conditions. However, in comparing stable transformants of the same A03 cell line with the *lac* repressor–NLS versus the EGFP–*lac* repressor–NLS constructs we still notice a small percentage (~5% or less) of highly expressing EGFP–repressor clones as compared to clones expressing the repressor–NLS fusion without the GFP. Future modifications incorporating additional mutations that yield improved thermal stability of the GFP (Siemering *et al.,* 1996) may be beneficial. Preliminary work in our laboratory has shown excellent expression of the EGFP–repressor–NLS fusion protein in transgenic *Drosophila* fly lines, suggesting again that the relatively poor expression in CHO cells may be related to the thermal instability of GFP.

VI. Gene Amplification and Cell Cloning

For transforming CHO cells we have used calcium phosphate precipitation, Lipofectamine (Gibco BRL), and electroporation. Our general experience is that for routine transformation the Lipofectamine liposome-mediated transformation method produces more consistent results than the calcium phosphate method, which is highly sensitive to the pH conditions used during precipitation. We have used electroporation when we were interested in obtaining single-copy insertions using linearized DNA; under appropriate conditions, ~70% of transformants correspond to integration of a single copy of the vector DNA (Boggs *et al.,* 1986).

For transformations with DHFR as the selectable marker, we use F12 medium that is deficient in thymidine and hypoxanthine for selection with dialyzed fetal bovine serum (Hyclone). Our experiments to date have been with CHO DG44 cells (Urlaub *et al.,* 1986), which contain a double deletion of the endogenous DHFR gene. These cells are propagated as normal CHO cells but are sensitive to EDTA treatment and need to be passaged using trypsin without EDTA added. Selection is imposed 3 days after transformation by changing from normal F12 media to selective media. Nontransformed cells show a highly flattened cell morphology and survive for 1 to 2 weeks after selection. Small colonies of stably transformed cells can be recognized within several days after selection by their more rounded, spindle-shaped morphology.

We select for cells that have undergone gene amplification using methotrexate (MTX) selection. Generally, initial selection can begin at higher methotrexate concentrations than those typically used for amplification of the endogenous DHFR locus. Originally we have started selection at 0.02 μM MTX, but more recently we have increased this starting concentration to 0.05 or 0.1 μM. We suspect that, due to the multicopy integration after typical transformation methods, these higher concentrations of MTX could be used to select stable transformants containing multiple-copy insertions even in cells containing the endogenous DHFR loci.

To select cell clones with specific HSRs we have mostly started with pools of stable transformants that have undergone MTX selection as a mixed population and then have subcloned these cells at different stages of amplification. A significant problem is isolating cell clones that are stable in HSR size and chromosome location. Previous work has indicated a large variation in HSR stability depending on the original integration event of the selectable marker (Wahl *et al.,* 1984). It therefore may be advantageous to start with individual stable clones prior to amplification and carry out gene amplification for each clone.

For transfection with the p3′SS vector and derivatives containing the GFP–repressor constructs we have used Lipofectamine for transformation with hygromycin B selection. Because of the limited half-life of hygromycin, media of cultures undergoing selection are changed every 3 or 4 days. Liquid media with hygromycin stored at 4°C are stable no longer than 3 or 4 weeks, so hygromycin should be added to smaller media reserves as needed.

Cloning of these stable transformants allowed generation of cell lines with characteristic expression levels of the fusion protein. Useful clones expressed *lac* repressor at sufficient levels to saturate *lac* operator sites but not high enough to cause excessive nuclear background staining. For cell lines containing a single-vector insertion containing the 256-copy direct *lac* operator repeat, the amount of GFP–*lac* repressor expressed was critical for successful detection. Empirically, we have found that optimal detection occurs for cells in which the nuclei have a low-level but obvious nuclear background of repressor staining.

As described earlier, problems with GFP–repressor expression levels in stable transformants still exist, leading typically to a low percentage of cells with bright fluorescence levels. We have tackled this problem in two ways. The first is by sequential cell cloning into 96-well microtiter dishes, using a pooled population approach. Cells are first plated at a density yielding 5 to 40 individual colonies per well; after cell growth and cell passaging, cells from each well are screened microscopically to select wells in which a suitable clone is present at an enriched percentage relative to the starting population. Then this procedure is repeated at a lower number of individual clones per well. Using several cycles we have isolated cell clones from starting populations in which desirable clones were present at less than 0.1%. A disadvantage of this method is that we have found a wide range of cell proliferation rates between clones. For this reason, the enrichment one sees at each stage may be less than that expected and may lead to diminishing returns depending on the exact dilution sequence used in this sequential cloning scheme.

A second approach uses direct, microscopy-based screening of individual colonies. Because of the small size of the HSR staining and the low nuclear background, it is not possible to visualize positive colonies except at high magnification with a high-numerical-aperture lens. This precludes direct screening through plastic tissue culture plates. Instead we cover the bottom of a plastic Petri dish with sterile, 12-mm-diameter glass coverslips. Cells are diluted at several concentrations such that coverslips contain a range of colony numbers, depending on the initial percentage of desirable cells in the original population. Typically we use 2 to 5 colonies per coverslip. After cell growth leads to visible colonies, coverslips are transferred, cell surface down under sterile conditions, to a larger Petri dish with a 4-cm-diameter, #1½ coverslip mounted in a hole in the Petri dish bottom. This Petri dish contains cell medium containing 20 m*M* HEPES buffer and is mounted on the UV light microscope. After direct visual screening to identify a colony with sufficient GFP expression, the coverslip is transferred to a new sterile, plastic Petri dish. A sterile, diamond pen is used to break the coverslip and select the piece containing the chosen colony. This piece is then trypsinized and the cells subcloned by serial dilution into microtiter dish wells. We find that under our usual conditions ~75–90% of cells after this initial trypsinization are derived from the single colony we have selected and are easily purified by a further, one-step subcloning.

VII. Repressor Staining and Immunodetection of the *lac* Operator Repeat

Direct, *in vivo* visualization of the *lac* operator repeats in cells using the GFP–*lac* repressor fusion is certainly an important and often ultimate goal of experiments with this system, but for a number of reasons it is advantageous to be able to stain the operator repeats using immunostaining methods. We have already mentioned some of the difficulties encountered with obtaining stable transformants expressing suitable levels of GFP–repressor. We have also mentioned the increase in instability of the HSRs encountered with long-term expression of *lac* repressor. We have found it important to be able to characterize cell clones containing *lac* operator repeats in the absence of expressed *lac* repressor. Also, a long-term direction of our work involves extending the *in vivo* light microscopy capabilities to ultrastructural analysis using the operator repeats as a tag for immunoEM (electron microscopy) approaches.

Detailed descriptions of protocols for staining cells using purified *lac* repressor, followed by immunostaining, or direct immunostaining of cells expressing *lac* repressor are found elsewhere (Robinett *et al.*, 1996; Belmont, 1997b). Originally our experience was that the sensitivity of immunodetection was not as high as that seen with *in vivo* expression of the GFP–repressor fusion, due to the higher background seen with repressor and immunostaining. In particular, detection of single-copy insertions of the 256-copy *lac* operator direct repeat was not possible. More recently, we have found two improvements that allow immunofluorescence detection of single copy insertions. The most important development has been that we have prepared our own mouse anti-*lac* repressor antibody with a much lower background staining than the commercial antibody we used previously. To further increase the signal we have used buffer conditions known to disrupt large-scale chromatin organization (Belmont *et al.*, 1989), such as PBS (phosphate buffered saline), in conjunction with detergent permeabilization. A significant increase in signal is seen under these conditions, presumably due to increased accessibility of antibodies. This approach is useful for light microscopy investigations but can not be used for ultrastructural analysis. Sensitivity levels for immunogold detection are still being optimized.

VIII. *In Vivo* Observation of GFP–Repressor Localization

For short-term visualization of GFP–repressor localization we have used Petri dishes that contain a hole over which a large #1½ coverslip is mounted with vacuum grease. Cells grown on smaller coverslips are placed cell surface down on top of the larger coverslip and examined with an inverted fluorescence microscope. We use normal culture media with 20 m*M* HEPES buffer added.

For long-term visualization of GFP–repressor localization, we have used two incubation chambers sold by Bioptechs (Butler, PA). Both are designed to heat the chamber by passing an electric current through a thin, transparent metallic coating of the coverslips used to assemble the chamber. This design is aimed at removing temperature gradients through the observation chamber. The deltaT system is designed around a disposable, 35-mm tissue culture dish containing a glass coverslip bottom. The cells grow directly on this coverslip, which also contains the heating element. This system is the most convenient to use, particularly for short time observations, and provides the closest temperature control over what the cells are actually experiencing. However, because the coverslip undergoes small changes in focus height as a function of heating, this creates difficulties for high-resolution, optical-sectioning approaches. It is possible to turn off the feedback control of heating and maintain a constant current through the heating circuit, but we have found it difficult to adjust this current to maintain a constant temperature. To minimize evaporation of medium, a layer of mineral oil is added to the culture medium in the dish before mounting on the microscope.

The second system we use is the Bioptechs FSC2 chamber, a closed system allowing perfusion of medium. This apparatus shows greater stability of focus, with the heating carried out on the nonobservation coverslip away from the objective lens. Using this apparatus we have been able to maintain cell proliferation and observations over several days. However, preparing the chamber for observations is more time-consuming and requires advance planning.

We used a 4-cm glass coverslip for the FSC2. Ideally this coverslip should have #1½ thickness to minimize spherical aberration for conventional lenses. Otherwise, an objective lens with a correction collar for variable-thickness glass coverslips should be used. A coordinate system was established by mounting a gold, EM finder grid in the center of the coverslip using vacuum grease. Care was taken that the grease was used only at the tab and grid edges. This coverslip was sterilized by autoclaving. Cells were plated onto the coverslip in a 60-mm tissue culture Petri dish 24 to 48 h before use.

The FSC2 chamber and connecting tubing was sterilized by pumping 70% EtOH through the chamber. The EtOH was then replaced by sterile water, and the chamber and associated tubing were washed extensively with ~100 ml of sterile water for 8 to 12 h. Finally, the chamber and tubing were rinsed with 5 to 10 ml of conditioned cell culture medium. A perfusion pump was used to pump liquid through the FSC2 chamber. The chamber was connected to a 100-ml bottle containing conditioned medium; this bottle had 5% CO_2 bubbled into it to maintain the pH of the medium. To prevent air bubbles from forming within the chamber, this reservoir was heated to 37°C. As an alternative, recently we have been using media within a syringe as reservoir, eliminating the need for bubbling of 5% CO_2 and allowing use of the oxygen scavenging system, Oxyrase (Oxyrase, Inc., Mansfield, Ohio).

The FSC2 chamber is a perfusion chamber that incorporates a grooved fluid distribution system to minimize shear forces created by fluid flow. However, we found that over a 12- to 24-h period of continuous flow, CHO cells changed morphology, rounding significantly into "sausage"-shaped cells, and eventually came off the coverslip, presumably due to cell death. Reducing the flow rate to the minimum (~0.5–1 ml/h) allowed by the Lambda Model 700 miniature perfusion pump (Instech Laboratories, Plymouth Meeting, PA) did not solve this problem. Instead, we used a timer to turn on the pump every 4 h for a 30-min period. This schedule allowed long-term growth and proliferation of CHO cells within the chamber over 2 to 3 days of observations.

In addition to the FSC2 chamber, a lens heater was used for our oil-immersion lens. The use of the lens heater is important to avoid large temperature gradients created by the large heat sink capability of the lens. To avoid repeated thermal stress from expansion and contraction of the lens, we keep the lens in a small incubator at the observation temperature when not in use on the microscope.

IX. Phototoxicity Issues

We were able to visualize large-scale chromatin dynamics in cells containing HSRs containing the *lac* operator repeats and expressing the GFP–repressor–NLS protein. However, we noticed a range of phototoxicity effects produced by long-term observations.

With prolonged observations we noticed that at a certain point cell movements would stop. We therefore explored the effects of long-term excitation in these cells. As a control we first illuminated control cells that did not express the GFP–repressor–NLS protein. Cells were illuminated continuously for variable times with blue excitation light; they were then examined 12–48 h later after growth in the chamber. The relative increase in the number of cells in the illuminated area was compared to that observed in areas that had not been exposed to light. We found that after 5- or 10-min exposures, the cells appeared normal and increased in number comparable to the surrounding areas after 48 h. At 20-min continuous exposure, however, cells appeared normal 12 h later but had all died and were lost from the coverslip after 48 h. It is not clear whether this toxicity was due to long-term exposure to blue light or to a leakage of UV light. However, because our exposures with GFP-expressing cells were limited to less than 1 min, we do not expect a significant contribution of our observed GFP phototoxicity from the exciting light.

This work was followed by experimentation with the A03 cell line containing a late-replicating, heterochromatic HSR and stably expressing the GFP–dimer *lac* repressor fusion. Cells were blocked in early S phase (Tobey and Crissman, 1972; Belmont and Bruce, 1994) and then released from this block and followed for the next 24 h. Because there was a wide range in brightness of the GFP signal, we kept track of the number of exposures generating minimal signals (peak intensity of 500–1000 out of the 4096 intensity levels of the CCD). We assumed that if the observed phototoxicity was due to GFP excitation that this measure might correlate better with phototoxicity as compared to total exposure. Typical exposure times for these minimal exposures varied from 0.1 to 0.5 s.

Cells with no obvious HSR fluorescence in the same field as cells with GFP fluorescence were used as internal controls. In general we found that with increasing exposure the first abnormal behavior observed was chromatid bridging and incomplete daughter chromatid segregation during anaphase. Further exposure resulted in cells failing to enter mitosis. Although HSR conformation changes typical of middle to late S phase were observed (Li *et al.,* 1998), the cells failed to divide as expected. In a number of cases by transmitted light the cells appeared to round up in a similar fashion as cells entering prophase; however, they then did not divide over a number of hours. In other examples the cells remained flat without further changes for more than 24 h. These results suggest the possibility that a mitotic checkpoint control responsive to DNA damage may be responsible for the observed lack of division.

Statistically, whereas after 60 exposures, 6/8 cells with no detectable GFP expression showed normal division within 24 h after release from the early S-phase block, 0/14 cells expressing GFP showed normal division. In this case division was considered normal only if the cells divided within 24 h after release and showed normal chromatid segregation of the labeled HSR. With less than 30 exposures, 2/2 cells not expressing GFP divided but only 3/10 of cells expressing GFP showed normal division.

Although these results are preliminary and not extensive, they suggest a problem related to GFP phototoxicity. The observed chromatid bridging and abnormal segregation of the labeled HSR points to either failure of DNA replication or problems with chromosome segregation. This bridging phenomenon has also been observed in time lapse experiments visualizing chromosome dynamics in *Drosophila* embryos using fluorochrome-labeled histones (Hiraoka *et al.,* 1989; Minden *et al.,* 1989) and suggests DNA damage possibly related to cross-linking as a result of phototoxicity effects. Experiments using Oxyrase (Mikhailov and Gundersen, 1995) to scavenge free oxygen are now in progress to see if the useful period of observation can be extended under these conditions.

X. Present Results and Future Directions

We have used the technology discussed in this chapter to describe the cell-cycle dynamics of a late-replicating, heterochromatic HSR (Li *et al.,* 1998). Previous work has described a large degree of heterogeneity when comparing large-scale chromatin structures within a single nucleus or between nuclei in a population of log phase cells. Although changes in overall chromatin structure as a function of the interphase cell cycle have been described, it has never been clear how tightly correlated these structural transitions are to cell cycle progression. Similarly, it has not been known how reproducible the large-scale chromatin folding of a specific chromosome region would be when comparing cells at the same cell cycle stage.

For this specific HSR, we now have demonstrated a precise choreography of changes in large-scale chromatin organization occurring with each cell cycle. Dramatic reorganization and changes in intranuclear positioning of this HSR occur at specific times during the interphase cell cycle, including events that are tightly linked with DNA replication of the HSR. This work was carried out using a combination of immunofluorescence and immunogold staining on fixed samples with direct *in vivo* visualization using the GFP–repressor system. What became apparent in this work was that pronounced conformational changes in structure could occur within short time periods and nonsynchronously within the population; for instance, an uncoiling of the telophase HSR into a linear chromatid of twice the telophase length occurred in early G1 within times as short as 10–20 min. This was followed by a recondensation into a compact ball over a comparable time period. The capability of direct *in vivo* visualization was essential in deducing a correct temporal sequence of events and establishing that all cells showed the same sequence of conformational changes.

Interestingly, a movement of the HSR from the nuclear periphery to the interior occurs within a 1- to 2-h time window in middle to late S phase in synchronized cells. *In vivo* visualization is now being used to analyze the nature of this motion, specifically whether it corresponds to directed motion versus random diffusion.

A second application of this methodology has been to confirm the existence of large-scale chromatin fibers, which we have termed chromonema fibers (Belmont, 1997a). Previous light and electron microscopy on fixed preparations had suggested a chromonema model of interphase and mitotic chromosome structure in which chromosome condensation occurred through the folding of these ~100-nm fibers (Belmont *et al.*, 1989; Belmont and Bruce, 1994; Li *et al.*, 1998). A controversial aspect of these results was their dependence on the analysis of fixed specimens and use of specific isolation buffers that maintained the condensed structure of these large-scale fibers. Using serial section and tomography reconstructions it was possible to unambiguously trace these fibers as distinct entities for distances of up to 2 μm. In contrast, using the selective tagging provided by the *lac* operator/repressor system, we were able to visualize extended large-scale chromatin fibers within living cells for distances exceeding 5 μm. Using a GFP–*lac* repressor fusion with the transcriptional activation domain of VP16 we were able to observe an unfolding of the A03 heterochromatic HSR into an extended large-scale fiber of ~30 μm total length (T. Tumbar, G. Sudlow, G. Li, and A. Belmont, unpublished data).

Work is now in progress examining the dynamics of extended chromonema fibers. Preliminary work suggests that over short time periods of several hours these fibers show elastic deformations but in general do not show significant long-range motions or rearrangements. Importantly, these fibers persist for long periods, demonstrating the stability of this large-scale organization of chromatin. It will be very interesting to examine the dynamics of these extended fibers as cells enter and leave mitosis as a means of understanding the condensation of mitotic chromosomes. An example of short-term observations of an HSR is shown in Fig. 4.

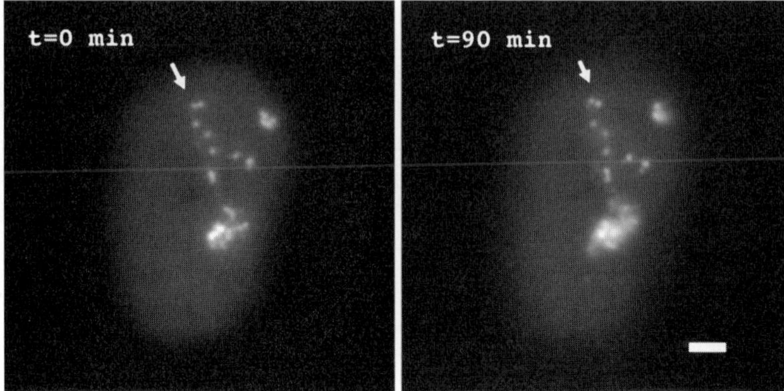

Fig. 4 *In vivo* visualization of an extended HSR in CHO cells. Two time points separated by 90 min are shown. Overall the location and conformation of this extended fiber changes little during this time period. However, the arrow points to a local segment that changes its orientation. Scale bar = 2 μm.

Future improvements in this technology will include improvements of GFP constructs to improve thermal stability in mammalian cells and to exploit different color GFPs. We have recently placed the GFP–repressor fusion protein under inducible control using the tetracycline system (T. Tumbar and A. Belmont, unpublished data). We expect this may be useful in improving the long-term stability of the *lac* operator repeats within the cells, particularly when expanded by gene amplification. Work in our laboratory is now aimed at extending this technology to the analysis of specific genes. An exciting application of this technology would be to combine it with methods for homologous recombination and gene targeting in mammalian cells as has been done with yeast (Straight *et al.,* 1996).

In this chapter we have concentrated on the applications of this technology in mammalian cells. In fact the *lac* operator/repressor system should be applicable to the visualization *in vivo* of specific chromosome regions in a variety of organisms. In general the methodology has potential in allowing visualization of mitotic chromosome dynamics in organisms without cytologically condensed and visible chromosomes, and in allowing visualization of interphase chromosome dynamics in all organisms. To date it has been successfully applied to visualize bacterial (Webb *et al.,* 1997) and yeast (Minshull *et al.,* 1996; Straight *et al.,* 1996; Straight *et al.,* 1997) chromosome dynamics, and many other applications by a number of laboratories are now in progress. In our own laboratory we have recently been successful in applying this system to *Drosophila melanogaster,* using P-element insertions to introduce the *lac* operator repeats. Significantly, although the *lac* operator array frequently undergoes recombination during the initial P-element insertion event, the operator array is highly stable in established fly lines over many generations. Using the Clontech enhanced GFP–repressor fusion under control by the UAS–Gal 4 system, we have been able to visualize the single, 256-copy operator array in both polytene and diploid tissues (C. Merli, C. Tolerico, G. Sudlow, and A. Belmont, unpublished data).

To date current applications have focused on exploiting this technology to allow improved cytological investigation of chromosome dynamics *in vivo.* Extensions of this work should be invaluable in exploring chromosome behavior during transcriptional activation, replication, and recombination. We also foresee future adaptation of this approach to facilitate genetic screens based on direct observation of chromosome dynamics, and possibly biochemical analysis of chromatin structure.

Acknowledgments

This work was supported by NIH grant GM42516 to A. Belmont. The authors thank all of the current and past members of the Belmont laboratory who have contributed to the project described in this chapter.

References

Belmont, A. S. (1997a). Large-scale chromatin organization. *In* "Genome Structure and Function," pp. 261–276. Kluwer Academic Publishers, Dordrecht.

Belmont, A. S. (1997b). Nuclear ultrastructure: transmission electron microscopy and image analysis. *In* "Methods in Cell Biology," pp. 99–124. Academic Press, San Diego.

Belmont, A. S., Braunfeld, M. B., Sedat, J. W., and Agard, D. A. (1989). Large-scale chromatin structural domains within mitotic and interphase chromosomes *in vivo* and *in vitro*. *Chromosoma* **98,** 129–143.

Belmont, A. S., and Bruce, K. (1994). Visualization of G1 chromosomes–a folded, twisted, supercoiled chromonema model of interphase chromatid structure. *J. Cell Biol.* **127,** 287–302.

Boggs, S. S., Gregg, R. G., Borenstein, N., and Smithies, O. (1986). Efficient transformation and frequent single-site, single copy insertion of DNA can be obtained in mouse erythroleukemia cells transformed by electroporation. *Exp. Hematol.* **14,** 988–994.

Brown, M., Figge, J., Hansen, U., Wright, C., Jeang, K.-T., Khoury, G., Livingston, D. M., and Roberts, T. M. (1987). Lac repressor can regulate expression from a hybrid SV40 early promoter containing a lac operator in animal cells. *Cell* **49,** 603–612.

Chao, M. V., Martinson, H. G., and Gralla, J. D. (1980). Lac operator nucleosomes. 2. lac nucleosomes can change conformation to strengthen binding by lac repressor. *Biochemistry* **19,** 3260–3269.

Chen, J., and Matthews, K. S. (1992). Deletion of lactose repressor carboxyl-terminal domain affects tetramer formation. *J. Biol. Chem.* **267,** 13843–13850.

Delidakis, C., Swimmer, C., and Kafatos, F. C. (1989). Gene amplification: an example of genome rearrangement. *Curr. Opin. Cell Biol.* **1,** 488–496.

Fieck, A., Wyborski, D. L., and Short, J. M. (1992). Modifications of the *E. coli* Lac repressor for expression in eukaryotic cells: effects of nuclear signal sequences on protein activity and nuclear accumulation. *Nucleic Acids Res.* **20,** 1785–1791.

Hartley, J. L., and Gregori, T. J. (1981). Cloning multiple copies of a DNA segment. *Gene* **13,** 347–353.

Helm, R., Cubitt, A. B., and Tsien, R. Y. (1995). Improved green fluorescence. *Nature* **373,** 663–664.

Hiraoka, Y., Minden, J. S., Swedlow, J. R., Sedat, J. W., and Agard, D. A. (1989). Focal points for chromosome condensation and decondensation revealed by three-dimensional *in vivo* time-lapse microscopy. *Nature* **342,** 293–296.

Hu, M. C.-T., and Davidson, N. (1987). The inducible lac operator-repressor system is functional in mammalian cells. *Cell* **48,** 555–566.

Lawrence, J. B., Singer, R. H., and McNeil, J. A. (1990). Interphase and metaphase resolution of different distances within the human dystrophin gene. *Science* **249,** 928–932.

Li, G., Sudlow, G., and Belmont, A. S. (1998). Interphase cell cycle dynamics of a late replicating, heterochromatic HSR: precise choreography of condensation/decondensation and intranuclear positioning. *J. Cell Biol.* **140,** 975–989.

Mikhailov, A. V., and Gundersen, G. G. (1995). Centripetal transport of microtubules in motile cells. *Cell Motil. Cytoskeleton* **32,** 173–186.

Minden, J. S., Agard, D. A., Sedat, J. W., and Alberts, B. M. (1989). Direct cell lineage analysis in *Drosophila melanogaster* by time-lapse, three-dimensional optical microscopy of living embryos. *J. Cell Biol.* **109,** 505–516.

Minshull, J., Straight, A., Rudner, A., Dernburg, A., Belmont, A., and Murray, A. W. (1996). Protein phosphatase 2A regulates MPF activity and sister chromatid cohesion in budding yeast. *Curr. Biol.* **6,** 1609–1620.

Rawn, R. D. (1989). "Biochemistry," p. 901. Neil Patterson Publishers, Burlington, NC.

Robinett, C. C., Straight, A., Li, G., Willhelm, C., Sudlow, G., Murray, A., and Belmont, A. S. (1996). *In vivo* localization of DNA sequences and visualization of large-scale chromatin organization using lac operator/repressor recognition. *J. Cell Biol.* **135,** 1685–1700.

Sasmor, H. M., and Betz, J. L. (1990). Specific binding of lac repressor to linear versus circular polyoperator molecules. *Biochemistry* **29,** 9023–9028.

Siemering, K. R., Golbik, R., Sever, R., and Haseloff, J. (1996). Mutations that suppress the thermosensitivity of green fluorescent protein. *Curr. Biol.* **6,** 1653–1663.

Straight, A. F., Belmont, A. S., Robinett, C. C., and Murray, A. W. (1996). GFP tagging of budding yeast chromosomes reveals that protein–protein interactions can mediate sister chromatid cohesion. *Curr. Biol.* **6,** 1599–1608.

Straight, A. F., Marshall, W. F., Sedat, J. W., and Murray, A. W. (1997). Mitosis in budding yeast: anaphase A but not metaphase plate. *Science* **277,** 574578.

Tobey, R. A., and Crissman, H. A. (1972). Preparation of large quantities of synchronized mammalian cells in late G1 in the pre-DNA replicative phase of the cell cycle. *Exp. Cell Res.* **75,** 460–464.

Urlaub, G., Mitchell, P. G., Kas, E., Chasin, L. A., Funanage, V. L., Myoda, T. T., and Hamlin, J. (1986). Effect of gamma rays at the dihydrofolate reductase locus: deletions and inversions. *Som. Cell Molec. Genet.* **12,** 555–566.

Wahl, G. M., de Saint Vincent, B. R., and DeRose, M. L. (1984). Effect of chromosomal position on amplification of transfected genes in animal cells. *Nature* **307,** 516–520.

Webb, C. D., Teleman, A., Gordon, S., Straight, A., Belmont, A., Lin, D. C., Grossman, A. D., Wright, A., and Losick, R. (1997). Bipolar localization of the replication origin regions of chromosomes in vegetative and sporulating cells of *B. subtilis. Cell* **88,** 667–674.

CHAPTER 14

Centrosome Dynamics in Living Cells

Aaron Young,[*] Richard Tuft,[†] Walter Carrington,[†] and Stephen J. Doxsey[*]

[*] Program in Molecular Medicine
University of Massachusetts Medical Center
Worcester, Massachusetts 01605
[†] Biomedical Imaging Group
University of Massachusetts Medical Center
Worcester, Massachusetts 01605

METHODS IN CELL BIOLOGY, VOL. 58
Copyright © 1999 by Academic Press. All rights of reproduction in any form reserved.
0091-679X/99 $25.00

I. Introduction

For centuries scientists have drawn conclusions about biological processes from static images of fixed and stained cells. It has become clear through an increasing number of observations made in living cells that static images acquired from fixed preparations do not accurately represent the dynamic environment of the living cell. Over the last several decades, a vast number of techniques have been developed for introducing labeled macromolecules into cells to study cellular dynamics, but none has gained widespread use due to technical difficulties. The recent discovery of a gene encoding green fluorescent protein (GFP, Prendergast, Chapter 1 this volume) has provided a simple and universal way to label proteins and study their behavior *in vivo.*

GFP is a bioluminescent protein encoded by a gene from the jellyfish, *Aequorea victoria* (Prasher *et al.,* 1992). Virtually any gene of interest can be cloned into one of several commercially available vectors containing GFP (or versions that produce other colors), introduced into cells by conventional methods, and expressed as a chimeric protein under a variety of promoters. The excitation and emission profile of GFP (Heim *et al.,* 1995) is similar to that of fluorescein, rendering a chimeric protein visible by standard epi-fluorescence microscopy; filter sets optimized for GFP spectra can be used to image proteins at low light levels (Heim *et al.,* 1995). The ability to produce fluorescent chimeric proteins in cells simply by molecular cloning techniques has revolutionized the study of proteins and cellular processes *in vivo.*

We used GFP-based technology to study centrosomes in live cells. Centrosomes are small organelles located at the focus of microtubules, where they play a key role in the organization of microtubule arrays for a number of cellular functions including cellular organization, morphogenesis, and mitosis, (see Kellogg *et al.,* 1994). Through their ability to nucleate the growth of microtubules, centrosomes control the geometry, polarity, and number of cellular microtubules, and to a large extent, the microtubule distribution in the cell. Although the importance of the centrosome as an organizer of cellular events has been known for over a hundred years (see Wilson, 1925), little is known about the composition, assembly, and function of this organelle. One approach we have taken to understand centrosome function is to study the assembly and *in vivo* dynamics of the organelle and relate these changes to microtubule nucleation. To this end, we fused GFP to pericentrin, a conserved centrosome protein involved in centrosome organization and microtubule nucleation (Doxsey *et al.,* 1994). Here, we describe how GFP–pericentrin can be used in combination with a recently developed system for image capture and a new algorithm for image restoration, to obtain high-resolution three-dimensional images of centrosomes in living cells.

II. Cloning and Expression of GFP–Pericentrin

A. Buffers

Buffer A. Phosphate-buffered saline (PBS): 130 mM NaCl, 2 mM KCl, 8 mM Na_2HPO_4, 2 mM KH_2PO_4.

Buffer B. Permeabilization buffer: 80 mM PIPES, pH 6.8, 5 mM EGTA, 1 mM $MgCl_2$, 0.5% Triton X-100.

Buffer C. Buffer A with 1% bovine serum albumin (BSA, Sigma) and 0.5% Triton X-100.

B. Cloning of GFP–Pericentrin

We used pericentrin to study centrosome dynamics for several reasons. It is present at the centrosome throughout the entire cell cycle, it is a component of the pericentriolar matrix, which nucleates microtubules, and it is required for centrosome organization and spindle assembly (Doxsey *et al.,* 1994). For these reasons, pericentrin serves an reliable marker for functional centrosomes.

We had previously determined that a full-length pericentrin tagged with the hemagglutinin peptide (Wilson *et al.,* 1984) localized to centrosomes and that a central coiled coil domain encoded by amino acids 766–1343 was sufficient for centrosome localization. To produce a GFP–pericentrin fusion, we used standard PCR-based amplification and cloning techniques (Dictenberg *et al.,* 1998). A version of GFP that fluoresces more brightly than the original molecule (S65T) was fused to the amino terminus of the truncated pericentrin construct (p 766–1343) and the chimeric molecule was introduced into a vector for expression in mammalian cells (pcDNA II, InVitrogen). Nucleotide sequences were confirmed by dideoxy sequencing methods (Biorad). Currently, it is more convenient to use GFP vectors and others that encode fluorescent proteins with various emission spectra because they have superior properties and require much simpler cloning strategies (e.g., Clonetech, Quantum, Biorad).

C. Expression of GFP–Pericentrin

The GFP–pericentrin cDNA is introduced into COS-7 cells by conventional procedures such as Lipofectamine transfection (InVitrogen) or electroporation (Gibco, BRL). Transient expression of the protein is monitored by morphological and biochemical assays (discussed later).

Lipofectamine Transfection (see InVitrogen for details)

1. Cells are plated on 25-mm coverslips in 35-mm plates (Costar) 24 h in advance so that they will be 30–40% confluent on the day of the transfection.
2. Cells are washed once in serum-free medium before adding DNA.

3. DNA (2 μg) in 100 μl of serum-free medium is mixed with 6.0 μl of Lipofectamine in 100 μl of serum-free medium and incubated for 35 min at room temperature.
4. Serum-free medium (800 μl) is added to the DNA–Lipofectamine mixture and immediately added to cells.
5. Cells are incubated with the DNA mixture for 5 h in a 37°C CO_2 incubator.
6. The DNA mixture is aspirated, 2 ml of complete medium are added, and the cells are returned to the incubator for various times.

Electroporation

1. 1 x 10^6 log phase cells are collected by trypsinization from culture plates and washed once in complete medium by pelleting cells and aspirating medium.
2. The cells are resuspended in complete medium at 10^6/ml.
3. DNA (2 μg) is added to 1 ml of cells in complete medium.
4. Well-suspended cells are placed in an electroporation chamber (Gibco) and electroporated at 1180 μfarad and 300 volts set at high resistance (Gibco #344528).
5. The cells are immediately transferred to complete medium using a 1-ml pipet and plated on coverslips to achieve a confluency of 50% in 24 h. Typically, 30% of the cells are plated on 100-mm plates.

Transfection efficiency is usually 50% for lipofectamine transfections and 15% for electroporation. Similar results can be obtained with CHO cells using similar parameters. The cells are not adversely affected by the presence of GFP–pericentrin because their mitotic rate is similar to control cells expressing GFP alone or other constructs (discussed later).

D. Confirmation of GFP–Pericentrin Expression

At various times after transfection, cells can be examined to confirm GFP expression. This is accomplished simply by inverting the coverslip on a slide in medium, sealing it with nailpolish, washing the noncell surface with water, and viewing the slide immediately. Alternatively, cells are fixed and coverslips are mounted in glycerol-based mounting medium (Vectasheild, Vector Laboratories) before viewing. GFP–pericentrin fluorescence is brightest in unfixed cells but is also preserved after fixation in 4% formaldehyde or 1% glutaraldehyde in PBS (Buffer A); methanol fixation significantly diminishes the GFP signal. Cells are examined on a fluorescence microscope using a conventional filter set for fluorescein. GFP-positive cells should fluoresce green under these conditions; occasionally, we observe material that fluoresces brightly in both green and red channels and is not related to GFP. Biochemical methods can also be used to monitor protein expression and to confirm the size of the chimeric protein. We routinely perform Western blot analysis and immunoprecipitations on GFP-expressing cells (Harlow and Lane, 1988) using antibodies to pericentrin (Babco) or GFP (Clontech).

E. Confirmation of GFP–Pericentrin Localization

The GFP–pericentrin fusion protein produces a bright spot in the cytoplasm of transfected cells that is typically adjacent to the nucleus and usually 1–3 μm in diameter. Centrosome localization is dependent on time. At early time points (6 h after transfection), little incorporation is detected and the protein is largely cytoplasmic. At 35–72 h, GFP–pericentrin is localized to centrosomes in ~35% of cells.

Several assays have been performed to confirm that GFP–pericentrin is localized to centrosomes prior to imaging live cells. Cells can be fixed and stained by immunofluorescence for endogenous pericentrin using a pericentrin antibody (Babco) that does not cross-react with the truncated domain fused to GFP or with antibodies to other centrosomal markers such as γ-tubulin (Sigma). In addition, GFP–pericentrin at the centrosome can be analyzed using high-resolution microscopy (Carrington *et al.,* 1995). We have shown that it forms a lattice structure similar to that shown for endogenous pericentrin and γ-tubulin (Dictenberg *et al.,* 1998)

Microtubule regrowth can also be used to verify centrosomal localization of GFP–pericentrin (Fig. 1). Microtubules are depolymerized with nocodazole (10 μg/ml for 60 min) in complete medium at 37°C, washed several times with serum-free medium, then returned to complete medium at 37°C for 5–7 min. Cells are then permeabilized in microtubule-stabilizing buffer (Buffer B), fixed in 1% glutaraldehyde in PBS, and processed for α-tubulin immunofluorescence (Pihan *et al.,* 1998). A microtubule aster should emanate from the GFP–

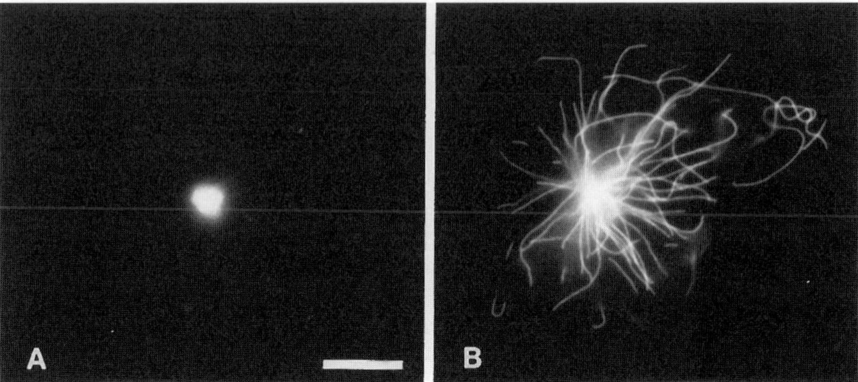

Fig. 1 GFP–pericentrin localizes to centrosomes and does not affect microtubule nucleation. GFP–pericentrin (A) is found at the center of a microtubule aster (B) formed by newly polymerized microtubules in a COS cell. Microtubules were depolymerized using nocodazole, repolymerized for 5 min at 37°C and processed for immunofluorescence using cy3 secondary antibodies as described in Section VI.C. Fluorescein (GFP) and rhodamine (microtubules) images were captured consecutively. Bar, 5 μm.

pericentrin staining focus, demonstrating that the protein localizes to the centrosome and does not impair its nucleating capacity.

In some cells, foci of GFP-staining material are present in the cytoplasm in addition to centrosomes. We believe that some of these structures represent intermediates in centrosome assembly whereas others are probably aggregated material that can also be found in some cells transfected with GFP alone. In cells with additional GFP-labeled structures, centrosomes can be distinguished by a variety of criteria before and after imaging (discussed later).

III. High-Speed Microscopy

Four-dimensional images of GFP–pericentrin in living cells (3-D images through time) are obtained on a novel custom-built high-speed wide-field epi-fluorescence microscope originally developed for the imaging of ionic signals in single isolated cells (Tuft *et al.,* 1997, Fig. 2). The maximum imaging speed of this instrument may be unnecessary for this application, but the highly dynamic changes in both the position and the conformation of the centrosome obtained by this method indicate that the emphasis on speed and sensitivity that guided its development are important. The general principal of the system is the rapid

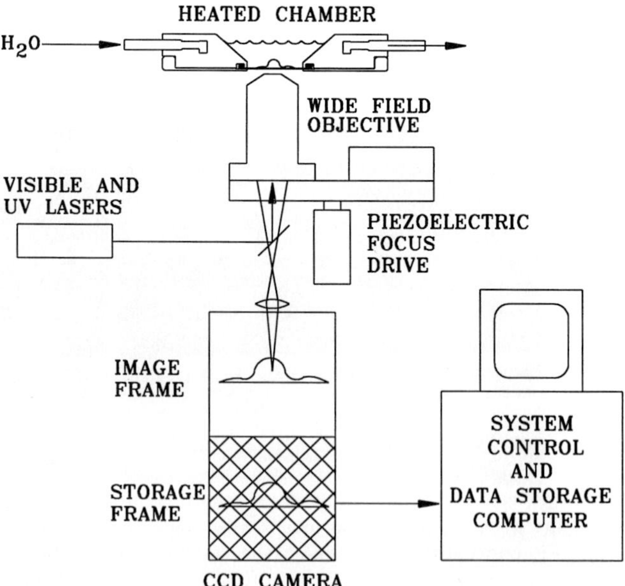

Fig. 2 System block diagram of custom-built, high-speed three-dimensional, epi-fluorescence microscope (see Section III.D for details).

capture of three-dimensional data sets of fluorescent probes in living cells through time with minimal photo damage. A system block diagram of the setup is shown in Fig. 2. The main components of the system are wide-field high-energy lasers to allow short light exposures, supersensitive charge-coupled-device (CCD) cameras to facilitate rapid image capture and image downloading, and a high-speed focus drive to permit rapid acquisition of 3-D data. Each of these components is described in detail later. High-resolution image restoration is described in the following section.

A. Wide-Field epi-Fluorescence Laser Illumination

Ultraviolet and visible lasers provide spectral lines (at 351, 364, 383, 488, 514, 521, 531, 568, and 647 nm) capable of exciting a number of common fluorophores and GFP mutants while providing sufficient excitation light for good imaging (i.e., signal-to-noise ratio $>10:1$) at normal dye loading or common GFP expression levels, with exposure times as short as 2 ms. For these experiments, the laser power and beam expansion optics were configured to provide a 488-nm laser flux of 36.5 watts/cm^2 at the specimen plane when used with a 100x objective. From the published S65T GFP extinction coefficient of 39,200 M^{-1} cm^{-1} at 489 nm (Meim *et al.,* 1995), it can be calculated that a single GFP molecule excited at the above flux will emit ~9 photons/ms. This excitation is adequate to produce the low-noise images (80-nm pixels with 100x objective) shown in Fig. 4 with 15-ms exposure times. It should be noted that the excitation flux can be easily changed depending on experimental requirements. For example, the laser power can readily be increased by a factor of 10 if exposures in the single millisecond range are needed. In any event, the rates of fluorophore cycling fall far below the saturation level for normal fluorophore transitions. The system is also configured to provide rapid changes between different excitation frequencies. This permits the simultaneous acquisition of signals from multiple fluorophores.

B. High-Speed Piezoelectric Focus Drive

To achieve rapid acquisition of three-dimensional data sets, we designed a drive capable of shifting focus, from dead stop to dead stop, at a maximum rate of 1 μm/ms. When used with high-numerical-aperture oil-immersion objectives, as in this application, the maximum practical rate of focus shift is limited by flexure of the coverslip as the oil layer is displaced during focus, and time must be allowed for the coverslip to return to its unflexed position. For relatively large unsupported cover slip areas, such as in the heated chamber used in these experiments, approximately 10–20 ms are required to complete small (100- to 200-nm) focal shifts. This rate of focus shift is attainable with commercially available focus drivers such as the PIFOC, available from Polytech PI, Inc. (Costa Mesa, CA and Auburn, MA). The combination of 15-ms exposure time and 10-ms time to shift focus allows the acquisition of a complete 30-plane three-dimensional data set in under 1 s.

C. Small-Format Four-Output-Port CCD Camera

In any imaging system that seeks to extract the maximum amount of information from fluorescent signals, the sensitivity of the detector is of paramount importance. The principal reason for this is the finite number of excitation–emission cycles a fluorophore can experience before photobleaching. For common fluorescent dyes, the number of fluorescent cycles is commonly between 10^{-4} and 10^{-6}. When imaging living cells, the need to follow cell dynamics over extended time periods requires that cell poisoning from the toxic by-products of fluorophore photobleaching, or even direct photodamage from the illumination acting on the cell itself, be minimized to the greatest possible extent. Careful design of the optics that collect and image the light is an extremely important feature of such imaging devices, but a major determinant of the success of such strategies is the detector itself.

The current standard of performance for sensitive imaging is the scientific CCD camera, which originally came to the attention of the scientific community as a detector for astronomical imaging. In this role, the readout time of these imagers, typically on the order of seconds, was of no significance compared with the several minutes to hours taken for typical astronomical imaging exposures. The imaging of cellular dynamics places different demands on the imager. For this application, we have collaborated with MIT Lincoln Laboratory in the development of a small-format, high-speed CCD imager. This device has a quantum efficiency approaching 70% at 550 nm and an intrinsic readout noise of 6 photoelectrons.

Because of the quantum nature of the detection process, in which each absorbed photon creates a photoelectron, there is an inherent uncertainty in the signal introduced by the process itself. This uncertainty, or Poisson noise, sets fundamental limits that cannot be bettered. For a perfect (i.e., no noise, 100% quantum efficiency detector), the repeated detection of a constant intensity signal of 100 photoelectrons will have a count uncertainty (standard deviation) of 10 photoelectrons. The Poisson-noise-limited signal-to-noise ratio would be 10 to 1 in this case. For the detector in our microscope, the ratio of 10:1 is reached at approximately 140 photoelectrons, or 200 incident photons, per pixel. This performance is comparable to that of current scientific CCDs, but is obtained at much higher imaging rates.

The maximum imaging rate of the camera is 543 images, or one image every 1.8 ms. This rate is achieved by virtue of the 128 × 128-pixel frame transfer format of the camera and four data output ports. The 128 × 128-pixel imaging area is sufficient to image, at high temporal and spatial resolution, well-defined and localized organelles such as the centrosome or localized signaling events such as calcium sparks. With the frame transfer geometry, the image (in the form of photoelectrons) on the image frame is transferred in 50 μs to the storage frame. While the next image is being formed, the stored image charge is read out through four output ports at a combined readout rate of 10 megapixels/s. This readout takes 1.8 ms, thus setting the maximum readout rate of the camera.

To maintain the detection efficiency gained by this detector, the optics in the camera imaging path are limited in number and have the highest transmission/reflection available. For this study, we used a 100× Nikon planapo oil objective lens. The 1.4 numerical aperture of this lens collects approximately 30% of the emitted fluorescence, the maximum attainable without complex multiple-lens collection schemes.

D. Heated Cell Chamber

To image live GFP–pericentrin expressing cells, we developed a chamber that attaches to the stage, allows cells to be maintained at various temperatures, and provides direct access to culture media (Fig. 2). Coverslips (see Section II.C) are secured between the top and bottom plates of the hinged chamber and sealed by a rubber O-ring mounted in the top plate. The bottom of the coverslip is washed with water and dried to remove residual salts and protein before lenses are brought into contact from below. The chamber is open to the atmosphere, facilitating access to the media for drug administration, immunofluorescence imaging, and other applications (discussed later). The medium is buffered with HEPES, obviating the need for bicarbonate buffers, which require CO_2 in the atmosphere. Cells are maintained at constant temperature (37°C) by water circulated from a heated waterbath through the interior of the top plate; temperature is monitored by a thermocouple immersed in the medium bathing the cells.

IV. Image Restoration by an Improved Deconvolution Method

Each set of three-dimensional images captured on the wide-field microscope is restored at high resolution as described previously (Carrington *et al.*, 1995) with adaptations for increased image resolution and image concatenation. The resolution that the light microscope can achieve is limited by the blurring caused by fundamental optical properties of the microscope. In the wide field microscope there is also substantial haze caused by light from out-of-focus parts of the cell, which substantially reduces contrast and the ability to distinguish fine details of the image. To circumvent this problem we obtain three-dimensional information by optical sectioning (i.e., by obtaining a series of two-dimensional images at different planes of focus through the cell). Because a cell is a three-dimensional object, each of these two-dimensional images or optical sections contains in-focus light from the in-focus plane of the cell and contains a blurred image of each out-of-focus part of the cell. The further a fluorescent molecule is from the in-focus plane, the more blurred its contribution to that optical section. To improve resolution and remove out-of-focus haze from these wide-field images, we use the computational method of image restoration, also known as deconvolution (Carrington *et al.*, 1995). The novel image restoration algorithm we have developed is designed to provide very high-resolution images with minimal photodamage and photobleaching. It provides three to four times the resolution of unprocessed confocal or wide-field microscope images; by making more effective use of the available light, it obtains higher spatial resolution than other image restoration methods without the very high magnification.

Figure 3 shows a quantitative description of the imaging process. At each plane of focus and each pixel of the camera, the light collected by that camera pixel is a weighted sum (integral) of the fluorescence emission throughout the cell. The further a fluorescent molecule is from the plane of focus or from the position of the camera pixel within that plane, the less light that camera pixel collects from that molecule. We characterize the blurring of the microscope by obtaining optical sections of a small fluorescent bead; from this we calculate the weighting functions $k_i(x, y, z)$ of Fig. 3. This procedure is analogous to other familiar calibration procedures: By quantitatively calibrating the microscope blurring we are able to calculate a more accurate image. There are several features of the imaging process that we explicitly include in our model of the imaging process. The number of pixels of data is finite. However, the cell does not consist of pixels, but is a continuum object; therefore, we model the cell's fluorescence dye density, $f(x,y,z)$, as a function in continuous space. We can then use the model in Fig. 3A to calculate a more accurate fluorescent dye distribution than the unprocessed image can provide. Figure 3B presents the mathematical expression that defines the mathematical algorithm we use. We

$$\text{A.}\quad g_i = \iiint_V k_i(x,y,z)f(x,y,z)\,dxdydz,\quad i=1,2,3,\ldots,N$$

$$\text{B.}\quad F(f) = \sum_{i=1}^{N} \left| \tilde{g}_i - \iiint_V k_i(x,y,z)f(x,y,z)\,dxdydz \right|^2 + \alpha \iiint_V \left| f(x,y,z) \right|^2 dxdydz.$$

Fig. 3 Deconvolution algorithm. The algorithm is based on the following formulation of the imaging process. At each optical section, each pixel collects light from a considerable volume of the cell, and the pixel's sensitivity to a point source depends on the location of the source. Thus, the light collected by pixel number i is a weighted sum of the fluorescence in the volume of the entire sample, as described by Equation (A), where $f(x, y, z)$ is the fluorescent dye density at (x, y, z), $k_i(x, y, z)$ is the response of a detector (camera pixel) to light from a point source located at (x, y, z) transmitted through the optics with the pixel focused onto the point (x_i, y_i, z_i), V is a volume containing the fluorescent specimen, and N is the number of sampled voxels of the image. The actual measurement will contain noise. The determination of the dye density $f(x, y, z)$ from Eq. (A) is an ill-posed problem to which we applied an L^2 regularization incorporating the powerful constraint that the dye density must be nonnegative. Thus we estimate the dye density as the nonnegative function, $f(x, y, z) > 0$, that minimizes the expression $F(f)$ in Eq. (B). The parameter α balances between the need to fit the data (first term) and the need to stabilize the estimate (second term).

compute the dye density as the nonnegative function $f(x, y, z)$ that minimizes the expression in Fig. 3B. This method allows us to compute the dye density, $f(x, y, z)$, on a more finely sampled grid than the sampling grid of the data. For example, in Fig. 4, camera pixels corresponded to squares 80 nm × 80 nm and the optical sections were spaced 100 nm apart. The image restoration was calculated on a grid spaced 40 × 40 × 50 nm. This is equivalent in size to doubling the optical magnification and doubling the number of optical sections. Our approach effectively zooms in on the sample computationally instead of optically. To achieve the same result by increasing the magnification would require eight times the total light exposure to the sample and would require a camera with four times the number of pixels and four times the readout rate.

Recent measurements of mechanically separated fluorescent beads have shown that we can resolve two point sources with a transverse separation of 100 nm. This is three to four times better resolution than is possible with confocal microscopes or unprocessed images from wide-field microscopes. Out-of-focus haze and quantitative fluorescence measurements are substantially improved.

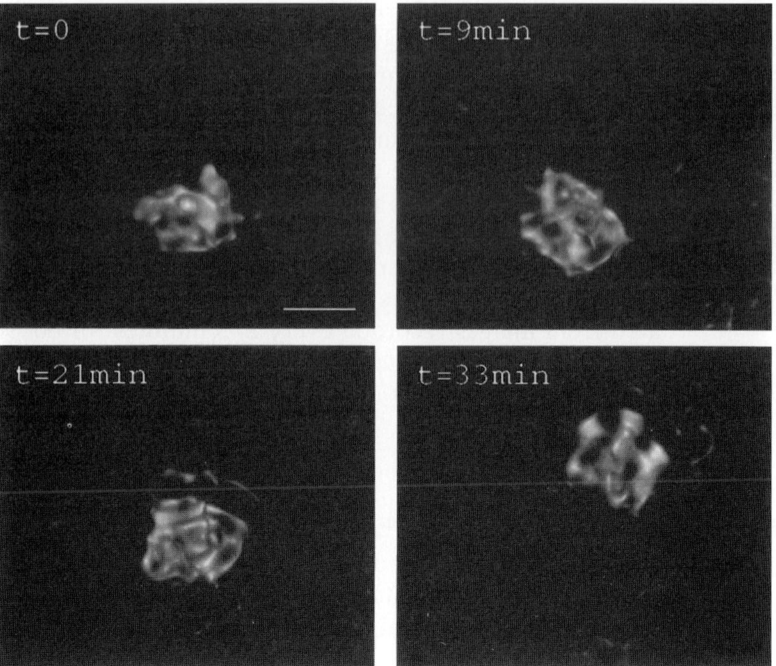

Fig. 4 Centrosome dynamics in a living cell. Time-lapse images of GFP–pericentrin in an interphase COS cell recorded every 9–12 min using high-speed, high-resolution three-dimensional imaging methods. Considerable spatial and morphological changes of the centrosome are observed over time demonstrating the dynamic nature of the organelle in living cells. Bar, 1 μm.

Once each three-dimensional data set of a series is restored at high resolution, the three-dimensional images are concatenated to produce a time-lapse series of images as shown in Fig. 4.

V. Imaging Centrosomes

A. Image Capture

Centrosomes labeled with GFP–pericentrin usually appear as small foci of staining (1–3 μm in diameter) that are brightly fluorescent. These properties permit acquisition of three-dimensional data sets of small volume at high signal-to-noise ratios even in low-light conditions. For each three-dimensional series, 30 optical sections (z-planes) are collected at 100-nm intervals in a 10 μm^2 region (x, y) centered on the centrosome. This acquisition area [10 μm (x) $\times$ 10 μm (y) x 3 μm (z)] is sufficient to image the entire centrosome even when the organelle undergoes translational movements in the cytoplasm (see Fig. 4); when required, additional z-planes can be obtained. Exposure times for each z-plane range from 15 to 30 ms depending on the intensity of GFP signal. In most cases, GFP signals at the centrosome are roughly 10-fold greater than the cytoplasmic staining that is present in virtually all cells. The minimum acquisition time of a complete three-dimensional series (30 z-planes), which includes times for exposure (15 ms) and objective movement and stabilization (10 ms) is 750 ms. Collection of three-dimensional images every three minutes is sufficient to capture most changes in centrosome movement and structure. Subtle changes in centrosome structure require image capture at more frequent intervals (30 s). If necessary, three-dimensional images of this size can be collected approximately every second using this system. Most time-lapse series of centrosomes are generated from images recorded every 2–3 minutes for 60–90 minutes (total, 20–45 three-dimensional images). Under these imaging conditions, there are no detectable deleterious effects of photodamage on centrosomes, microtubules, or cells (discussed later).

B. Properties of Centrosomes in Living Cells

The ability to image centrosomes in living cells has revealed several features that were not appreciated in fixed specimens. Changes in centrosome position, orientation, and internal structure are consistently observed in interphase centrosomes (Fig. 4). In images restored at high resolution, the GFP–pericentrin signal at the centrosome reveals a unique reticular lattice. This structure is remarkably similar to that observed for endogenous pericentrin in fixed cells stained by immunofluorescence (Dictenberg *et al.*, 1998). Changes in the structure of the lattice occur over time as is clearly demonstrated in the 9- to 12-min intervals shown in Fig. 4. When imaged at shorter time intervals, more fluid and continuous changes in lattice structure are observed. Similar dramatic structural changes are observed with endogenous pericentrin in fixed cells, thus confirming the observations with GFP–pericentrin in living cells. For example, when endogenous pericentrin is imaged by indirect immunofluorescence, predictable changes in lattice structure occur at specific cell cycle stages (Dictenberg *et al.*, 1998). However, only in live cells is it possible to observe small aggregates of GFP–pericentrin that undergo vectorial and centripetal movements toward the centrosome. This material ultimately joins with the GFP–pericentrin lattice in a manner suggesting protein recruitment to the centrosome. Many of these changes in centrosome movement and structure are detectable, although less well defined, when images are not restored at high resolution. All of these dynamic changes in centrosomes occur in the presence of an intact microtubule cytoskeleton (discussed later).

VI. Postimaging Confirmation of Centrosome Integrity and Function

The conditions that we have described for imaging centrosomes do not appear to have any detectable deleterious effects on centrosome structure or function. However, we periodically perform several assays following the imaging procedure to confirm that the imaging procedure (e.g., photodamage) does not affect the parameters under study. It is especially important to check these parameters when any aspect of the procedure is modified.

A. Centrosome Localization

Centrosomal GFP–pericentrin usually appears as a single dot 1–3 μm in diameter that is adjacent to the nucleus (which is darker than the cytoplasm because it excludes the fluorescent protein). Further confirmation of centrosome localization of the GFP–pericentrin can be accomplished by rapid indirect immunofluorescence staining for other centrosome proteins. This method is performed on unfixed cells and can be completed in ~15 min while the coverslip remains on the stage at 37°C. Centrosome integrity is maintained remarkably well in these unfixed preparations. Easy access to cells in the open chamber is one of the major advantages of this system. Solutions are kept at room temperature (22°C) before use in this assay.

1. Immediately after the imaging procedure the growth medium is aspirated from cells.
2. PBS (Buffer A, 1 ml) is added to cells and immediately aspirated.
3. Permeabilization buffer (Buffer B, 1 ml) is added to cells, incubated for 30 s then aspirated.
4. Anti-γ-tubulin antibody (200 μl) is added at 25 μg/ml in Buffer C and incubated for 5 min. Other anti-centrosome antibodies can also be used.

5. Cells are washed 10 times in rapid succession with Buffer C (1 ml, total time = 1 min).
6. Secondary antibody (cy3 or cy5 conjugates are used, Jackson Labs, 200 μl) in Buffer C is added at 25 μg/ml and incubated for 3 min.
7. Cells are washed as in step 5.
8. A 3-D image of GFP–pericentrin is immediately captured.
9. The laser excitation wavelength is changed to excite cy3 or cy5.
10. A 3-D image of the cy3-labeled centrosome antigen is captured.

Coincidence of GFP and cy3 signals confirms localization of GFP–pericentrin at the centrosome.

B. Centrosome Structure

The structure of the GFP–pericentrin after image restoration at high resolution can be compared with features of centrosomes imaged in fixed cells. The most reliable measure of structural integrity is the diameter of the rings of the lattice (see Fig. 4). Rings are consistently ~280 nm in diameter in fixed centrosomes (Dictenberg *et al.,* 1998). Similar dimensions are typically observed in GFP–pericentrin images at the end of the imaging procedure. A more stringent analysis of the GFP–pericentrin structure can be made by comparing signals from GFP–pericentrin and endogenous pericentrin in the same cells. Cells are processed for indirect immunofluorescence to label the endogenous protein in the absence of fixation as described above. Complete three-dimensional data sets are acquired, resolved at high resolution, and analyzed for coincidence of the two signals using software developed for this purpose (Lifschitz *et al.,* 1994).

The following assays are performed on fixed cells by one of two methods: (*a*) Cells can be rapidly fixed and processed on the stage in a manner similar to that described in Section VI.A. Great care must be taken to ensure that there are no leaks in the chamber that could allow fixatives to penetrate and modify the cement in the objective lenses. This can be prevented by changing O-rings after three to five uses, because they become less effective at sealing the coverslip after multiple exposures to fixatives. (*b*) An alternative method can be used to perform these assays that involves the use of gridded coverslips (Eppendorf), which facilitate localization of the same cell after removing the coverslip from the chamber of the microscope.

C. Centrosome Function

Perhaps the best test of centrosome function after imaging is its ability to nucleate the growth of new microtubules (Kellogg *et al.,* 1994). We developed a rapid procedure to analyze microtubule nucleation in cells based on the ability to rapidly and reversibly depolymerize microtubules (Blomberg and Doxsey, 1998). This procedure can be performed either on the microscope stage or on gridded coverslips removed from the stage. All solutions are prewarmed to 37°C unless otherwise stated.

1. After the last image is captured, cells are incubated with 100 μg/ml nocodazole (Sigma) in complete medium.
2. After 15 min, the medium is aspirated.
3. Cells are rapidly washed 10 to 15 times in Buffer A (total time, 1–2 min).
4. Complete medium is added and cells are incubated for 7–10 min.
5. Cells are permeabilized in Buffer B for 30 s.
6. Cells are fixed in 1% glutaraldehyde in Buffer A for 1–2 min.
7. Fixed cells are processed for indirect immunofluorescence by conventional methods using anti-α-tubulin antibodies to stain microtubules as described (Doxsey *et al.,* 1994).

The presence of microtubules at the focus of GFP staining, similar to that shown in Fig. 1, indicates that the imaging procedure has little effect on microtubule nucleation. Under optimal conditions of fixation and staining, microtubules are well resolved and can be counted to obtain a more accurate measure of the nucleating capacity of the centrosome.

D. Microtubule Integrity

Given the dramatic changes in centrosome position, orientation, and structure observed in living cells, it is important to confirm that the microtubule cytoskeleton is intact throughout the imaging period. To accomplish this, cells are simply fixed and stained for microtubules at various times after GFP–pericentrin is imaged using the method described in Section VII.C, steps 5–7. The general organization and number of microtubules in the cells are compared with cells that have not been imaged. We have not detected any changes in the microtubule cytoskeleton by these criteria. A future alternative method for analyzing microtubules after centrosome imaging is to use cells in which microtubules are labeled, for example by injection of rhodamine-labeled tubulin. In this way, microtubule images can be captured after a series of centrosome images are acquired. This system will also allow simultaneous imaging of microtubules and centrosomes in the same cell.

E. Cell Viability

The postimaging viability of cells can be monitored by testing for the ability of cells to undergo cell division. After imaging, cells plated on gridded coverslips are returned to the 37°C incubator for 24 h. This is sufficient time for most cells in culture to divide. The presence of two GFP–labeled progeny indicates that cell division has taken place.

F. Cell and Cytoplasmic Movements

Changes in the position of the centrosome could be due to centrosome-specific movements in the cytoplasm or more general types of motion including cytoplasmic movement (e.g., streaming) or cell motility. We believe that centrosomes are undergoing specific movements in the cell because they occur while other organelles remain stationary. For example, the nucleus can be used as a frame of reference because it is dark relative to the fluorescent cytoplasm (it excludes GFP–pericentrin), or it can be labeled with vital dyes that bind chromatin (Hoechst 33342, Sigma). Using either detection method, nuclei appear to remain stationary in nearly all cells analyzed thus far (>200) whereas changes in the cytoplasmic position of the centrosome are consistently observed (similar to that shown in Fig. 4). These observations strongly suggest that centrosome-specific movements occur during the imaging period. Centrosome dynamics seem to be independent of cell motility, as little to no change in cell movement is detected by phase contrast microscopy while significant centrosome movements are recorded. Finally, centrosome dynamics can be observed when immobile phase dense material in the cell was used as a frame of reference.

References

Blomberg, M., and Doxsey, S. (1998) Rapid isolation of centrosomes, *Meth. Enzymol.*,**298,** 228–238.

Carrington, W., Lynch, R. M., Moore, E. D., Isenberg, G., Fogarty, K. E., and Fay, F. S. (1995). Superresolution three-dimensional images of fluorescence in cells with minimal light exposure. *Science* **268,** 1483–1486.

Dictenberg, J., Zimmerman, W., Sparks, C., Young, A., Carrington, Zheng, Y., Vidair, C., W. Fay, F., and Doxsey, S. (1998). Pericentrin and gamma tubulin form a protein complex and are organized into a novel lattice at the centrosome. *J. Cell Biol.* **141,** 163–174.

Doxsey, S. J., Stein, P., Evans L., Calarco, P., and Kirschner. (1994). Pericentrin, a highly conserved protein of centrosomes involved in microtubule organization. *Cell* **76,** 639–650.

Harlow, E., and Lane, D. (1988) "Antibodies: A Laboratory Manual." Cold Spring Harbor Laboratory Press, Cold Spring Harbor, New York.

Heim R., Cubitt, A., and Tsien, R. (1995). Improved green fluorescence. *Nature* **373,** 663–664.

Kellogg, D. R., Moritz, M., and Alberts, B. M. (1994) The centrosome and cellular organization. *Annu. Rev. Biochem.* **63,** 639–674.

Lifschitz, L., Collins, J., Moore, E., and Gauch, J. (1994). "Computer Vision and Graphics in Fluorescence Microscopy," vol. 166. IEEE Computer Soc. Press. pp. 166–175 Los Alamos, CA.

Pihan, G., Purohit, A., Knecht, H., Woda, B., Quesenberry, P., and Doxsey, S. J. (1998). Centrosome defects and genetic instability in malignant tumors. Cancer Research, in press.

Prasher, D. C., Eckenrode, V. K., Ward, W. W., Prendergast, F. G., Cormier, M. J. (1992). Primary structure of the *Aequorea victoria* green fluorescent protein. *Gene* **111**, 229–233.

Tuft, R. A., Bowman, D. S., Carrington, W. A., Fogarty, K. E., and Fay, F. S. (1997). High-speed digital-imaging-microscopy with continuous focus scanning. *Biophys. J.* **72,** A215.

Wilson, E. B. (1925). "The Cell in Development and Heredity." Macmillan, Inc., New York.

Wilson, I. A., Niman, H. L., Houghton, R. A., Cherenson, A. R., Connolly, M. L., and Lerner, R. A. (1984). The strucuture of an antigenic determinant in a protein. *Cell* **37,** 767–778.

CHAPTER 15

Transfections of Primary Muscle Cell Cultures with Plasmids Coding for GFP Linked to Full-Length and Truncated Muscle Proteins

Guissou A. Dabiri, Kenan K. Turnacioglu, Joseph C. Ayoob, Jean M. Sanger, and Joseph W. Sanger

Department of Cell and Developmental Biology
University of Pennsylvania School of Medicine
Philadelphia, Pennsylvania 19104

Copyright © 1999 by Academic Press. All rights of reproduction in any form reserved.
0091-679X/99 $25.00

I. Introduction

The actin/myosin cytoskeleton of muscle and nonmuscle cells is a dynamic structure that assembles and disassembles reversibly in response to a variety of cellular signals. When nonmuscle cells and embryonic cardiomyocytes enter mitosis, their stress fibers and myofibrils disassemble and the component proteins are recruited to form a transitory contractile apparatus for cytokinesis (Sanger *et al.,* 1989, 1994). After completion of cell division, the proteins then re-form the stress fibers of the nonmuscle cells and the myofibrils of the cardiomyocytes. Localized protrusions of the cell membrane for motility (Theriot and Mitchison, 1992) or in response to certain microorganisms (Higley and Way, 1997) also involve localized recruitment of actin filaments, although on a smaller scale. Analyses of these and related forms of cell motility depend, in part, on observations of individual cytoskeletal proteins in live cells.

The introduction of fluorescently labeled probes into live cells has served that purpose and provided important information in studies of cell motility in both muscle and nonmuscle cells (reviewed in Wang and Taylor, 1989). For many probes, particularly cytoskeletal proteins, fluorescent dyes have been coupled to purified proteins, followed by microinjection of trace amounts of the labeled proteins into live cells. This requires expertise in biochemistry and microinjection techniques, and is limited to proteins that are sufficiently abundant and soluble, and cells that are amenable to microinjection. The development of expression plasmids for linking the cDNA encoding a protein of interest, or a fragment of one, to the cDNA coding for green fluorescent protein (GFP) has supplied the potential for examining a much wider range of proteins and cells during a variety of motile activities.

During the assembly of a myofibril, filaments of actin and myosin, and the many proteins associated with them, become aligned with one another in a series of steps that have not yet been defined (Epstein and Fischman, 1991; Dabiri *et al.,* 1997). Time-lapse observations in live muscle cells of the proteins involved would provide a framework for analyzing the process of myofibril assembly. Introduction of high levels of mutated or truncated forms of a protein could also provide evidence of the role of the protein in myofibril formation. Expression of fragments of a protein can indicate the region of the protein required for its proper localization in the myofibril. Transfection of muscle cells with plasmids expressing GFP-linked proteins provides the means for addressing such topics.

The present chapter describes results obtained with GFP-linked α-actinin, myosin light chain, and the Z-band portion of the titin molecule.

II. Construction of GFP-Linked Muscle Proteins

A. Choosing the N-Terminal or C-Terminal Site for the Addition of the GFP

GFP is a relatively large protein (238 amino acid residues) with a specific tertiary structure, which is required for its autofluorescence (Ormo *et al.,* 1996; Yang *et al.,* 1996). Despite its large size, GFP has been successfully fused to proteins of varying size without affecting GFP fluorescence or activity of the fused protein of interest (Chalfie *et al.,* 1994; Gerisch *et al.,* 1995; Rizzuto *et al.,* 1995; Stearns, 1995; Moores *et al.,* 1996). However, in some cases, the function of the fused protein may be affected by the proximity of GFP to an active site. For example, addition of GFP to the C terminus of α-actinin resulted in a fluorescent GFP–α-actinin fusion protein that specifically localized to the Z-bands of muscle sarcomeres (Fig. 1); whereas addition of GFP to the N terminus, which is near the actin-binding site (Arimura *et al.,* 1988), interfered with the proper localization of α-actinin in cells and resulted in diffusely fluorescent cells

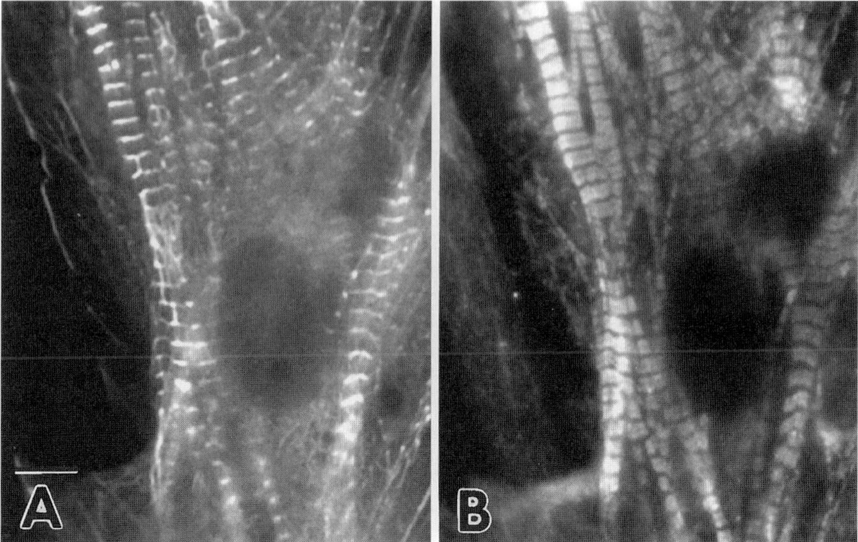

Fig. 1 Cardiomyocyte transfected with GFP–α-actinin and then fixed and stained with an antimuscle myosin II antibody. (A) Note the presence of the labeled α-actinin in the Z-bands of the cardiac muscle cell. The GFP was coupled to the carboxyl terminus of the cDNA coding for α-actinin. (B) The myosin antibody stains the broad A-bands that are localized between the Z-bands in (A). Scale = 10 microns

(Dabiri *et al.,* 1997). Similar effects of GTP placement with relation to the cDNA of interest have been observed with a fragment of titin (Turnacioglu *et al.,* 1996).

The regulatory light chain of myosin is a small protein (a sequence of 171 amino acids) (Zavodny *et al.,* 1988) that localized in the A-bands of muscle as a GFP fusion protein when the GFP was placed near the N terminus of the protein (Fig. 2). Although we did not try GFP fusion at the C terminus of the light chain, we predict that, like α-actinin, the location of GFP would be critical to the proper localization of the myosin light chain because the actin-binding site of the regulatory light chain is near its C terminus.

An alternative approach to avoiding inhibition of protein function due to GFP is to introduce a short amino acid linker between GFP and the protein of interest. In our GFP–myosin light chain construct there is a 12–amino acid linker between GFP and the myosin light chain. The effects of linker length with GFP fusion to yeast actin (Act1p) have been shown by Doyle and Botstein (1996).

B. Designing the Primers

As with construction of any fusion protein, it is important to have start and stop codons for proper protein expression. Most commercially available GFP vectors (e.g., Clontech and Quantum) have engineered these codons for expres-

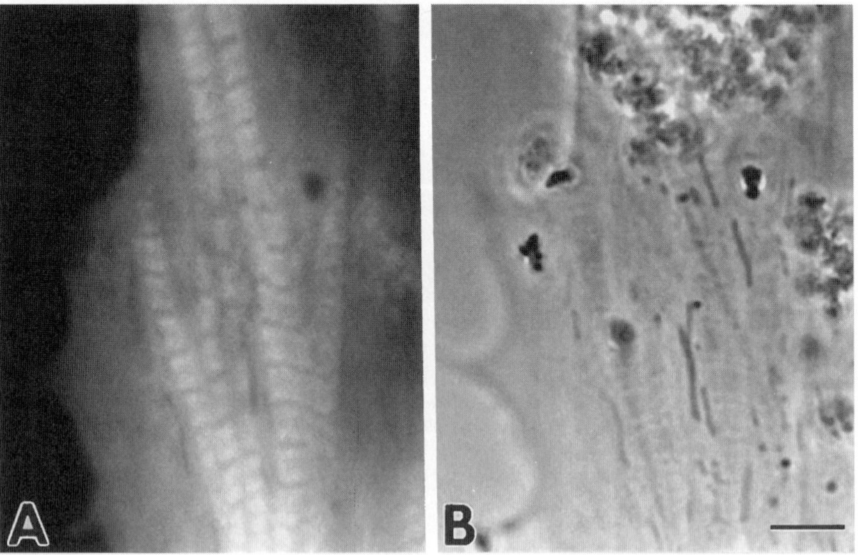

Fig. 2 (A) Fluorescent and (B) phase-contrast images of the same living cardiomyocyte transfected with a plasmid coding for GFP coupled to the cDNA for myosin light chains. Note that the GFP–myosin light chain is concentrated in the A-bands of the sarcomeres. The less bright area in the center of the A-bands reflects the absence of myosin light chains. Scale = 5 microns.

sion of GFP itself. When expressing a full-length cDNA with GFP, it is essential to remove the stop codon when fusion is to the 5′ end of GFP. Similarly, the stop codon in the GFP sequence should be removed if the fusion is at its 3′ end. Addition or elimination of start and stop codons can be achieved easily by PCR techniques with the use of specific primers. In this manner, convenient restriction sites could be introduced as well for cloning purposes.

III. Preparation of Embryonic Avian Cardiomyocytes and Skeletal Muscle Myoblasts

Cells isolated from cardiac and skeletal muscle tissue of avian embryos will grow in tissue culture and form myofibrils, allowing the process of myofibrillogenesis to be analyzed in fixed or living single cells with a variety of probes (Sanger *et al.,* 1984, 1986; Dome *et al.,* 1988; Epstein and Fischman, 1991; Rhee *et al.,* 1994).

Embryonated eggs provide the starting material for the preparation of cultures of cardiomyocytes and skeletal muscle myotubes. The eggs are usually obtained a few days younger than the age used for the preparation of cells, and are best kept in a slowly rocking, humidified egg incubator at 37°C until use. The muscle cultures derived from the embryos should be kept in a tissue culture incubator at 37°C in 5% CO_2 with enough humidity in the incubator to slow the loss of fluid in the culture dishes. If the investigator is preparing muscle cells for the first time, it is highly recommended that the cells be isolated and cultured once or twice before any attempt at transfection studies is made. The embryonic avian cardiomyocytes spread and beat in culture over several days and occasionally undergo cell divisions (Chacko, 1973; Sanger 1977). It is difficult to see myofibrils in these live cells with phase-contrast microscopy because of the density of the cytoplasm. The avian skeletal muscle cultures contain mononuclear myoblasts that will begin to fuse on the second and subsequent days of culture to form long multinucleated myotubes that contract. The myofibrils are easier to see in the myotubes, particularly in the quail myotubes, because they are aligned across the cell at the level of their Z-bands. The procedures that have worked for us in the study of spreading cardiomyocytes are presented in a step by step procedure listed in Table I. A similar approach for the isolation of myoblasts from embryonic avian breast muscle is presented in Table II. These procedures have been modified over the years from procedures described in the following papers: (for cardiomyocytes) Cavanaugh (1955), DeHaan (1968), and Sanger (1977); (for skeletal muscle cells) Bischoff and Holtzer (1968), Yaffee (1968), Fischbach (1972), and Sanger *et al.* (1986).

IV. Methods of Transfection of Cross-Striated Cells in Culture

We have found that, in general, DNA transfection of primary cells is very inefficient compared to cell lines. In this section we describe conditions that

Table I
Chick Embryonic Heart Myocytes Preparation

Necessary material: (have ready ahead of time)

a. Heart medium
b. Two 10-ml tubes of 0.05% Trypsin–0.53 m*M* EDTA
c. 100-mm and 35-mm Petri dishes
d. 100-mm tissue culture dish
e. Ten 7-day-old fertilized chick eggs
f. Waste beaker
g. Sterile forceps and irrecdectomy knives

To make 100 ml of heart medium:

Fetal bovine serum (10 ml)
Fill to total volume of 100 ml with MEM (w/o L-glutamine).
Filter through a 200-ml, 0.22-micron filter system

1. Rinse eggs with a flow of 70% ethanol and air dry.
2. Break open egg, remove the embryo and collect in a 100-mm Petri dish.
3. Dissect out the beating red hearts and place them in a single pile in a 35-mm petri dish.
4. Cut the pile with fine, sharp irredectomy knives. At the end of the operation, each heart should be in about eight pieces.
5. Add 10 ml of trypsin solution to the dish and place the tissue into a 14-ml plastic tube.
6. Place the tube in a 37°C incubator (5% CO_2) for 15 min.
7. Invert the tube to disturb settled cells.
8. Let pieces settle again for about 2 min. Slowly remove trypsin solution above the pellet with pipette. This removes most of the nucleated red blood cells.
9. Add fresh trypsin (1 ml/heart) to the minced heart pieces.
10. Place the tube for another 15 min in the incubator.
11. Spin for 5 min at a setting of 5 in the table top centrifuge (we use an IEC Clinical centrifuge).
12. Gently resuspend the pellet in 10 ml of the normal heart medium (1ml/heart). Titurate the pellet (i.e., draw the pieces up and down until they have separated into cells).
13. Preplate the cell suspension of the 100-mm tissue culture dish for at least 1 h in the incubator. The cardiac fibroblasts will attach readily to the dish.
14. Gather medium from the preplate dish. Wash this dish lightly several times with the previously collected medium to release any loosely attached cardiac muscle cells.
15. Gently wash the preplate dish with a fresh 10 ml of avian muscle culture medium. Collect and add this medium to the previous wash collected in step 14.
16. Count the collected cells in the hemocytometer.
17. A sterile glass coverslip should be placed in the middle of 35-mm plastic dishes. These glass cover slips are 22 × 22 mm and 1.5 mm thick. The glass coverslips can be made sterile by overnight UV illumination. Sterile plastic dishes are also available that have glass bottoms (MatTek, Ashland, MA), which are particularly useful for observations of live cells on an inverted microscope.
18. Plate the cells onto the 35-mm dishes at a density of 0.5×10^6 cells per dish. The cardiac muscle cells attach to the glass surfaces as round balls and gradually spread on the surfaces after 24 h. The cardiomyocytes are often beating in culture. Fibroblasts will also be present in the cultures. The lack of glutamine in the cultures slows the growth of the fibroblasts. (Inhibitors of DNA synthesis are not used because the embryonic cardiomyocytes can divide).
19. Feed the cardiac muscle cultures every other day. The muscle cells are healthy for about a week under these conditions.

Table II
Avian Myotube Preparation

Necessary material: (have ready ahead of time)

a. Collagen-coated 35-mm dishes (glass-bottom or coverslips)
b. Avian myotube medium
c. Sterilized Sweeney filters
d. Sterilized forceps
e. 10-day-old quail or chick eggs
f. 100-mm Petri and tissue culture dishes
g. Waste bucket
h. Hank's solution without $CaCl_2$, $MgCl_2$, or $MgSO_4$ (alternate solution: two 10-ml tubes of 0.05% Trypsin–0.53 m*M* EDTA)

To make 100 ml of medium:

10 ml Chick Embryo Extract
10 ml Horse Serum
1.5 m L-glutamine
Fill up to 100 ml with MEM (without L-glutamine)

1. Dishes containing coverslips or glass-bottom dishes (MatTek, Ashland, MA) should be coated with collagen several hours before the start of the skeletal muscle preparation. The collagen-coated dishes should be placed in a tissue culture hood with the UV light on. The lids of the dishes are removed from the dishes but left in the hood. This allows the collagen-coated dishes to be dried and sterilized at the same time. (We usually do this the night before the prep.)

2. Rinse eggs with 70% ethanol and air dry.

3. Break open eggs, remove the embryos and collect into a 100-mm Petri dish.

4. Decapitate the embryo, spread the legs to stabilize, and gently peel off skin tissue from chest area.

5. Scoop out the breast tissue by pinching the muscle with the curved ends of the forceps. Place the excised tissue into a 35-mm Petri dish. Make sure there are no bones in the muscle sample.

6. Add 10 ml (or 1 ml/embryo) of Hank's balanced salt solution without $CaCl_2$, $MgCl_2$, or $MgSO_4$ to this dish. (An alternate dissociating solution is the use of trypsin–EDTA. Ten-day-old and younger avian muscle readily disassociates in Ca–Mg-free solutions. (One must use trypsin solutions to dissociate myoblasts from muscles of older embryos.) Suck up the suspension of tissue and empty it into a 14-ml tube.

7. Place the test tube with tissue in the tissue culture incubator (37°C, 5% CO_2) for 15 min.

8. After 15 min, change the Hank's solution and incubate for another 15 min in the incubator.

9. Spin for 5 min at a setting of 6 in the tabletop centrifuge.

10. Gently remove the supernatant without disturbing the pellet. Gently resuspend the pellet in the 10 ml of Avian Myotube medium (1 ml/embryo). Let the resuspended pellet stand for a minute to allow any large pieces of tissue to settle.

11. Carefully remove the supernatant and filter it through a Sweeney filter using a 10-ml syringe.

12. Preplate the filtrate of cells for 1 h in a 100-mm tissue-culture-grade Petri dishes to remove excess fibroblasts.

13. Collect the medium from the preplate dish. Wash this dish lightly several times with the previous collected medium to release any loosely attached myoblasts.

(*continues*)

Table II *continued*

14. Gently wash the preplate dish with a fresh 10 ml of avian muscle culture medium. Collect and add this medium to the previous wash collected in step 13.

15. Count the collected cells in a hemocytometer.

16. Plate the cells out onto the collagen coated dishes at a density of 0.5×10^6 cells per dish.

17. Change the medium after two days of culture in the incubator. Many of the myoblasts have exited the mitotic cycle by this time. These cells begin to fuse with one another to form elongated multinucleated myotubes. To inhibit the growth of the fibroblasts in these cultures, cytosine arabinoside (ara-c) is added at a concentration of 10^{-5} *M* on day 3 of culture.

18. Feed the skeletal muscle cultures every other day. The muscle cells are healthy for about a week under these conditions. Myofibrils form in these myotubes on the third day of culture. The contractions of the myotubes begin to pull the cells from the surface of the dish by the end of the week.

allow transfection of 5–10% and 50% of a primary cell population of embryonic cardiomyocytes and skeletal myotubes, respectively, with GFP probes, which are sufficient for microscopic examination (Dabiri *et al.,* 1997).

A. Lipid-Based Methods

Despite the basic similarities of their myofibrils, avian embryonic cardiomyocytes and skeletal muscle myotubes react quite differently to the standard techniques of DNA transfection. In our experience, embryonic chick cardiomyocytes (see Table I) can be transfected with DNA only with lipid-based reagents such as Lipofectamine (Gibco-BRL). Optimally, cardiomyocytes isolated from 7-day-old chicken embryos should be plated one day before transfection at 50% confluency ($\sim 0.5 \times 10^6$ cells/35-mm dish) (Dabiri *et al.,* 1997). For each 35-mm dish of cells to be transfected, approximately 1–2 μg DNA is added to 100 μl of serum-free culture medium in one tube (tube A), and 10 μl Lipofectamine is added to 100 μl of serum-free medium in a second tube (tube B). The contents of tube A are mixed with tube B, and the combined mixture is incubated at room temperature for 15 to 30 min following the manufacturer's protocol. At the end of the incubation, 800 μl of serum-free medium is added to the mixture and the mixture is added to a dish of cells that has been rinsed previously with serum-free medium. The DNA–lipid mixture remains on the cells for 5–6 h before the addition of 1 ml of culture medium containing serum. One day after transfection this medium is replaced with medium containing serum. Between 18 and 48 h after transfection, GFP fluorescence becomes apparent. The procedures presented here are optimized for chick embryonic cardiomyocytes and the DNA constructs described in this chapter; however, it is highly recommended that the conditions be optimized when using different constructs and cells.

B. Calcium Phosphate Method

Unlike cultured embryonic cardiomyocytes, cultured myotubes are unable to take up DNA with lipid-based reagents. However, they are easily transfected with standard calcium phosphate methods (Sambrook *et al.,* 1989). Chicken myotubes, from 10- to 11-day-old eggs (see Table II for details of skeletal muscle preparation), are plated at 0.5×10^6 cells per 35-mm dish one day before transfection. The next day, when the cells reach a 50–80% confluency, the medium is replaced with 2 ml of fresh medium, approximately 2 h before transfection. For each 35-mm dish of cells, calcium phosphate–DNA precipitate is then prepared as described later.

In one tube, 2 μg DNA in 125 μl sterile distilled water is mixed with 125 μl of 2x HBS (HEPES buffered saline: 250 m*M* NaCl, 10 m*M* KCl, 1.5 m*M* $Na_2HPO_4 \cdot 2H_2O$, 12 m*M* dextrose, 50 m*M* HEPES). To this mixture, 15.5 μl of 2 *M* $CaCl_2$ is slowly added in a dropwise manner while vortexing vigorously. The mixture is incubated at room temperature for 30 min, at which time the medium of each dish is replaced with 2 ml of fresh medium. Subsequently, 200 μl of the calcium phosphate–DNA mixture is added directly to the 2 ml of freshly added culture medium in each dish of cells and transfection is allowed to occur overnight at 37°C. In each case, the medium is exchanged the next day with fresh culture medium and the cells are allowed to incubate for another 24 h. This procedure has resulted in approximately 50% transfection efficiency of myotubes with α-actinin-GFP (G. A. Dabiri, J. M. Sanger, and J. W. Sanger, unpublished data). Similar procedures have been used to transfect myotubes with other non-GFP probes (Schultheiss *et al.,* 1991).

C. Microinjection Method

An alternate approach to the DNA transfection methods described is the direct microinjection of DNA into cells (Capecchi, 1980; Mittal *et al.,* 1992). This method, though technically more difficult (Pochapin *et al.,* 1983; Sanger *et al.,* 1985), allows for introduction of DNA into specific cells of interest, potentially valuable for time-lapse microscopic examination of single living cells. An added advantage of microinjection is that it allows two different agents to be introduced simultaneously into a single cell (Sanger *et al.,* 1995), such as an antibody or a rhodamine-labeled protein along with a GFP fusion probe.

V. Transfection of Cross-Striated Muscle Cells with Full-Length cDNA for Sarcomeric Proteins

A. α-Actinin

Embryonic chick cardiomyocytes transfected with a plasmid coding for α-actinin–GFP exhibited fluorescently labeled Z-bands within 24 hours of transfection (Fig. 1A). When the transfected cells were fixed, permeabilized, and stained with various antibodies (sarcomeric α-actinin or muscle-specific myosin II) and rhodamine-labeled secondary antibodies, or with rhodamine phalloidin, the GFP fluorescence persisted, allowing confirmation that the α-actinin–GFP was indeed located in the Z-bands. There was a colocalization of α-actinin antibodies with the α-actinin–GFP in the Z-bands (Dabiri *et al.,* 1997), and a complementary staining pattern was present along the myofibril with α-actinin–GFP and myosin II antibodies (Fig. 1B). α-Actinin–GFP was also incorporated into all structures of cardiomyocytes known to contain α-actinin: focal adhesions, ruffles, and intercalated disks (Dabiri *et al.,* 1997). In living transfected cells, contractions of transfected cells were not affected by the presence of α-actinin–GFP in their Z-bands. Similar localization of α-actinin–GFP was obtained with transfected skeletal muscle cells (Fig. 4).

To demonstrate that the localization of GFP-α–actinin was not due to a nonspecific reaction of the GFP with Z-bands, cardiomyocytes were transfected with a plasmid encoding only GFP. The fluorescent protein was diffusely distributed throughout the cell, especially in the nucleus (Fig. 3A). Contraction was unaffected and rhodamine phalloidin staining showed a normal distribution of myofibrils in the GFP-transfected cardiomyocytes (Fig. 3B). When GFP was linked to the N terminus, near the actin-binding region of α-actinin, a similarly diffuse distribution of GFP fluorescence was seen (Dabiri *et al.,* 1997).

The real potential of GFP-labeled probes lies in their use in studies of processes in live cells. Expressing GFP probes for cytoskeletal proteins in muscle cells allows for real-time observations of the assembly, distribution, and dynamics

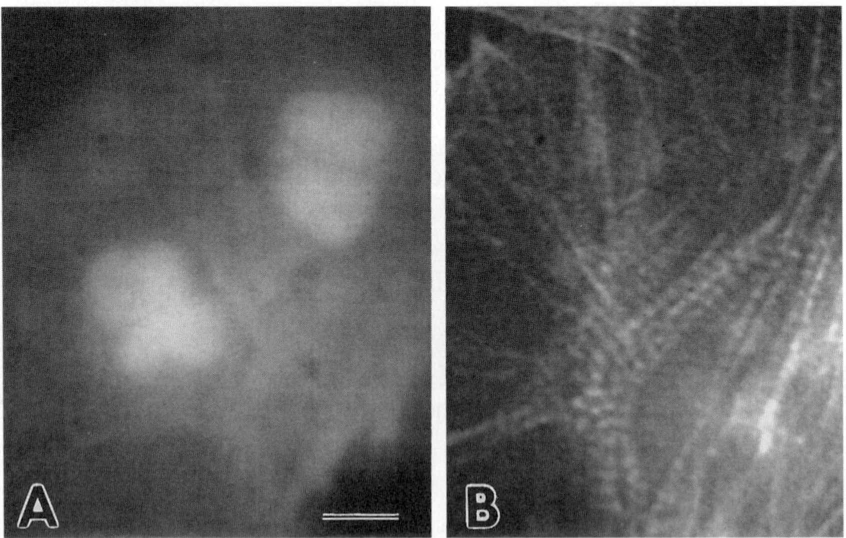

Fig. 3 (A) Image of a living cardiomyocyte transfected with a plasmid expressing GFP alone. There is a diffuse concentration of the GFP in the cell with the highest concentration in the nucleus. The normal contractions of the muscle were unaffected by the diffuse soluble GFP. (B) The cell was subsequently fixed and stained with rhodamine-labeled phalloidin. Scale = 10 microns.

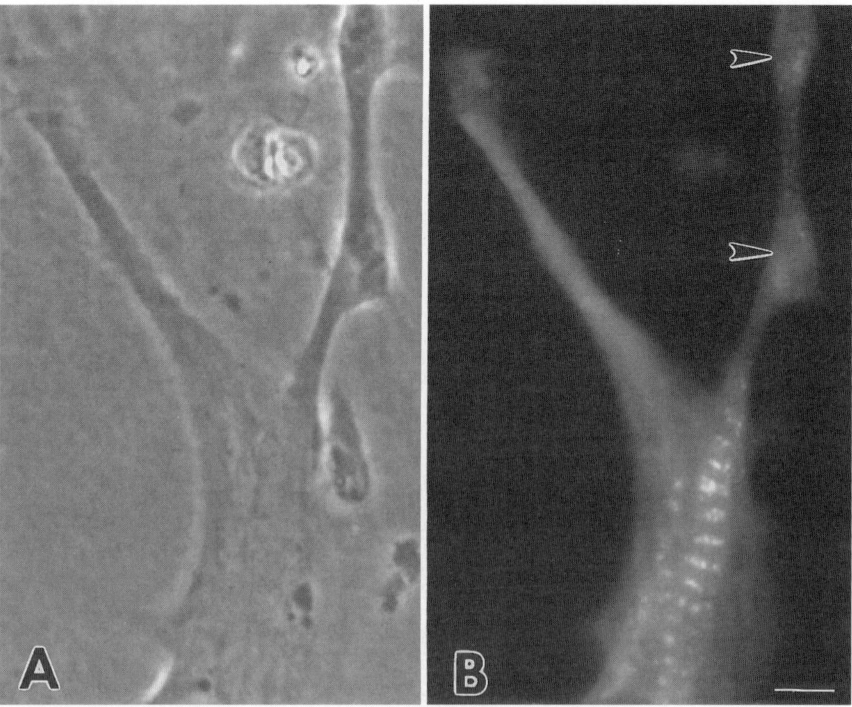

Fig. 4 (A) Phase contrast image and (B) fluorescent image of living skeletal muscle cells transfected with a plasmid coding for α-actinin coupled to GFP. α-actinin–GFP is localized in the Z-bands of the myotube, and is diffusely distributed in myoblasts (arrowheads) that have fused with the myotube. Scale = 10 microns.

of the different elements of the myofibrils. We chose α-actinin for a study of myofibrillogenesis in embryonic cardiomyocytes because its concentration in the Z-bands marks the edges of each sarcomere. Figure 5 illustrates the result of one such experiment, in which α-actinin–GFP was used to follow the assembly of myofibrils in areas of a cell initially lacking any such structures. As the cardiomyocyte spread laterally and increased in length, α-actinin–GFP was first deposited in small aligned dense bodies, termed Z-bodies, near the cell periphery. It had been proposed from observations of fixed cardiomyocytes and stained with various cytoskeletal antibodies at different stages of spreading, that linear arrays of Z-bodies (termed premyofibrils) aligned laterally with other premyofibrils and then fused together at the level of the Z-bodies to form the Z-bands of the mature myofibrils (Rhee *et al.*, 1994). Time-lapse observations of living spreading cardiomyocytes transfected with α-actinin–GFP (Dabiri *et al.*, 1997) support this premyofibril hypothesis for the assembly of myofibrils in cardiomyocytes and also indicate that existing mature myofibrils can elongate by the lateral coalescing of adjacent myofibrils (Fig. 5).

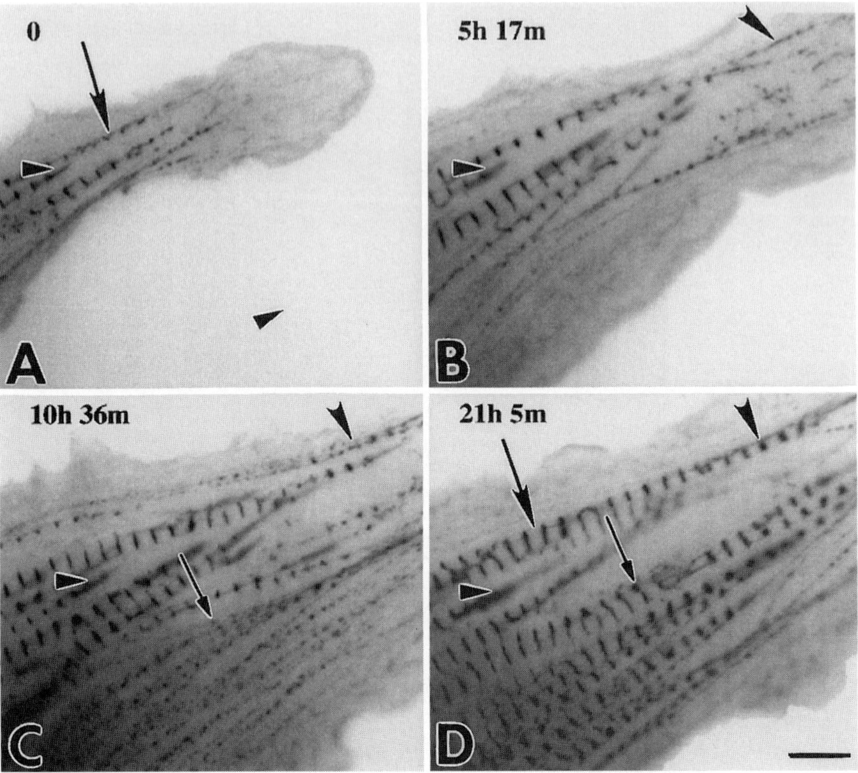

Fig. 5 Time-lapse observations of an embryonic chick cardiomyocyte transfected with α-actinin–GFP. The cell, shown in reverse contrast with the time of acquisition relative to (A) indicated on each micrograph (B–D) was beating as it spread in culture. The same prominent adhesion plaque is marked in each panel as a reference point (A–D, horizontal small arrowheads). Note the extent of spreading with respect to this plaque. As the cell spreads toward the lower part of the image, punctate Z-bodies assemble into linear arrays that mature into Z-bands of myofibrils. One myofibril indicated by a large arrow (A) appears to double in length over 21 h (D). This appears to occur by lateral coalescence of myofibrils (B to D, larger oblique arrowheads). Adjacent Z-bodies (B, small arrow), increase in diameter (C, small arrow) and fuse (D, small arrow) into a mature Z-band. Scale = 5 microns. Images reproduced with permission from the *Proceedings of the National Academy of Sciences.*

Embryonic cardiomyocytes can undergo mitosis and cytokinesis in culture (Chacko, 1973; Sanger, 1977). Mitotic cardiomyocytes were also detected in transfected cultures. In these cells as the chromosomes separated, α-actinin–GFP assembled in the cleavage furrow that formed in the region of the former metaphase plate (Fig. 6A). The concentration of α-actinin–GFP in the furrow was composed of small fluorescent patches similar to those detected in cleaving cardiomyocytes that had been stained with sarcomeric α-actinin antibodies (Sanger *et al.*, 1994a). These patches were also similar to structures detected

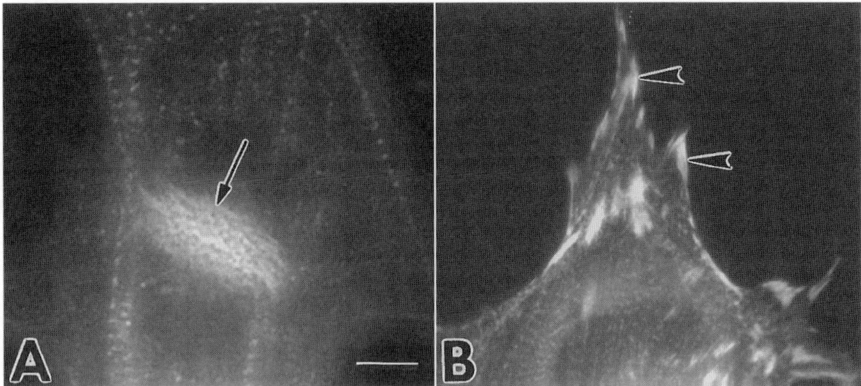

Fig. 6 (A) Living cardiomyocyte and (B) cardiac fibroblast transfected with a plasmid coding for the cDNA for α-actinin coupled to that of GFP. Note the periodic distribution of α-actinin in the cleavage furrow (arrow) of the mitotic cardiomyocyte (A). In the fibroblast (B), α-actinin–GFP is concentrated in the focal adhesions and in the periodic stress fibers. These α-actinin–GFP–staining dense bodies marked the sarcomeric units of the stress fibers. Scale = 10 microns.

in cleaving PtK2 cells previously microinjected with fluorescently labeled talin (Sanger *et al.*, 1994b), suggesting that these patches in the cleavage furrows may be involved in the membrane attachment of the actin filaments.

Because fibroblasts are also present in primary cultures of cardiomyocytes, it is not surprising to find transfected fibroblasts expressing α-actinin–GFP (Figs. 6B and 7A). The fluorescent probe is found in all areas of these nonmuscle cells known to contain concentrations of α-actinin: the dense bodies of stress fibers, focal adhesions, and ruffles. It is known from studies of nonmuscle cells, previously injected with rhodamine-labeled α-actinin, that there is a pool of the α-actinin that can exchange with α-actinin in the focal contacts (Kreis *et al.*, 1984).

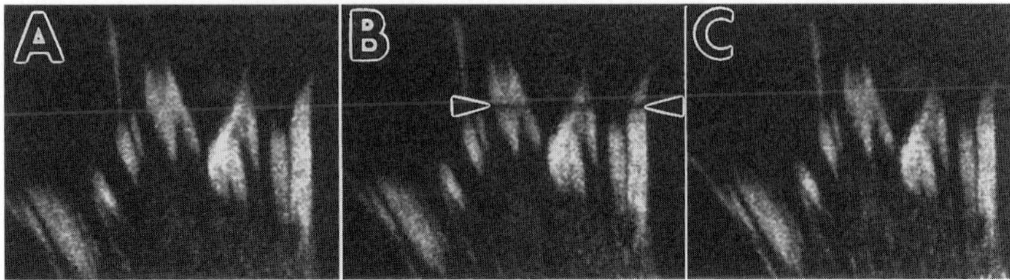

Fig. 7 Fibroblast transfected with α-actinin-GFP from a primary culture of cardiac muscle. Confocal micrographs illustrate: (A) the distribution of the α-actinin-GFP in the focal adhesions; (B) the loss of fluorescence in the focal adhesion plaques where a laser line, about 0.2 microns in width, was used to photobleach a band across the cell (arrowheads); (C) the recovery of fluorescence in the former photobleached area five minutes after the photobleaching indicating that a pool of α-actinin–GFP in this area exchanges with α-actinin–GFP in the focal adhesions.

To demonstrate that α-actinin–GFP molecules are also capable of similar dynamics, the focal contacts in these transfected fibroblasts were photobleached with a narrow slit of intense laser light (Fig. 7), resulting in a dark narrow band across several focal contacts (Fig. 7B). This photobleached area was gradually replaced in the focal adhesions with α-actinin–GFP from the cytoplasmic pool (Fig. 7C). These experiments are an example of how α-actinin–GFP can be used to follow the dynamics of α-actinin in both muscle and nonmuscle cells.

B. Myosin Light Chains

In cardiomyocytes transfected with plasmids encoding for GFP linked to the N terminus of the regulatory light chain, the expressed construct was incorporated into the phase-dense A-bands of the myofibrils (Fig. 2). Transfections of these cultures with GFP–myosin light chain did not affect their contractions or the stability of their myofibrils, which was similar to previous results following microinjection of rhodamine-labeled myosin light chains into cardiomyocytes (Mittal *et al.,* 1987). Like injected rhodamine light chains, the GFP–myosin light chains were distributed initially in nuclei as well as the cytoplasm; but whereas the injected light chains gradually left the nuclei, the GFP–myosin light chains remained in nuclei over the many hours of observation. We presume that the small size of the myosin light chains allows them to diffuse into nuclei (Mittal *et al.,* 1987), and in transfected cells continual synthesis of the GFP–myosin light chains may produce more light chains than can be incorporated into the myosin molecules in the A-bands.

C. Fragments of Titin

Titin is the largest known protein, with a molecular weight of about 3 million daltons (Labeit and Kolmerer, 1995). The N-terminal region of the molecule is embedded in the Z-bands of both cardiac and skeletal muscle cells, and the C terminus, about 1 micron away from the N terminus, is associated with the middle of the thick filaments. Zeugmatin, originally identified as a Z-band protein, was discovered to be part of the N-terminal region of titin (Turnacioglu *et al.,* 1996). A 1.1-kb fragment of zeugmatin (Z1.1), or titin, coupled to GFP at the fragment's C terminus, was localized to the Z-bands of transfected cardiac and skeletal muscle cells (Fig. 8; Turnacioglu *et al.,* 1997a,b). If the plasmid was constructed so that GFP was ligated to the N terminus of this titin fragment, there was no incorporation into any structures of the transfected muscle cells (Turnacioglu *et al.,* 1997b), indicating that GFP at its N-terminal region interferes with the targeting site of the Z1.1 fragment.

Two different long-term effects on the stability of the myofibrils were observed in cells transfected with the titin fragment–GFP (Turnacioglu *et al.,* 1997b). In cells expressing low amounts of the titin fragment, the Z-bands remained uniformly fluorescent and the myofibrils were stable (Fig. 9A and B). In transfected

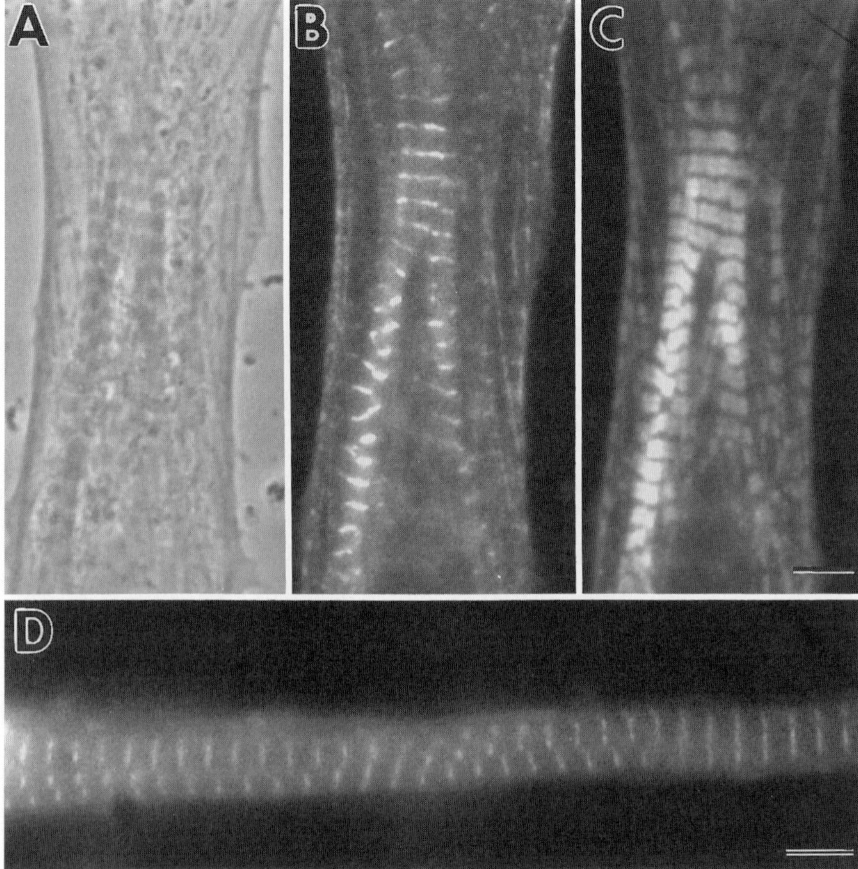

Fig. 8 Cardiac (A–C) and skeletal (D) muscle cells. Phase-contrast (A) and fluorescent images (B and C) of a cardiomyocyte transfected with a plasmid expressing a truncated region of titin (1.1 kb) coupled to GFP (Z1.1–GFP) (B and C), then stained with a muscle-specific myosin antibody. Note the incorporation of the truncated titin–GFP into the Z-bands and the myosin antibody staining of the intervening A-bands (C). (D) Skeletal muscle cell transfected with a plasmid expressing the entire Z-band region of titin (1000 amino acids) coupled to GFP. The GFP probe is incorporated into all of the Z-bands of the transfected myotube. Scale = 5 microns.

cells expressing higher levels of the 1.1-kb titin–GFP construct, the Z-bands and myofibrils disassembled (Fig. 9C and D). These observations could be quantitated by measuring the intensities of fluorescence in the Z-bands and between them in the middle of the A-bands. The ratio of the two intensities was plotted over the 40 to 50 h of observation (Fig. 10). In cells expressing lower amounts of the protein, the ratio remained constant, whereas in overexpressing cells, the fluorescence between the Z-bands steadily increased with respect to the Z-band fluorescence. We interpreted these measurements as indicating that very high

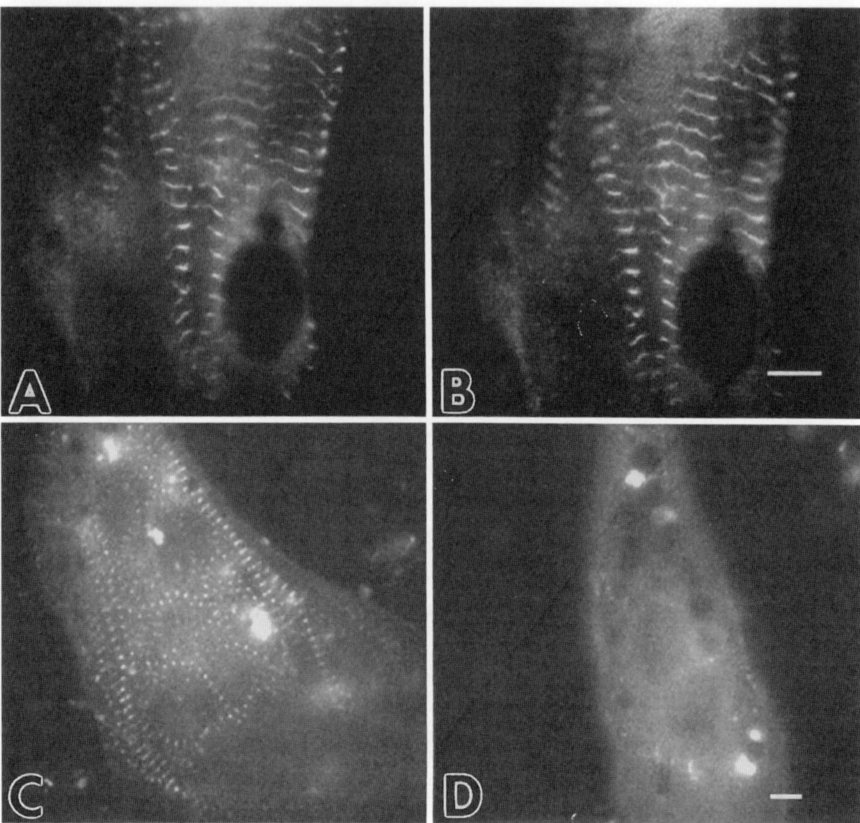

Fig. 9 Time-lapse observations of two different embryonic cardiomyocytes transfected with a plasmid coding for part of the Z-band region of titin (Z1.1) responsible for targeting titin to Z-bands. In the cell in (A), there was no discernible change in the myofibril arrangement and contraction over 36 h. This cell is shown 36 h later in (B), still contracting with intact labeled Z-bands. In the second cell, with a high level of background fluorescence (C), there was a loss of Z-bands and contractions 12 h later (D). The GFP tag demonstrates that initially the truncated titin targets to the Z-bands with no apparent effect on the structure and beating of the myofibrils in the cardiomyocytes (C). However, as the product to continues to be produced, it leads to the disassembly of the myofibrils and the loss of contractions (D). Images reproduced by permission from "Molecular Biology of the Cell." (A and B) Scale = 10 microns. (C and D) Scale = 6 microns.

levels of the GFP–titin fragment were competing for the same site in the Z-band as full-length endogenous titin, and with increasing synthesis of the GFP–titin fragment, intact titin was unable to bind the Z-bands (Turnacioglu *et al.*, 1997b). Because titin attaches the A-bands to the Z-bands, overexpression of the same Z-band fragment of titin might be expected to interfere with A-band alignment as well, which was also observed to be the case (Turnacioglu *et al.* 1997b).

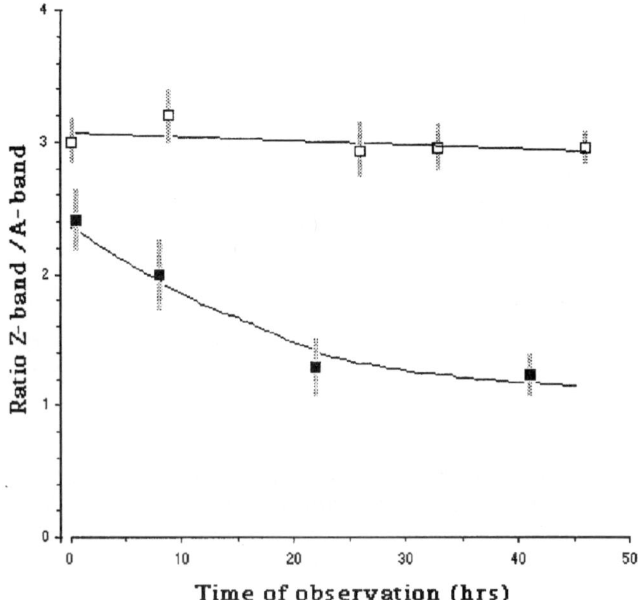

Fig. 10 Graph of ratios of Z-band fluorescence intensities to A-band intensities plotted over time of observation for two different cardiomyocytes illustrated in Fig. 9. The pixel intensity of the same Z-band and a region between the Z-bands (i.e., A-band region) was measured, after background subtraction, in cCCD images that were acquired every 2 h. The cardiomyocyte in Fig. 9 (A and B), which shows no loss of Z-bands or contractility over 50 h of observation, yielded a linear plot that showed no increase in background fluorescence (open squares). The cardiomyocyte in Fig. 9 (C and D) showed Z-band disassembly concomitant with a rise in background fluorescence. Measurements from this cell yielded a tapering curve (filled squares) as background fluorescence rose to Z-band levels. Four pairs of pixel intensities were acquired for each time point shown, and the gray bars are equal to 1 SD unit. Figure reproduced by permission from "Molecular Biology of the Cell."

VI. Microscopic Observations of Live Cells

A. Culture Chambers

1. Culture Dishes for Inverted Microscope

For time-lapse observations of live cardiac or skeletal myocytes, cells were grown on glass-bottom dishes (MatTek, Ashland, MA). These are 35-mm-diameter plastic culture dishes that have a glass coverslip glued over a 1-cm circular opening in the bottom dish. Embryonic cardiomyocytes and fibroblasts, as well as a number of tissue culture cell lines (PtK2, REF-52, Caco-2, LLC-PK) grow readily on the glass bottom of these dishes. Collagen (Type I rat tail collagen, 5 mg/ml; Collaborative Biomedical Products, Bedford, MA) was applied to the coverslip for the skeletal myotube culture. The dishes can be directly transferred from the incubator to the stage of an inverted microscope, in which the cells can be microinjected and observed with low-power and oil-immersion lenses without being mounted in special chambers.

2. Maintaining Temperature and CO_2

Temperature was maintained at 37°C on the microscope stage with a heat curtain (ASI 400 Air Stream Incubator; NevTek, Burnsville, VA). A homemade plastic cover was fashioned from the bottom of a 100-mm plastic Petri dish to contain an atmosphere of 5% CO_2 around the glass-bottom plastic culture dish. A needle hole was made in the side of the top of a Petri dish by pushing a heated 22-gauge needle through it, and 5% CO_2 was perfused into the perforated Petri dish through tubing connecting a tank of 5% CO_2 to a disposable 22-gauge needle inserted into the Petri dish. Plastic wrap was placed over the stage, with a hole for the objective lens, and the perforated Petri dish was inverted over the MatTek culture dish on the microscope stage. A weight such as an annular microscopic stage insert for larger culture dishes was placed over the Petri chamber to press it onto the plastic wrap and slow the escape of the CO_2 and water vapor. The CO_2 was bubbled through distilled water to humidify it, and in addition, drops of distilled water were placed around the culture dish on the plastic wrap surface under the Petri chamber. The Petri dish can be easily lifted to change the culture medium in the dish when desired. For long-term observations medium was changed about every 12 h.

B. Observation and Imaging of the Transfected Cells

1. Cameras

Cells were observed with phase-contrast and epi-fluorescence optics, with pairs of images acquired sequentially with a cooled CCD camera (Photometrics AT 200; Photometrics, Tuscon, AZ) controlled by Metamorph software (Universal Imaging, West Chester, PA). For living cells, exposure to light from the mercury lamp was as brief as possible to avoid photodamage to the cells. Exposures of 500 ms were usually sufficient. Thick cells required images to be acquired in several focal planes to visualize all myofibrils in the area of the cell being followed.

2. Time-Lapse Acquisition and Image Processing

With Metamorph software, images were collected in time-lapse sequences at intervals from as short as every 2 h to as long as every 12 h. When fluorescent images of several planes were collected for one time point, they were combined digitally to reveal all the fibrils present. In some cases, the contrast of GFP fluorescence was inverted to more clearly illustrate the fluorescent structures. Sharpening filters were sometimes applied to the fluorescent and phase-contrast digital images (Inoué and Spring, 1997).

VII. Postprocessing of Transfected Cells

A. Antibody Staining

Muscle and nonmuscle cells transfected with GFP plasmids can be fixed, permeabilized, and stained with different specific antibodies (Figs. 1 and 8). GFP-transfected cells have been stained with antibodies directed against α-actinin, muscle myosin II, and nonmuscle myosin IIB (Dabiri *et al.,* 1997; Turnacioglu *et al.,* 1997a). The immunofluorescent images are identical to those of similarly stained control cardiomyocytes (Rhee *et al.,* 1994; Turnacioglu *et al.,* 1997b). The staining of Z-bands with sarcomeric α-actinin antibodies in cells transfected with either α-actinin–GFP or 1.1-kb titin–GFP indicates that the large size of the GFP probe does not block the access of the antibody.

B. Rhodamine Phalloidin

GFP transfected cells can also be fixed and stained with the F-actin staining agent, rhodamine-labeled phalloidin (Fig. 2; Turnacioglu *et al.,* 1997b). The presence of α-actinin–GFP along certain regions of the actin filaments does not interfere with the ability of rhodamine phalloidin to bind to the actin in these areas.

VIII. Problems Encountered in Cells Transfected with GFP–Sarcomeric Proteins

A. Under- and Overexpression

1. Underexpression

The underexpression of the GFP–cytoskeletal protein in transfected cultures is an indication that one needs to increase the concentrations of plasmids to the culture. Normally, 1–2 μg of DNA per 35-mm dish was used to successfully transfect cardiomyocytes and nonmuscle cells (Dabiri *et al.,* 1997; Turnacioglu *et al.,* 1997b). A 10-fold lower concentration of DNA was used in microinjection studies (Turnacioglu *et al.,* 1997b). It is a good idea to transfect one or two dishes of muscle cells with a plasmid coding for just GFP along with the experimental fusion GFP constructs to check culturing and transfection techniques.

2. Overexpression

Muscle cells are able to incorporate large amounts of rhodamine-labeled α-actinin (Sanger *et al.,* 1986). Similarly, we have found that α-actinin–GFP is readily incorporated into Z-bands in cardiomyocytes. The Z-bands appear to be able to bind high levels of α-actinin–GFP. We have not tried to increase the level of DNA (i.e., greater than 2 μg/ml/10^6 cells) to determine the effect of very high levels of α-actinin–GFP on myofibrillar stability.

It is clear that muscle cells do not retain their myofibrils in the presence of large amounts of a 1.1-kb fragment of the Z-band region of titin linked to GFP (Figs. 9 and 10). We did deliberately overexpress this titin fragment in culture. There were populations of cells expressing various levels of the GFP–protein. We used this dominant negative effect of truncated titin to examine possible roles for titin in myofibril assembly (Turnacioglu *et al.,* 1997b).

B. Photobleaching

To reduce photobleaching of the GFP signal, exposure of transfected cells to light should be minimized to prevent diminution of the fluorescence. Although epi-fluorescence is needed to locate a transfected cell, the cell should be positioned in the field of view and focused with phase-contrast microscopy before final imaging of the GFP signal. Neutral density filters should be inserted in the pathway of the exciting light to diminish the light's intensity. Exposures should be kept as short as possible (less than a second) to prevent photobleaching of the signal. In contrast to cells injected with a given volume of dye-labeled proteins, transfected cells produce increasing levels of product. Thus in experiments lasting many hours, the fluorescence intensity of the signal is usually replenished. Nevertheless, photobleaching of a defined region of a GFP-labeled structure can be analyzed by quantitation of the fluorescence recovery after photobleaching (FRAP) to measure exchange of the GFP protein within structures in the cell or movement of the protein within the structure (Cole *et al.,* 1996).

IX. Overview

The large size of the GFP tag does not appear to interfere with the incorporation of α-actinin, myosin light chains, and titin fragments into the appropriate sites of the myofibrils. The continued synthesis of the GFP-labeled cytoskeletal probes in the transfected muscle cells allows dynamic processes such as myofibrillogenesis to be studied in living cells over several days. Similar transfection studies with truncated parts of sarcomeric proteins linked to GFP should permit studies of the functions of different protein domains during myofibrillogenesis.

References

Arimura, C., Suzuki, T., Yanagisawa, M., Immura, M., Hamada, Y., and Masaki, T. (1988). Primary structure of chicken muscle and fibroblast alpha-actinins deduced from cDNA sequences. *Eur. J. Biochem.* **177,** 649–655.

Bischoff, R., and Holtzer, H. (1968). The effect of mitotic inhibitors on myogenesis *in vitro. J. Cell Biol.* **36,** 111–127.

Capecchi, M. R. (1980). High efficiency transformation by direct microinjection of DNA into cultured mammalian cells. *Cell* **22,** 479–488.

Cavanaugh, M. W. (1955). Pulsation, migration and division in dissociated chick embryo heart cells *in vitro. J. Exp. Zool.* **128,** 573–590.

Chacko, S. (1973). DNA synthesis, mitosis and differentiation in cardiac myogenesis. *Develop. Biol.* **35,** 1–18.

Chalfie, M., Tu, Y., Euskirchen, G., Ward, W. W., and Prasher, D. C. (1994). Green fluorescent protein as a marker for gene expression. *Science* **263,** 802–805.

Cole, N. B., Smith, C. L., Sciaky, N., Terasaki, M., Edidin M., and Lippincott-Schwartz, J. (1996). Diffusional mobility of Golgi proteins in membranes of living cells. *Science* **273,** 797–801.

Dabiri, G. A., Turnacioglu, K. K., Sanger, J. M., and Sanger, J. W. (1997). Myofibrillogenesis visualized in living embryonic cardiomyocytes. *Proc. Natl. Acad. Sci. U.S.A.* **94,** 9493–9498.

DeHaan, R. L. (1968). Emergence of form and function in the embryonic heart. *Develop. Biol.* **Suppl. 2,** 208–250.

Dome, J. S., Mittal, B., Pochapin, M. B., Sanger J. M., and Sanger, J. W. (1988). Incorporation of fluorescently labeled actin and tropomyosin into muscle cells. *Cell Differ.* **23,** 37–52.

Doyle, T., and Botstein, D. (1996). Movement of yeast cortical actin cytoskeleton *in vivo. Proc. Natl. Acad. Sci. U.S.A.* **93,** 3886–3891.

Epstein, H. F., and Fischman, D. A. (1991). Molecular analysis of protein assembly in muscle development. *Science* **251,** 1039–1044.

Fischbach, G. D. (1972). Synapse formation between dissociated nerve and muscle in low density cell cultures. *Develop. Biol.* **28,** 407–429.

Higley, S., and Way, M. (1997). Actin and cell pathogenesis. *Curr. Opin. Cell Biol.* **9,** 62–69.

Gerisch, G., Albrecht, R., Heizer, C., Hodgkinson, S., and Maniak, M. (1995). Chemoattractant-controlled accumulation of coronin at the leading edge of *Dictyostelium* cells monitored using a green fluorescent protein–coronin fusion protein. *Curr. Biol.* 5, 1280–1285.

Inoué, S., and K. R. Spring. (1997). "Video Microscopy: The Fundamentals," pp. 1–741. Plenum Press, New York.

Kreis, T. E., Avnur, Z., Schlessinger, J., and Geiger, B. (1984). Dynamic properties of cytoskeletal proteins in focal contacts. *In* "Molecular Biology of the Cytoskelton" (Borisy, G. G., Cleveland, D., and Murphy, D. B., Eds.), pp. 45–57. Cold Spring Harbor Press, Cold Spring Harbor, NY.

Labeit, S., and Kolmerer, B. (1995). Titins: giant proteins in charge of muscle ultrastructure and elasticity. *Science* **270,** 293–296.

Mittal, B., Sanger, J. M., and Sanger, J. W. (1987). Visualization of myosin in living cells. *J. Cell Biol.* **105,** 1753–1760.

Mittal, B., Danowski, B. A., Sanger, J. M., and Sanger, J. W. (1992). Expression of desmin cDNA in PtK2 cells results in assembly of desmin filaments from multiple sites throughout the cell. *Cell Motil. Cytoskel.* **23,**188–200.

Moores, S. L., Sabry, J. H., and Spudich, J. A. (1996). Myosin dynamics in live *Dictyostelium* cells. *Proc. Natl. Acad. Sci. U.S.A.* **93,** 443–446.

Ormo, M., Cubitt, A. B., Kallio, K., Gross, L. A., Tsien, R. Y., and Remington, S. J. (1996). Crystal structure of the *Aequorea victoria* green fluorescent protein. *Science* **273,** 1392–1395.

Pochapin, M. B., Sanger J. M., and Sanger, J. W. (1983). Microinjection of Lucifer Yellow CH into sea urchin eggs and embryos. *Cell Tissue Res.* **234,** 309–318.

Rhee, D., Sanger, J. M., and Sanger, J. W. (1994). The premyofibril: evidence for its role in myofibrillogenesis. *Cell Motil. Cytoskel.* **28,** 1–24.

Rizzuto, R., Brini, M., Pizzo, P., Murgia, M., and Pozzan, T. (1995). Chimeric green fluorescent protein as a tool for visualizing subcellular organelles in living cells. *Curr. Biol.* **5,** 635–642.

Sambrook, J., Fritsch, E. F., and Maniatis, T. (1989). "Molecular Cloning: A Laboratory Manual." Cold Spring Harbor Laboratory Press. Cold Spring Harbor, NY.

Sanger, J. W. (1977). Mitosis in beating cardiac myoblasts treated with cytochalasin-B. *J. Exp. Zool.* **201,** 463–469.

Sanger, J. M., Pochapin, M. B., and Sanger, J. W. (1985). Midbody sealing after cytokinesis in embryos of the sea urchin, *Arbacia punctulata. Cell Tissue Res.* **240,** 287–292.

Sanger, J. M., Mittal, B., Pochapin, M. B., and Sanger, J. W. (1986). Myofibrillogenesis in living cells microinjected with fluorescently labeled alpha-actinin. *J. Cell Biol.* **102,** 2053–2066.

Sanger, J. M., Mittal, B., Dome, J. B., and Sanger, J. W. (1989). Analysis of cell division using fluorescently labeled actin and myosin in living cells. *Cell Motil. Cytoskel.* **14,** 201–219.

Sanger, J. M., Rhee, D., Leonard, M., Price, M., Zhukarev, V., Shuman, H., and Sanger, J. W. (1994a). Assembly of myofibrils and cleavage furrows in cardiomyocytes. *Mol. Biol. Cell* **5**: 165a.

Sanger, J. M., Dome, J. S., Hock, R. S., Mittal, B., and Sanger, J. W. (1994b). The occurrence of contractile fibers and their association with talin in the cleavage furrow in living PtK2 cells. *Cell Motil. Cytoskel.* **27,** 26–40.

Sanger, J. M., Golla, R., Safer, D., Choi, J. K., Yu, K., Sanger, J. W., and Nachmias, V. T. (1995). Increasing intracellular concentrations of thymosinbeta4 in PtK2 cells: effects on stress fibers, cytokinesis and cell spreading. *Cell Motil. Cytoskel.* **31,** 307–322.

Schultheiss, T., Lin, Z., Ishikawa, H., Zamir, I., Stoeckert C. J., and Holtzer, H. (1991). Desmin/vimentin intermediate filaments are dispensable for many aspects of myogenesis. *J. Cell Biol.* **114,** 953–966.

Stearns, T. (1995). The green revolution. *Curr. Biol.* **5,** 262–264.

Theriot, J. A., and Mitchison, T. J. (1992). The nucleation-release model of actin filament dynamics in cell motility. *Trends Cell Biol.* **2,** 219–222.

Turnacioglu, K. K., Mittal, B., Sanger, J. M., and Sanger, J. W. (1996). Partial characterization and DNA sequence of zeugmatin. *Cell Motil. Cytoskel.* **34,** 108–121.

Turnacioglu, K. K., Mittal, B., Dabiri, G. A., Sanger, J. M., and Sanger, J. W. (1997a). Connectin (titin) and the formation of myofibrils. Cell Struct. Funct. **22,** 73–82.

Turnacioglu, K. K., Mittal, B., Dabiri, G. A., Sanger, J. M., and Sanger, J. W. (1997b). An N-terminal fragment of titin coupled to Green Fluorescent Protein localizes to the Z-Bands in living muscle cells: overexpression leads to myofibril disassembly. *Mol. Biol. Cell* **8,** 705–717.

Wang, Y.-L and Taylor, D. L. (Eds.) (1989). Fluorescence microscopy of living cells in culture part a. Fluorescent analogs, labeling cells and basic microscopy. *Methods Cell Biol.* **29,** 1–328.

Yaffee, D. (1968). Retention of differentiation potentialities during prolonged cultivation of myogenic cells. *Proc. Natl. Acad. Sci. U.S.A.* **61,** 477–483.

Yang, F., Moss, L. G., and Phillips, G. N. (1996). The molecular structure of green fluorescent protein. Nat. Biotechnol. **14**: 1246–1251.

Zavodny, P. J., Petro, M. E., Kumar, C. C., Daily, S. H., Lonial, H. K., Narula, S. K., and Leibowitz, P. J. (1988). The nucleotide sequence of chicken smooth muscle myosin light chain two. *Nucleic Acids Res.* **16,** 1214.

Zhukarev, V. Sanger, J. M., Sanger, J. W., Goldman, Y. E., and Shuman, H. (1997). Distribution and orientation of rhodamine–phalloidin bound to thin filaments in skeletal and cardiac myofibrils. *Cell Motil. Cytoskel.* **37,** 363–377.

CHAPTER 16

Monitoring the Dynamics and Mobility of Membrane Proteins Tagged with Green Fluorescent Protein

J. Lippincott-Schwartz,* J. F. Presley,† K. J. M. Zaal,† K. Hirschberg,‡ C. D. Miller,† and J. Ellenberg†

‡Supported by the Human Frontiers Science Program
*Cell Biology and Metabolism Branch, NICHD, NIH
†Howard Hughes Medical Institute Scholar at NIH
Bethesda, Maryland 20982

METHODS IN CELL BIOLOGY, VOL. 58
Copyright © 1999 by Academic Press. All rights of reproduction in any form reserved.
0091-679X/99 $25.00

I. Introduction

Eukaryotic cells contain an elaborate internal membrane system comprised of distinct membrane-bound compartments (including the endoplasmic reticulum, nuclear envelope, Golgi complex, lysosomes, and endosomes) that intercommunicate by vesicle/tubule-mediated membrane transport pathways. Many important cellular functions are thereby served, including protein processing and secretion, nuclear compartmentalization, regulation of ion gradients, and the uptake and degradation of extracellular materials. Studies of the localization and dynamics of proteins within this membrane system until recently have been restricted primarily to fixed specimens labeled with a limited repertoire of antibodies. This has provided only static snapshots of the distribution of molecules, so progress toward understanding temporal events involved in membrane sorting, retention, and transport processes has been hampered.

The use of green fluorescent protein (GFP) chimeras overcomes many of the limitations inherent to studies of fixed specimens labeled with antibodies (Prasher *et al.,* 1992; Chalfie *et al.,* 1994). Not only can GFP chimeras be visualized in the unperturbed environment of a living cell, virtually any protein can be tagged with GFP, producing fusion proteins that are intrinsically fluorescent and photostable, and that usually retain parent molecule targeting and function. GFP fusion proteins have been used in a wide variety of applications, including time-lapse-imaging, double-labeling, and photobleaching experiments. As such, they have become powerful tools in the analysis of membrane trafficking pathways, membrane protein mobility, and the biogenesis of secretory and endocytic organelles (Kaether and Gerdes, 1995; Lippincott-Schwartz and Smith, 1997; Rizzuto *et al.,* 1995).

In this chapter we discuss methods that can be used to study intracellular membrane dynamics using GFP-tagged membrane proteins expressed in living cells. We begin by discussing approaches for optimizing GFP chimera brightness and expression levels within cells, so that intracellular membranes expressing GFP constructs can be readily resolved. Next, we discuss techniques for acquiring time-lapse images and methods to analyze them. We then describe approaches for quantitating the number of GFP molecules being imaged in a cell, so that kinetic steps of membrane transport can be analyzed. Finally, we describe methods of photobleaching that can be used to measure the diffusional mobility of GFP chimeras and the extent of their movement between membrane-bound compartments.

II. Constructing and Expressing GFP Fusion Proteins: Strategies for Optimizing Brightness and Assessing Chimera Function

Many experiments using GFP-tagged membrane proteins are likely to require multiple exposures of the same cell, including time-lapse imaging or serial sectioning to reconstruct three-dimensional images. In general, therefore, the brightest, most photostable GFP variant available should be used in constructing a chimera. Several engineered GFP mutants with altered excitation and emission characteristics are significantly brighter than wild-type GFP, and therefore are recommended. These include S65T and F64L, S65T mutants (Cormack *et al.,* 1996; Heim and Tsien 1996), which are commercially available from Clontech (Palo Alto, CA) (e.g., EGFP), Quantum (Montreal, Canada), and other suppliers. In addition to being intrinsically brighter, these variants fold more efficiently at 37°C and are more photostable than wild-type GFP.

The intrinsic brightness of the GFP molecule is not the only factor involved in producing a bright intracellular signal from a GFP chimera. The expression level of the chimera in the cell is also important. Using vectors with strong promoters to enhance transcription levels and proper codon usage to optimize translation can greatly increase the expression levels of the chimera and thereby enhance its overall brightness within the cell (Yang *et al.,* 1996a; Zolotukhin *et al.,* 1996). This is important when cellular autofluorescence makes distinguishing GFP-chimera-derived fluorescence from background difficult and when sensitive camera systems [i.e., confocal and cooled charge-coupled-device (CCD) systems] are not being used.

Even when using the brightest GFP variant, the signal of the GFP chimera within cells may be unsatisfactory. When this occurs, a variety of procedures can be tried to further enhance the expression level and brightness of the chimera. If a stable cell line expressing the GFP chimera is used, the addition of sodium butyrate to the culture medium (1–5 μm for up to 12 h) will increase overall gene expression levels, and frequently will lead to enhanced GFP chimera expression and a brighter signal (Shima *et al.,* 1997). Use of transiently transfected cells rather than stably transfected cell lines can also help, because the former usually give much higher expression levels. As a last resort to enhance GFP chimera expression, construction of chimeras containing double or triple GFP molecules in tandem has been shown to boost signal (K. Zaal, unpublished observations).

Once a GFP-tagged membrane protein is optimally expressed within cells, the next question is whether it has targeted correctly. The easiest way to address this is to compare the distribution of the chimera with the distribution of the parent molecule detected with fluorescently labeled antibodies in unexpressing cells that are fixed and permeabilized. If the overall pattern of the two fluorescent labels is not similar, then addition of the GFP moiety to the parent molecule has prevented it from targeting correctly or the expression level of the GFP chimera is so high that the machinery for targeting the chimera has been saturated. In the latter case, lowering the expression level of the chimera by decreasing the amount of DNA used during transfection, by selecting stable cell lines or using different promoters can be helpful.

The preservation of parental function by the GFP chimera may be demonstrated in a number of ways, including the ability of the chimera to rescue a phenotype in cells with mutations or deletions of the parent molecule or incorporation of the chimera into a functional macromolecular structure (e.g., actin or microtubule filaments) (Olson *et al.,* 1995; Shelby *et al.,* 1996). When a GFP chimera fails to target or function correctly within expressed cells, one solution is to change the position where GFP is attached to the parent molecule. Many functional GFP membrane fusion proteins have been generated with GFP placed at the N or C terminus of the membrane protein (Cole *et al.,* 1996; Ellenberg *et al.,* 1997).

III. Practical Guidelines for the Preparation and Imaging of GFP-Expressing Cells

Microscope objectives are designed for observing specimens through a glass coverslip of specific thickness, so cells expressing GFP fusion proteins must be grown on or attached to a glass coverslip prior to imaging. This can be accomplished cheaply and easily by sterilizing coverslips in ethanol and then flaming. The sterile coverslips are then placed in standard tissue culture dishes and medium containing GFP-expressing cells in suspension is added to the dish. If the cells fail to grow well, the coverslips can be cleaned with chromic acid and/or treated with poly-D-lysine (1 mg/ml solution for 15 min) to enhance cell attachment. Care should be taken that the correct glass coverslip thickness (e.g., #1 glass coverslips for 63× and 100× objectives) is used for imaging cells. When a coverslip differs from the thickness recommended for a specific objective by only 5%, significant spherical aberration can result, giving poor microscopic images.

After GFP-expressing cells are grown on coverslips, they can be visualized live on the fluorescence microscope by mounting them in a chamber that contains buffered medium. A simple chamber can be made by cutting a small hole in a sheet of silicon rubber, sealing the rubber sheet to a glass slide with grease to prevent leaking, and filling the newly created well with buffered medium. The coverslip is then inverted on top with the attached cells placed cell side down. Alternatively, it is possible to grow cells in commercially available coverslip-bottom open dishes (Lab-Tek chambers, Nunc, Rochester, NY). This allows drugs or other materials to be easily added during an experiment. Such chambers tend to be relatively expensive and can only be used with an inverted microscope, however.

Cells can be kept warm during imaging on the microscope using a hair dryer or an air stream incubator (Nevtek, Burnsville, VA). This method will work with either an inverted or upright microscope setup and has the advantage of low cost and extreme simplicity. The apparatus used to warm the microscope stage will likely produce vibrations and should therefore be isolated from the table the microscope is on. To monitor the temperature, a probe or thermometer can be taped close to the coverslip.

To acquire optimal images of GFP-expressing cells, the overall goal is to maximize the GFP signal over existing background fluorescence. It is essential, therefore, to avoid using fluorescent materials such as phenol red in the medium. One must also check the basic elements of the fluorescent microscope (including the excitation light source, objective lens, filter cube, and detector) to be sure that they are optimized for detecting GFP fluorescence. The best objective lenses to use are those that have a high NA. By collecting more light than lower NA objectives, they make the specimen appear brighter. A mercury lamp or argon laser is an excellent light source for viewing GFP because both give off bright blue light, which maximally excites S65T variants (e.g., EGFP). A standard fluorescein filter cube set is usually suitable for obtaining a good GFP signal. When the GFP signal is difficult to distinguish from background fluorescence, however, the use of narrower bandpass filters to maximally collect light from the GFP signal rather than background fluorescence is recommended (Niswender *et al.,* 1995; Endow and Komma, 1996).

IV. Time-Lapse Imaging of GFP Chimeras: Critical Parameters

A. Illumination and Focal Control

An important application of GFP chimeras expressed in living cells is time-lapse imaging, in which a cell is viewed over extended periods of time. The simplest way to record time-lapse images is to attach an electronic camera to a fluorescence microscope and then to acquire images manually at fixed intervals. This approach is not suitable for imaging of GFP chimeras, however, because the prolonged illumination will photobleach the fluorophore, decreasing the fluorescence signal. The exposure time can be reduced by placing a shutter between the light source and the microscope when images are not being collected. Ideally, the shutter should be computer controlled and therefore automatic, but it can be as simple as moving a piece of cardboard by hand in and out of the light path. Photobleaching can also be minimized if illumination levels are lowered. A graded series of neutral density filters (with transmittance of 100, 30, 10, 3, and 1%) will allow selection of the minimum illumination compatible with good image quality for a particular specimen. These can be installed in a computer-controlled filter wheel, but manual placement of filters in the excitation light path is also satisfactory and does not require expensive modifications to the microscope.

Some specimen photobleaching will always occur during time-lapse imaging experiments, even when illumination times and levels are minimized. To determine the extent of photobleaching during an experiment, a good method is to acquire a time-lapse series of fixed GFP-expressing cells using the same imaging parameters as in living cells. Loss of fluorescence under these conditions will be due only to GFP photobleaching and not to biological degradative processes. The photobleaching rate, which is generally an exponential decay of intensity, can then be used to correct the images in the experiments with living cells.

Sometimes a specimen will drift out of focus during time-lapse imaging. If this is a problem one can manually refocus during the time-lapse sequence or use a motorized autofocusing objective that regains the original focus position.

B. Image Acquisition Devices

Many structures of interest labeled with GFP-tagged membrane proteins are likely to have only small numbers of fluorophore molecules (i.e., vesicles and tubules). It is therefore essential to use an image acquisition system during time-lapse imaging that can capture small numbers of photons. Three such systems can be used for GFP imaging: intensified video cameras, charge-coupled-device (CCD) cameras, and the confocal laser scanning microscope.

Intensified video cameras and silicon-intensified-target (SIT) cameras are sensitive light detectors whose main advantage is in acquiring low-light images at video rate (and onto videotape) using a relatively simple setup (see Inoue, 1986). The camera plugs directly into a port in the microscope. Usually there is a beam splitter or removable mirror that directs light from the specimen to the microscope eyepiece or to the camera. The signal from the camera is generally filtered before being directed to a videotape recorder, monitor screen, or computer where the image can be digitized. The signal from the video/SIT camera can be manipulated by adjusting an offset (this adds or subtracts a constant value to the signal) and a gain (which controls amplification of the signal). When faint signals are being imaged it is useful to average frames rather than increase the gain to levels that may amplify noise. Averaging frames will also significantly increase image quality. Because of their sensitivity, video/SIT camera systems are suitable for time-lapse imaging of GFP chimeras expressed at moderately low levels within cells. However, because such systems do not guarantee a linear response to light detection, quantitative measurements are difficult.

A second image acquisition device for quantitative time-lapse imaging GFP chimeras is the CCD camera. It contains light-sensitive silicon chips whose surface electrons are excited by photons emitted by the GFP chimera. The excited electrons are recorded and converted to a digital image. The signal is linear with increasing photon count, so the pixel values can be used as a direct measure of fluorophore intensity. The image can be optimized by acquiring images over longer times ("on-chip integration") because signal strength is proportional to acquisition time. Additionally, cooling the CCD chip (usually to −40°C for fluorescence applications) can improve the resolution and dynamic range (up to 16-bit = 2^{16} grayscales and 40–60% detection efficiency). Under these conditions, it is more sensitive than a video/SIT camera. The major disadvantage of CCD cameras compared to video systems is that frames are acquired over longer periods of time, so imaging in real time is essentially impossible. Nevertheless, one can easily perform satisfactory time-lapse imaging with moderately bright GFP chimeras on a cooled CCD camera with frames captured at 1-s intervals. A caveat to imaging with a CCD camera is that if the object of interest moves during the imaging period, the imaged object will appear elongated or "streaked" relative to its normal appearance. If this occurs, capturing CCD images at shorter exposure times is necessary.

The confocal laser scanning microscope (CLSM) is a third method for image acquisition suitable for quantitative time-lapse imaging. Composed of a laser and scan box attached to a conventional microscope, the CLSM produces sharp images of structures by selectively collecting light from a thin focal plane within the specimen. This is in contrast to CCD and video/SIT camera systems, which cannot discriminate between in-focus and out-of-focus fluorescence without deconvolution software. The confocal laser is focused to a point inside the sample that is imaged by the objective lens. The point is scanned over the sample at a fixed depth and used to excite the fluorophore. Emission from the fluorophore is acquired by the objective lens and passed to a photomultiplier tube (PMT) in the scan box. En route to the PMT the light passes through a small adjustable aperture (pinhole), which excludes emitted light from above or below the plane of focus. This arrangement causes structures within the focal plane to appear more sharply defined then they would with a conventional microscope because there is essentially no light from out-of-focus areas. When imaging confocal specimens with an open pinhole, objects are brighter because more out-of-focus fluorescence is collected. A closed pinhole eliminates out-of-focus fluorescence (as well as some in-focus fluorescence), resulting in a dimmer signal. Thus, the pinhole opening determines the acquired section thickness. The PMT turns the emission light into a voltage, which is then digitized, typically into a number from 0 (black) to 255 (white) for each point scanned. The confocal microscope's dynamic range (usually 8-bit = 2^8 grayscales) and detection efficiency (usually 15–30%) are usually lower than those of a cooled CCD camera.

Time-lapse imaging of GFP chimeras with a confocal microscope presents some special problems. Normally a single depth within a cell is imaged over time because collecting a series of *z*-sections at each time point is usually too time-consuming (i.e., several minutes per stack of images). Vertical movement of GFP-containing structures during imaging is therefore a concern because these structures will disappear as they move out of the focal plane. Use of flat cell types and focusing on objects in the cell periphery (where cell thickness is less than 2 μm) can alleviate this problem. Another solution is to image with the pinhole partly or entirely open and to use objectives with a lower NA (for example, 25× with 0.8 NA). Maximum focal depth into a sample increases with lower NA.

The zoom setting used on a confocal microscope is important for optimal imaging. It determines the size of the scan region and the apparent magnification of the image. For example, a zoom factor of 2 will scan an area half as large in each dimension as that of a zoom factor of 1, so pixels in a zoom-2 image have areas half the width and length of those in a zoom-1 image. A rule of thumb is to always use a pixel size that is one-half to one-third the optical resolution of the objective. If the pixel size is made too large by using a lower than optimal zoom factor, minimal structures will not be resolved on the display monitor. Likewise, if the pixel size is made too small by using a higher than optimal zoom factor, the specimen will be exposed to too much irradiation and be photobleached. This is because the rate of photobleaching increases proportionately to the square of the zoom factor (Centonze and Pawley, 1995).

Whatever light detection system is used, images should always be acquired with minimal illumination levels so that emitted light from the object of interest is never saturated. Detection of fluorescence emission increases linearly with light intensity only until emission is saturated. Thereafter, only detection of fluorescence from background sources, which usually are not saturated, rise as the illumination level is increased. This decreases the signal-to-noise ratio of the image.

V. Analysis of Time-Lapse Imaging Data

One important application of time-lapse imaging is to track a fluorescent object so that its path and velocity can be determined. This requires converting the distance between two image locations in pixels to an actual distance. The calibration can be done by imaging a calibration grid (Yona Microscope, Rockville, MD) using brightfield illumination and the same objective lens and camera system used for fluorescence imaging. The distance traversed by an object of interest between two successive frames can then be calculated. Velocity is simply this distance divided by elapsed time between the frames. The path followed by the object can be determined by plotting the x- and y-coordinates at each time point. An example in which this approach has been used to track intracellular membrane structures is shown in Fig. 1.

We have written a macro for NIH Image (NIH Image can be obtained free from the NIH Web site at http://hrsb.info.nih.gov/nih-image/) that simplifies this procedure. The macro allows one to use a computer mouse to click on the object of interest in the first frame in a stack and then in subsequent frames. As output, it produces text with complete distance and velocity information in a form that is readable by a graphing program. It also will produce a graph that describes the path followed by the object of interest. This macro is available via the Internet at http://dir.nichd.nih.gov/CBMB/pb6labob.htm.

It is often useful to determine whether the number of GFP chimeras associated with an intracellular structure changes over time, due to transport, sorting, and/or degradative processes. To accomplish this start by identifying the object of interest and a nearby region that gives an appropriate estimate of the background contribution of fluorescence intensity of the object. Measure the average total

Fig. 1 Tracking the path and velocity of a GFP-tagged membrane protein chimera. COS cells expressing the viral glycoprotein, VSVG, tagged with GFP (VSVG–GFP) were imaged upon temperature shift from 15°C (at which VSVG–GFP accumulates in pre-Golgi structures) to 32°C (at which pre-Golgi structures move to the Golgi region). (A) The left panel shows the distribution of VSVG–GFP in pre-Golgi structures at the time of warm-up, and the right panel shows the distribution 9 min later in the same cell. The paths of pre-Golgi structures (middle panel) were obtained by mapping the movement of these structures. Images were collected at 3.5-s intervals. Bar = 20 μm. (B) Path of a single pre-Golgi structure as it moves toward the Golgi region. The position of the structure at 3.5-s intervals is shown. Arrow points to the direction of motion. (C) Average velocity of the same structure.

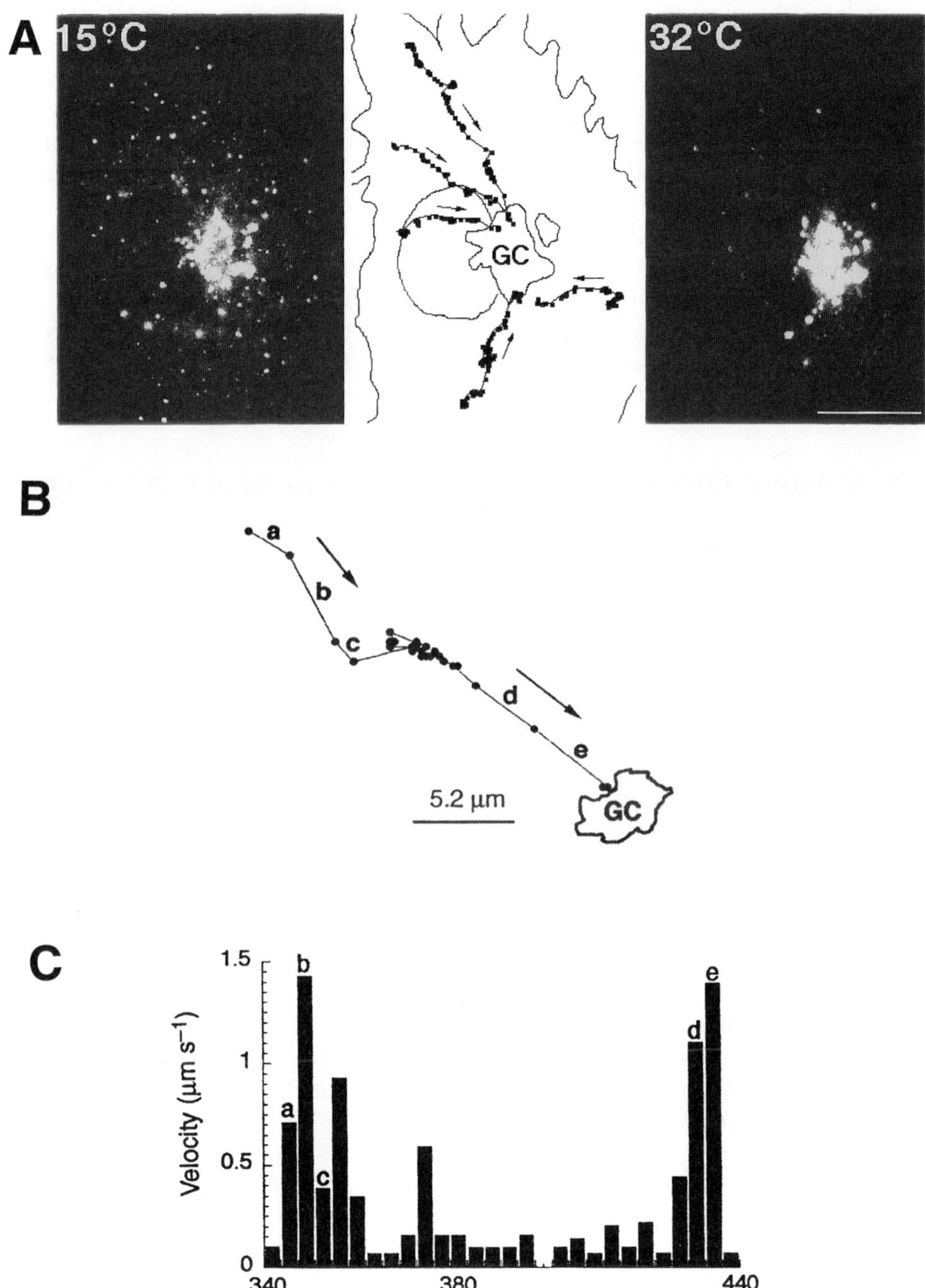
A
15°C
GC
32°C
B
a
b
c
d
e
5.2 μm
GC
C
Velocity (μm s^{-1})
1.5
1
0.5
0
340
380
440
Time (s)
a
b
c
d
e

intensity of the two regions in each frame of the time-lapse sequence. Make sure that the object of interest does not move out of the plane of focus during the experiment and that the background does not change in unexpected ways. Next, subtract background intensity from the intensity associated with the object of interest. If total intensities are measured, the two regions should contain identical numbers of pixels or the background intensity should be appropriately normalized to account for the difference in the areas. Plot the raw values, the background values, and the background-corrected values as a function of time on a graph. It is also useful to plot total fluorescence associated with the images (or just in the entire cell of interest) as a function of time on a log-linear plot. This will reveal the extent of photobleaching (which should be a simple exponential decay) and any problems due to focal drift (which will tend to show up as abrupt or oscillating changes in total fluorescence).

VI. Relating GFP Chimera Fluorescence to Actual Numbers of GFP Molecules

Because the number of GFP molecules within a GFP chimera is fixed (usually one GFP per chimera, unless tandem GFPs have been attached), fluorescence microscopy can be used as a quantitative tool to determine the number of GFP chimeras being imaged in a single cell. This provides a powerful method for describing and analyzing intracellular protein dynamics. Following is a simple method for determining the number of GFP molecules being imaged within a cell or within different subcellular structures. For a more detailed discussion of quantitative imaging of GFP see Chapter 3.

The aim of this method is to compare total cellular GFP fluorescence with a standard curve generated by imaging GFP in aqueous solutions of known concentrations and from this comparison to calculate the number of cellular GFP molecules expressed. The standards consist of solutions of GFP (Clontech, Palo Alto, CA) in TE buffer in concentrations of 0.37 μM to 37 μM placed in coverslip chambers. Images of the standard solutions are collected with a confocal laser scanning microscope using imaging parameters identical to those used in experiments with living cells. In choosing the objective and pinhole settings for the images, it is important to select a combination that will ensure that the total fluorescence of the cellular structures will be detected. For example, in experiments measuring total GFP chimera fluorescence in COS cells, we have used a 25× objective (NA 0.8) with a pinhole setting of 150 on the Zeiss LSM410.

The mean pixel values for images of the standards are then processed using NIH image software, and the background signal is subtracted from the mean pixel values prior to further calculations. The fluorescence in a 100-μm^2 region of interest is then plotted against the number of GFP molecules estimated to be within the volumetric region of interest (VOI), determined by the product of the 100-μm^2 area and the FWHM (Full Width Half Maximum). The FWHM is the distance in the z-direction between planes where the intensity is half that at the plane of focus. It can be calculated as in the equation (obtained from Zeiss confocal microscope manual),

$$\text{FWHM } (\mu\text{m}) = (1.41)\,\frac{\lambda_1}{\lambda_2}\left[\left(\frac{2n}{\text{M}\cdot\text{NA}}\right)\cdot P + \frac{n\lambda_2}{\text{NA}^2}\cdot \exp\left(-\frac{2\cdot\text{NA}\cdot P}{\lambda_2\cdot M}\right)\right], \qquad (1)$$

where λ_1 = emission wavelength (nm); λ_2 = excitation wavelength (nm); NA = numerical aperature; M = magnification; n = refractive index of immersion medium; P = pinhole setting in digital units; FWHM = Full Width Half Maximum (μm). In our experience, we have been able to fit plots of GFP molecules versus total fluorescence (the sum of the mean pixel values in an area) with a linear function in the range of GFP concentrations stated earlier.

In generating the standard curve, we have assumed that the FWHM approximates the z-dimensional thickness of the sample, which contributes to the fluorescence signal, and that the efficiency of detection of fluorescence within the volume is constant. The validity of these assumptions can be demonstrated by imaging small spherical droplets of GFP solution in oil (with diameter-FWHM) and comparing the number of GFP molecules obtained from the calibration curve to that derived from the product of GFP concentration and sphere volume, which can be accurately determined by measuring the dimensions in the x–y plane. In applying the standard curve to living cells, it is necessary to make the additional assumptions that the quantum yield, detection efficiency, and proportion of properly folded (and therefore active) GFP molecules is similar for GFP chimeras in cells and for GFP in aqueous solution.

VII. Fluorescence Recovery after Photobleaching: FRAP

Although photoinduced destruction of the GFP fluorophore by photobleaching is undesirable in most types of imaging experiments, it is desirable in FRAP experiments. In FRAP, a small region of a specimen is exposed to photobleaching radiation by intense laser light. Fluorescence recovery into the area being photobleached is then measured. This usually occurs by diffusional exchange between bleached and unbleached fluorescent molecules. The rate of recovery can be used to calculate the diffusion constant, D, of the fluorescent protein. The fraction of labeled proteins that participate in recovery, called the mobile fraction, can also be measured. This technique allows the lateral mobility of fluorescent proteins to be monitored in very small or large membrane areas within cells.

GFP chimeras are excellent probes for use in photobleaching studies because when bleached they cause much less damage to their surrounding environment than other fluorophores. This is probably due to the protection of the GFP fluorophore by the tightly packed beta barrel structure of GFP (Ormö et al., 1996; Yang *et al.,* 1996b). Because GFP chimeras are endogenously expressed within cells, FRAP can be used to study the mobility of intracellular membrane components. Later, we describe two types of FRAP experiments, qualitative and quantitative, that can be used for studying lateral mobility of GFP chimeras. These protocols do not require custom-designed equipment and can be adapted to most commercial CLSMs.

In qualitative FRAP experiments, the entire cell with the structure of interest is imaged with the confocal microscope before the bleach and during recovery. Images are collected with a comparatively long time interval between the recovery images (i.e., ~10 s to 1 min). This approach allows recovery in areas within and outside the bleached zone to be monitored with good morphological resolution and provides an assessment of the overall effect of photobleaching on the cell, including potential photodamage. Although D is not accurately determined using qualitative FRAP, the mobile fraction of molecules can be measured by comparing the ratio of fluorescent intensities in regions inside and outside the photobleaching box before photobleaching and after recovery (Cole *et al.*, 1996). For many intracellular organelles, total fluorescence is frequently lower after photobleaching than before photobleaching because a significant portion of fluorescence is lost in the photobleach. This method provides a means for correcting this error.

The main purpose of quantitative FRAP experiments is to determine an accurate D of the GFP chimera. For this, recovery is only imaged within the bleached area, typically a 2- to 4-μm wide strip across the cells. This allows fluorescence intensities to be acquired very rapidly (i.e., every 0.5 s), which is crucial for an accurate determination of D. With GFP chimeras expressed within intracellular membranes, fluorescence within the bleached strip rarely returns to its initial value even when there is 100% recovery because a significant proportion of total fluorescence is usually lost in the photobleach. The entire cell should therefore be imaged before the bleach and after recovery is complete to determine the mobile fraction of fluorescent molecules. This will also allow the overall effect of photobleaching on the cells to be assessed.

Following is a general description of the protocol for qualitative and quantitative FRAP. Most confocal microscope software systems (e.g., Zeiss LSM410/510, BioRad MRC series, Leica TCS) allow programming of the mentioned functions within a command macro using the microscope's specialized macro language. The system used in our laboratory is based on a Zeiss LSM 410 with an Omnichrome KrAr 488/586/647 Series 43 laser. Program codes for the described protocols are available via the Internet at http://dir.nichd.nih.gov/CBMB/pb2labob.htm.

VIII. Qualitative FRAP Experiments

When the cell of interest has been identified and focused on the confocal microscope, prepare the command macro for initiating the photobleaching protocol. Once started, the macro will automatically record the image of the cell before photobleaching, execute the bleach and record it after the photobleach and during recovery. See Fig. 2 for an example of a FRAP experiment using this macro.

In preparing the command macro, the photobleaching and imaging radiation must be experimentally determined for every specimen depending on specimen size and brightness as well as other optical parameters. The imaging radiation should be 300- to 1000-fold lower than the maximum laser power available for the photobleach to ensure that bleaching does not occur during the monitoring step. This means using low power, high attenuation, low zoom, and minimal averaging for imaging. For imaging radiation in the macro, set laser power, attenuation (neutral density filters), zoom, and averaging to imaging intensity levels that allow a high-resolution image of the entire cell or structure of interest. Scan the image at the imaging laser intensity level and then save the image.

Next, define the region of interest (ROI) for the photobleach at the highest possible zoom. A high zoom will increase the time the laser prevails on the bleached region per line scan and thus increase the bleaching radiation per area. Remember that photobleaching increases proportionally to the square of the zoom factor. If the desired bleach area is larger than the whole field scanned at maximum zoom, lower the zoom until it just accommodates the bleach ROI. Some microscope softwares allow one to define the ROI interactively in a dialog window. Alternatively, the bleach ROI can be defined before the start of the macro. The coordinates in the macro command that define it then have to be altered each time. One can also define a conveniently sized ROI close to the center of the screen, mark it on the monitor surface, and position the cell prior to the experiment so that the bleach ROI hits the structure intended to be bleached.

To photobleach the ROI, set the laser power at maximum and attenuation (i.e., neutral density filter) to zero for bleaching intensity. Scan the bleach ROI at the bleach intensity and bleach zoom until fluorescence is reduced to background levels. It is not necessary to save the image. An optimum photobleach should be sufficient to reduce the intensity in the bleached area to background levels in as short a time as possible (typically one scan of the ROI without averaging). If that is not sufficient to reduce fluorescence levels to background intensity (at imaging laser settings), try averaging scans or increasing the number of scans. If a complete photobleach takes longer than 20 s, a more powerful laser may be required. The bleach intensity should not result in significant photobleaching outside the bleach target by light scattering or draining of material due to diffusion into the bleached region during the course of the bleach.

To acquire recovery images, switch the laser power, attenuation, zoom, and averaging back to imaging intensity settings. Take a time series of images of the entire cell starting immediately after the photobleach and until recovery is judged complete. Images should be taken and saved every 5 to 30 s. This allows time for acquisition of averaged images that have good resolution. The time for full recovery depends on the diffusion constant of the probe and the size of the bleached area. A good range is 1–10 min.

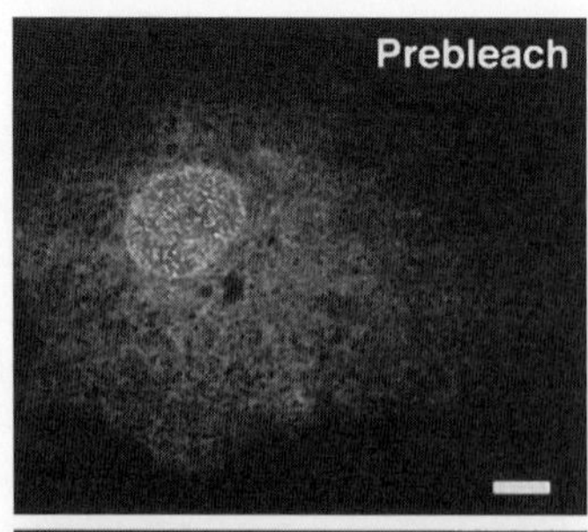
Prebleach
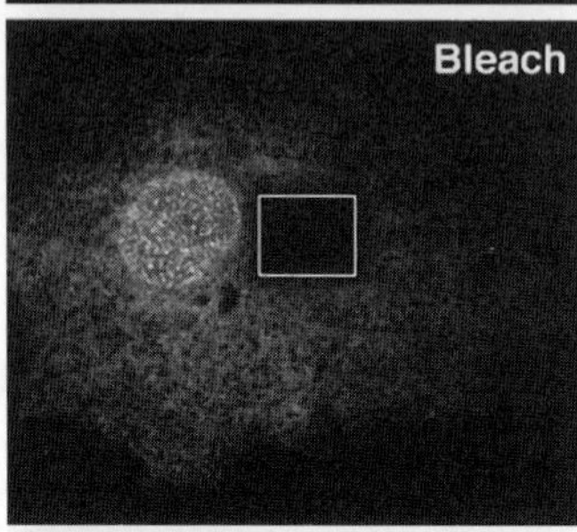
Bleach
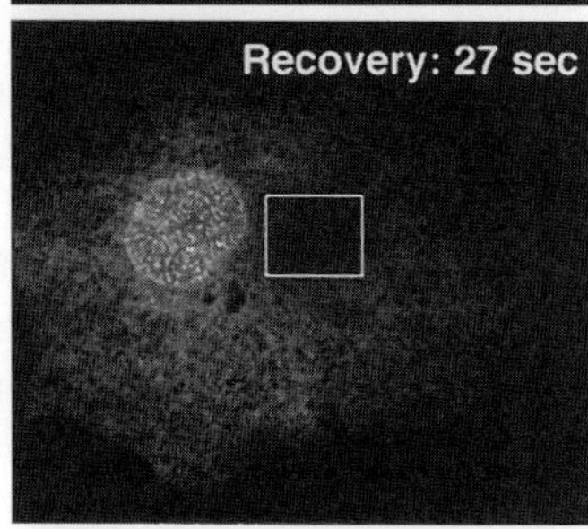
Recovery: 27 sec
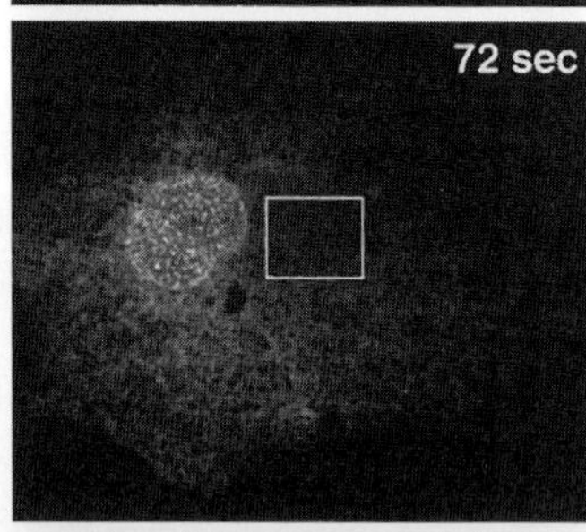
72 sec
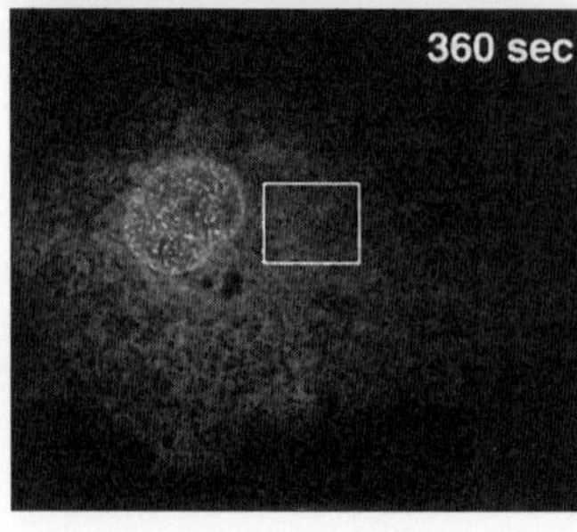
360 sec

IX. Quantitative FRAP

In contrast to qualitative FRAP, the quantitative FRAP protocol is designed to collect fluorescence intensities within the bleached area with high temporal resolution to allow accurate determination of *D*. As a result, it does not provide morphological information of the whole cell during recovery. Identify the cell of interest and focus it on the confocal microscope. Next, prepare the command macro to allow settings to be changed quickly during photobleaching and recovery. The macro will record the image of the entire cell before photobleaching, execute the bleach, record the bleached area after the photobleach and during recovery, and record the entire cell after recovery. See Fig. 3 for an example of quantitative FRAP using this method.

In the macro under imaging radiation, set laser power, attenuation (neutral density filters), zoom, and averaging to imaging intensity levels that allow a high-resolution image of the whole cell or structure of interest. Scan the entire cell or structure of interest at imaging laser intensity and then save the image. This image allows the overall prebleach image morphology to be recorded. Together with the image collected at the end of recovery, it can serve as a control for obvious photodamage and enables one to calculate mobile fractions using the method described for qualitative FRAP.

Scan the strip that is going to be bleached (usually a 2- to 4-μm-wide box that spans the entire cell) at imaging intensity. Save this image. This conveniently measures the prebleach intensity in the bleach strip, even though the same information is contained in the first entire cell image. It is useful to calibrate a number of different strips for the respective zooms, mark them on plastic overlays that can be placed on the monitor, and position the cell until the strip covers the desired area. An accurate measurement of the size of the bleach strip should be made on a fixed specimen for which there is no recovery during or after the photobleaching episode. This measurement is crucial for the determination of *D* and usually has an uncertainty of 20% (Edidin, 1994).

Define the bleach ROI at the highest possible zoom. Set the laser power to maximum and the attenuation to zero (bleaching intensity). Scan the bleach ROI at bleaching intensity and zoom until fluorescence is reduced to background levels. It is crucial that the cell be bleached completely across its surface and through its entire depth. Otherwise recovery will not follow two dimensional

---

Fig. 2 Qualitative FRAP. A COS cell expressing lamin B receptor tagged with GFP (LBR–GFP) in the ER and nuclear envelope was photobleached in the outlined box. Recovery of fluorescence was monitored every 9.1 s for 6 min, when recovery was complete. Note the rapid recovery of fluorescence of this chimera within ER membranes. Bar = 10 μm.

kinetics. The depth of the bleach depends on the power and NA of the objective used (a lower NA gives greater depth of field). For structures with a significant depth, the bleach dimensions of each objective need to be calibrated on a fixed specimen.

Switch the laser power, attenuation, zoom, and averaging back to imaging intensity settings. Scan and record the bleached ROI immediately after the photobleach at imaging intensity and continue at short intervals until recovery is judged complete. Save each image in the series. The time for full recovery depends on the diffusion constant of the probe and the width of the bleached strip. A good range is 30 s to 5 min.

Scan the whole cell after recovery is complete and save the image. This image can be compared to the prebleach image of the whole cell to check for any photodamage and loss of focus. The comparison of the prebleach and recovery entire cell images can also be used to determine the mobile fraction. This enables one to assess overall loss of fluorescence due to the bleach, which should not exceed 20%. If it does exceed this value, try narrowing the width of the strip.

To make sure that the bleaching fluorescence does not damage the sample, it is important to repeat a FRAP experiment on the same area. *D* should remain the same, but the mobile fraction should now be close to 100%. This is because the immobile fraction was bleached in the previous photobleach. If *D* is changed or the mobile fraction is less than 100%, it is possible that the specimen has become damaged. In that case, the experimental data is flawed and the photobleaching protocol needs to be changed.

X. Calculating D

The virtue of the quantitative FRAP protocol is that it allows an accurate determination of *D* based simply on fluorescent intensities within the bleached strip before the bleach and during recovery. The following steps to determine *D* can be automated as a batch program or command macro with most image analysis programs (NIH Image, IP Lab Spectrum, Zeiss LSM software, etc.).

Background-subtract all the images taken within the bleached strip. Quantitate the mean fluorescence intensity per area of the region within the bleached strip that contains fluorescent cellular material in the prebleach image. Because the strip reaches all the way across the field of view, it often includes nonfluorescent areas at its edges. To measure mean pixel value per unit area, narrow the ROI down to the region that actually contained fluorescence before the bleach.

Correct the postbleach intensities for overall loss of fluorescence determined from total fluorescence in the entire cell images taken at the beginning and end of the experiment. Overall loss of fluorescence is given by $(I_{\Sigma 0} - I_{\Sigma\infty})/I_{\Sigma 0}$, with $I_{\Sigma 0}$ = total cumulative pixel values in prebleach whole cell image and $I_{\Sigma\infty}$ = total cumulative pixel values in postbleach whole cell image. Due to this loss, postbleach fluorescence intensities in the strip can never reach 100% of the prebleach value and must be corrected by multiplying by $I_{\Sigma 0}/I_{\Sigma\infty}$.

Plot the mean fluorescence intensities per area over time (see Fig. 3 for an example output). Fit the experimental data to a formula for one-dimensional diffusion. The empirical formula we use is:

$$I_{(t)} = I_{\text{final}}\,(1 - (w^2(w^2 + 4\pi\,\text{D}t)^{-1})^{1/2}) \qquad (2)$$

where $I_{(t)}$ = fluorescence intensity as a function of time; zero of time t is taken as the midpoint of the bleach; $I_{(\text{final})}$ = final intensity reached after complete recovery, w = strip width (i.e., 2 μm), and D = effective one-dimensional diffusion constant. This equation assumes one-dimensional recovery because the membranes are bleached all across their length and entire depth. Because intracellular membranes are frequently convoluted, the true motion of molecules occurs over a longer path than that observed in a projected image of the cell (Wey *et al.*, 1981). The effective D measured with our method, therefore, is generally underestimated with a maximum error of 1/3 (Sciaky *et al.*, 1997).

The diffusion constant D can be obtained directly from the preceding fit. The fitting can be performed by most spreadsheet programs using a least squares algorithm [e.g., Kaleidagraph (Abelbeck Software, Reading, PA)]. To assess the effects of geometry, as well as the nonuniform fluorescence density in the cell,

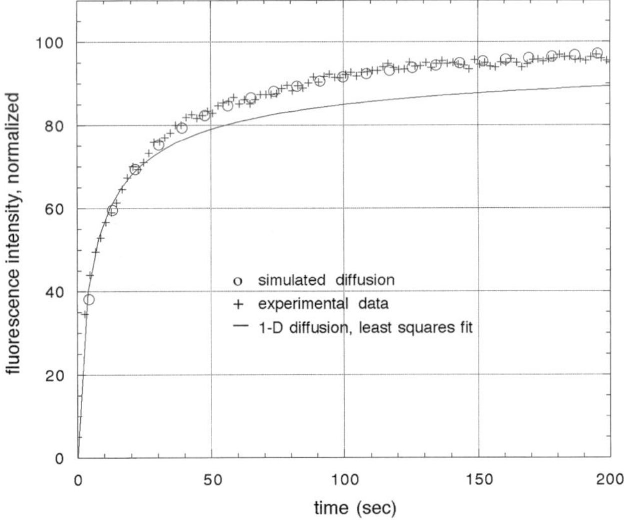

Fig. 3 Quantitative FRAP. A COS cell expressing LBR–GFP was photobleached over a 4-μm-wide strip covering the ER (similar to Fig. 2). Mean fluorescence intensity in the strip is plotted versus time (crosses). The prebleach intensity was normalized arbitrarily to 100%. A least squares fit to Eq. (2) is shown as a line, computer-simulated diffusion from a prebleach whole cell image as circles. Note the excellent correspondence between simulated diffusion and experimental data.

D calculated from the Eq. (2) has been checked against a numerical simulation that uses the prebleach intensity of the entire cell as input to simulate diffusive recovery into the bleached strip (as shown in Fig. 3). A detailed description of such a simulation is found in Sciaky *et al.* (1997).

If the prebleach intensity is normalized to 100%, the final intensity reached in recovery is a direct measure for the mobile fraction. This is frequently an underestimate, however, because the total cellular fluorescence is usually lowered during the photobleaching process. A lack of or slow recovery after photobleaching can be due to nondiffusional factors such as discontinuities within the structure the fluorophore is localized to. This is especially relevant to the endomembrane systems of cells. Membranes of the ER and the cell surface are completely interconnected, which means that recovery of fluorescence of a GFP fusion protein in these membranes is only limited by diffusional constraints on its lateral mobility. In contrast, membrane systems such as endosomes and lysosomes consist of isolated membrane structures that lack permanent connections. A FRAP experiment performed on such structures might show little or no recovery not because the GFP chimeras are immobilized, but because the membranes are not interconnected. Photobleaching experiments can therefore be used to assess fragmentation of a formerly connected membrane system by perturbations of their integrity (Subramanian *et al.,* 1997), as well as fusion of formerly isolated membranes, as described later.

XI. Fluorescence Loss in Photobleaching (FLIP) Using a Confocal Microscope

This technique provides a powerful means for examining the connectivity of a membrane-bound compartment, whether it be the ER, Golgi complex, nuclear envelope or other membranes. FLIP measures the extent to which regions outside a photobleached box contribute to fluorescence recovery in a bleached site. Using this method, a region of interest is bleached repeatedly and fluorescence loss outside the region of interest is monitored over time. The extent to which areas outside the region of interest lose fluorescence over time describes the dimensions of a compartment within which GFP chimeras diffuse. Figure 4 shows an example of a FLIP experiment.

To perform a FLIP experiment, first identify the cell of interest and get it in focus. Set laser power, attenuation (neutral density filters), zoom, and averaging to imaging intensity levels that allow a high-resolution image of the whole cell

---

Fig. 4 FLIP. A COS cell in mitosis expressing LBR–GFP is photobleached repetitively in the outlined box at one tip of the cell. Fluorescence is drained entirely over a period of 800 s, indicating that the mitotic membranes enriched in LBR–GFP are completely interconnected. Bar = 10 μm.

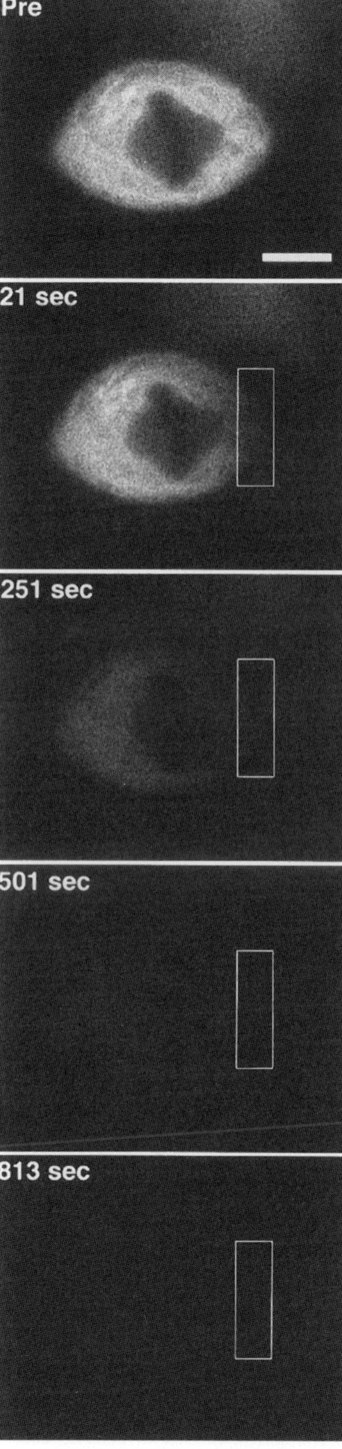
Pre
21 sec
251 sec
501 sec
813 sec

or structure of interest. Scan the whole cell (or structure of interest) at imaging laser intensity. Save the image.

Define the region of interest of interest to be photobleached (bleach ROI) at the highest possible zoom. Because the aim of a FLIP experiment is to deplete fluorescence in regions outside the bleached area in order to access the continuity of a structure, it is desirable to bleach only a small portion (5–20%) of the structure of interest. Set the laser power to maximum and the attenuation to zero (bleaching intensity). Scan the bleach ROI at bleaching intensity and zoom for several times within a prolonged time period (e.g., 30 s). Do not save the image. Bleaching for a long period will not only deplete the bleach ROI of fluorescence but will also deplete areas in direct continuity with it. To avoid excess photodamage the individual scans can be separated by pauses to allow exchange of fluorescent molecules into the bleached area that are then bleached in the next scan. Switch the laser power, attenuation, zoom, and averaging back to imaging intensity settings. Scan the whole cell (or structure) once at imaging intensity. Save the image. Repeat the procedure. This loop should be executed until loss of fluorescence outside the photobleached area is no longer detected.

Several controls should be performed to assess the FLIP results. These include FLIP of fixed cells to ensure that fluorescence loss is not due to light leakage during FLIP or photobleaching during the imaging step. It is also important to fix cells after the FLIP experiment and then stain them with antibodies to the structures that have been photobleached. This will allow one to determine whether structures inside the photobleached zone have been damaged or dramatically changed in their morphology by the repetitive bleaching episodes.

XII. Other Applications of Photobleaching

In addition to measuring the diffusional properties of GFP fusion proteins and assessing the extent of continuity of membrane structures using photobleaching methods, these techniques can be used to enhance the imaging of dim structures within cells or in areas of the cell next to very bright objects. Excellent examples of this application include the photobleaching of background fluorescence in axons in order to visualize trafficking of GFP-tagged post-Golgi membrane transport intermediates (Nakata *et al.,* 1998) and the photobleaching of bright Golgi stacks in order to study targeting to and fusion of GFP-tagged pre-Golgi structures with this compartment (Presley *et al.,* 1997).

References

Centonze, V., and Pawley, J. (1995). Tutorial on practical confocal microscopy and use of the confocal test specimen. *In* "Handbook of Biological Confocal Microscopy," 2nd ed. (Pawley, J., Ed.), pp. 549–570. Plenum, New York.

Chalfie, M., Tu, Y., Euskirchen, G., Ward, W. W., and Prasher, D. C. (1994). Green fluorescent protein as a marker for gene expression. *Science* **263,** 802–805.

Cole, N. B., Smith, C. L., Sciaky, N., Terasaki, M., Edidin, M., and Lippincott-Schwartz, J. (1996). Diffusional mobility of Golgi proteins in membranes of living cells. *Science* **273,** 797–801.

Cormack, B. P., Valdivia, R. H., and Falkow, S. (1996). FACS-optimized mutants of the green fluorescent protein (GFP). *Gene* **173,** 33–38.

Edidin, M. (1994) . Fluorescence photobleaching and recovery, FPR, in the analysis of membrane structure and dynamics. *In* "Mobility and Proximity in Biological Membranes" (Damjanovich, S., Edidin, M., Szollosi, J., and Tron, L., Eds.), pp. 109–135. CRC, Boca Raton, FL.

Ellenberg, J., Siggia, E. D., Moreira, J. E., Smith, C. L., Presley, J. F., Worman, H. J., and Lippincott-Schwartz, J. (1997). Nuclear membrane dynamics and reassembly in living cells: targeting of an inner nuclear membrane protein in interphase and mitosis. *J. Cell Biol.* **138,** 1193–1206.

Endow, S. A., and Komma, D. J. (1996). Centrosome and spindle function of the *Drosophila* Ncd microtubule motor visualized in live embryos using Ncd-GFP fusion proteins. *J. Cell Sci.* **109,** 2429–2442.

Heim, R., and Tsien, R. Y. (1996). Engineering green fluorescent protein for improved brightness, longer wavelengths and fluorescence resonance energy transfer. *Curr. Biol.* **6,** 178–182.

Inoue, S. (1986). "Video Microscopy," pp. 191–260. Plenum Press, New York.

Kaether, C., and Gerdes, H.-H. (1995). Visualization of protein transport along the secretory pathway using green fluorescent protein. *FEBS Lett.* **369,** 267–271.

Lippincott-Schwartz, J., and Smith, C. L. (1997). Insights into secretory and endocytic membrane traffic using green fluorescent protein chimeras. *Curr. Opin. Neurobiol.* **7,** 631–639.

Nakata, T., Terada, S., and Hirokawa, N. (1998). Visualization of the dynamics of synaptic vesicle and plasma membrane proteins in living axons. *J. Cell Biol.* **140,** 659–674.

Niswender, K. D., Blackman, S. M., Rohde, L., Magnuson, M. A., and Piston, D. W. (1995). Quantitative imaging of green fluorescent protein in cultured cells: comparison of microscope techniques, use in fusion proteins and detection limits. *J. Microsc.* **180,** 109–116.

Olson, K. R., McIntosh, J. R., and Olmsted, J. B. (1995). Analysis of MAP4 function in living cells using green fluorescent protein (GFP) chimeras. *J. Cell Biol.* **130,** 639–650.

Ormö, M., Cubitt, A. B., Kallio, K., Gross, L. A, Tsien, R. Y., and Remington, S. J. (1996). Crystal structure of the *Aequorea victoria* green fluorescent protein. *Science* **273,** 1392–1395.

Prasher, D. C., Eckenrode, V. K., Ward, W. W., Prendergast, F. G., and Cormier, M. J. (1992). Primary structure of the *Aequorea victoria* green fluorescent protein. *Gene* **111,** 229–233.

Presley, J., Zaal, K, Schrorer, T., Cole, N. B., and Lippincott-Schwartz, J. (1997). ER to Golgi transport visualized in living cells. *Nature* **389,** 81–85

Rizzuto. R., Brini, M., Pizzo, P., Murgia, M., and Pozzan, T. (1995). Chimeric green fluorescent protein as a tool for visualizing subcellular organelles in living cells. *Curr. Biol.* **5,** 635–642.

Sciaky, N., Presley, J., Smith, C., Zaal, K., Cole, N., Terasaki, M., Siggia, E., and Lippincott-Schwartz, J. (1997). Golgi tubule traffic and the effects of brefeldin A visualized in living cells. *J. Cell Biol.* **139,** 1137–1155.

Shelby, R. D., Hahn, K. M., and Sullivan, K. F. (1996). Dynamic elastic behavior of alpha-satellite DNA domains visualized in situ in living human cells. *J. Cell Biol.* **135**, 545–557.

Shima, D. T., Haldar, K., Pepperkok, R., Watson, R., and Warren, G. (1997). Partitioning of the Golgi apparatus during mitosis in living HeLa cells. *J. Cell Biol.* **137,** 1211–1228.

Subramanian, K., and Meyer, T. (1997). Calcium-induced restructuring of nuclear envelope and endoplasmic reticulum calcium stores. *Cell* **89,** 963–971.

Wey, C.-L., Edidin, M., and Cone, R. A. (1981). Lateral diffusion of rhodopsin in photoreceptor cells measured by fluorescence photobleaching and recovery. *Biophys J.* **5,** 387–394.

Yang, T. T., Cheng, L., Kain, S. R. (1996a). Optimized codon usage and chromophore mutations provide enhanced sensitivity with the green fluorescent protein. *Nucleic Acids Res.* **24**, 4592–4593.

Yang, F., Moss, L. G., Phillips, G. N. Jr. (1996b). The molecular structure of green fluorescent protein. *Nat. Biotechnol.* **14,** 1246–1251.

Zolotukhin, S., Potter, M., Hauswirth, W. W., Guy, J., and Muzyczka, N. (1996). A humanized green fluorescent protein cDNA adapted for high-level expression in mammalian cells. *J. Virol.* **70,** 4646–4654.

CHAPTER 17

Synchronous Real-Time Reporting of Multiple Cellular Events

Jeffrey D. Plautz and Steve A. Kay
Department of Cell Biology
The Scripps Research Institute
La Jolla, California 92037

I. Introduction

From whole-animal physiology to unicellular metabolism, life is by necessity a complex coordination of biological processes. Biology as a science strives to understand both the processes that make up life and the interplay among these processes. Recent advances in live-cell imaging have pushed the frontiers of our understanding, especially at the cellular level (Sabri *et al.*, 1997). Real-time imaging permits the observation of dynamic cellular events (Wang and Hazelrigg, 1994), and multiply stained cells allow the (co)localization of multiple subcellular features to be visualized (Stichel and Muller, 1991; Ochi *et al.*, 1993). Moreover, genetic incorporation of reporter genes into nearly every major experimental system has allowed the reliable recording of transcriptional, translational, and trafficking events, in some cases in live cells and/or without any fixing, staining, or additional cofactors.

METHODS IN CELL BIOLOGY, VOL. 58
Copyright © 1999 by Academic Press. All rights of reproduction in any form reserved.
0091-679X/99 $25.00

This chapter will describe the next step in the evolution of cellular monitoring: the simultaneous real-time reporting of multiple cellular processes in living tissue. Specifically, we will describe our experimental system for using green fluorescent protein (GFP) as an *in vivo* spatial marker along with luciferase as a real-time reporter of gene transcription. We have used these reporters to study the expression of the circadian clock-gene *period* (*per*) in *Drosophila* (Brandes *et al.,* 1996; Plautz *et al.,* 1997b). GFP, a bright, stable marker, is ideal for identifying cells in which *per* is expressed (Prasher, 1995; Plautz *et al.,* 1997b). Luciferase, on the other hand, is very dim compared to GFP, but its short half-life makes it useful for monitoring transcription (Millar *et al.,* 1992; Brandes *et al.,* 1996). Each of these reporters is practical for addressing a subset of our specific questions, and each has its own special requirements for detection.

II. Green Fluorescent Protein

GFP was cloned from *Aequorea victoria* (Prasher, 1995). It is an extremely bright and stable 231–amino acid monomer that is active in both live and fixed tissue (Plautz *et al.,* 1996). It is easily used either as a transcriptional reporter (Plautz *et al.,* 1996) or in fusion proteins (Wang and Hazelrigg, 1994). In most cases the addition of the GFP sequence does not interfere with native protein function. The main advantage of GFP is that it will fluoresce without any additional cofactors, and there are many reports of genetic manipulations that make this reporter brighter or better translated. It has been used in a wide range of experimental systems, with much of its success coming in studies of cell labeling or protein localization and movement (reviewed in Prasher, 1995; Welsh and Kay, 1997). Although fluorescence is detectable soon after the GFP is transcribed, the long half-life of the reporter (at least 24 h) makes it unsuitable for determining when a gene is down-regulated (Wood, 1995). In short, GFP is useful in labeling cells and/or proteins, but it is unsatisfactory for all but the most basic measures of transcriptional dynamics.

Our early studies in optimizing GFP detection in *Drosophila* used GMR–GFP flies (Fig. 1A, see color plate; Plautz *et al.,* 1996). These transgenic flies drove wild-type GFP from a synthetic *glass*-responsive multimer. Because the *glass* gene is expressed in photoreceptor cells of the eyes and ocelli (photoreceptive structures on top of the head), GFP can be detected in these tissues. This work went on to show that GFP can be visualized in both live and fixed *Drosophila* tissue, and brought out the difficulties arising from the yellowish cuticle autofluorescence in the animal.

In our current studies we have used *per*-GAL4; UAS–GFP transgenic flies (Fig. 1B, see color plate; Plautz *et al.,* 1997b). These flies fluoresce due to *trans*-activation of the yeast upstream activating sequence (UAS) via GAL4 driven from the *per* promoter. Thus, wherever the promoter is activated, GAL4 is produced; the GAL4 protein can then activate the UAS promoter to produce

GFP. The GFP in this case is the S65T variant (Heim *et al.,* 1995), which is ~100× brighter than wild-type GFP. Although position effects in the transgenic *Drosophila* result in an imperfect fluorescent replication of *per* immunostaining (Plautz *et al.,* 1997b), the vast majority of the GFP-expressing cells do truly express *per.*

Working with GFP in *Drosophila* has two main difficulties: photobleaching and autofluorescence. Although GFP is more resistant to photobleaching than most other fluorescent molecules, bleaching can occur. Along with this is the related issue of nonspecific photodamage to live cells from exposure to the excitation light. Therefore, to keep the tissue living and fluorescing for as long as possible, a minimum amount of exciting light should be used.

Autofluorescence is perhaps the biggest difficulty encountered with GFP in *Drosophila.* The cuticle of the fly, when excited by light that will excite GFP, emits a characteristic yellow autofluorescence (see Fig. 1). The peak of the yellow autofluorescence is separated by ~50 nm from the peak of green S65T GFP fluorescence. With a monochrome detector, this can lead to the detection of many false-positive tissues. Although appropriate filter cubes exist that block yellow light, these generally dim the desired signal, necessitating longer exposure to the potentially photodamaging light. A color detector, though, collects all of the green and yellow light, then allows the user to visually separate the two.

III. Luciferase

Luciferase is a perfect counterpart to GFP in dual-reporter studies. Whereas GFP is very stable, making it excellent for *in vivo* cell labeling, luciferase with its 3- to 5-h half-life (Wood, 1995; Plautz *et al.,* 1997a) is ideally suited for temporal studies. Fast activation and degradation of activity, along with bioluminescence proportional to transcriptional rates, have allowed luciferase to be used as a quantitative temporal marker in systems such as plants (Millar *et al.,* 1992), flies (Brandes *et al.,* 1996), and mammals (Geusz *et al.,* 1997). It is a monomer that biolumineses upon exposure to luciferin (Ow *et al.,* 1986). O_2 and ATP are also needed for the reaction, which means that *in vivo* bioluminescence can only come from living tissue with an adequate oxygen supply.

In vivo optimization work in our lab used transgenic flies in which luciferase was expressed from the heat-shock promoter (Fig. 2A, see color plate; Kay *et al.,* 1994; Stanewsky *et al.,* 1997). Although these flies express some luciferase constitutively, production is dramatically increased upon heat shock. Bioluminescence from these flies was detectable *in vitro* and *in vivo,* but the extremely dim light can only be detected with sensitive instruments; it is too dim for human eyes. In these experiments, live heat-shocked flies were anesthetized with ether (another common anesthetic, CO_2, dims the luciferase reaction because it displaces O_2; likewise, cold slows the enzymatic reaction) and luciferin in solution was administered topically.

Currently we use luciferase to track *per* gene transcription. We generated a series of transgenic flies that drive luciferase from the *per* promoter (Fig. 2B, see color plate; Brandes *et al.,* 1996). When live transgenic flies are fed a constant supply of food containing luciferin, they bioluminesce rhythmically with a 24-h period (Fig. 3A, see color plate). We have also extended this technique to cultured pieces of the fly; these (clock-containing) pieces likewise show rhythmic bioluminescence (Fig. 3, see color plate; Plautz *et al.,* 1997a). Although most of our rhythm studies have been carried out with temporal (but not spatial) detection systems to allow for greater throughput, we have recently shown that luciferase can be imaged *in vivo* with both spatial and temporal resolution (Fig. 6, see color plate).

The main difficulties of luciferase detection come from keeping the tissue alive, permitting sufficient exposure to O_2 (the oxygen in room air is sufficient for bioluminescence), and collecting sufficient amounts of light. Circumventing these problems involves proper animal mounting/cell culture techniques, as well as a light path and detector optimized to capture every available photon. A side effect of such a sensitive detection system is the need for complete darkness; light leaks from a single LED in a room with an open detector can completely saturate the detection system.

IV. Instrumentation and Techniques

A. Nonspatial Detection

Several methods are used to capture emitted bioluminescence or fluorescence. These have varying degrees of sensitivity and spatiotemporal resolution. The simplest method for capturing bioluminescence is through exposure to photographic film. This method is semiquantitative and gives some spatial resolution (although not at the microscopic level) but is not especially sensitive. Specialized bioluminescence detectors (luminometers) are useful for quantitative bioluminescence recording; these detectors are also very sensitive. Similarly, fluorescence can be detected and measured with a fluorometer. Both of these machines are excellent for quantifying emitted light, but they are nonspatial and not well suited for high-throughput assays. Scintillation counters can be used to detect bioluminescence but, like luminometers, are nonspatial. Generally, scintillation counters are not as sensitive as dedicated luminometers.

We have used a modified scintillation counter in our research. The Packard TopCount Multiplate Scintillation and Bioluminescence Counter allows automated single-photon counting. Although nonspatial, this machine is ideal for high-throughput, repeated-measure quantification, in that bioluminescence from hundreds of samples can be recorded hourly for weeks with no user intervention (Brandes *et al.,* 1996). Several manufacturers are also beginning to introduce instruments that can measure both bioluminescence and fluorescence from a single sample in an automated manner. Improvements in sensitivity, speed, and detection versatility will continue to expand the possibilities of multiplexed reporter genes; however, although these machines excel in high-throughput assays, they still lack spatial detection capabilities.

B. Spatial Detection—The Microscope

For spatial detection and quantitation of fluorescence and bioluminescence, we have turned to the digital microscope. Bioluminescence and fluorescence cannot be detected absolutely synchronously because fluorescence (which is extremely bright compared to bioluminescence) quickly washes out any bioluminescent signal. Also, different detection devices, exposure times, and microscope components are optimal for detection of GFP and luciferase. Therefore, although both reporters may be expressed in a single organism, tissue, or cell, detection of each reporter is a very different task.

1. GFP Detection

Because GFP is a fluorescent molecule, it absorbs light of one wavelength and emits light of a longer wavelength. In most cases, detection is accomplished with a filter set that allows illumination of the sample with light near the excitation peak and blocks light that is not near the emission peak. Although FITC filter sets are sufficient for S65T emission and excitation, GFP-specific filter sets are more useful. A comparison of three filter sets for visualizing S65T GFP in *Drosophila* is shown in Fig. 4 (see color plate).

GFP is far easier to detect than bioluminescence simply because it is so bright. In cases in which there is high autofluorescence, especially spectrally close to the true GFP signal, a color camera is indispensable. Our early GFP work captured images with a simple 35-mm camera mounted directly to the microscope (see Fig. 1A; Plautz *et al.,* 1996). Although some of the exposures were long (60 s or more), this relatively simple technology easily allows a true GFP signal to be distinguished from the yellow-green autofluorescence. More recently, though, we have moved to a computerized image acquisition system (Plautz *et al.,* 1997b). An integral part of this is a Hamamatsu C5810 3-chip color CCD camera (see Fig. 1B). Although this camera does not have the resolution of traditional 35-mm film, the resolution is still adequate for our needs and the advantages of digital image processing and computerized acquisition control compared to a 35-mm camera are immense.

In some circumstances, autofluorescence is known not to be a problem. For example, visual inspection of the sample may confirm that some internal tissues that express GFP do not exhibit autofluorescence; alternately, an appropriate filter set may pass sufficient green signal for detection. In these cases monochromatic detectors can be of use. These could range from black-and-white 35-mm film to video cameras to CCD devices.

Perhaps the most intriguing monochromatic detector, though, is the confocal microscope. The optical sectioning power of this instrument allows signal clarity and resolution that may not be achievable with traditional fluorescent microscopes. Although some autofluorescence is present in samples of interest to our lab, we have still been able to detect fluorescence expression *in vivo* in the animal during development (Fig. 5). Multiphoton confocal microscopy offers two advantages over traditional single-photon microscopy: there is less photobleaching and photodamage, and signal can be captured from deeper within the tissue than with single-photon microscopy (Denk *et al.,* 1995). Although the reported benefits in photobleaching are not of tremendous import to our GFP imaging, the increased depth of excitation light penetration in the sample is extremely useful, especially in thick, living samples.

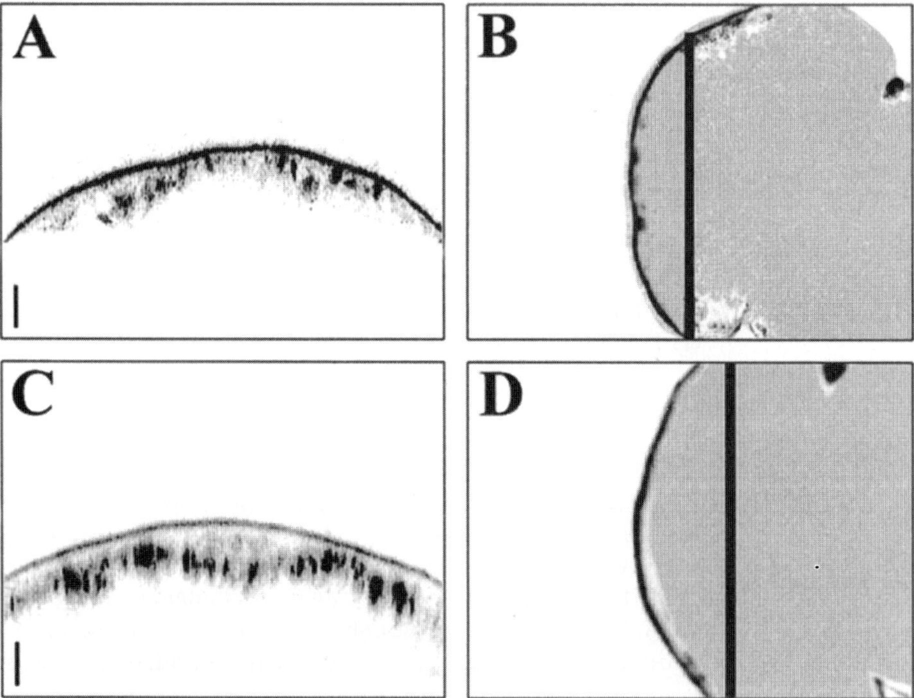

Fig. 5 Confocal GFP imaging. Confocal microscopy allows optical sectioning and 3-D reconstruction of fluorescent samples. In this figure, imagine planes of a pupal *per*–GAL4; UAS–GFP animal were captured from the dorsal side (panels B and D); coronal sections (panels A and C) were subsequently generated through computer reconstruction using Metamorph software. (A) The depth of detection using an Olympus single-photon confocal microscope; (B) the plane from the whole animal from which (a) was constructed. (C) and (D) Corresponding data (from a different animal) collected with a Bio-Rad two-photon confocal microscope. As measured with aid of the 50-micron scale bar in panels A and C, using multiple photons approximately doubles the imaging depth, from 40 to 80 μm.

2. Bioluminescence Detection

Because bioluminescence is so dim, maximum detection sensitivity is critical. This necessitates long exposure times, maximum light transmittance through the microscope system, and most important, a single-photon-capable detector. Our early work used a Hamamatsu VIM camera system coupled to a microscope (see Fig. 2; Brandes *et al.,* 1996). This camera detects single emitted photons, then electrically amplifies them before ouputting electrons to a CCD chip. Recently we have used a Photometrics liquid-nitrogen-cooled CCD camera (Fig. 6). Although the brightness per pixel is generally less with the Photometrics camera, resolution is greater, so the two systems are comparable overall in terms of sensitivity. However, in our experience the Hamamatsu camera has higher background counts than the Photometrics, as well as a lower bit depth. Both systems, though, are more than adequate for low-light imaging.

The requirement for complete darkness in the microscope room cannot be overstressed. The cameras used for bioluminescence detection are incredibly sensitive, so any light leak will at best wash out the image and at worst ruin the camera. Because samples are generally imaged on an open microscope stage, the entire microscope will ideally be placed in a light-tight box within a darkroom. Minimally, the microscope should be in a darkroom in which all light leaks are sealed. If possible, the microscope itself should be covered separately during imaging. A good way to test the imaging environment is to capture two consecutive background images, the first with the light path to the camera closed, and the second with it open (but no sample on the stage). If the environment is sufficiently dark, quantitative image analysis should give identical overall counts for each image.

3. Optimizing the Light Path and Collection Parameters

The overall goal of imaging is the collection of photons. Thus, a good imaging system will allow the most efficient collection of light emitted from the sample. In reality, techniques used to boost the signal also generally increase noise; therefore, detection is a balance that must be optimized for each particular experimental situation. For integrating detection devices (in which photons are collected and summed over a given period of time), such as 35-mm film and CCD chips, more photons can be collected by exposing the sample for longer periods of time. However, background levels also increase over time. In fluorescence imaging, more photons are emitted (up to a certain point) with increasing amounts of excitation light; however, such increases also lead to photobleaching and potential photodamage to the sample itself. With some detection devices, the gain can be adjusted to make the camera more sensitive. This also leads to increased background levels. Because every photon matters in bioluminescence detection even more than in fluorescence, the imaging setup and collection parameters are especially important. These factors will need to be balanced according to each particular experimental situation. When detecting either luciferase activity or fluorescence, the entire light path must be clean and in perfect alignment; otherwise some of the photons (and therefore data) are lost before the experiment even begins.

In a computerized system, signal will ideally be collected so that it covers the full dynamic range of the detector. This means that background is detected; no "empty" pixels should be present in the image. Likewise, no pixel should be fully saturated. In both of these cases, extremes in detection indicate that some data either above or below the working sensitivity range of the detector has been lost. It is better to collect data in the greatest possible range initially, then subtract background and/or optimize the range of values in postprocessing.

4. Postprocessing

In this age of digital image processing and quantitative image analysis, collecting images is only a fraction of the experiment. Common techniques for visualizing multiple reporter genes include imagine optimization, noise reduction, quantification, and visualization (Sabri *et al.,* 1997). The key to image processing is information—without sufficient information, there is nothing to process. Therefore, as much data as possible should be collected for each sample. This may involve collecting multiple images under identical conditions and images showing emitted as well as reflected light; both of these should be collected with high resolution. A good rule of thumb is to collect more information than will ever be needed. More data allows better and more flexible image processing, and once a sample is discarded it is nearly impossible to collect data from it again. When dealing with images, especially with multiple reporters in a single sample, an entire experiment can be rendered useless for want of one additional image. "Excess" data can be discarded, but missing data is missing forever.

V. Future Directions

We have shown here that GFP and luciferase may be used concurrently in *Drosophila.* GFP, a stable fluorescent molecule, is ideal for labeling cells of interest; luciferase is better suited for temporal measures of gene transcription within those same cells. As reporter genes and detector technologies become more advanced, we can expect to see further advances in multiplexed reporter systems. Variants of GFP and luciferase that express in a range of colors already exist (Prasher, 1995; Ohmiya *et al.,* 1996). These could conceivably be used for stably marking several types of cells with the synchronous monitoring of multiple transcriptional events. Presently, though, a thorough understanding of the advantages and limitations of reporters and monitoring technologies is essential for recording multiple outputs from a single living tissue.

Acknowledgments

The authors would like to thank James Pawley and the staff of the 1997 UBC 3-D Microscopy of Living Cells course for the opportunity to directly compare different confocal microscopes, and Olympus Microscopes and Chroma Technologies for providing microscopy equipment.

References

Brandes, C., Plautz, J. D., Stanewsky, R., Jamison, C. F., Straume, M., Wood, K. V., Kay, S. A., and Hall, J. C. (1996). Novel features of *Drosophila* transcription revealed by real-time luciferase reporting. *Neuron* **16,** 687–692.

Denk, W., Piston, D. W., and Webb, W. W. (1995). Two-photon molecular excitation in laser-scanning microscopy. *In* "Handbook of Biological Confocal Microscopy" (Pawley, J. B., ed.), pp. 445–458. (Plenum Press, New York).

Geusz, M., Fletcher, C., Block, G. D., Straume, M., Copeland, N. G., Jenkins, N. A., Kay, S. A., and Day, R. N. (1997). Long-term monitoring of circadian rhythms in *c-fos* gene expression from suprachiasmatic nucleus cultures. *Curr. Biol.* **7,** 758–766.

Heim, R., Cubitt, A. B., and Tsien, R. Y. (1995). Improved green fluorescecnce. *Nature* **373,** 663–664.

Kay, S. A., Millar, A. J., Smith, K. W., Anderson, S. L., Brandes, C., and Hall, J. C. (1994). Video imaging of regulated firefly luciferase activity in transgenic plants and *Drosophila. Promega Notes* **49,** 22–27.

Millar, A. J., Short, S. R., Hiratsuka, K., Chua, N.-H., and Kay, S. A. (1992). Firefly luciferase as a reporter of regulated gene expression in higher plants. *Plant Mol. Biol. Rep.* **10,** 324–337.

Ochi, S., Lim, J. Y., Rand, M. N., During, M. J., Sakatani, K., and Kocsis, J. D. (1993). Transient presence of GABA in astrocytes of the developing optic nerve. *Glia* **9,** 188–198.

Ohmiya, Y., Hirano, T., and Ohashi, M. (1996). The structural origin of the color differences in the bioluminescence of firefly luciferase. *FEBS Lett.* **384,** 83–86.

Ow, D. W., Wood, K. V., DeLuca, M., de Wet, J. R., Helinski, D. R., and Howell, S. H. (1986). Transient and stable expression of the firefly luciferase gene in plant cells and transgenic plants. *Science* **234,** 856–859.

Plautz, J. D., Day, R. N., Dailey, G., Welsh, S. B., Hall, J. C., Halpain, S. L., and Kay, S. A. (1996). Green fluorescent protein and its derivatives as versatile markers for gene expression in living *Drosophila,* plant and mammalian cells. *Gene* **173,** 83–87.

Plautz, J. D., Straume, M., Stanewsky, R., Jamison, C. F., Brandes, C., Dowse, H., Hall, J. C., and Kay, S. A. (1997a). Quantitative analysis of *Drosophila* period gene transcription in living animals. *J. Biol. Rhythms* **12,** 204–217.

Plautz, J. D., Kaneko, M., Hall, J. C., and Kay, S. A. (1997b). Independent photoreceptive circadian clocks throughout *Drosophila. Science* **278,** 1632–1635.

Prasher, D. C. (1995). Using GFP to see the light. *Trends Genet.* **11,** 320–323.

Sabri, S., Richelme, F., Pierres, A., Benoliel, A. M., and Bongrand, P. (1997). Interest of image processing in cell biology and immunology. *J. Immunol. Methods* **208,** 1–27.

Stanewsky, R., Jamison, C. F., Plautz, J. D., Kay, S. A., and Hall, J. C. (1997). Multiple circadian-regulated elements contribute to cycling *period* gene expression in *Drosophila. EMBO J.* **16,** 5006–5018.

Stichel, C. C., and Muller, H. W. (1991). Dissociated cell culture of rat cerebral cortical neurons in serum-free, conditioned media: GABA-immunopositive neurons. *Dev. Brain Res.* **64,** 145–154.

Wang, S., and Hazelrigg, T. (1994). Implications for *bcd* mRNA localization from spatial distribution of *exu* protein in *Drosophila. Nature* **369,** 400–403.

Welsh, S., and Kay, S. A. (1997). Reporter gene expression for monitoring gene transfer. *Curr. Opin. Biotechnol.* **8,** 617–622.

Wood, K. V. (1995). Marker protein for gene expression. *Curr. Opin. Biotech.* **6,** 50–58.

CHAPTER 18

Visualizing Protein Interactions in Living Cells Using Digitized GFP Imaging and FRET Microscopy

Ammasi Periasamy* **and Richard N. Day**†

*Department of Biology
Center for Cellular Imaging
University of Virginia
Charlottesville, Virginia 22903
† Department of Medicine and Cell Biology
NSF Center for Biological Timing
University of Virginia Health Sciences Center
Charlottesville, Virginia 22908

METHODS IN CELL BIOLOGY, VOL. 58
Copyright © 1999 by Academic Press. All rights of reproduction in any form reserved.
0091-679X/99 $25.00

I. Introduction

Determining when and where specific proteins associate with one another in the living cell is of considerable importance to many biologists. Through the use of conventional fluorescence microscopy, proteins labeled with different fluorophores can be localized within fixed or living cell preparations. The fluorophores absorb light at one wavelength and emit light at another, longer wavelength. By using appropriate filters, it is possible to detect several different labeled proteins in the same preparation. However, the optical resolution of the light microscope limits determination of protein proximities to $\geq 0.2 \mu m$. Resolving the relative proximities of proteins that exceed the optical limit of the microscope is necessary to reveal the physical interactions between protein partners. This degree of spatial resolution can only be achieved in light microscopy using the technique of fluorescence resonance energy transfer (FRET). FRET is a process by which radiationless transfer of energy occurs from a fluorophore in the excited state to an acceptor molecule in close proximity. As will be discussed in Section II, the range over which resonance energy transfer can occur is limited to ~0.01 micrometers (100 Å) and the efficiency of energy transfer is extraordinarily sensitive to the distance between fluorophores. The measurement of FRET in the microscope provides a noninvasive approach to visualize the spatiotemporal dynamics of the interactions between protein partners in the living cell.

A major obstacle to implementation of FRET microscopy in living cells has been the lack of suitable methods for specifically labeling intracellular proteins with the appropriate fluorophores. The cloning of the jellyfish green fluorescent protein (GFP) by Prasher and colleagues (1992) and its expression in a wide variety of cell types has proven this fluorescent protein to be a versatile marker for both gene expression and protein localization in living cells (Chalfie *et al.,* 1994; Cubitt *et al.,* 1995; Gerdes and Kaether, 1996; Plautz *et al.,* 1996). When illuminated by blue light, the jellyfish GFP yields a bright green fluorescence that does not require any cofactors, substrates, or additional gene products. GFP retains its fluorescent properties when fused to other proteins, allowing fluorescence microscopy to be used to visualize dynamic changes in protein localization in intact cells (Carey *et al.,* 1996; Ludin *et al.,* 1996; Lang *et al.,* 1997; Barak *et al.,* 1997). Recently, several mutant variants of the GFP with emission in both the green and the blue spectrum (blue fluorescent protein, BFP) have become available (Heim *et al.,* 1994, 1995; Heim and Tsien, 1996; Ormö *et al.,* 1996). The expression of genetic vectors encoding protein fusions with these mutant forms of GFP provides a general method for simultaneously labeling two different proteins within the same cell (Rizzuto *et al.,* 1996). Moreover, as will be discussed later, the excitation and emission spectra for the mutant GFP and BFP proteins are compatible with the FRET approach.

The use of GFP- and BFP-labeled proteins, coupled with recent advances in microscope optics and computer imaging technology, provide the tools necessary for the application of FRET to visualize the dynamic interactions of protein partners within the living cell. This chapter reviews the theoretical basis of FRET microscopy, with emphasis on its application to the living cell. We describe the characteristics of the GFPs that make them useful for FRET imaging studies and outline the important characteristics of an imaging system suitable for detecting FRET signals. We use the expression of genetic vectors encoding protein fusions to GFP and BFP to demonstrate the practical implementation of FRET imaging. Finally, we describe the future direction of FRET imaging technology in biological applications.

II. The Theory of FRET

When a fluorophore absorbs light it goes from a ground state (S0) to a higher vibrational level in the excited state (S2). Within picoseconds the fluorophore decays to the lowest of the vibrational levels of the excited state without emitting a photon. The fluorophore can then decay over a few nanoseconds to one of the ground states by emission of a photon (fluorescence), whose wavelength is longer than the exciting wavelength. In the late 1940s, Förster proposed the theory of FRET, which described how energy could be transferred directly from a fluorophore in the excited state (the donor, D) to a nonidentical acceptor (A) fluorophore (Förster 1948, 1960, 1965). The transfer of excited-state energy occurs without the production of heat and does not require that a collision occur between D and A. FRET is a quantum mechanical process by which radiationless transfer of excitation energy from the D fluorophore to an A occurs by way of a long-range dipole–dipole interaction. A simplified Jablonski energy-level diagram demonstrating the process of energy transfer from a donor to an acceptor fluorophore is shown in Fig. 1.

There are several essential requirements for FRET to occur between proteins labeled with D and A fluorophores. First, there must be a substantial overlap of the emission spectrum of the D with the absorption spectrum of A (J_{DA}, the overlap integral), and the quantum yield of the donor should be high (ϕ_D, the ratio of the number of photons emitted to the number absorbed). Second, the D and A transition dipoles must be in alignment relative to one another in space. The factor k^2 is used to describe the orientation of the D and A transition dipoles and, as will be seen later, the k^2 value is required to determine the precise distance between D and A. The k^2 value can be estimated by taking polarization measurements for both D and A fluorophores. However, slow readout of the signals from some of the detectors described in Section VII makes it difficult to precisely measure the decay of emission anisotropy. Further, uncertainty in the exact positions of the fluorophores is introduced by flexibility of the linkers that attach the probes to the interacting proteins. However, this flexibility does provide dynamic averaging of the orientation factor. Because of this, the k^2 value is often assumed to equal ⅔, which is the value for D and A fluorophores that

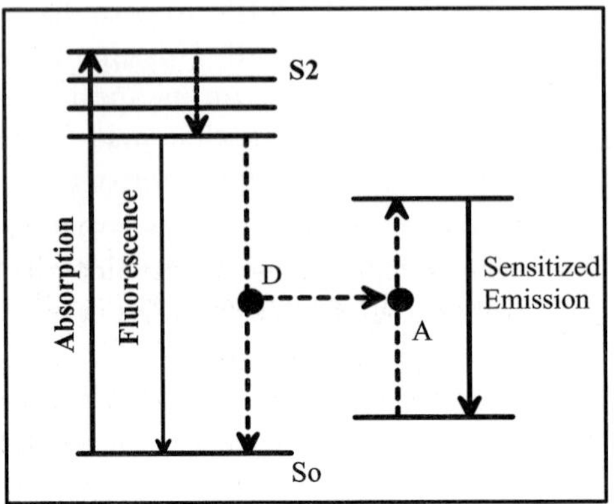

Fig. 1 A simplified illustration of a Jablonski energy-level diagram demonstrating the process of energy transfer. In the presence of a suitable acceptor (A) and donor (D) pair, energy can be transferred directly from D to A without emitting a photon. This nonradiative process of sensitized emission can be measured using FRET microscopy.

are in fast isotropic motion (Dale *et al.,* 1975, 1979; dos Remedios and Moens, 1995). This assumption may lead to an error of ~20% in the calculation of distance between D and A and in the calculation of Förster critical distance. Although the uncertainity of orientation limits the use of FRET for precise distance measurements, the approach is a powerful tool for determining the relative distance between probes. The third essential requirement for FRET to occur is that the distance between D and A fluorophores fall within the appropriate range for energy transfer. Depending on the properties of the fluorophores, energy can be transferred from D to A by a resonance dipole–dipole mechanism over distances of ~10 to ~100 Å. This is because energy transfer is a quantum mechanical effect that is dependent upon the electric field produced by the D fluorophore. Therefore, the efficiency of energy transfer varies inversely with the sixth power of the distance separating D and A, limiting FRET to the range of ~100 Å. At distances closer than ~10 Å, many other modes of energy transfer or electron transfer can come into play (Dexter, 1953; Ernsting *et al.,* 1988). However, this lower limit is of little consequence for studies using the GFPs as probes due to the size of the fluorescent protein (see Section VI).

Förster derived the energy transfer rate equation that related the distance between D and A fluorophores and the spectroscopic properties of the fluorophores. The distance between D and A at which the average efficiency of energy transfer is 50% is known as the Förster distance (R_0). Consider a D–A pair that are separated by a distance *r*. The rate of energy transfer from the D to the acceptor (K_T) is given by

$$K_T = (1/\tau_D)\,(R_0/r)^6 \qquad (1)$$

$$R^6{}_0 = (8.8 \times 10^{-25})\,(k^2 n^{-4} \phi_D\, J_{DA})\ (\text{in cm})\,, \qquad (2)$$

where τ_D is the fluorescence lifetime of the D in the absence of A, $8.8 \times 10^{-25} = 9000(\ln 10)/128\pi^5 \times$ Avogadro's number (6.02×10^{23} per mole), k^2 is the orientation factor; n is the refractive index of the medium, which is generally assumed to be 1.4 for proteins, ϕ_D is the quantum yield of the donor in the absence of acceptor, and J_{DA} is the overlap integral of the donor emission with the acceptor absorption spectrum and is defined as

$$J_{DA} = \int F_D(\lambda)\varepsilon_A(\lambda)\lambda^4 \, d\lambda \,, \tag{3}$$

where F_D is the peak-normalized fluorescence spectrum of the donor, ε_A is the molar absorption coefficient of the acceptor (M^{-1} cm^{-1}).

The energy transfer efficiency (E_T) is defined by the ratio of the transfer rate (K_T) to the sum of the rate processes that depopulate the excited state of the donor. Because E_T varies as the inverse of the sixth power of the distance separating the D and A fluorophores, E_T is related to r by the expression

$$r = R_0 \, (1/E_T - 1)^{1/6} \tag{4}$$

$$E_T = 1 - (\tau_{DA}/\tau_D) \,, \tag{5}$$

where τ_{DA} is the lifetime of the donor in the presence of the acceptor. Thus, by measuring the lifetime of the donor with and without an acceptor, one can calculate the distance between D and A using Eqs. (4) and (5). The practical importance of R_0 is that it provides a means of estimating the range over which FRET can be achieved for any given fluorophore pair. The efficiency of energy transfer (E_T) decreases dramatically in the range 0.5 R_0, to 1.5 R_0 and it is difficult to accurately determine distance beyond this range. The application of FRET microscopy to measure the intermolecular distances in living cells was described by Ludwig *et al.,* (1992). They examined the chromatin structure in the cell nucleus using energy transfer between two DNA-specific dyes, Hoechst 33342 and acridine orange. The Förster distance for this dye pair was determined to be 45 Å, which is the approximate distance of one helical turn of DNA.

III. Review of the FRET Literature

Energy transfer has been applied as a spectroscopic ruler in the fields of biophysics and biochemistry for the last 30 years. A number of excellent review articles deal with theory and application of FRET spectroscopy (Förster, 1965; Schiller, 1975; Fairclough and Cantor, 1978; Stryer, 1978; Jovin, 1979; Botts *et al.,* 1984; Lakowicz, 1986; dos Remedios *et al.,* 1987; Szöllösi *et al.,* 1987; Cheung, 1991; Dewey, 1991; Miki *et al.,* 1992; Clegg *et al.,* 1994; Wu and Brand, 1994; dos Remedios and Moens, 1995; Clegg, 1996). As will be discussed later, the approach of FRET microscopy provides the cell biologist with a method to resolve the spatial and temporal dynamics of protein–protein interactions in the living cell. A number of articles in the literature focus on this specific biological application

(Herman, 1989; Jovin *et al.,* 1989; Jovin and Arndt-Jovin, 1989a,b; Uster and Pagano, 1989; Cardullo *et al.,* 1991; Bottiroli *et al.,* 1992; Dunn *et al.,* 1994). This chapter deals with the application of FRET microscopy to the living cell using a relatively new class of fluorescent probes, the jellyfish GFPs. At present there are few articles reporting the use of GFPs and FRET (Heim and Tsien, 1996; Mitra *et al.,* 1996; Mahajan and Herman, 1997; Periasamy and Day, 1997; Romoser *et al.,* 1997).

IV. Why Use FRET Microscopy?

Because measurements on the scale of angstroms can be obtained under physiological conditions using FRET spectroscopy, this approach has been widely applied to determination of protein and nucleic acid structure. It serves as an alternative to other, more destructive techniques such as X-ray diffraction, nuclear magnetic resonance and electron microscopy. Importantly, intramolecular distances determined by these techniques have validated distance information determined by FRET (dos Remedios and Moens, 1995). For example, distances calculated by FRET for the monomeric actin cocrystallized with Dnase I, profilin, and gelsolin segment 1 were in close agreement with distances determined by X-ray diffraction (Kabsch *et al.,* 1990; McLaughlin *et al.,* 1993; Schutt *et al.,* 1993). Where the FRET approach significantly departs from these other techniques is in its application to the living cell environment. Average energy transfer efficiencies have been obtained from living cell populations using the spectrofluorimeter, and similar information can be acquired using dual-laser flow cytometry (reviewed in Jovin and Arndt-Jovin, 1989; Mátyus, 1992). It is the measurement of FRET from single living cells viewed through the fluorescent microscope, however, that provides the most powerful tool to the cell biologist. FRET microscopy allows the visualization of the spatial distribution of energy transfer efficiency in different locations within a single living cell (Herman, 1989; Jovin and Arndt-Jovin, 1989; Clegg, 1992; Gadella and Jovin, 1995; Bastiaens *et al.,* 1996). Recent advances in light detectors and digital imaging technology (see Section VII) have provided the means to acquire computer processed images to establish energy transfer efficiencies on a pixel-by-pixel basis. The efficiency of energy transfer at a particular location within the digital image provides information that surpasses the optical limit of the light microscope, allowing the measurement of relative proximities of interacting proteins within $\sim 0.04\ \mu m^2$.

V. Why Use the GFPs for FRET?

Many studies utilizing FRET microscopy have relied upon fluorescein (donor)– and rhodamine (acceptor)–labeled probes, typically in the form of antibodies or as labeled receptor ligands that interact with proteins at the cell surface (Gadella and Jovin, 1995; Guo *et al.,* 1995; Jurgens *et al.,* 1996). Until recently, FRET analysis of protein–protein interactions within the intracellular environment relied upon covalent attachment of fluorescent probes to purified proteins, which were then microinjected into individual cells (Bacskai *et al.,* 1993). It is now possible, by transfection of cells with genetic vectors encoding proteins fused to GFP, to fluorescently tag any protein that can be expressed in cells. The jellyfish GFP is a single-chain protein of 238 amino acids, and it owes its fluorescent properties to the autocatalytic formation of a peptide chromophore (Prasher *et al.,* 1992; Brejc *et al.,* 1997; Reid and Flynn, 1997). The crystal structure of GFP revealed that the chromophore, consisting of the cyclic tripeptide sequence serine^{65}-dehydrotyrosine^{66}-glycine^{67} (numbers denote amino acid position), is buried in the center of a nearly perfect cylinder formed by an 11-stranded β-barrel (Ormö *et al.,* 1996). The folding of this intricate barrel structure, complete with top and bottom, is necessary for formation of the chromophore. This explains the observation that nearly the entire protein sequence is required for fluorescence (Cubitt *et al.,* 1995).

The wild type GFP has a complex excitation spectrum with a maximum absorption at 395 nm and a minor peak absorbance at 475 nm. GFP emits green light maximally at 508 nm. Mutagenesis of GFP has yielded a variety of spectral variants with peak fluorescent emission ranging from blue to yellowish green (Heim *et al.,* 1994; Cubitt *et al.,* 1995; Heim and Tsien, 1996; Ormö *et al.,* 1996). Mutation to change the chromophore serine^{65} to cysteine, leucine, alanine, or threonine simplified the excitation spectra to a single peak in the 471- to 489-nm range (Cubitt *et al.,* 1995). The reason for the loss of the ultraviolet excitation was recently determined by comparison of the crystal structure of wild-type GFP with the serine^{65} to threonine (S65T) mutant. In the GFP^{S65T} structure, the hydrogen bonding network in the chromophore is stabilized, resulting in a permanently ionized form of the fluorophore absorbing at 489 nm (Brejc *et al.,* 1997). GFP^{S65T} has proven to be an exceptionally useful spectral variant. It has a four- to six-fold improvement in the intensity of green light emission when compared to wild-type. The posttranslational cyclization and oxidation to produce the chromophore also occurs more rapidly, allowing fluorescence to be detected at earlier time points after cell transfection (Heim *et al.,* 1995). Further improvements in the expression of GFP^{S65T} were obtained by optimizing the codon usage to facilitate translation of the messenger RNA encoding the protein in plant or mammalian cells (Chiu et al., 1996; Yang *et al.,* 1996; Zolotukhin *et al.,* 1996). In addition, the introduction of mutations that improve the folding of the protein at 37°C have resulted in increased expression of the protein under cell culture conditions (Siemering *et al.,* 1996; Kimata *et al.,* 1997). Taken together, these mutations have dramatically improved the fluorescence signal obtained from GFP fusion proteins expressed in living cells.

A second group of interesting spectral variants were obtained by mutation of the chromophore tyrosine^{66} to histidine, phenylalanine, or tryptophan. These changes resulted in fluorescent proteins that, when excited by ultraviolet light, emit light in the blue spectrum (the BFPs). The tyrosine^{66} to histidine mutant has an excitation maximum at 382 nm and a peak emission at 447 nm (Heim *et al.,* 1994). Because of its ultraviolet excitation, BFP^{Y66H} is sensitive to photobleaching and its quantum yield is reduced when compared to GFP. Further mutation to change the tyrosine^{154} to phenylalanine (BFP$^{Y66H,\ Y154F}$) improved the quantum efficiency and also reduced the time required for post-translational processing of the fluorophore (Heim and Tsien, 1996). The crystal structure for this BFP$^{Y66H,\ Y154F}$ was determined, and it indicated that the hydrogen bonding around the chromophore differed considerably from that of GFP (Wachter *et al.,* 1997). Because the peak emission for BFP$^{Y66H,\ Y154F}$ and GFP^{S65T} are substantially different, the two fluorophores can be readily distinguished within the same cell when viewed sequentially with a fluorescence microscope (Rizzuto *et al.,* 1996). Thus, expression of GFP^{S65T} and BFP$^{Y66H,\ Y154F}$ fusion proteins in the same cell provides a general method to determine the relative positions of two different proteins at the optical resolution of the microscope.

As was discussed in Section II, FRET microscopy provides a method for determining the relative distance between proteins on the scale of angstroms. FRET requires that the emission energy of a D fluorophore overlap with the energy need to excite an A. Because living cells tolerate longer wavelength light better than they do near ultraviolet, GFP variants for FRET studies would be optimal if GFP^{S65T} could serve as a D for a A fluorophore emitting in orange or red wavelengths. However, extensive mutagenesis of the GFP protein has failed to shift peak excitation further than 513 nm (Ormö *et al.,* 1996). Thus, in the absence of a "red" fluoresecent protein, it has been necessary to rely on the BFP spectral variants as Ds of excitation energy to GFP^{S65T} for FRET studies. Based upon the significant overlap in the emission spectrum of BFP$^{Y66H,\ Y154F}$ with the excitation spectrum of GFP^{S65T}, it was predicted that these fluorophores would be suitable for FRET (Cubitt *et al.,* 1995). The Förster distance, R_0, was estimated by the energy transfer rate equation to be 40 Å for BFP$^{Y66H,\ Y154F}$ and GFP^{S65T}, assuming a k^2 value of ⅔ (Heim and Tsien, 1996). Energy transfer between these fluorescent proteins was verified in experiments using protein fusions in which BFP and GFP were physically coupled by a flexible polypeptide linker. Excitation of this purified fusion protein with ultraviolet light caused green emission from GFP. Proteolytic cleavage within the linker to separate the two probes substantially decreased green emission and enhanced blue emission due to the reduction in excited state quenching by GFP (Heim and Tsien, 1996; Mitra *et al.,* 1996). We used FRET microscopy to demonstrate energy transfer signals from single living cells expressing GFP fused to BFP by a 29–amino acid linker (Periasamy *et al.,* 1997). Recently, Romoser and colleagues (1997) extended this approach to demonstrate that FRET signals from GFP linked to BFP by a 17–amino acid calmodulin-binding domain peptide could be used as an indicator of spatiotemporal changes in free calcium within living cells. These studies confirm that BFP^{Y66H} can transfer its excitation energy to GFP^{S65T} when the two proteins are physically tethered. As will be described in Section VIII, we have successfully applied FRET microscopy to image the interaction of GFP and BFP fusions with the pituitary specific transcription factor Pit-1 in the nucleus of single living cells (Periasamy *et al.,* 1997; Periasamy and Day, 1997). We will describe how digital deconvolution FRET imaging using donor and acceptor fluorescence images acquired at different optical sections through the cell can localize with high resolution the association of the GFP– and BFP–Pit-1 proteins within the living cell nucleus.

VI. Some Considerations for the Use of GFPs in FRET Imaging

A. The Fusion Proteins

There are several important considerations when using the GFPs to label proteins for FRET studies. Protein fusions with the GFPs come in only two flavors; with the fluorescent protein positioned at either the amino- or the carboxy-terminal end of the protein of interest. An educated guess about which fusion will work best may be made based upon the known domain structure of the protein of interest. However, in many cases it may be useful to prepare both the amino- and the carboxy-terminal fusions. If the GFPs are to be positioned at the amino-terminal end, it is important to utilize GFPs encoded by cDNAs with optimal codon usage. Another important consideration is whether a linker sequence is to be used to separate the protein of interest from the fluorescent proteins. In some cases this may help by allowing both the tagged protein and the fluorophore to fold properly. However, it should be remembered that 15 amino acids forming an alpha helix are approximately 20 Å in length, which is 0.5 R_0 for the GFPs. Given the extreme sensitivity of FRET to distance around R_0, insertion of a linker may adversely impact detection of energy transfer. In this regard, the insertion of a proline residue into a linker sequence to disrupt the secondary structure may be beneficial. Alternatively, it may be possible to dispense with some of the sequence encoding the protein of interest, thereby moving the fluorescent protein closer to important interacting domains. In many cases, it may be necessary to prepare several different GFP fusion proteins and test each one to find a chimera with optimal characteristics.

It is critical that the protein of interest be fully functional when fused to the 27-kDa fluorescent proteins. Likewise, it is necessary that the protein fusion not interfere with the ability of the GFPs to fold properly to form the fluorophore. The function of fusion proteins can be assessed in a number of ways. Fluorescence microscopy of transfected cells will quickly reveal the level of expression and localization of the fusion protein. In these pilot studies, control cells expressing GFP alone may be a useful comparison for signal intensity and localization. Further, when transiently transfected cells are imaged, some cells in the microscope field will not be expressing the fusion protein, providing an internal control for the autofluorescence background of the sample. Confirmation of expression of full-length fusion proteins can be obtained by Western blotting of proteins extracted from the transfected cells. Antibodies detecting the protein of interest or GFP can be used sequentially to positively identify the expressed protein. Antibodies directed against GFP that are suitable for Western analysis are commercially available. Provided that the experiments described earlier indicate that full-length fusion proteins are being expressed, it is important to identify a method that directly demonstrates that the function of the protein is not impaired by the GFPs. For example, the activity of a kinase fused to GFP can be determined by its ability to phosphorylate a specific substrate. Similarly, the activity of a GFP fusion to a transcription factor can determined by its ability to influence transcription of a reporter gene construct in transient transfection assays. Further, nuclear extracts prepared from these transfected cells can be used in electrophoretic mobility shift studies to assess the ability of a fusion protein to bind to an appropriate DNA element.

B. Expression of GFP Fusion Proteins

A variety of methods are available to introduce purified fusion proteins or expression vectors into cells. The method of microinjection provides a means to introduce purified proteins, *in vitro* transcribed mRNA, or DNA into cells in culture. This method can result in very high expression levels and may be the method of choice for primary cell cultures. However, the approach is technically challenging and the number of cells that can be injected for each experiment is limited. Transient transfection techniques, on the other hand, can yield high numbers of cells expressing the proteins of interest. Although a high efficiency of transfection is important for assessing fusion protein function by the methods described earlier, it is less critical for actual imaging studies. Individual cells expressing GFP representing less than 1% of the total cell population can be easily detected by fluorescence microscopy. The method of transfection is dictated by the cell type used, and the conditions need to be optimized. HeLa cells were chosen for the imaging studies described in Section VIII. The cells are transfected by the method of electroporation using 0.2-cm gap cuvettes and a 250-volt pulse at total capacitance of 1200 microfarads. Typical transfection efficiencies range from 25 to 60%. After transfection, the cells are used to inoculate coverglass in 35-mm culture dishes and maintained in phenol-red-free medium to reduce the background autofluorescence. When using GFPs that are not optimized for folding at 37°C (see Section V), culturing transfected cells at 33°C will increase the fluorescence signals compared to cells grown at the higher temperature. For fluorescence microscopy the coverglass is inserted into a chamber designed specifically for the microscope stage. The culture medium is replaced with a medium that is buffered to maintain pH in room air. For the relatively short time duration studies described here, the stage was maintained at room temperature. However, longer time course experiments or studies following the dynamics of protein interactions may necessitate maintaining the stage at 37°C, and temperature-controlled chambers are commercially available (Periasamy and Herman, 1994).

C. Evaluation of a FRET Imaging System with GFPs

Before initiating energy transfer experiments using proteins tagged with $BFP^{Y66H, Y154F}$ and GFP^{S65T}, it is important to design both positive and negative controls to calibrate the microscope system. Several examples of positive control proteins in which GFP was physically linked to BFP were mentioned in Section V. The expression of a tethered GFP–BFP fusion protein under the precise experimental conditions provides a way to optimize the optics, filters, and detectors for the discrimination of fluorescence signals and energy transfer. Of equal importance is a negative control to verify that these signals result from energy transfer and are not due to channel overlap (Periasamy and Day, 1997). For the studies described in Section VIII, we used the expression of noninteracting GFP and BFP fusion proteins that were targeted to the same location in the cell as a negative control for FRET imaging.

Transient cotransfection of cells with GFP- and BFP-fusion-protein expression vectors will result in a heterogeneous population displaying a range of fluorescent protein levels. We have found that in general there is good agreement in the levels for GFP and BFP fusion proteins, and selection of a cell expressing a particular level of GFP fluorescence is often a good predictor of BFP fluorescence. Therefore, cells expressing the fusion proteins can be first identified using GFP fluorescence, so as not to photobleach the $BFP^{Y66H, Y154F}$-fusion-protein partner. It is important to use some neutral density filter to reduce the spectral scattering from the excitation light source and to control photobleaching of the BFP partner. This will also help to reduce autofluorescence from the cells and culture medium, which can be substantial at near ultraviolet wavelengths. Autofluorescence from living cells arises from intracellular NADH, riboflavin, flavin coenzymes, and flavoproteins bound in the mitochondria (Aubin, 1979). To facilitate the use of neutral density filter and to maximize the BFP signal over the autofluorescence background, it is important to use genetic vectors that are optimized for BFP-fusion-protein expression (see Section V).

VII. Some Considerations for Designing a FRET Imaging System

A. The Microscope

Conventional fluorescence microscopes (inverted or upright) can be used for FRET imaging. A number of research and review articles have been written on the various types of optical microscopes implemented in biological applications

(Inoué, 1987; Kam, 1987; Shotton, 1989; Taylor *et al.,* 1992; Periasamy and Herman, 1994). A schematic diagram of the FRET imaging system used in the experiments presented here is shown in Fig. 2. The system consists of an inverted microscope equipped with epi-fluorescence and transmitted illumination (IX-70, Universal infinity system; Olympus America Inc., Melville, NY). The microscope stage and the focus system were coupled to a stepper motor and interfaced with SGI computer, allowing acquisition of optical sections in the *z*-axis of a living specimen. The selection of the objective lens depends on the specimen under investigation. In general, the higher the numerical aperture of the objective, the better the resolution. Moreover, water-immersion lenses provide better resolution at deeper optical sections due to decreased spherical aberration. A Planapo 60× magnification, 1.2–numerical aperture (NA), water-immersion objective lens was used for the experiments presented here. The epi-fluorescent light source used was a 100-W mercury–xenon arc lamp (Hamamatsu Corp., Middlesex, NY). The Ludl (Ludl Electronic Products Ltd., Hawthorne, NY) excitation, emission,

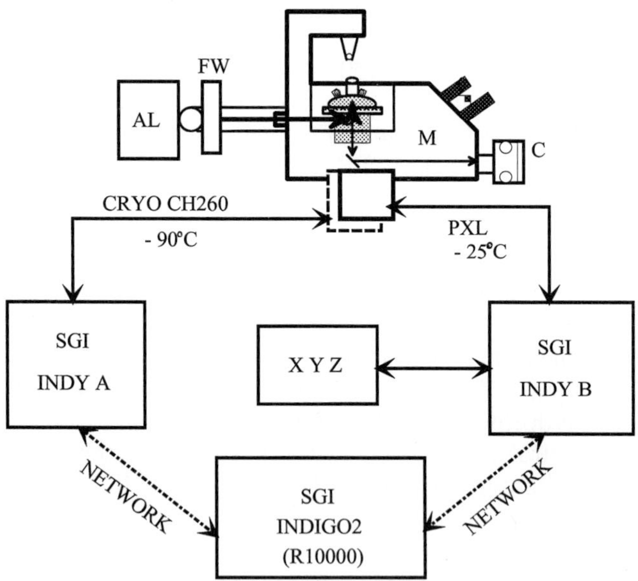

Fig. 2 Schematic diagram of the FRET imaging system used in the studies presented in Section VIII. The Olympus infinity-corrected inverted microscope (M) has four output ports. The liquid-nitrogen-cooled CCD camera (CRYO CH260) was coupled to the side port and the PXL CCD camera to the bottom port, just below the objective. The emission filter wheel was coupled in front of the camera (not shown). The *x, y, z* stage controller (XYZ) was interfaced to a SGI Indy B computer for the acquisition of optical sections from the living cells. An SGI Indigo2 (R10000) was used for image analysis and digital deconvolution of the optical sectioned images for three-dimensional reconstruction. C, 35-mm camera; FW, excitation and neutral density filter wheels; AL, mercury–xenon arc lamp.

and neutral density filter wheels were interfaced to the SGI computer and were driven by the Inovision Corporation ISEE software (Raleigh, NC). The dichroic mirrors were installed in the filter cube.

B. The Filters

FRET microscopy using the GFPs requires the discrimination of fluorescent signals from BFP (D) and GFP (A) fusion proteins that overlap in the D emission spectrum and the A absorption spectrum. We used the expression of both the positive and negative control proteins described in Sections VI.C and VIII to characterize the channel overlap of several different filter combinations for the imaging system and detectors described later. Due to the broad nature of the excitation and emission spectra for the $BFP^{Y66H,\ Y154F}$ and GFP^{S65T} variants (Heim and Tsien, 1996), it is critical to quantify any excitation of GFP^{S65T} when using the A (FRET) filter combination. If a substantial signal is detected, changing the excitation filter to slightly lower wavelengths or narrowing the band pass to reduce coincidental excitation of GFP^{S65T}, may improve detection for FRET signals. Based upon these evaluations (Periasamy *et al.,* 1997), we selected the excitation, dichroic, and emission filters (Omega Optical Inc., Brattleboro, VT) shown in Table I for detection of GFP^{S65T}, $BFP^{Y66H,\ Y154F}$ (D) and FRET (A) images.

C. The Detectors

The major limitation to the imaging of FRET from living cells is the detection of signals above background noise. Fluorescence imaging of living cells, especially when using BFP, requires the detection of very low levels of light, typically on the order of 10^{-4} to 10^{-8} foot-candles. The video imaging detector needs to be sufficiently sensitive to detect these low-level signals. Further, the detector is where the majority of signal amplification occurs, and it is important that the dark current noise of the detector be minimized. For FRET microscopy it is especially important to select a detector with high quantum efficiency (less noise), high sensitivity, and a fast readout rate. These are characteristics that can be found in both charge-coupled-device (CCD) cameras and photomultiplier tubes (PMTs). The slow-scan, cooled CCD cameras have reduced readout noise and

Table I
Filter Sets for Fluorescence Imaging

Purpose	Excitation filter (nm)	Dichroic mirror (nm)	Emission filter (nm)
GFP image	485/22	510	535/50
Donor image (BFP)	365/15	390	460/50
Acceptor image (FRET)	365/15	390	520/40

are capable of prolonged exposures to detect low photon fluxes. Alternatively, the PMTs have high sensitivity, stability, low noise, rapid response (on the order of subnanoseconds) and very large dynamic range (>1 million–fold). A number of articles have reviewed the characteristics of various detectors used in fluorescence microscopy (Bright and Taylor, 1986; Hiraoka *et al.,* 1987; Aikens *et al.,* 1988; Bookman, 1990; Csorba, 1990; Tsay *et al.,* 1990; Spring, 1991; Periasamy and Herman, 1994).

The camera that we have used for some of the FRET imaging described here was a slow-scan, liquid-nitrogen-cooled CCD camera with a back-thinned, back-illuminated imaging chip (CH260, Photometrics, Ltd., Tucson, AZ). This camera has 35% quantum efficiency in the blue spectrum, compared to ~10% for conventional CCD cameras. When cooled to −90°C, this camera has extremely low dark current at 4× gain (~4.0e^-/h) and very high sensitivity. In low-light-level detection the read noise, associated with the on-chip output amplifier of the CCD camera, exceeds the photon shot noise (quantum effect) to improve the signal level. These characteristics are especially important when imaging the relatively low light signals from BFP fusion proteins. However, because of the relatively low spatial resolution (512 × 512 pixels, 16 bits) and slow readout rate (40 KHz), this camera is not suitable for obtaining optical sections to be used for digital deconvolution imaging. This technique, described later, allows three-dimensional reconstruction of the optical sections to localize FRET signals within the cell. For this purpose we have used the high-spatial-resolution (1317 × 1035 pixels, 12 bits), fast-readout (2 MHz) PXL camera (Photometrics).

D. Digital Image Acquisition and Processing

Digital images of single or multiple cells expressing the GFP and BFP fusion proteins can be obtained depending upon the objective and detector used. The larger the digital image, the more computer memory is required to store the images. Each 512 × 512 pixel image requires 750 KB of disk space. For the studies described here (see Section VIII), we have used the SGI-based ISEE software to acquire and process digital images. Once a field of cells with the desired characteristics has been identified by green fluorescence, a reference image is obtained using the GFP filter set (Table I). A second image is acquired for D fluorescence (BFP) from the same field of cells using donor filter set. A third image of the same field is then acquired using the acceptor filter set (Table I). The camera gain, ND filter, and image acquisition time must be kept constant for the D and A images. Camera dark current noise can be digitally subtracted from D and A images by acquiring an image with shutter closed. Similarly, an image of a microscope field with no cells can be acquired under identical conditions as the D and A images and digitally subtracted to remove the background contribution of the media. Using the ISEE software, a pixel-by-pixel ratio of the background-subtracted A to D images will yield the FRET image (I_A/I_D). The ratioing process also serves to remove specimen autofluorescence at the D and A wavelengths.

E. Digital Deconvolution Image Processing

Images collected at the focal plane using wide-field fluorescence microscopy contain out-of-focus light from above and below the focal plane. When the optical axis is changed in the *z*-direction, fine image details are lost in the background of out-of-focus light resulting in a substantial decrease in the contrast. This out-of-focus information can be removed in two ways: (*a*) by using a pinhole aperture in the emission path to reject out-of-focus light, as in confocal microscopy, or (*b*) by use of a digital deconvolution algorithm that is applied to each optical section collected in the *z*-axis. There are number of articles on digital deconvolution methods in the literature (Agard *et al.,* 1989; Shaw, 1993) but here we focus the usage of digital deconvolution system to acquire the FRET signals at different optical sections (see Section VIII).

VIII. The Practical Application of FRET to Visualize Protein–Protein Interactions

We have applied the techniques described earlier to study protein–protein associations involving the pituitary specific transcription factor Pit-1. The Pit-1 protein is a member of the POU homeodomain transcription factor family, which includes the octamer binding proteins Oct-1 and Oct-2 (Herr *et al.,* 1988). Pit-1 is expressed in several different anterior pituitary cell types, where it functions as an important determinant of pituitary-specific gene expression (Andersen and Rosenfeld, 1994). The protein was cloned based upon its binding to cell-specific DNA elements in the promoter regions of the rat prolactin (PRL) and growth hormone genes (Bodner *et al.,* 1988; Ingraham *et al.,* 1988). During anterior pituitary development, the appearance of the Pit-1 protein precedes the emergence of the PRL and growth-hormone-producing cells, and it is required for ontogeny of these cell types (Andersen and Rosenfeld, 1994). The Pit-1 protein generally interacts with DNA elements in the PRL gene promoter as a dimer, although it is thought to exist as monomers when it is not complexed with DNA (Ingraham *et al.,* 1990; Holloway *et al.,* 1995; Chang *et al.,* 1996). We made both GFP–and BFP–Pit-1 infusions to begin to characterize the use of FRET to monitor dimerization of this protein in the living cell nucleus. These fluorescent proteins were positioned at the amino terminus of Pit-1, and the fusion proteins were demonstrated to be fully competent to *trans*-activate the PRL gene promoter linked to luciferase. Further, electrophoretic mobility shift assay demonstrated that the fusion proteins interacted with Pit-1 DNA-binding elements but not the unrelated interferon responsive element.

We applied FRET imaging to HeLa cells expressing the nuclear localized GFP–Pit-1 and BFP–Pit-1 fusion proteins. Dimerization of the BFP–Pit-1 (D) and GFP–Pit-1 (A) proteins would be expected to result in nonradiative transfer of the D excitation energy to the A provided the fluorophores are within 100 Å and the D and A transition dipoles are aligned. This would be evidenced by an increase in the A_A/I_D ratio, whereas no change or a decrease in the ratio would indicate that the fluorophores are not within the range of 10 to 100 Å, or not aligned. Cells coexpressing the BFP–Pit-1 and GFP–Pit-1 proteins were first identified by green fluorescence using the GFP filter set. Images of these same cells were then obtained with the donor (BFP–Pit-1) filter set and background-subtracted (I_D, Fig. 3B, see color plate). As described earlier, camera gain, ND, and acquisition time were kept constant to acquire the background-subtracted acceptor fluorescence (I_A, Fig. 3A, see color plate) image from this cell. The ratio of the acceptor to donor fluorescence was used to generate the processed FRET image (I_A/I_D, Fig. 3C, see color plate). The red color within localized regions of the cell nuclei indicates maximum grayscale intensity corresponding to an increase in the I_A/I_D ratio. This result is consistent with resonance energy transfer from the D– to A–Pit-1 proteins, implicating the close association (<100 Å) or dimerization of these fusion proteins within the cell nucleus.

To demonstrate that simple colocalization of the blue and green fluorophores is *not* sufficient to yield energy transfer, we coexpressed a GFP with a nuclear localization signal (GFP–NLS) and a BFP–Pit-1 fusion protein in the same HeLa cell. Expression of GFP–NLS and BFP–Pit-1 in the same cells resulted in colocalization of the green and blue fluorophores within the nuclear compartment at levels similar to those in the previous experiment (Fig. 3D and E, see color plate). The resulting processed FRET image indicated a decrease in the I_A/I_D ratio, consistent with the lack of energy transfer between these noninteracting but colocalized, proteins (Fig. 3F, see color plate). These results demonstrated that a FRET signal can be detected in single living cells using digital imaging and the selected filter combinations, and that simple colocalization of the fluorophores was not sufficient for resonance energy transfer.

The FRET signals from the dimerized GFP– and BFP–Pit-1 fusion proteins shown in Fig. 3C were obtained from these two cells at the focal plane. Variations in FRET signal strength from different optical planes within the living cell can be used to determine with high resolution the location of protein–protein interations. To precisely localize where dimerization of the GFP– and BFP–Pit-1 proteins was occuring within the cell nucleus, both D and A fluorescence images were acquired at different optical sections using the computer driven stage. The focal plane for the cells shown in Fig. 3A–C was first established and then A and D (BFP–Pit-1) images were sequentially acquired at 0.4-μm steps through the cells. The Delta Vision digital deconvolution algorithm was then used to process both the I_D and the I_A stacks of optical sections. Each of the paired I_A and I_D optical sections was then ratioed to generate the processed FRET image for that section. The resulting FRET images for each optical section through the cell nuclei demonstrates the complex pattern of Pit-1 dimer localization (Fig. 4, see color plate), providing much better resolution of the FRET signal than image shown in Fig. 3C. This is because the digital deconvolution algorithm corrects each pixel in the image for optical noise and out-of-focus information and can correct problems due to photobleaching at different optical sections. Once the digitally deconvolved optical stacks have been obtained, the images can be volume rendered to yield three-dimensional projections. Here, the A, D, and FRET images at one of the optical sections for the cells shown in Fig. 4 were volume rendered to obtain three-dimensional projections. These images were then superimposed to obtain the three-dimensional view of the FRET signals emanating from the living cell nuclei shown in Fig. 5 (see color plate).

IX. Overview and Conclusion

In this chapter we have reviewed the application of FRET microscopy to study protein–protein interactions in the living cell using the spectral variants of the jellyfish GFP. We outlined some of the important considerations in using the GFPs and in the design of an imaging system suitable for detecting FRET. Finally, we demonstrated how the application of digital deconvolution imaging can improve the resolution of localized FRET signals within the cell. As was mentioned earlier, the factors that limit the detection of the weak fluorescent signals associated with FRET include the contribution of autofluorescence, photobleaching, light scattering, and fluctuation in excitation light intensity relative to the image background. These factors can be especially significant for the acquisition of images using ultraviolet light excitation of BFP. The contribution of these factors could be considerably reduced by using two-photon excitation fluorescence imaging microscopy (TEFIM) (see Chapter 4; Denk *et al.,* 1990, 1994; Piston *et al.,* 1994). Two-photon excitation using a 780-nm femtosecond laser light would result in fluorescence emission from BFP. The lower energy red light would greatly reduce sample absorbance and light scattering, minimizing photobleaching and damage to the living cells. The use of TEFIM would limit excitation of the D fluorophore to the focal volume and reduce the time necessary to obtain images. Further improvements in detecting the dynamics of localized FRET may be obtained using fluorescence lifetime imaging microscopy (FLIM). Because FRET is a excited-state process, the application of FLIM would provide subnanosecond temporal resolution of protein interactions (Lakowicz *et al.,* 1992; Periasamy *et al.,* 1995a,b; 1996). Fluorescence lifetime measurements are independent of excitation intensity and fluorophore concentration, and would provide improved precision in the determination of the distance between fluorophores. By combining the TEFIM and FLIM, it should be possible to significantly improve the resolution of the spatiotemporal dynamics of protein interactions within living cells.

Acknowledgments

R.N.D. acknowledges the support of the National Institute of Health (RO1 DK 43701), the National Science Foundation (SGER IBN9528526), and the National Science Foundation Center for Biological Timing (DIR 8920162). A.P. acknowledges the support of National Science Foundation (DBI-9604709) and an award from the Academic Enhancement Program of the University of Virginia.

References

Agard, D. A., Hiraoka, Y., Shaw, P., and Sedat, J. W. (1989). Fluorescence microscopy in three dimensions. *Methods Cell Biol.* **30,** 353–377.

Aikens, R., Agard, D., and Sedat, J. (1988). Solid state imagers for optical microscopy. *Methods Cell Biol.* **29,** 291–313.

Aubin, J. E. (1979). Autofluorescence of viable cultured mammalian cells. *J. Histochem. Cytochem.* **27,** 36–43.

Andersen, B., and Rosenfeld, M. G. (1994). Pit-1 determines cell types during development of the anterior pituitary gland. A model for transcriptional regulation of cell phenotypes in mammalian organogenesis. *J. Biol. Chem.* **269,** 29335–29338.

Bacskai, B. J., Hochner, B., Mahaut-Smith, M., Adams, S. R., Kaang, B. K., Kandel, E. R., and Tsien, R. Y. (1993). Spatially resolved dynamics of cAMP and protein kinase A subunits in Aplysia sensory neurons. *Science* **260,** 222–226.

Barak, L. S., Ferguson, S. S., Zhang, J., Martenson, C., Meyer, T., and Caron, M. G. (1997). Internal trafficking and surface mobility of a functionally intact beta$_2$-adrenergic receptor-green fluorescent protein. *Mol. Pharmacol.* **51,** 177–184.

Bastiaens, P. I., Majoul, I. V., Verveer, P. J., Soling, H. D., and Jovin, T. M. (1996). Imaging the intracellular trafficking and state of the AB5 quaternary structure of cholera toxin. *EMBO J.* **15,** 4246–4253.

Bodner, M., Castrillo, J. L., Theill, L. E., Deerinck, T., Ellisman, M., and Karin, M. (1988). The pituitary-specific transcription factor GHF-1 is a homeobox-containing protein. *Cell* **55,** 267–275.

Bookman, R. (1990). Temporal response characterization of video cameras. *In* "Optical Microscopy for Biology" (Herman, B., and Jacobson, K., Eds.), pp. 235–250. Wiley-Liss, New York.

Brejc, K., Soxma, T. K., Kitts, P. A., Kain, S. R., Tsien, R. Y., and Remington, S. J. (1997). Structural basis for dual excitation and photoisomerization of the *Aequorea victoria* green fluorescent protein. *Proc. Natl. Acad. Sci. U. S. A.* **94,** 2306–2311.

Bottiroli, G., Croce, A. C., and Ramponi, R. (1992). Fluorescence resonance energy transfer imaging as a tool for in situ evaluation of cell morphofunctional characteristics. *J. Photochem. Photobiol. B: Biol.* **12,** 413–416.

Botts, J., Takashi, R., Torgerson, P., Hozumi, T., Mhlrad, A., Mornet, D., and Morales, M. F. (1984). On the mechanism of energy transduction in myosin subfragment 1. *Proc. Natl. Acad. Sci. U.S.A.* **81,** 2060–2064.

Bright, G. R., and Taylor, D. L. (1986). Imaging at low light level in fluorescence microscopy. *In* "Applications of Fluorescence in the Biomedical Sciences" (Taylor, D. L., Lanni, F., Waggoner, A. S., Murphy, R. F., and Birge, R. R., Eds.), pp. 257–288. Alan R. Liss, New York.

Cardullo, R. A., Mungavon, R. M., and Wolf, D. E. (1991). Imaging membrane organization and dynamics. *In* "Biophysical and Biochemical Aspects of Fluorescence Spectroscopy" (Dewey, T. G., Ed.), pp. 231–260. Plenum, New York.

Carey, K. L., Richards, S. A., Lounsbury, K. M., and Macara, I. G. (1996). Evidence using a green fluorescent protein-glucocorticoid receptor chimera that the Ran/TC4 GTPase mediates an essential function independent of nuclear protein import. *J. Cell Biol.* **133,** 985–996.

Chalfie, M., Tu, Y., Euskirchen, G., Ward, W. W., and Prasher, D. C. (1994). Green fluorescent protein as a marker for gene expression. *Science* **263,** 802–805.

Chang, W., Zhou, W., Theill, L. E., Baxter, J. D., Schaufele, F. (1996). An activation function in Pit-1 required selectively for synergistic transcription. *J. Biol. Chem.* **271,** 17733–17738.

Cheung, H. C. (1991). Resonance energy transfer. *In* "Topics in Fluorescence Spectroscopy" (Lakowicz, J. R., Ed.), vol. 3, pp. 127–176. Plenum, New York.

Chiu, W., Niwa, Y., Zeng, W., Hirano, T., Kobayashi, H., Sheen, J. (1996). Engineered GFP as a vital reporter in plants. *Curr. Biol.* **6,** 325–330.

Clegg, R. M. (1992). Fluorescence resonance energy transfer and nucleic acids. *Methods Enzymol.* **211,** 353–388.

Clegg, R. M., Murchie, A. H., and Lilley, D. M. J. (1994). The solution structure of the four-way DNA junction at low-salt conditions: a fluorescence resonance energy transfer analysis. *Biophys. J.* **66,** 99–109.

Clegg, R. M. (1996). Fluorescence resonance energy transfer. In "Fluorescence Imaging Spectroscopy and Microscopy" (Wang, X. F., and Herman, B., Eds.), Chemical Analysis Series, vol. 137, pp. 179–251. John Wiley & Sons, New York.

Csorba, I. P. (1990). "Selected Papers on Image Tubes," SPIE vol. 20. Optical Engineering Press, Bellingham, WA.

Cubitt, A. B., Heim, R., Adams, S. R., Boyd, A. E., Gross, L. A., and Tsien, R. Y. (1995). Understanding, improving and using green fluorescent proteins. *Trends Biochem. Sci.* **20,** 448–455.

Dale, R. E., and Eisinger, J. (1975). Polarized excitation energy transfer. *In* "Biochemical Fluorescence: Concepts" (Chen, R. F., and Edelhoch, H., Eds.), vol. 1, pp. 115–284. Marcel Dekker, New York.

Dale, R. E., Eisinger, J., and Blumberg, W. E. (1979). The orientational freedom of molecular probes. The orientation factor in intramolecular energy transfer. *Biophys. J.* **26,** 161–194.

Denk, W., Piston, D. W., and Webb, W. W. (1994). Two-photon molecular excitation in laser scanning microscopy. In "Handbook of Biological Confocal Microscopy" (Pawley, J., Ed.), 2nd ed., chapter 28. Plenum, New York.

Denk, W., Strickler, J. H., and Webb, W. W. (1990). Two photon laser scanning fluorescence microscopy. *Science* **248,** 73–76.

Dexter, D. L. (1953). A theory of sensitized luminescence in solids. *J. Chem. Phys.* **21,** 836–850.

Dewey, T. G. (1991). Fluorescence energy transfer in membrane biochemistry. *In* "Biophysical and Biochemical Aspects of Fluorescence Spectroscopy" (Dewey, T. G., Ed.), pp. 197–230. Plenum, New York.

dos Remedios, C. G., Miki, M., and Barden, J. A. (1987). Fluorescence resonance energy transfer measurements of distances in actin and myosin. A critical evaluation. *J. Muscle. Res. Cell Motil.* **8,** 97–117.

dos Remedios, C. G., and Moens, P. D. J. (1995). Fluorescence resonance energy transfer spectroscopy is a reliable "ruler" for measuring structural changes in proteins. Dispelling the problem of the unknown orientation factor. *J. Struct. Biol.* **115,** 175–185.

Dunn, K. W., Mayor, S., Myers, J. N., and Maxfield, F. R. (1994). Applications of ratio fluorescence microscopy in the study of cell physiology. *FASEB J.* **8,** 573–582.

Ernsting, N. P., Kaschke, M., Kleinschmidt, J., Drexhage, K. H., and Huth, V. (1988). Sub-picosecond time-resolved intramolecular electronic energy transfer in bi-chromophoric rhodamine dyes in solution. *Chem. Phys.* **122,** 431–442.

Fairclough, R. H., and Cantor, C. R. (1978). The use of singlet-singlet energy transfer to study macromolecular assemblies. *Methods Enzymol.* **48,** 347–379.

Förster, T. (1948). Intermolecular energy migration and fluorescence. *Ann. Phys.* (*Leipzig*) **2,** 55–75.

Förster, T. (1960). Transfer mechanisms of electronic excitation energy. *Radiat. Res. Suppl.* **2,** 326–339.

Förster, T. (1965). Delocalized excitation and excitation transfer. *In* "Modern Quantum Chemistry" (Sinanoglu, O., Ed.), vol. 3, pp. 93–137. Academic Press, New York.

Gadella, T. W. Jr., Jovin, T. M. (1995). Oligomerization of epidermal growth factor receptors on A431 cells studied by time-resolved fluorescence imaging microscopy. A stereochemical model for tyrosine receptor activation. *J. Cell. Biol.* **129,** 1543–1558.

Gerdes, H. H., and Kaether, C. (1996). Green fluorescent protein: applications in cell biology. *FEBS Lett.* **389,** 44–47.

Guo, C., Dower, S. K., Holowka, D., Baird, B. (1995). Fluorescence resonance energy transfer reveals interleukin (IL)-1-dependent aggregation of IL-1 type I receptors that correlates with receptor activation. *J. Biol. Chem.* **270,** 27562–27568.

Heim, R., Prasher, D. C., and Tsien, R. Y. (1994). Wavelength mutations and posttranslational autoxidation of green fluorescent protein. *Proc. Natl. Acad. Sci. U. S. A.* **91,** 12501–12504.

Heim, R., Cubitt, A. B., and Tsien, R. Y. (1995). Improved green fluorescence. *Nature* **373,** 663–664.

Heim, R., and Tsien, R. Y. (1996). Engineering green fluorescent protein for improved brightness, longer wavelengths and fluorescence resonance energy transfer. *Curr. Biol.* **6,** 178–182.

Herman, B. (1989). Resonance energy transfer microscopy. *Methods Cell Biol.* **30,** 216–243.

Herr, W., Sturm, R. A., Clerc, R. G., Corcoran, L. M., Baltimore, D., Sharp, P. A., Ingraham, H. A., Rosenfeld, M. G., Finney, M., Ruvkin, G., and Horvitz, H. R. (1988). The POU domain: a large conserved region in the mammalian pit-1, oct-1, oct-2, and *Caenorhabditis elegans* unc-86 gene products. *Genes Dev.* **2,** 1513–1516.

Hiraoka, Y., Sedat, J. W., and Agard, D. A. (1987). The use of a charge-coupled device for quantitative optical microscopy of biological structures. *Science* **238,** 36–41.

Holloway, J. M., Szeto, D. P., Scully, K. M., Glass, C. K., and Rosenfeld, M. G. (1995). Pit-1 binding to specific DNA sites as a monomer or dimer determines gene-specific use of a tyrosine-dependent synergy domain. *Genes Dev.* **9,** 1992–2006.

Ingraham, H. A., Chen, R., Mangalam, H. J., Elsholtz, H. P., Flynn, S. E., Lin, C. R., and Simmons, D. M., Swanson, L., Rosenfeld, M. G. (1988). A tissue specific transcription factor containing a homeodomain specifies a pituitary phenotype. *Cell* **55,** 519–529.

Ingraham, H. A., Flynn, S. E., Voss, J. W., Albert, V. R., Kapiloff, M. S., Wilson, L., and Rosenfeld, M. G. (1990). The POU-specific domain of Pit-1 is essential for sequence specific, high affinity DNA binding and DNA dependent Pit-1-Pit-1 interactions. *Cell* **61,** 519–529.

Inoué, S. (1987). "Video Microscopy." Plenum Press, New York.

Jovin, T. M. (1979). Fluorescence polarization and energy transfer: theory and application in flow systems. *In* "Flow Cytometry and Sorting" (Melamed, M. R., Mullaney, P. F., and Mendelsohn, M. L., Eds.), pp. 137–165. Wiley, New York.

Jovin, T. M., and Arndt-Jovin, D. J. (1989a). FRET microscopy: digital imaging of fluorescence resonance energy transfer: application in cell biology. *In* "Cell Structure and Function by Microspectrofluorometry" (Kohen, E., and Hirschberg, J. G., Eds.), pp. 99–117. Academic Press, San Diego.

Jovin, T. M., and Arndt-Jovin, D. J. (1989b). Luminescence digital imaging microscopy. *Annu. Rev. Biophys. Chem.* **18,** 271–308.

Jovin, T. M., Marriott, G., Clegg, R. M., and Arndt-Jovin, D. J. (1989). Photophysical processes exploited in digital imaging microscopy: fluorescence resonance energy transfer and delayed luminescence. *Ber. Bunsenges. Phys. Chem.* **93,** 387–391.

Jurgens, L., Arndt-Jovin, D., Pecht, I., and Jovin, T. M. (1996). Proximity relationships between the type I receptor for Fc epsilon (Fc epsilon RI) and the mast cell function-associated antigen (MAFA) studied by donor photobleaching fluorescence resonance energy transfer microscopy. *Eur. J. Immunol.* **26,** 84–91.

Kabsch, W., Mannherz, H. G., Suck, D., Pai, E. F., and Holmes, K. C. (1990). Atomic structure of the actin-Dnase-I complex. *Nature* **347,** 37–44.

Kam, Z. (1987). Microscopic imaging of cells. *Quant. Rev. Biophys.* **20,** 201–259.

Kimata, Y., Iwaki, M., Lim, C. R., and Kohno, K. (1997). A novel mutation which enhances the fluorescence of green fluorescent protein at high temperatures. *Biochem. Biophys. Res. Commun.* **232,** 69–73.

Lakowicz, J. R. (1986). "Principles of Fluorescence Spectroscopy." Plenum Press, New York.

Lakowicz, J. R., Szmacinski, H., Nowaczyk, K., and Johnson, M. L. (1992). Fluorescence lifetime imaging of calcium using Quin-2. *Cell Calcium* **13,** 131–147.

Lang, T., Wacker, I., Steyer, J., Kaether, C., Wunderlich, I., Soldati, T., Gerdes, H. H., Almers, W. (1997). Ca^{2+}-triggered peptide secretion in single cells imaged with green fluorescent protein and evanescent-wave microscopy. *Neuron* **18,** 857–863.

Ludin, B., Doll, T., Meili, R., Kaech, S., and Matus, A. (1996). Application of novel vectors for GFP-tagging of proteins to study microtubule-associated proteins. *Gene* **173,** 107–111.

Ludwig, M., Hensel, N. F., and Hartzman, R. J. (1992). Calibration of a resonance energy transfer imaging system. *Biophys. J.* **61,** 845–857.

Mahajan, N., and Herman, B. (1997). Alterations in the molecular interaction of Bcl-2 and Bax during apoptosis assessed using fluorescence resonance energy transfer (FRET) microscopy and green fluorescent protein (GFP)-Bax and blue fluorescent protein (BFP)-Bcl-2 expressing proteins. *In* "Proceedings of Microscopy and Microanalysis" (Baily, G. W., Dimlich, R. V. W., Alexander, K. B., McCarthy, J. J., and Pretlow, T. P., Eds.), vol. 3, pp. 135–136. Springer-Verlag, New York.

Mátyus, L. (1992). Fluorescence resonance energy transfer measurements on cell surfaces. A spectroscopic tool for determining protein interactions. *J. Photochem. B: Biol.* **12,** 323–337.

McLaughlin, P. J., Gooch, J. T., Mannherz, H. G., and Weeds, A. G. (1993). Structure of gelsolin segment-1-actin complex and the mechanism of filament severing. *Nature* **364,** 685–692.

Miki, M., O'Donoghue, S. I., and dos Remedios, C. G. (1992). Structure of actin observed by fluorescence resonance energy transfer spectroscopy. *J. Muscle Res. Cell Motil.* **13,** 132–145.

Mitra, R. D., Silva, C. M., and Youvan, D. C. (1996). Fluorescence resonance energy transfer between blue-emitting and red-shifted excitation derivatives of the green fluorescent protein. *Gene* **173,** 13–17.

Ormö M., Cubitt, A. B., Kallio, K., Gross, L. A., Tsien, R. Y., and Remington, S. J. (1996). Crystal structure of the *Aequorea victoria* green fluorescent protein. *Science* **273,** 1392–1395.

Periasamy, A., and Herman, B. (1994). Computerized fluorescence microscopic vision in the biomedical sciences. *J. Comput. Assist. Microsc.* **6,** 1–26.

Periasamy, A., Siadat-Pajouh, M., Wodnicki, P., Wang, X. F., and Herman, B. (1995a). Time-gated fluorescence microscopy in clinical imaging. *Microsc. Anal.* **March,** 33–35.

Periasamy, A., Wang, X. F., Wodnicki, P., Gordon, G. W., Kwon, S., Diliberto, P. A., and Herman, B. (1995b). High speed fluorescence microscopy: lifetime imaging in the biomedical sciences. *J. Microsc. Soc. Amer.* **1,** 13–23.

Periasamy, A., Wodnicki, P., Wang, X. F., Gordon, G. W., and Herman, B. (1996). Time resolved fluorescence lifetime imaging microscopy using picosecond pulsed tunable dye laser system. *Rev. Sci. Instrum.* **67,** 3722–3731.

Periasamy, A., and Day, R. N. (1997). Pit-1 protein localization at different optical sections in a single living cell using FRET microscopy and green fluorescent proteins. *In* "Proceedings of Microscopy and Microanalysis" (Bailey, G. W., Dimlich, R. V. W., Alexander, K. B., McCarthy, J. J., and Pretlow, T. P., Eds.), vol. 3, pp. 133–134. Springer-Verlag, New York.

Periasamy, A., Kay, S. A., and Day, R. N. (1997). Fluorescence resonance energy transfer (FRET) imaging of a single living cell using green fluorescent protein. *SPIE* **2983,** 58–66.

Piston, D. W., Kirby, M. S., Cheng, H., Lederer, W. J., and Webb, W. W. (1994). Two-photon-excitation fluorescence imaging of three dimensional calcium-ion activity. *Appl. Opt.* **33,** 662–669.

Plautz, J. D., Day, R. N., Dailey, G. M., Welsh, S. B., Hall, J. C., Halpain, S., and Kay, S. A. (1996). Green fluorescent protein and its derivatives as versatile markers for gene expression in living *Drosophila melanogaster,* plant and mammalian cells. *Gene* **173,** 83–87.

Prasher, D. C., Eckenrode, V. K., Ward, W. W., Prendergast, F. G., and Cormier, M. J. (1992). Primary structure of the *Aequorea victoria* green-fluorescent protein. *Gene,* **111,** 229–233.

Reid, B. G., Flynn, G. C. (1997). Chromophore formation in green fluorescent protein. *Biochemistry* **36,** 6786–6791.

Rizzuto, R., Brini, M., De Giorgi, F., Rossi, R., Heim, R., Tsien, R. Y., and Pozzan, T. (1996). Double labeling of subcellular structures with organelle-targeted GFP mutants *in vivo. Curr. Biol.* **6,** 183–188.

Romoser, V. A., Hinkle, P. M., and Persechini, A. (1997). Detection in living cells of Ca^{2+}-dependent changes in the fluorescence emission of an indication composed of two green fluorescent protein variants linked by a calmodulin-binding sequence: a new class of fluorescent indicators. *J. Biol. Chem.* **272,** 13270–13274.

Schiller, P. W. (1975). The measurement of intramolecular distances by energy transfer. *In* "Biochemical Fluorescence: Concepts" (Chen, R. F., and Edelhoch, H., Eds.), pp. 285–303. Dekker, New York.

Schutt, C. E., Myslik, J. C., Rozycki, M. D., Goonesekere, N., and Lindberg, U. (1993). The structure of crystalline profilin-α-actin. *Nature* **365,** 810–816.

Shaw, P. J. (1993). Computer reconstruction in three-dimensional fluorescence microscopy. *In* "Electronic Light Microscopy" (Shotton, D., Ed.), pp. 211–230. Wiley-Liss, New York.

Shotton, D. M. (1989). Confocal scanning optical microscopy and its applications for biological specimens. *J. Cell. Sci.* **94,** 175–206.

Siemering, E. M., Golbik, R., Sever, R., and Haseloff, J. (1996). Mutations that suppress the thermosensitivity of green fluorescent protein. *Curr. Biol.* **6,** 1653–1663.

Spring, K. R. (1991). Detectors for fluorescence microscopy. *Scanning Microsc.* **5,** 63–69.

Stryer, L. (1978). Fluorescence energy transfer as a spectroscopic ruler. *Annu. Rev. Biochem.* **47,** 819–846.

Szabò, G., Jr., Scott Pine, P., Weaver, J. L., Kasari, M., and Aszalos, A. (1992). Epitope mapping by photobleaching fluorescence resonance energy transfer measurements using a laser scanning microscope system. *Biophys. J.* **61,** 661–670.

Szöllösi, J., Damjanovich, S., Mulhern, S. A., and Trón, L. (1987). Fluorescence energy transfer and membrane potential measurements monitor dynamic properties of cell membranes: a critical review. *Prog. Biophys. Mol. Biol.* **49,** 65–87.

Taylor, D. L., Nederlof, M., Lanni, F., and Waggoner, A. S. (1992). The new vision of light microscopy. *Am. Sci.* **80,** 322–335.

Tsay, T.-T., Inman, R., Wray, B., Herman, B., and Jacobson, K. (1990). Characterization of low-light-level cameras for digitized video microscopy. *J. Microsc.* **160,** 141–159.

Uster, P. S., and Pagano, R. E. (1989). Resonance energy transfer microscopy: visual colocalization of fluorescent lipid probes in liposomes. *Methods Enzymol.* **171,** 850–857.

Wachter, R. M., King, B. A., Heim, R., Kallio, K., Tsien, R. Y., Boxer, S. G., Remington, S. J. (1997). Crystal structure and photodynamic behavior of the blue emission variant Y66H/Y145F of green fluorescent protein. *Biochemistry* **36,** 9759–9765.

Wu, P., and Brand, L. (1994). Review—resonance energy transfer: methods and applications. *Anal. Biochem.* **218,** 1–13.

Yang. T. T., Cheng, L., and Kain, S. R. (1996). Optimized codon usage and chromophore mutations provide enhanced sensitivity with the green fluorescent protein. *Nucleic Acids Res.* **24,** 4592–4593.

Zolotukhin, S., Potter, M., Hauswirth, W. W., Guy, J., and Muzyczka, N. (1996). A "humanized" green fluorescent protein cDNA adapted for high-level expression in mammalian cells. *J. Virol.* **70,** 4646–4654.

CHAPTER 19

Flow Cytometric Analysis and FACS Sorting of Cells Based on GFP Accumulation

David W. Galbraith[*], Michael T. Anderson[†], and Leonard A. Herzenberg[†]

[*] Department of Plant Sciences
University of Arizona
Tucson, Arizona 85721

[†] Department of Genetics
Stanford University Medical School
Beckman Center, B007
Stanford, California 94305-5125

I. General Introduction

A. Principles of Flow Cytometry and Cell Sorting

The technique of flow cytometry involves the optical analysis of biological particles and cells, constrained to flow within a fluid stream through the focus

Copyright © 1999 by Academic Press. All rights of reproduction in any form reserved.
0091-679X/99 $25.00

of an intense source of light, thereby absorbing and scattering light. The resultant scatter and fluorescence signals are detected as intensity-versus-time waveforms using photomultipliers (PMTs) and photodiodes. These pulse waveforms are digitized, and the values are stored either on a cell-by-cell basis or in the form of population frequency distributions. The data produced by flow cytometry preserves optical differences between cells, and subpopulations of cells having different optical characteristics can be recognized and their numbers determined.

The optical characteristics of cells are determined in part by physical parameters, which affect the intensities of light scattered both at angles close to the axis of illumination (forward-angle light scatter, FALS) and orthogonally to the axes of illumination and fluid flow (90° light scatter). Fluorescence emission by cells reflects either endogenous autofluorescence or application of exogenous fluorochromes. A great variety of fluorochromes are available either as synthetic chemicals or as purified natural products. These either have intrinsic specificity, for example, the various dyes that bind to DNA, or are covalently linked to carriers with particular binding characteristics (the prototypical example being the linking of fluorescein to antibodies). Based on the specificities of these fluorescent molecules, flow cytometry can therefore provide a sensitive measure of a large variety of cellular parameters.

From a practical viewpoint, the type of illumination that can be applied to the cells depends on the resources available to the flow cytometric facility. Commercial instruments, under increasing cost constraints, are minimally equipped with an air-cooled argon laser emitting light of 488 nm. With increasing fiscal resources, flow cytometric facilities can be equipped with instruments having water-cooled argon and krypton lasers. These provide tuned illumination either in the UV or at various visible wavelengths from about 400 to 800 nm. Fully equipped, flow cytometric facilities have instruments with multiple independent sources of laser illumination, each tuned to a specific wavelength.

From a technical viewpoint, the point of illumination and analysis of the fluid stream either is done in air or within an enclosed cuvette. In either case, emergence of the fluid stream into air involves passage through an orifice of precise diameter, typically 40–400 μm). The technique of cell sorting derives from the principle that fluid streams in air are unstable and necessarily decay into spherical droplets as a consequence of minimizing free energy. Imposition of a harmonic oscillation on the fluid stream precisely constrains the point of droplet formation, which therefore corresponds to a precise point in time after a given cell has been analyzed. Application of a voltage to the fluid stream at the time this cell occupies the last-attached droplet imposes a corresponding charge on the surface of the newly forming droplet. The detached droplet then can be deflected by passage through a fixed HV electrostatic field. Because the process of fluid charging can be rapidly and reversibly switched, cell sorting at high rates is readily achieved.

B. Green Fluorescent Protein (GFP)

GFP, the endogenous fluorochromatic protein found in the marine invertebrate *Aequorea victoria,* is a single polypeptide of 238 amino acids having a molecular mass of 26 kDa. The cloning of the full-length cDNA (Prasher *et al.,* 1992) led to the discovery that GFP expression, resulting in cytoplasmic fluorescence, can be readily achieved for a variety of prokaryotic and eukaryotic species (Chalfie *et al.,* 1994). Fluorochrome formation involves a series of posttranslational intramolecular reactions, involving cyclization and autoxidation of amino acids 65–67 (Ser-Tyr-Gly) to generate a *p*-hydroxybenzylidine–imidazolidinone fluorescent chromophore (Heim *et al.,* 1994). This reaction requires neither cellular cofactors nor prosthetic groups. At the moment, GFP occupies a unique position as an *in vivo* marker of gene expression and has been extensively employed in studies of transcription, translation, and protein targeting.

C. Fluorescence Characteristics of GFP

The fluorophore of wild-type GFP has an excitation spectra that is complex with absorption peaks at 395 nm (major), and 475 nm (minor). In contrast, its emission spectrum is unimodal, peaked at approximately 509 nm. Biochemical analysis indicates the bimodal excitation spectrum is a consequence of two distinct states of the fluorophore (Cody *et al.,* 1993). When electrically neutral, the fluorophore exhibits maximal absorbance at 395 nm, whereas when the fluorophore is negatively charged, this peak of absorbance shifts to 475 nm. In its native context, the ionization state of the fluorophore is determined by a hydrogen bond network formed between the fluorophore and the side chains of the surrounding amino acids (Brejc *et al.,* 1997). The extent to which this network forms determines the ratio of the excitation maxima at these two wavelengths. Amino acid substitutions either in the fluorophore or in the surrounding sequences can influence the extent to which this network forms and thereby alters the excitation peak ratio.

Variants that are altered in their 395/475-nm excitation ratios or those that possess new wavelengths of excitation and/or emission form the spectral class of GFP variants. The mutations found in this class have significantly contributed to the construction of several distinct GFP reporters suitable for use with the flow cytometer. The first subclass of spectral variants is characterized by a unimodal absorption spectrum peaked at 395 nm. Members of this subclass have been designated Vex, reflecting their violet excitation. The two Vex variants thus far described share an threonine to isoleucine substitution at amino acid 203 (Heim *et al.,* 1994; Ehrig *et al.,* 1995). This substitution disrupts the hydrogen bond network formed with the fluorophore (Ormo *et al.,* 1996; Brejc *et al.,* 1997). As a consequence, tyrosine 66 is fully protonated, leading the fluorophore to be electrically neutral. For flow cytometric applications, the recently introduced Innova 300 krypton laser, which produces a powerful 407-nm line, is particularly suited for excitation of GFP variants of the Vex subclass (Anderson *et al.,* 1996).

A second subclass of GFP spectral variants is characterized by unimodal excitation spectra peaked at wavelengths near the 475-nm excitation peak of wild-type GFP. Variants of this subclass are designated as Bex due to their excitation in the blue. To date, nearly a dozen different Bex variants have been isolated (Delagrave *et al.,* 1995; Heim *et al.,* 1995; Cormack *et al.,* 1996). Among these is the popular "red-shift" variant, possessing a serine to threonine substitution at amino acid 65 (Heim *et al.,* 1995). The amino acid substitutions in the Bex variants stabilize the hydrogen-bonding network formed with the fluorophore (Brejc *et al.,* 1997). This strengthened network promotes dissociation of a proton from the phenolic hydroxyl on tyrosine 66, leaving this amino acid negatively charged. The excitation peaks of the Bex variants tend to be at slightly longer wavelengths than the 475-nm peak of wild-type GFP. This shift partly accounts for the increase in brightness found with Bex variants (up to sevenfold higher), as compared to wild-type GFP, when they are excited by the 488-nm argon line (Heim *et al.,* 1995). This increase also results from a greater proportion of the Bex fluorophores being in an ionization state that is excitable in the blue. The extinction coefficients and quantum yields of both Bex and Vex are at least as great as those of wild-type GFP and comparable to those of fluorescein.

Members of the third subclass of GFP spectral variants have excitation and/or emission spectra that differ in their peak wavelengths from those found for wild-type GFP. One member of this subclass has an excitation peak in the UV at 382 nm; its emission peak is at 488 nm. It possesses a tyrosine to histidine substitution at amino acid 66 (Heim *et al.,* 1994). Application of a consistent nomenclature to this variant designates it as Uex. Although readily distinguishable from wild-type GFP and from the Vex and Bex variants, the Uex variant has a lower extinction coefficient and quantum yield (Heim and Tsien, 1996). Its relative dullness in terms of flow cytometric analysis is compounded by the absence of an available strong laser line coincident with the Uex excitation peak (this lies midway between the 350-nm argon and the 407-nm krypton laser lines) (Ropp *et al.,* 1995, 1996). Recently another member of this subclass has been described; it has an excitation peak at 513 nm and an emission peak at 527 nm (Ormo *et al.,* 1996). It was constructed by substituting glycine for serine at amino acid 65 and histidine for serine at amino acid 203. The histidine substitution is thought to increase the excitation and emission wavelengths by extending the electron conjugation system of the fluorophore. Applying a consistent nomenclature to this variant designates it as Gex due to its excitation peak in the green.

In addition to the class of spectral variants, two other classes of GFP variants have provided important contributions in for the construction of reporters suitable for fluorescence-activated cell sorting (FACS). One of these classes is characterized by greater than wild-type thermostability in terms of establishment of the native three-dimensional structure of the protein molecule. Whereas a significant proportion of wild-type GFP molecules fail to properly fold and cyclize when synthesized at 37°C, variants of this class fold correctly and therefore are significantly brighter. This effect, not seen at 22°C, is conferred by several different amino acid substitutions, and in some instances these mutations can function additively (Anderson *et al.,* 1996; Crameri *et al.,* 1996; Heim and Tsien, 1996; Siemering *et al.,* 1996; Kimata *et al.,* 1997). One member of this class of variants has a valine to alanine substitution at amino acid 163, which increases its fluorescence by about fivefold when expressed in mammalian cells (Anderson *et al.,* 1996). Combinatorial incorporation of multiple mutations of this class of mutations can lead to an overall tenfold increase in fluorescence (Anderson *et al.,* 1998b).

The final class comprises GFP variants that provide enhanced levels of intracellular fluorescence over that of wild-type GFP, but without changes in the amino acid sequence. Variants of this class produce mRNA species that are more efficiently translated than mRNA produced from the wild-type coding sequence. This appears largely a consequence of codon bias; some codons found in the GFP cDNA from the jellyfish *Aequorea victoria* are infrequently utilized in yeast, plants, insects, and mammals. Replacement of these codons with ones preferred within the organism of interest increases GFP production (Chiu *et al.,* 1996; Ha *et al.,* 1996; Pang *et al.,* 1996; Yang *et al.,* 1996; Zhang *et al.,* 1996; Zolotukhin *et al.,* 1996; Cormack *et al.,* 1997; Kimata *et al.,* 1997; Muldoon *et al.,* 1997; Rouwendal *et al.,* 1997). For mammalian cells, the increase is approximately four- to fivefold.

Because independent mechanisms underly the increases in fluorescence seen in these three classes of variants, combining mutations from these different classes has resulted in the production of several spectrally distinct GFP reporters with greatly enhanced levels of fluorescence (up to 100-fold higher than that of wild-type GFP). This is important in terms of sensitivity because at most only a single fluorophore can be formed for each GFP molecule that is synthesized. In flow cytometry, a rule of thumb is that at least 500 fluorescein molecules are required for a particle or cell to produce a fluorescent signal that is above the background derived from cellular autofluorescence and instrument noise. It is considered that optimized S65T GFP and fluorescein produce roughly equivalent amounts of fluorescence, which renders GFP approximately 100-fold less sensitive than the FACS–GAL enzymatic assay. However, the versatility of GFP (at least three spectrally distinct variants are available for simultaneous flow cytometric analysis) provides obvious theoretical and practical advantages.

Considerable interest is emerging in the use of flow cytometry and sorting for the analysis of GFP fluorescence in prokaryotic and eukaryotic systems. This chapter does not claim to comprehensively cover all species, tissues, and cell types, and the reader is recommended to consult the following references for alternative methods (Ropp *et al.,* 1995, 1996; Lybarger *et al.,* 1996; Subramanian and Srienc, 1996; Bierhuizen *et al.,* 1997; Gervaix *et al.,* 1997; Kain and Kitts, 1997; Mosser *et al.,* 1997; Muldoon *et al.,* 1997).

II. Methods and Specific Applications

We have grouped the methods according to organism type, separately describing specific applications for higher eukaryotes (animal and plants) and for prokaryotes.

A. Preparation of GFP-Expressing Cells of Higher Eukaryotes

1. Mammalian Cells

Several methods are available to introduce GFP expression constructs into mammalian cells. The most widely used method is to transiently transfect the cells using lipofectin, calcium phosphate, or electroporation. In the majority of the transfected cells, the DNA construct is maintained as an extrachromosomal episome. In the rare cell, the construct stably integrates into the chromosome. Those cells with stable integrants can be selected by cotransfection of a drug resistance marker. However, retroviral infection is a more efficient method of producing cells with constructs stably integrated into their chromosomes.

In this chapter, we detail a calcium phosphate–mediated transfection protocol. It is a generally useful protocol for transiently introducing GFP expression constructs into adherent cells. A more specific application of this protocol is to introduce defective retroviral constructs into either the BOSC 23 or the Phoenix producer lines (Pear *et al.,* 1993; Kinsella and Nolan, 1996). These lines are derivatives of the highly transfectable 293 human cell line. Each of these producer lines has two constructs integrated within its chromosomes. These constructs function to complement the genes absent from defective Moloney retroviruses. One of these integrants constitutively expresses the viral *gag* and *polymerase* genes; the other constitutively expresses the viral *envelope* gene. When either producer line is transiently transfected with defective retroviral constructs, it can rapidly produce high-titer retrovirus, which can be used for infection of susceptible cells (Pear *et al.,* 1993; Kinsella and Nolan, 1996).

The marrying of GFP reporters with this method of retroviral production is likely to gain considerable popularity as biology moves toward efficient assay of the function of genes through complementation analysis. Together, these methodologies permit the selection of cells expressing the gene of interest within 4 days of preparation of the retroviral construct.

a. Biological Materials

We have used the Moloney retrovirus derivative MFG (Anderson *et al.,* 1996; Krall *et al.,* 1996) as a retroviral vector. Within many cell lines, the MFG LTR directs high levels of expression from genes that are inserted within this vector. Other useful retroviral vectors include the pBABE series of SIN retroviruses, which lack functional LTRs and express inserted genes from internally placed *cis*-acting sequences (Hofmann *et al.,* 1996; Naviaux *et al.,* 1996). The transcripts from either series of retroviruses can be made dicistronic by inclusion of an internal ribosomal entry site (IRES) within the transcript (Hofmann *et al.,* 1996; Mosser *et al.,* 1997). This allows two distinct proteins to be expressed from the same transcript, one of which can be a FACS-selectable GFP reporter.

b. Transfection Procedure

This procedure is detailed for the Phoenix or the BOSC 23 cell lines, but is applicable to many different mammalian cell lines.

a. Producer lines are cultured in supplemented Dulbecco's Minimal Essential Medium (sDMEM), supplemented to 10% (v/v) with a 1:1 mixture of fetal calf to horse serum, 100 U/ml each of penicillin and streptomycin, 2 m*M* glutamine, and 50 μM β-mercaptoethanol. The day before transfection, 2.5×10^6 cells of the producer line are plated onto each 60-mm tissue Falcon culture dish (Becton Dickinson, Franklin Lakes, NJ). The plate of cells is then placed overnight in a 37°C incubator at 5% CO_2. Immediately prior to transfection, chloroquine is added to the plates at a final concentration of 25 μM.

b. To prepare the transfection mixture, 10 μg of the DNA construct is pipetted into a FALCON 12- × 75-mm tube. Sufficient H_2O is added to bring the final volume to 438 μl. Then, 62 μl of 2*M* $CaCl_2$ is added followed by 500 μl of 2× HBS (50 m*M* HEPES, pH 7.05, 10 m*M* KCl, 12 m*M* dextrose, 280 m*M* NaCl, 1.5 m*M* Na_2HPO_4).

c. This solution is mixed by bubbling air from a 2-ml pipette through it for 10–20 s, and then the solution is added dropwise to the plate of producer cells. After this addition, the cells are returned to the incubator.

d. Between 8 and 12 h after transfection, the fluid in the plate is replaced with 3 ml of fresh sDMEM. This process is repeated at 24 h after transfection.

e. At 48 h after transfection, the retrovirus containing supernatant is decanted from the plate of cells and then centrifuged at 500 × *g* at 4°C for 5 min. The supernatant is then ready for use in infecting susceptible cells.

This protocol routinely results in transfection rates of 60–100%; such high rates are necessary to produce high-titer retrovirus. To achieve these rates, several aspects of the transfection process are critical. One of these aspects is the pH of the culture medium. This pH is established by the CO_2 concentration of the incubator. Therefore, the cell culture plates should spend only minimal amounts of time outside of the incubator. In practice, only the particular dish being transfected is transferred from the incubator to the tissue culture transfer hood. A second critical aspect is the pH of the HBS solution; although we define a pH of 7.05 as optimal for the 2x HBS solution, it is necessary to routinely prepare several different batches of this medium differing stepwise from pH 7.00 to 7.10. Each batch should then be evaluated as to which confers the highest transfection efficiency. A final critical aspect of this protocol is the length of time that transfection mixture is bubbled with air. Each individual bubbles this mixture with a slightly different level of vigor. Consequently, the length of bubbling time that produces the highest transfection efficiencies varies between individuals.

c. Viral Infection of Cells

a. For the retroviral infection of adherent murine cell lines such as NIH 3T3 with retrovirus, approximately 5×10^5 cells are seeded onto a 60-mm Falcon plate on the day before they are to be infected. To infect the cells, an aliquot of viral supernatant (titer $0.5–2.0 \times 10^6$ pfu/ml is added to the plate of cells and then the liquid on the plate is adjusted to 4 μg/ml in polybrene.

b. To infect nonadherent murine cells such as lymphocyte cell lines grown in suspension, we have incorporated a modification of the preceding protocol described by Bunnell *et al.* (1995). For the infection of cells, the virus and the cells are pelleted together by centrifugation at 1000 x *g* at 22°C.

In general we have been able to detect by flow cytometry GFP expression in as little as 18 h following retroviral infection.

2. Plant Cells

For flow cytometric analysis and sorting of plant cells according to GFP accumulation, we require production of single-cell suspensions that are expressing GFP. Two strategies are available: the first is the production of protoplasts, which are subsequently transfected, then examined for GFP expression; the second is the production of transgenic plants expressing GFP, which are then reduced to protoplasts prior to flow analysis and sorting. The first strategy is detailed in this chapter. This is followed by an analysis of transgenic accumulation of GFP, specifically employing nuclear targeting as a means for generating fluorescent nuclei that are suitable for FACS analysis.

a. GFP Expression in Transfected Plant Protoplasts

Preparation of protoplasts for different plant species follows defined protocols optimized for the different species and tissue types that are under study. The reader is referred to comprehensive references (see, for example, Gamborg and Philips, 1995).

i. Protoplast Preparation. Protoplasts are most commonly prepared from leaf tissues, which provide adequate supplies of starting materials that can be obtained easily in axenic form. The following protocol works well for two model plant systems: *Arabidopsis thaliana* and *Nicotiana tabacum. Arabidopsis* is accepted as a model for plant molecular genetic analysis, and tobacco is particularly amenable to tissue culture and transformation. The given protocol illustrates the important general concepts of protoplast preparation: use of an osmoticum, selective solubilization of the cell wall using polysaccharidases, recovery and purification of the protoplasts, and viability determinations.

a. Macerase, Cellulysin, and propidium iodide were obtained from Calbiochem, Inc. (La Jolla, CA), and fluorescent microspheres from the Coulter Corporation (Miami, FL). MS (Murashige and Skoog) medium was from Gibco (Grand Island, NY). All remaining chemicals were obtained from the Sigma Chemical Co. (Saint Louis, MO).

b. Plants are grown within standard growth chambers (Conviron) under sterile conditions in Magenta boxes on MS medium containing 3% sucrose. Tobacco (*Nicotiana tabacum* L. cv. Xanthi) plants are maintained at 22°C under continuous light (Harkins *et al.,* 1990). *Arabidopsis* plants (*Arabidopsis thaliana* L. ecotype Columbia) are maintained at 22°C under continuous illumination.

c. All procedures are done using standard sterile technique. Fully expanded leaves are excised from axenic tobacco or *Arabidopsis* plants (approximately 600 mg wet weight), and are transferred into 20 ml of digestion medium, contained in a sterile plastic Petri dish. (*Note:* Repeated harvesting of leaves from individual plants is not recommended. Molecular changes to the cell wall and plasma membrane structure occur, probably as a consequence of wound-induced signal transmission through the body of the plant. Thus, second and subsequent harvests of leaves do not produce protoplasts at yields and viabilities comparable to those from the leaves of the first harvest.) The digestion medium comprises 0.1% driselase, 0.1% macerase, and 0.1% cellulysin, dissolved in a buffer containing 0.5 *M* mannitol, 10 m*M* $CaCl_2$, and 3 m*M* MES, pH 5.7. This medium is sterilized by Millipore filtration (GSWP 047). Incubation is continued at 22°C overnight (18–20 h) in darkness without agitation.

d. The protoplast suspension is filtered through two layers of sterile cheesecloth into a 50-ml sterile centrifuge tube, and centrifuged at $50 \times g$ for 8 min.

e. Protoplasts are purified by centrifugal flotation using a step sucrose gradient. The pelleted protoplasts are gently resuspended in 20 ml of 25% (w/v) sucrose dissolved in modified TO medium (Harkins *et al.,* 1990). This is overlaid with 5 ml of W5 medium (Negrutiu *et al.,* 1987) and centrifuged at $50 \times g$ for 10 min. Viable protoplasts, found at the interface, are carefully removed using a wide-bore Pasteur pipet and are diluted with two volumes of W5 medium. Protoplasts are counted using a hemocytometer and are recovered by centrifugation at $50 \times g$ for 5 min.

ii. Preparation of Recombinant GFP Constructions. For transient expression, recombinant GFP constructs are prepared in high-copy-number vectors. Features of the constructs include promoter sequences for regulated or constitutive expression [the Cauliflower Mosaic Virus (CaMV) 35S promoter is frequently used for constitutive expression], the GFP coding region, and a transcriptional terminator. For expression in *Arabidopsis,* elimination of a cryptic splice site is critical (Haseloff *et al.,* 1997). Use of targeted gene fusions (translation fusions that direct GFP to specific cellular compartments) can also enhance fluorescence emission (Grebenok *et al.,* 1997a,b; Haseloff *et al.,* 1997). Preparation of these recombinant constructions is done according to routine procedures of molecular biology (Sambrook *et al.,* 1989).

iii. Protoplast Transfection. Protoplasts can be transfected by addition of polyethylene glycol (PEG).

a. For PEG-mediated transfection, protoplasts are resuspended in 600 μl W5 to a concentration of $\sim 1.6 \times 10^6$ protoplasts/ml. Plasmid DNA and carrier (calf thymus) DNA are then added to a concentration of 20 and 50 μg/ml, respectively.

b. Tap the side of the tube to ensure that the protoplasts are well resuspended, then immediately add 1.5 volumes of PEG solution, comprising 40% w/v PEG 4000 (Baker Chemical Co.), 0.4 *M* mannitol, and 0.1 *M* $Ca(NO_3)_2 \cdot 4H_2O$, pH 7.0. Gently mix by resuspension using a 1-ml disposable Pipetman pipet tip.

c. Incubate at room temperature for 25 min.

d. Add eight volumes of culture medium [medium NTTO (Harkins *et al.,* 1990)], modified to contain 0.18 *M* glucose and 0.15 *M* mannitol, and supplemented with 75 mg/ml ampicillin).

e. Incubate for 18 to 24 h in darkness at room temperature.

Protoplasts can also be transfected by electroporation (for details of maize protoplast preparation, see Galbraith *et al.,* 1995).

a. For electroporation, the protoplast pellet is resuspended in an electroporation buffer comprising 0.6 *M* mannitol, 20 m*M* KCl, 4 m*M* MES, and 5 m*M* EGTA, pH 5.7.

b. The protoplasts are pelleted once by centrifugation at 100 x *g* for 10 min. Protoplasts (1.5×10^5) are resuspended in 0.3 ml of electroporation buffer and transferred into a plastic electroporation cuvette (gap width 4 mm; BTX, San Diego, CA).

c. Plasmid DNA is added (25–50 μg in 30–60 μl TE). Electroporation conditions involve a 10-ms pulse length, a field strength of 400–500 V/cm, a 200-mF capacitance setting, and a total of three pulses.

d. The protoplasts are transferred into plastic 6-well culture plates (Falcon 3046), kept on ice for 10 min after electroporation, and diluted by addition of electroporation solution (0.7 ml/well).

e. The tissue culture plates are covered with foil and incubated at room temperature for 15 to 18 h.

b. GFP Expression in Transgenic Plants

Transgenic tobacco plants expressing GFP are prepared via agrobacterium-mediated transformation. The methods are well established and routine, and need not be reiterated here (see, for example, Rogers *et al.,* 1986, and for the GFP constructions described in this chapter, Grebenok *et al.,* 1997a,b).

B. Preparation of GFP-Expressing Bacteria

When GFP is expressed in bacteria from appropriate transcriptional/translational controls, it generally produces a FACS-detectable fluorescent signal. This results from GFP folding properly within the reducing environment of the bacterial cytosol. To express GFP in bacteria, we and others have used the tight transcriptional/translational controls of the T7 bacteriophage (Heim *et al.,* 1994). These controls are present within the pET vector, which is available from Novagen Inc. (Madison WI). To express GFP from this vector, the GFP coding sequence is inserted between the BamHI site downstream of the T7 promoter/lac operator and the Xho I site upstream of the T7 terminator; this typically requires PCR-based manipulation to introduce flanking restriction sites. The resulting GFP expression vectors are then transfected into the BL21(DE3) strain of bacteria by standard methodologies such as electroporation (Dubendorff and Studier, 1991). This bacterial genotype contains a lambda DE lysogen, which expresses the T7 polymerase under the control of the lacUV5 promoter. Growth of transfectants under noninductive conditions (high glucose, no IPTG) leads to little or no GFP expression. In contrast, significant levels of fluorescence reflecting increases in GFP expression are obtained under inducing conditions (removal of glucose and addition of IPTG).

C. Flow Cytometry and Sorting

a. Analysis and Sorting of Mammalian Cell Lines Expressing GFP

The following methods are platform independent but are detailed for a custom flow cytometer and FACS constructed in the Herzenberg laboratory. It is capable of excitation with three laser lines and possesses sufficient detectors to assay the fluorescent signal from Bex, Vex, and six immunofluorescence parameters. It consists of a FACStar Plus optical bench (Becton-Dickinson, San Jose, CA), MoFlo electronics (Cytomation, Fort Collins, CO), which collect and digitize the signals from the optical bench, and custom-designed electronics designed and built in the Herzenberg laboratory that control the PMT voltages and that handle data transfer to the computers.

a. The 488-nm line and the 595-nm dye laser line are both powered by a 5-watt Innova-90 argon-ion laser (Coherent, Sunnyvale, CA), run in "all lines" mode. The light from this laser is split by a dichroic mirror into its shorter and longer wavelength components. The shorter wavelengths are bandpass filtered to provide only 488-nm illumination. The longer wavelength lines are directed to a dye head tuned to 595 nm. The third laser line comes from an Innova-300 krypton laser tuned to produce light at 407 nm.

b. The signals collected include forward-angle light scatter (FALS), 90° light scatter (SSC), 488-nm excited GFP–Bex fluorescence (495–535 nm) and 407-nm excited GFP–Vex fluorescence (495–535 nm). A more detailed description of the configuration of this optical bench including the optical filters used to collect the immunofluorescence signals can be found in Roederer *et al.* (1998) and in Anderson *et al.* (1997a). The components of the optical bench are aligned with 3.2-μm multidye polystyrene test particles obtained from Spherotech Inc. (Libertyville, IL). PMT voltages are selected that result in the cells with and without GFP expression both being on scale.

c. Cells are analyzed at a flow rate of between 200 and 1000 events/s in a stream formed from a nozzle with a 70- to 80-μm orifice. Generally PBS is used as the sheath fluid; however, when cells are to be sorted for high viability, RPMI-deficient medium (Hyclone Laboratories, Logan, UT) lacking phenol red and biotin, is substituted.

d. To sort cells, a nozzle pressure of 12 psi is established. The electronics are set for a drop drive frequency of 30 kHz, the drop delay is set between 14 and 16, and the sorting setting is set at the two drops mode.

b. Analysis and Sorting of Plant Protoplasts Expressing GFP

The following two methods are also platform independent but are in this case described for the instrument currently available in the Galbraith laboratory (Elite, Coulter Electronics, Miami Lakes, FL). This is equipped with a 20-mW 488-nm argon-ion laser and a 100-μm-diameter sense-in-quartz flow tip.

a. Signals are collected for forward-angle light scatter (FALS), 90° light scatter (PMT1), and fluorescence [PMTs 2 (green; 505–545 nm) and 4 (red; 670–680 nm)]. Typical high-voltage and amplification settings are FALS, 260/10; PMT1, 350/7.5; PMT2, 850/10; and PMT4, 950/5. The FALS discriminator is set to 50, and all other discriminators are turned off.

b. The cytometer is aligned using fluorescent calibration particles (DNA Check, Coulter Electronics). Biparametric histograms are accumulated [log 90° light scatter versus log green fluorescence, or log red (chlorophyll) fluorescence versus log green fluorescence] to a total count of 100,000. Uniparametric histograms can also be acquired.

c. A sample flow rate of 100–200/s is employed for analysis of the protoplasts. The sheath fluid comprises 0.47 *M* mannitol, 50 m*M* KCl, 10 m*M* $CaCl_2$, and 4 m*M* MES, pH 5.7, filtered through a 0.22-μm filter (Millipore GSWP047) prior to use.

d. For sorting, the Elite is operated at a sheath pressure of 8.0 psi, a sample pressure of 7.3 psi, a drive frequency of 15 kHz, and a drive amplitude of 11.5%. This typically gives a sort stream having six free drops above the deflection plate assembly. Delay settings are systematically optimized by sorting batches of 25 particles onto a microscope slide for various delay settings. The proportion of recovered particles is then determined under the microscope. DNA Check calibration particles (diameter 10 μm) and Lycopodium spores (diameter 28 μm, Polysciences, Warrington, PA) are used for this procedure. Delay settings of 17 to 19 gave stable sort streams, using a 70% deflection amplitude. Samples were sorted at a rate of 25 to 50/s into 12- × 75-mm plastic tubes.

c. Analysis of Plant Nuclei Accumulating GFP

a. Nuclei are isolated from transgenic plants by chopping (Galbraith *et al.*, 1983). Leaf material (250 mg) is transferred into 2 ml of ice-cold homogenization medium and homogenized by chopping for approximately 1 min using a single-edged razor blade. The homogenization medium comprises 0.225 *M* mannitol, 50 m*M* potassium acetate, 20 m*M* HEPES–KOH, pH 7.3, 5 m*M* sodium acetate, 2 m*M* magnesium acetate, 2 m*M* PMSF, 0.125 m*M* spermine, 0.125 m*M* spermidine, and 5 mg/ml each of aminocaproic acid, aprotinin, leupeptin, and pepstatin A (Hicks *et al.*, 1996).

b. The homogenate is filtered through nylon mesh (pore size 60 μm, Tetko, NY).

c. Flow cytometry is done based on detection of FALS, 90° light scatter (PMT1), 505- to 545-nm fluorescence (green; PMT2), 555- to 595-nm fluorescence (orange; PMT3), and 670- to 680-nm fluorescence (red; PMT4), at a sample rate of 100/s. Typical high-voltage and amplification settings are FALS, 260/10; PMT1, 350/7.5; PMT2, 900/7.5; PMT3, 900/7.5; PMT4, 750/7.5. The PMT3 integral signal discriminator is set to 50, and all other discriminators are turned off.

d. Triggering is done on PMT3 fluorescence, enabling detection of signals both from nuclei containing GFP and from chloroplasts. Triggering should not be done on based on FALS, due to the excessive proportion of nonfluorescent, light-scattering particles in the homogenate. Uniparametric and biparametric histograms are collected to a total count of $5–10 \times 10^5$ particles, or for defined periods of time.

e. For sorting of nuclei using the 100-μm flow tip, the sheath and sample pressures are set to 12 and 11.5 psi. Typical settings are an analysis rate of 100/s, a drive frequency of 16.9 kHz and amplitude of 70%, and a delay setting of 16.4. Sorted nuclei are allowed to settle prior to observation under the microscope.

d. Analysis and Sorting of GFP-Expressing Bacteria

To analyze bacteria with the flow cytometer, the bacteria must be prepared as a single-cell suspension. To prepare a suspension from clonal colonies present on 100- × 150-mm Petri plates, 10 ml of LB are pipetted onto the plate, which is then incubated at 22°C, with gentle rocking for 10–20 min. An important consideration in this process is that none of the agar from the plate contaminate the suspension, because it can clog the fluidic lines of the flow cytometer. Immediately prior to FACS analysis, the suspension is diluted with phosphate buffered saline (PBS), to an extent that leads to a data acquisition rate on the flow cytometer of about 200 to 1000 events/s.

The analysis of bacteria on the flow cytometer is complicated by their small size. However, in contrast to debris, bacteria have a characteristically high side scatter, allowing this parameter to be employed as the discriminator. Bacteria can then be further distinguished with a forward-scatter gate set with a log gain. In general, the analysis of bacteria is done under a similar instrumentation set up to that used for mammalian cells. In the studies shown here, a standard dual-laser FACStar$^+$ was used. Excitation was from a Innova 70 argon laser producing 200 mW at 488 nm. In all other respects, except for the use of side scatter as the discriminator, the optical filters, analysis conditions, and sorting conditions were identical to those used with mammalian cells.

The analysis of bacteria and other tissue culture contaminants on the flow cytometer often raises concerns about the ability to sterilize the instrument for subsequent use with mammalian cells. After the analysis of bacteria on the flow cytometer, we sterilize it by passing through the fluidic lines diluted dish soap followed by a 70% solution of alcohol.

III. Typical Results

A. Mammalian Cell Systems

The original GFP gene (WT-GFP) of *Aequorea victoria* produces too little fluorescent signal for it to be useful in FACS either as a selectable marker or as a transcriptional reporter. This is illustrated in Fig. 1, which employs FACS to compare the fluorescence emissions of uninfected 3T3 cells with those of 3T3 cells infected at a high multiplicity of infection (MOI) with MFG–WT-GFP; at this MOI, nearly every cell is productively infected. Comparison of their fluorescence intensity distributions shows that these distributions overlap, their median values differing by only a factor of two. This is in part due to the majority of the WT-GFP being improperly folded and in part due to only a minority of the productively folded GFP having fluorophores in a state excitable by the 488-nm line. It is also in part due to a high background from autofluorescence. For lymphocytes, this background signal is similar in intensity to that of at least 500 fluorescein molecules (Alberti *et al.,* 1988). GFP that is properly folded and that has a fluorophore in a state optimally excitable by the 488-nm line produces a fluorescent signal comparable to that of fluorescein (Cubitt *et al.,* 1995). Consequently, many more than 500 molecules of WT-GFP need to be expressed in order to exceed the background due to autofluorescence.

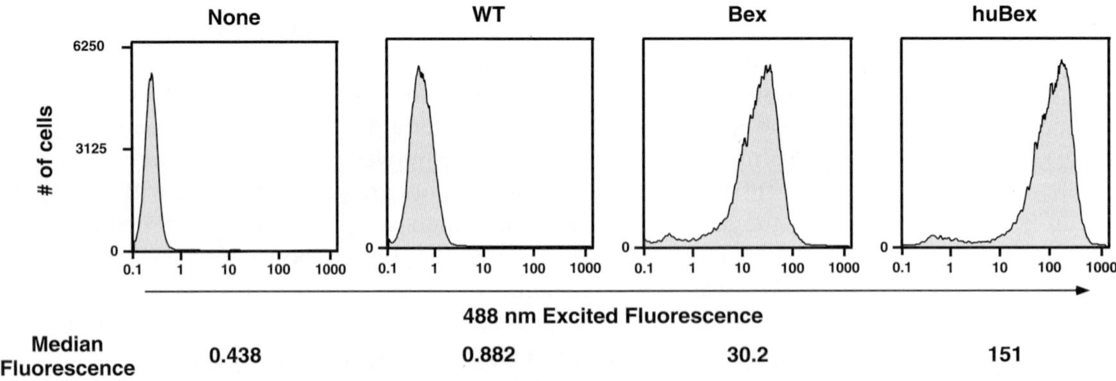

Fig. 1 FACS histograms of NIH/3T3 cells infected with either WT-GFP or improved GFP variants. NIH/3T3 cells were infected with (A) WT-GFP, (B) Bex–GFP, or (C) HuBex–GFP retrovirus and then analyzed by flow cytometery 48 h after infection. The MOI of each virus was similar and sufficient to infect nearly every cell. Emission data were collected with 488-nm excitation and emissions were collected between 495 nm and 535 nm. Representative histograms are shown.

This difficulty has been largely overcome with brighter versions of the GFP reporter obtained through either genetic screens or directed mutagenesis. These brighter versions of GFP produce adequate levels of signal to be used as selectable markers with flow cytometry and sorting. For example, 3T3 cells infected with MFG–GFP–Bex are completely distinguished from controls as a consequence of their GFP expression (Fig. 1). In these experiments, the increased brightness of the Bex variant derives from improved fluorochrome excitation at 488 nm because of the S65T mutation and an increase in the proportion of correctly folded GFP molecules because of the V163A mutation. When the Bex variant coding sequence is further modified to incorporate only those codons that are frequently used in mammalian cells (huGFP-Bex), even greater levels of fluorescence are produced by the infected cells (Fig. 1).

Questions do remain as to whether these bright GFP variants will turn out to be generally applicable as transcriptional reporters. We have shown their utility as transcriptional reporters in transient transfection analysis using flow cytometry (Anderson *et al.,* 1996). They have also proven useful as reporters of inducible constructs stably integrated into the chromosome (Gervaix *et al.,* 1997). However, even the brightest GFP reporters are likely to be less sensitive than other reporters used in flow cytometry, such as β-galactosidase and β-glucuronidase; these GFPs all lack the property of signal amplification inherent to substrate to product conversion by enzymatic reporters (Nolan *et al.,* 1988; Lorincz *et al.,* 1996). The overall sensitivity of GFP reporters is also a function of the stability of the GFP protein and its mRNA. Both the GFP mRNA and GFP protein appear relatively stable, because the GFP signal progressively increases until 36–48 h after infection (Fig. 2).

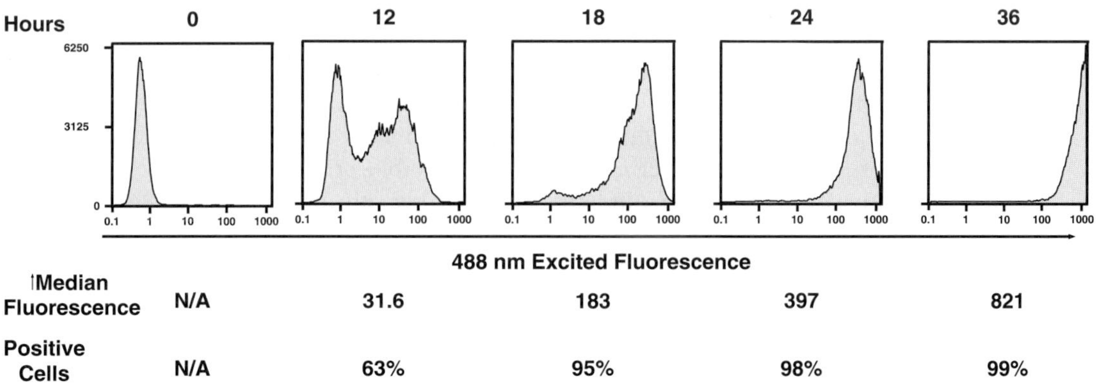

Fig. 2 FACS histograms depicting the increased huGFP–Bex fluorescent signal with increases in time after infection. Aliquots of NIH/3T3 cells were infected identically with huGFP–Bex retrovirus. Individual aliquots were analyzed at 0, 12, 18, 24, and 36 h after infection. MOI of infection were sufficient to result in nearly every cell becoming infected. Emission data were collected with 407-nm excitation, and emissions were collected between 495 and 535 nm. Representative histograms are shown. Cells positive for GFP expression were defined as those cells with fluoresence levels two standard deviations above the median fluorescence levels of the uninfected cells.

Despite these questions, the GFP variants provide unique capabilities, particularly the ability to simultaneously detect the activities of two different reporters within a single living cell. This makes it possible to acquire data simultaneously about two processes or pathways—for example, different transcriptional responses to given stimuli within single cells—and is a result of the availability of the Bex and Vex series of GFP variants. These variants differ from each other in excitation spectra but not in their emission spectra. Thus, Bex is optimally excited by the 488-nm argon line but only negligibly by the 407-nm krypton line. In contrast, Vex is well excited by the 407-nm krypton line but not by the 488-nm argon line. It is not intuitively obvious that flow cytometric instrumentation can be configured to analyze fluorochromes differing only in excitation spectra. Nonetheless, this is the case, and the fluorescent signal from fluorophores differing only in excitation can be as distinctly and quantitatively measured as the signal from fluorophores differing only in emission (Fig. 3). In this case, biparametric population frequency distributions are presented for 3T3 cells infected singly or simultaneously with the MFG–GFP–Bex and MFG–GFP–Vex retroviruses. Only 3% crosstalk between the fluorescent signals of these variants is observed for the uncompensated data. This is small enough to allow ready and unambiguous recognition of the various populations of cells expressing either one or both of these variants (Fig. 4).

As a demonstration of the power and unique information obtainable with the dual-GFP-reporter system, we tested the relative susceptibility to infection of members of a 3T3 cell population and whether infection with one retrovirus precluded infection with a second. The methodology necessary to answer these questions required discrimination of multiply infected cells. This is best done with multiple distinguishable reporters. To test this, we infected NIH/3T3 cells with equal amounts of MFG–GFP–Vex1 and MFG–GFP–Bex1 at 3 different MOIs (Fig. 4). FACS analysis of the 3T3 cells infected with the lowest MOI, 0.064 of each variant, reveals three populations of cells: one with low levels of fluorescence characteristic of uninfected cells and two with higher levels of fluorescence. The latter clearly distinguishable populations were excited exclusively with 488- or 407-nm light, representing cells infected by a single retrovirus of either the GFP–Bex1 or GFP–Vex1 type, respectively. Infection with a twofold greater amount of retroviral supernatant yields the predicted MOI of 0.13 for each virus and generates, in addition to the two singly infected population, a new "doubly positive" population showing bright fluorescence with both 488-nm and 406-nm excitation, consistent with cells expressing both GFP variants due to coinfection. With a fivefold further increase in the amount of retrovirus used for the infection of the 3T3 cells, the number of double positives increases. These results are consistent with those predicted by the Poisson distribution with an MOI of 0.65. Therefore, with the dual-reporter system we were able to show that each of the cells is equally infectable with retrovirus and that infection with one retrovirus does not alters the cell's infectability by a second retrovirus.

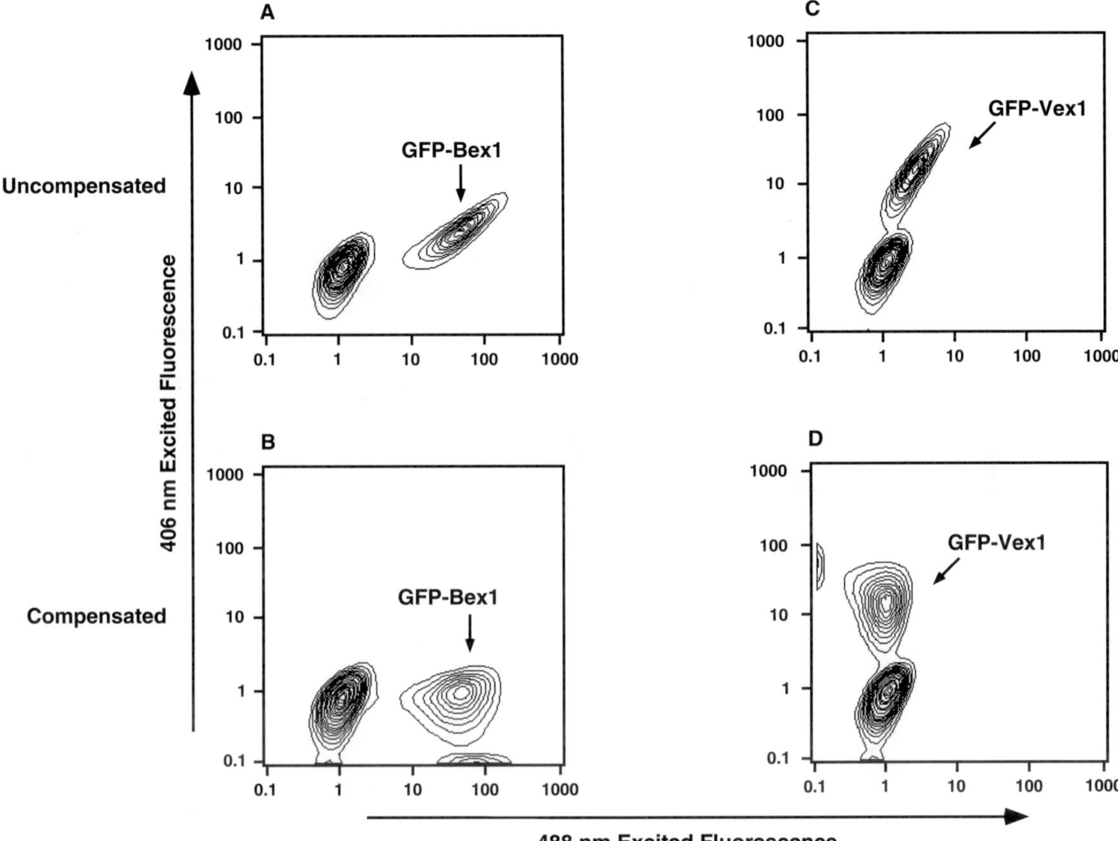

Fig. 3 Raw and compensated data of GFP–Bex1 and GFP–Vex1 fluorescence in NIH/3T3 cells. NIH/3T3 cells were infected with MFG–GFP–Bex1 (A and B) or MFG–GFP–Vex1–GFP (C and D) retrovirus and analyzed by flow cytometry 48 h after infection. Emission data collected with 407-nm and 488-nm excitation are displayed on the abscissa and ordinate, respectively, in uncompensated (A and C) and compensated (B and D) form. Representative plots from three independent experiments are shown. (Reproduced from Anderson *et al.,* 1996, with permission.)

B. Plant Cell Systems

1. Analysis and Sorting of WT-GFP-Expressing Protoplasts

Plant protoplasts in general have larger diameters than mammalian cells and consequently have much larger cytoplasmic volumes. For this reason, perhaps, it is relatively easy to accumulate levels of cytoplasmic GFP to give fluorescent signals that are detectable through flow cytometry. Expression of levels of GFP sufficient for flow analysis is illustrated for transfected maize protoplasts in Fig. 5. In this biparametric analysis, nontransfected protoplasts occupy a single cluster

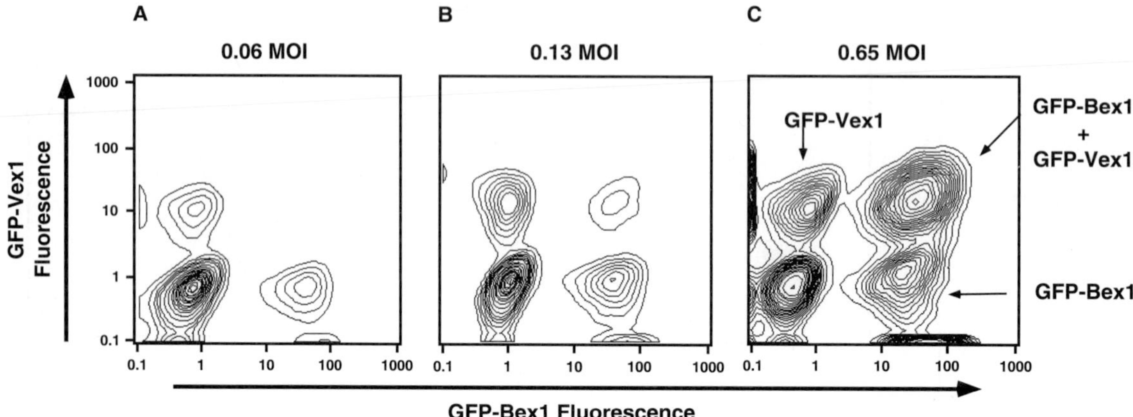

Fig. 4 Simultaneous detection of GFP–Bex1 and GFP–Vex1 fluorescence. Increasing amounts of a 1:1 mixture of the GFP–Bex1 and GFP–Vex1 viruses were used to infect NIH/3T3 cells. Cells were infected with an MOI for each of the two viruses of (A) 0.06 MOI, (B) 0.13 MOI, or (C) 0.65 MOI, as determined by multiplying the MOI measured in (A) by the relative amount of supernatant used in (B) and (C). FACS analysis was conducted as described for Fig. 2B and D. Representative plots from three independent experiments are shown. (Reproduced from Anderson *et al.,* 1996, with permission.)

having a very low level of green autofluorescence and a slightly higher level of red autofluorescence. This autofluorescence derives from the cellular pigments other than GFP (Fig. 5A). Flow analysis of protoplasts 10–18 h after transfection with GFP reveals a second cluster, characterized by greatly enhanced green

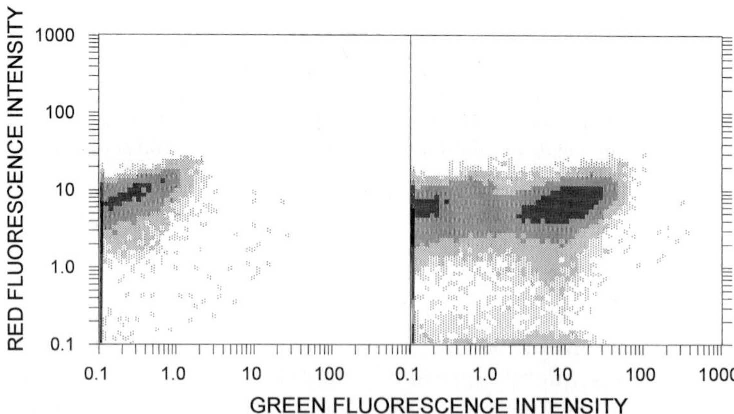

Fig. 5 Biparametric flow analysis of GFP expression in maize protoplasts, based on green and red fluorescence emission. (A) Control (nontransfected) protoplasts, (B) Protoplasts 24 h after transfection. (Reproduced from Galbraith *et al.,* 1995, with permission.)

fluorescence but unaltered red fluorescence (Fig. 5B). This corresponds to those protoplasts expressing GFP, which, in this experiment, approximates 14% of the population. The variability in transfection efficiencies between experiments was quite high, with the proportion of protoplasts expressing GFP ranging from 4 to 34% of the total. The fluorescence emission intensity from GFP-expressing protoplasts averaged about 80-fold more than that from nontransfected cells.

The maize protoplasts are about 25 μm in diameter and, being prepared from etiolated tissues, lack mature chloroplasts. Similar results were obtained following expression of the S65T humanized variant of GFP in protoplasts prepared from tobacco leaves (diameters about 40 μm), although the proportion of GFP-expressing protoplasts was lower, ranging from 2 to 10%. It is not possible to directly compare these results to extract information about the GFP constructions most appropriate for all situations because the experiments differ in protoplast types and sizes, in their relative biosynthetic capacities, and particularly in transfection efficiencies. Different plant species also appear to differ in their abilities to excise the cryptic intron found in the WT-GFP coding sequence. A systematic comparison of those that are most suitable for flow cytometric detection for specific situations should, however, be relatively straightforward.

We have previously described conditions for the sorting of intact maize protoplasts according to GFP expression (Galbraith *et al.,* 1995). Sort conditions were optimized using *Lycopodium* spores, which are of approximately the same size as the protoplasts. After placing a bit-mapped sort window around the cluster corresponding to the GFP-expressing protoplasts and enabling the sorting circuitry, protoplasts were sorted that remained intact and exhibited green fluorescence.

2. Analysis and Sorting of GFP-Accumulating Nuclei

Levels of GFP sufficient for detection using flow cytometry can be accumulated within subcellular organelles (Fig. 6). In this example, nuclear accumulation of GFP [the S65T "humanized" version (Chiu *et al.,* 1996)] requires an effective nuclear localization signal. It also requires enlargement of the molecular size of GFP, which in the native state is too small to be prevented from passive diffusion into and out of the nucleus (Grebenok *et al.,* 1997). Identification of nuclei accumulating GFP requires biparametric flow analysis in which green fluorescence emission is combined with 90° light scatter or with orange fluorescence emission. In the latter case, the population of GFP-accumulating nuclei falls on a distinct diagonal, due to spillover of the GFP-derived fluorescence into PMT3.

C. Bacteria

GFP was first heterologously expressed in *Escherichia coli.* Subsequently, because of the superior genetic tractability of this organism, it has been used as the expression vehicle in the selections for brighter GFPs. Many of the brighter

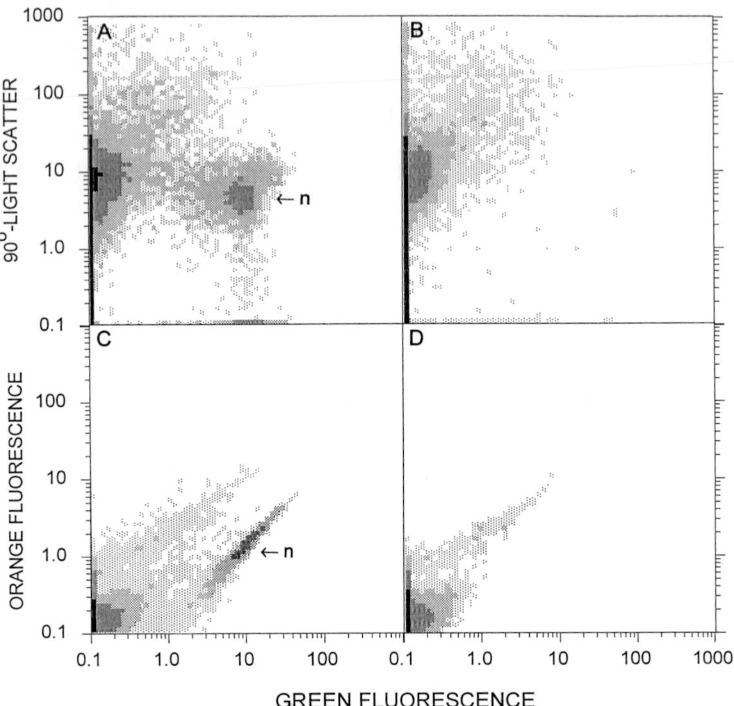

Fig. 6 Flow cytometric analysis *in vitro* of GFP targeting to nuclei. Homogenates from leaves of transgenic plants (A and C) or nontransgenic controls (B and D) were subjected to biparametric analysis of 90° light scatter versus green fluorescence (A and B) or of orange versus green fluorescence (C and D). The regions corresponding to nuclei containing GFP (key: n, nuclei) are absent from the controls. (Reproduced from Grebenok *et al.,* 1997b, with permission.)

GFP variants were selected by visual examination, but some were selected by FACS, allowing greater quantitation of their signals and greater efficiency in their selection. When huWT-GFP is expressed in *E. coli,* it is, as expected, dull. As shown in Fig. 7, the induced expression of huWT-GFP leads to only a 3.5-fold increase in the fluorescent signal compared to the signal produced with uninduced bacteria. This relative dullness is partly due to the time (approx. 3 h) required for this protein to fold and cyclize (Heim *et al.,* 1995). It is also partly due to the majority of GFP folding improperly at 37°C as well when properly folded, most of its fluorophores being in a state not well excited by the 488-nm line.

However, the brighter GFP variants produce signal levels markedly higher than those obtained with WT-GFP. For example, the induced expression of huBex–GFP leads to 80-fold higher fluorescence levels than are obtained in the absence of induction. In addition to humanized codons, this variant has the S65T spectral mutation and the V163A thermal stability mutation. The S65T mutation

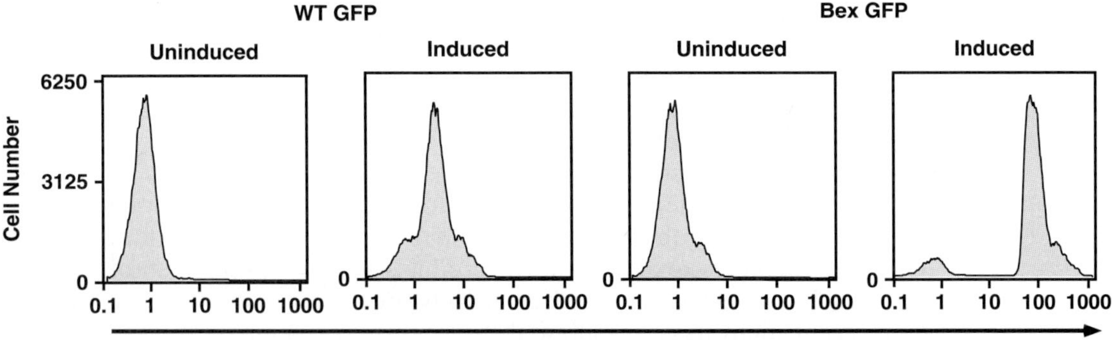

Fig. 7 Inducible expression of huWT-GFP and huBex–GFP *E. coli.* The BL21(DE3) strain of *E. coli* was transformed with pET expression vectors containing either huWT-GFP (A and B) or huBex–GFP (C and D). Suspensions of these transfectants were grown for 3 h at 37°C under induced conditions (2 m*M* IPTG, no glucose) or uninduced conditions (no IPTG, 2% glucose). Suspension cultures were diluted with PBS and analyzed on the flow cytometer. Emission data were collected with 488-nm excitation, and emissions were collected between 495 nm and 535 nm. Representative histograms are shown.

converts the fluorophore to a state better excited by the 488-nm laser line (Fig. 6) and causes GFP to more rapidly fold correctly (average time less than 30 min) (Heim *et al.,* 1995). The length of time required for folding influences the GFP signal levels because, for example with WT-GFP, this time significantly exceeds the cell doubling time. The influence of the other spectral mutations and thermal stability mutations on the rate of folding still remains to be explored. A particular advantage in the use of flow cytometer in the selection of new GFP variants is that this instrument can quantitatively detect not only the magnitude of the GFP fluorescent signal but also spectral changes in its excitation and emission (Cormack *et al.,* 1996; Crameri *et al.,* 1996).

IV. Discussion and Conclusions

The use of GFP as a reporter in flow cytometry and cell sorting is in its infancy. A major benefit provided by GFP is that it can be measured noninvasively. Alterations to the primary structure of the molecule have provided variants that are readily detectable in both prokaryotic and eukaryotic cells and within subcellular organelles. Further spectral variants that can be combined for multiparametric analyses will find considerable future utility in flow cytometry.

In animal cell systems, GFP is particularly suited as an indicator of viral infection. When expressed in viruses, either as a gene fusion or when substituted for a nonessential viral gene, GFP fluorescence provides a sensitive and accurate measure of the proportions of infected cells and the progress of viral infection (Anderson *et al.,* 1996; Balague *et al.,* 1997; Bierhuizen *et al.,* 1997; Chen *et al.,* 1997; Lee *et al.,* 1997). Coupling this approach for viral detection with concerted mutagenesis of virus provides a screen for the roles of the various viral components within the infection process.

Flow cytometric detection of GFP also finds practical application in gene therapy. By marking the viral vectors containing therapeutic gene products with GFP, the members expressing these products within a population of cells can be identified, selected, and used exclusively for the reconstitution of organisms. One recent innovation has allowed this approach to employed with retroviral vectors. It involves the insertion of an IRES element within the retroviral transcript allowing both the therapeutic gene product and GFP to be expressed from the transcript (Mosser *et al.,* 1997). Cells infected with such dicistronic viruses can be not only stochastically selected for GFP expression but also dynamically selected for specific levels of GFP expression. These levels are anticipated to correlate with the levels of therapeutic product within the cells. Although this strategy is not unique for the GFP reporter, the noninvasive measure of its level and the small size of its gene greatly aid its implementation.

It can be anticipated that the methodologies used in the gene therapeutic approaches can also provide a generalized approach for the basic study of cell function. In this approach, exogenous genes introduced by virus are tested not only for their effects on cell viability and cell proliferation but for their effects upon less dramatic end points of cellular processes, the transcription of gene products. This can be measured as the activity of *cis*-acting elements linked to a reporter such as Bex–GFP or β-galactosidase integrated into the genome (Nolan *et al.,* 1988; Anderson *et al.,* 1996). This activity should be measurable either with GFP–Bex or with the FACS–GAL assay of β-galactosidase activity. GFP–Vex, which is spectrally distinct from the fluorescence derived from these two reporters, would then mark those cells within the population that have been infected with the retrovirus genetically complemented by virus and that are therefore expressing the protein being tested (Mosser *et al.,* 1997). This approach and similar ones made possible by pairs of FACS-detectable markers should greatly expand the genetic analysis of cell function, which is particularly relevant in an age of gene discovery driven by large-scale genome sequencing projects.

Recently, it has been demonstrated that pairs of GFP variants can be coupled through energy transfer and, when appropriately engineered, provide real-time readouts of changes in cellular physiology (Romoser *et al.,* 1997). This is specifically seen when Uex and Bex variants are brought into close proximity. The light absorbed by the Uex molecule transfers nonradiatively to the Bex molecule, which subsequently emits its characteristic fluorescence. Changes in the proximity of the two GFP molecules are detected as alterations in the proportions of fluorescence produced at the Uex and Bex emission maxima. Such changes are measurable continuously within single cells through ratiometric fluorescence microscopy, but they can also be surveyed in entire cell populations using flow cytometry, which is well suited for the simultaneous detection of fluorescence at two different wavelengths. Thus far, GFP energy transfer couples have been successfully employed as assays for proteolytic cleavage of a peptide linker linking the two GFP domains in a single fusion protein and for calcium-dependent three-dimensional rearrangements in a calmodulin domain arranged to link the two GFP domains. Further developments of this novel technology, particularly in assays designed to monitor protein–protein interactions (such as the two-hybrid system) are expected to emerge.

In higher plants, transient expression of GFP is readily detected in transfected protoplasts and these protoplasts can be recovered in purified form through FACS sorting. One of our particular interests is the development of methods for the analysis of global patterns of gene expression in plants. One possible way to achieve this involves the use of GFP as a means to specifically highlight cell types having developmentally interesting patterns of gene expression, thence to employ FACS as a means to purify protoplasts prepared from these cells or tissues. Although we and others have previously established that protoplast-derived reports can be representative of the tissues from which the protoplasts are prepared (Harkins *et al.,* 1990; Sheen *et al.,* 1995), it is not a simple task to prepare and sort protoplasts. Further, although reports of high-level expression of GFP within transgenic plants have emerged (Chiu *et al.,* 1996; Pang *et al.,* 1996; Haseloff *et al.,* 1997), a systematic study employing flow cytometry will be needed to establish the dynamic range over which GFP expression within transgenic protoplasts is suitable for the study of cell-specific gene expression.

We have been pursuing an alternative strategy to protoplast sorting. This is based on our observation that, through construction of translational fusions, GFP chimeric proteins can be used to monitor the process of targeting to the nucleus (Grebenok *et al.,* 1997a,b). Our work indicates it is possible to employ FACS for the isolation of nuclei that have accumulated GFP. This leads to the idea of examining the RNA transcripts within sorted nuclei as a way to monitor gene expression. Coupling nuclear accumulation of GFP to cell-, tissue-, or developmentally specific promoters would allow characterization of patterns of gene expression within these subpopulations of cells. The strength of the approach relies on the observation that nuclei are intrinsically nonautofluorescent, hence give a low background, and are highly homogeneous in terms of physical dimensions. This renders them easy to analyze and sort using FACS, as has been well established for analyses of genome sizes (Galbraith *et al.,* 1983). In principle, this approach is applicable to any eukaryotic system and may prove particularly important in the study of the interactions of different organisms, such as in infections, or in disease states characterized by abnormal patterns of gene expression and/or abnormal genome sizes.

GFP as a transcriptional reporter in bacteria should prove particularly important in the development of therapeutics against bacterial pathogens. In the past, it has been difficult to clone virulence factors from invasive bacterial pathogens.

However, because GFP is an intrinsically fluorescent reporter, it allows the identification of bacterial transcription units up-regulated in response to engulfment. This has allowed Valdivia and Falkow (1997) to screen a library of *Salmonella typhimurium* with GFP gene fusions to identify genes upregulated in this organism as a response to engulfment by host macrophages (Valdivia and Falkow, 1997). From this genetic screen they were able to clone four of the virulence factors of *S. typhimurium.* This strategy should find wide applicability to a range of host/pathogen interactions.

Acknowledgments

D.G. would like to thank Georgina Lambert for valuable technical assistance. This work was supported by grants to D.G. from the U.S.D.A.-N.R.I. Competitive Grants Program (Plant Genome), the Instrumentation and Instrument Development Program of N.S.F., and the N.S.F./U.S.D.A./D.O.E. Triagency Program in Plant Biology.

L.A.H. and M.T.A. would like to thank the Stanford Shared FACS Facility (David Parks, Director) for the custom instrumentation and custom software used in many of these analysis. In addition, we greatly appreciate the assistance of Iwan Tijoe in the construction of the GFP variants. This work was supported by a grant CA-42509 to L.A.H.

References

Alberti, S., Parks, D. R., and Herzenberg, L. A. (1988). A single laser method for subtraction of cell autofluorescence in flow cytometry. *Cytometry* **8,** 114–119.

Anderson, M. T., Tjioe, I. T., Lorincz, M. C., Parks, D. R., Herzenberg, L. A., Nolan, G. P., and Herzenberg, L. A. (1996). Simultaneous fluorescence-activated cell sorter analysis of two distinct transcriptional elements within a single cell using engineered green fluorescent proteins. *Proc. Natl. Acad. Sci. U. S. A.* **93,** 8508–8511.

Anderson, M. T., Gerstein, R., Haugland, R., Tjioe, I., Herzenberg, L. A., and Herzenberg, L. A. (1998a). Three violet-excited fluorophores for immunofluorescence analysis on the flow cytometer (in preparation).

Anderson, M. T., Baumgarth, N., Haugland, R. P., Gerstein, R. M., Tjioe, I. T., Herzenberg, L. A., and Herzenberg, L. A. (1998b). Submitted.

Balague, C., Kalla, M., and Zhang, W. W. (1997). Adeno-associated virus Rep78 protein and terminal repeats enhance integration of DNA sequences into the cellular genome. *J. Virol.* **71,** 3299–3306.

Bierhuizen, M. F., Westerman, Y., Visser, T. P., Wognum, A. W., and Wagemaker, G. (1997). Green fluorescent protein variants as markers of retroviral-mediated gene transfer in primary hematopoietic cells and cell lines. *Biochem. Biophys. Res. Commun.* **234,** 371–375.

Brejc, K., Sixma, T. K., Kitts, P. A., Kain, S. R., Tsien, R. Y., Ormo, M., and Remington, S. J. (1997). Structural basis for dual excitation and photoisomerization of the *Aequorea victoria* green fluorescent protein. *Proc. Natl. Acad. Sci. U. S. A.* **94,** 2306–2311.

Bunnell, B. A., Muul, L. M., Donahue, R. E., Blaese, R. M., and Morgan, R. A. (1995). High-efficiency retroviral-mediated gene transfer into human and nonhuman primate peripheral blood lymphocytes. *Proc. Natl. Acad. Sci. U. S. A.* **92,** 7739–7743.

Chalfie, M., Tu, Y., Euskirchen, G., Ward, W. W., and Prasher, D. C. (1994). Green fluorescent protein as a marker for gene expression. *Science* **263,** 802–805.

Chen, B. K., Feinberg, M. B., and Baltimore, D. (1997). The kappaB sites in the human immunodeficiency virus type 1 long terminal repeat enhance virus replication yet are not absolutely required for viral growth. *J. Virol.* **71,** 5495–5504.

Chiu, W., Niwa, Y., Zeng, W., Hirano, T., Kobayashi, H., and Sheen, J. (1996). Engineered GFP as a vital reporter in plants. *Curr. Biol.* **6,** 325–330.

Cody, C. W., Prasher, D. C., Westler, W. M., Prendergast, F. G., and Ward, W. W. (1993). Chemical structure of the hexapeptide chromophore of the *Aequorea* green-fluorescent protein. *Biochemistry* **32,** 1212–1218.

Cormack, B. P., Bertram, G., Egerton, M., Gow, N. A., Falkow, S., and Brown, A. J. (1997). Yeast-enhanced green fluorescent protein (yEGFP) a reporter of gene expression in *Candida albicans. Microbiology* **143,** 303–311.

Cormack, B. P., Valdivia, R. H., and Falkow, S. (1996). FACS-optimized mutants of the green fluorescent protein (GFP). *Gene* **173,** 33–38.

Crameri, A., Whitehorn, E. A., Tate, E., and Stemmer, W. P. (1996). Improved green fluorescent protein by molecular evolution using DNA shuffling. *Nat. Biotechnol.* **14,** 315–319.

Cubitt, A. B., Heim, R., Adams, S. R., Boyd, A. E., Gross, L. A., and Tsien, R. Y. (1995). Understanding, improving and using green fluorescent proteins. *Trends Biochem. Sci.* **20,** 448–455.

Delagrave, S., Hawtin, R. E., Silva, C. M., Yang, M. M., and Youvan, D. C. (1995). Red-shifted excitation mutants of the green fluorescent protein. *Bio/Technology* **13,** 151–154.

Dubendorff, J. W., and Studier, F. W. (1991). Creation of a T7 autogene. Cloning and expression of the gene for bacteriophage T7 RNA polymerase under control of its cognate promoter. *J. Mol. Biol.* **219,** 61–68.

Ehrig, T., O'Kane, D. J., and Prendergast, F. G. (1995). Green-fluorescent protein mutants with altered fluorescence excitation spectra. *FEBS Lett.* **367,** 163–166.

Galbraith, D. W., Harkins, K. R., Maddox, J. R., Ayres, N. M., Sharma, D. P., and Firoozabady, E. (1983). Rapid flow cytometric analysis of the cell cycle in intact plant tissues. *Science* **220,** 1049–1051.

Galbraith, D. W., Lambert, G. M., Grebenok, R. J., and Sheen, J. (1995). Flow cytometric analysis of transgene expression in higher plants: green-fluorescent protein. *Methods Cell Biol.* **50,** 3–14.

Gamborg, O. L., and Philips, G. C. (1995). "Plant Cell, Tissue, and Organ Culture: Fundamental Methods." Springer-Verlag, New York.

Gervaix, A., West, D., Leoni, L. M., Richman, D. D., Wong-Staal, F., and Corbeil, J. (1997). A new reporter cell line to monitor HIV infection and drug susceptibility *in vitro. Proc. Natl. Acad. Sci. U. S. A.* **94,** 4653–4658.

Grebenok, R. J., Pierson, E., Lambert, G. M., Gong, F. C., Afonso, C. L., Haldeman-Cahill, R., Carrington, J. C., and Galbraith, D. W. (1997a). Green-fluorescent protein fusions for efficient characterization of nuclear targeting. *Plant J.* **11,** 573–586.

Grebenok, R. J., Lambert, G. M., and Galbraith, D. W. (1997b). Characterization of the targeted nuclear accumulation of GFP within the cells of transgenic plants. *Plant J.* **12,** 685–696.

Ha, D. S., Schwarz, J. K., Turco, S. J., and Beverley, S. M. (1996). Use of the green fluorescent protein as a marker in transfected *Leishmania. Mol. Biochem. Parasitol.* **77,** 57–64.

Harkins, K. R., Jefferson, R. A., Kavanagh, T. A., Bevan, M. W., and Galbraith, D. W. (1990). Expression of photosynthesis-related gene fusions is restricted by cell type in transgenic plants and in transfected protoplasts. *Proc. Natl. Acad. Sci. U. S. A.* **87,** 816–820.

Haseloff, J., Siemering, K. R., Prasher, D. C., and Hodge, S. (1997). Removal of a cryptic intron and subcellular localization of green fluorescent protein are required to mark transgenic *Arabidopsis* plants brightly. *Proc. Natl. Acad. Sci. U. S. A.* **94,** 2122–2127.

Heim, R., Prasher, D. C., and Tsien, R. Y. (1994). Wavelength mutations and posttranslational autoxidation of green fluorescent protein. *Proc. Natl. Acad. Sci. U. S. A.* **91,** 12501–12504.

Heim, R., Cubitt, A. B., and Tsien, R. Y. (1995). Improved green fluorescence [letter]. *Nature* **373,** 663–664.

Heim, R., and Tsien, R. Y. (1996). Engineering green fluorescent protein for improved brightness, longer wavelengths and fluorescence resonance energy transfer. *Curr. Biol.* **6,** 178–182.

Hicks, K. A., Millar, A. J., Carre, I. A., Somers, D. E., Straume, M., Meeks-Wagner, D. R., and Kay, S. A. (1996). Conditional circadian dysfunction of the *Arabidopsis* early-flowering 3 mutant. *Science* **274,** 790–792.

Hofmann, A., Nolan, G. P., and Blau, H. M. (1996). Rapid retroviral delivery of tetracycline-inducible genes in a single autoregulatory cassette [see comments]. *Proc. Natl. Acad. Sci. U. S. A.* **93,** 5185–5190.

Kain, S. R., and Kitts, P. (1997). Expression and detection of green fluorescent protein (GFP). *Methods Mol. Biol.* **63,** 305–324.

Kimata, Y., Iwaki, M., Lim, C. R., and Kohno, K. (1997). A novel mutation which enhances the fluorescence of green fluorescent protein at high temperatures. *Biochem. Biophys. Res. Commun.* **232,** 69–73.

Kinsella, T. M., and Nolan, G. P. (1996). Episomal vectors rapidly and stably produce high-titer recombinant retrovirus. *Hum. Gene Ther.* **7,** 1405–1413.

Krall, W. J., Skelton, D. C., Yu, X. J., Riviere, I., Lehn, P., Mulligan, R. C., and Kohn, D. B. (1996). Increased levels of spliced RNA account for augmented expression from the MFG retroviral vector in hematopoietic cells. *Gene Ther.* **3,** 37–48.

Lee, A. H., Han, J. M., and Sung, Y. C. (1997). Generation of the replication-competent human immunodeficiency virus type 1 which expresses a jellyfish green fluorescent protein. *Biochem. Biophys. Res. Commun.* **233,** 288–292.

Lorincz, M., Roederer, M., Diwu, Z., Herzenberg, L. A., and Nolan, G. P. (1996). Enzyme-generated intracellular fluorescence for single-cell reporter gene analysis utilizing *Escherichia coli* beta-glucuronidase. *Cytometry* **24,** 321–329.

Lybarger, L., Dempsey, D., Franek, K. J., and Chervenak, R. (1996). Rapid generation and flow cytometric analysis of stable GFP-expressing cells. *Cytometry* **25,** 211–220.

Mosser, D. D., Caron, A. W., Bourget, L., Jolicoeur, P., and Massie, B. (1997). Use of a dicistronic expression cassette encoding the green fluorescent protein for the screening and selection of cells expressing inducible gene products. *Biotechniques* **22,** 150–154.

Muldoon, R. R., Levy, J. P., Kain, S. R., Kitts, P. A., and Link, C. J. Jr. (1997). Tracking and quantitation of retroviral-mediated transfer using a completely humanized, red-shifted green fluorescent protein gene. *Biotechniques* **22,** 162–167.

Naviaux, R. K., Costanzi, E., Haas, M., and Verma, I. M. (1996). The pCL vector system: rapid production of helper-free, high-titer, recombinant retroviruses. *J. Virol.* **70,** 5701–5705.

Negrutiu, I., Shillito, R., Potrykus, I., Biasini, G., and Sala, F. (1987). Hybrid genes in the analysis of transformation conditions. *Plant Mol. Biol.* **15,** 363–373.

Nolan, G. P., Fiering, S., Nicolas, J. F., and Herzenberg, L. A. (1988). Fluorescence-activated cell analysis and sorting of viable mammalian cells based on beta-D-galactosidase activity after transduction of *Escherichia coli* lacZ. *Proc. Natl. Acad. Sci. U. S. A.* **85,** 2603–2607.

Ormo, M., Cubitt, A. B., Kallio, K., Gross, L. A., Tsien, R. Y., and Remington, S. J. (1996). Crystal structure of the *Aequorea victoria* green fluorescent protein [see comments]. *Science* **273,** 1392–1395.

Pang, S. Z., DeBoer, D. L., Wan, Y., Ye, G., Layton, J. G., Neher, M. K., Armstrong, C. L., Fry, J. E., Hinchee, M. A., and Fromm, M. E. (1996). An improved green fluorescent protein gene as a vital marker in plants. *Plant Physiol.* **112,** 893–900.

Pear, W. S., Nolan, G. P., Scott, M. L., and Baltimore, D. (1993). Production of high-titer helper-free retroviruses by transient transfection. *Proc. Natl. Acad. Sci. U. S. A.* **90,** 8392–8396.

Prasher, D. C., Eckenrode, V. K., Ward, W. W., Prendergast, F. G., and Cormier, M. J. (1992). Primary structure of the *Aequorea victoria* green-fluorescent protein. *Gene* **111,** 229–233.

Roederer, M., DeRosa, S., Gerstein, R., Anderson, M., Bigos, M., Stovel, R., Nozaki, T., Parks, D., Herzenberg, L., and Herzenberg, L. (1997). 8 color, 10-parameter flow cytometry to elucidate complex leukocyte heterogeneity. *Cytometry* **29,** 328–339.

Rogers, S. G., Bisaro, D. M., Horsch, R. B., Fraley, R. T., Hoffmann, N. L., Brand, L., Elmer, J. S., and Lloyd, A. M. (1986). Tomato golden mosaic virus A component DNA replicates autonomously in transgenic plants. *Cell* **45,** 593–600.

Romoser, V. A., Hinkle, P. M., and Persechini, A. (1997). Detection in living cells of Ca2+-dependent changes in the fluorescence emission of an indicator composed of two green fluorescent protein variants linked by a calmodulin-binding sequence. A new class of fluorescent indicators. *J. Biol. Chem.* **272,** 13270–132704.

Ropp, J. D., Donahue, C. J., Wolfgang-Kimball, D., Hooley, J. J., Chin, J. Y., Hoffman, R. A., Cuthbertson, R. A., and Bauer, K. D. (1995). *Aequorea* green fluorescent protein analysis by flow cytometry. *Cytometry* **21,** 309–317.

Ropp, J. D., Donahue, C. J., Wolfgang-Kimball, D., Hooley, J. J., Chin, J. Y., Cuthbertson, R. A., and Bauer, K. D. (1996). *Aequorea* green fluorescent protein: simultaneous analysis of wild-type and blue-fluorescing mutant by flow cytometry. *Cytometry* **24,** 284–288.

Rouwendal, G. J., Mendes, O., Wolbert, E. J., and Douwe de Boer, A. (1997). Enhanced expression in tobacco of the gene encoding green fluorescent protein by modification of its codon usage. *Plant Mol. Biol.* **33,** 989–999.

Sambrook, J., Fritsch E. F., and Maniatis, T. (1989). "Molecular Cloning, a Laboratory Manual." Cold Spring Harbor Laboratory Press, Cold Spring Harbor, NY.

Sheen, J., Hwang, S., Niwa, Y., Kobayashi, H., and Galbraith, D. W. (1995). Green-fluorescent protein as a new vital marker in plant cells. *Plant J.* **8,** 777–784.

Siemering, K. R., Golbik, R., Sever, R., and Haseloff, J. (1996). Mutations that suppress the thermosensitivity of green fluorescent protein. *Curr. Biol.* **6,** 1653–1663.

Subramanian, S., and Srienc, F. (1996). Quantitative analysis of transient gene expression in mammalian cells using the green fluorescent protein. *J. Biotechnol.* **49,** 137–151.

Valdivia, R. H., and Falkow, S. (1997). Fluorescence-based isolation of bacterial genes expressed within host cells. *Science* **277,** 2007–2011.

Yang, T. T., Cheng, L., and Kain, S. R. (1996). Optimized codon usage and chromophore mutations provide enhanced sensitivity with the green fluorescent protein. *Nucleic Acids Res.* **24,** 4592–4593.

Zhang, G., Gurtu, V., and Kain, S. R. (1996). An enhanced green fluorescent protein allows sensitive detection of gene transfer in mammalian cells. *Biochem. Biophys. Res. Commun.* **227,** 707–711.

Zolotukhin, S., Potter, M., Hauswirth, W. W., Guy, J., and Muzyczka, N. (1996). A "humanized" green fluorescent protein cDNA adapted for high-level expression in mammalian cells. *J. Virol.* **70,** 4646–4654.

CHAPTER 20

GFP Biofluorescence: Imaging Gene Expression and Protein Dynamics in Living Cells

Design Considerations for a Fluorescence Imaging Laboratory

Paul C. Goodwin
Instrumentation Resource
Fred Hutchinson Cancer Research Center
Seattle, Washington 98109

Copyright © 1999 by Academic Press. All rights of reproduction in any form reserved.
0091-679X/99 $25.00

I. Introduction

In the past ten years, substantial improvements have been made in scientists' ability to visualize, document, follow, and quantitate fluorescent molecules in biology. The discovery of autofluorescent proteins comes at a very opportune time. While chemists and molecular biologists have converged with cell biologists to create whole new classes of chimeric proteins for the study of cellular structure and function, imaging technology has experienced a concomitant growth that enables these novel proteins to be studied in vital cells and tissues. Ten years ago, few biologists knew what computer imaging was and even fewer had access to it. Today, virtually every institution has some digital imaging capabilities. However, creating a successful facility for imaging fluorescence, in particular, autofluorescent proteins, requires a good deal more than merely assembling the pieces. A successful fluorescence imaging laboratory is created by carefully considering the entire imaging environment and not merely the individual pieces. Indeed, the imaging laboratory behaves more as a complex system than it does an assemblage of systems. With this in mind, it is surprising that there are very few publications that have attempted to describe such a complex system. A search of biomedical literature for papers written on creating an imaging laboratory reveals only a single such attempt, and that was in the field of electron microscopy (Willis, 1969). An excellent source of information specifically addressing the needs of imaging fluorescence in living cells was produced by Wang and Taylor (1989). But even in this series, there was no apparent deliberate attempt to describe a laboratory for imaging fluorescence. The description here is derived from experience garnered in creating the Image Analysis Laboratory at the Fred Hutchinson Cancer Research Center. As this system evolved through the 1990s, attempts were made to carefully integrate all aspects of the imaging environment so that the facility could handle the changing needs of the scientific projects at the Hutchinson Center and throughout the Pacific Northwest. Certainly, there are many ways to build such a facility, and few will have the luxury of being able to create such as facility at a single point in time. The intent of this chapter not so much to reveal a single solution to the problem of creating an imaging laboratory but rather to raise the issues that need to be considered in creating such a facility.

As systems evolve and as technology advances, it is easy to assume that every new technology should be embraced. The difficulty that most scientists face, however, is very limited resources. The effort necessary for most scientists to obtain funding for sophisticated imaging equipment necessitates that purchasing decisions be made carefully. Poor purchasing decisions severely hamper scientific discovery and in most cases reduce the chance of having the funds to make the decision again in the future. At the same time, there are strong competitive pressures to make quick decisions on technology. Undo hesitation in adopting effective technology can also slow scientific discovery and significantly handicap the ability of scientists to acquire new funding. This careful balance forces scientists either to become experts in the technology themselves or to create a resource that is carefully staffed by scientists and technicians who can follow advancements in technology and either create the equipment themselves or assist in obtaining funds to acquire commercial systems. These shared resources often prove to be efficient uses of instrumentation funds and can easily be justified in a multiuser environment. Whether you are creating a resource for hundreds of users or for a single lab, the issues raised here should assist in the creation of an effective fluorescence imaging laboratory.

II. Facilities

In designing a laboratory for digital fluorescence microscopy, it is essential to carefully consider the space in which the equipment is to exist. In the ideal world, the scientists will be afforded the luxury of being able to sit down with facilities engineers and work together to create the ideal working space and resources to optimally support the laboratory. In reality, the scientist is well served by being prepared to handle many of the design issues for the lab. The more information the scientist can give to design engineers, the better the engineer will be able to accommodate the needs of the scientist. In this section, we will consider the facilities issues of the laboratory including the size of the room, power requirements, lighting, heating/ventilation/air-conditioning (HVAC), and ergonomics.

A. Room Size

Fluorescence microscopes and associated computers and accessories needed for live cell experiments tend to be large. They also usually require access to all sides of the equipment. Each equipment vendor can give you specifications for the actual equipment that they supply. They can also tell you what access you need to the sides and back of the instrument. At the Fred Hutchinson Cancer Research Center (FHCRC), our confocal and deconvolution microscope systems are in rooms that are 9–10 square meters. This is sufficient for 1–3 people to work together along with the instrument. The philosophy at that we have adopted at FHCRC is that the instrument itself is generally viewed as an acquisition device. Whenever possible, computations, analysis, visualization, and file management are performed on a remote computer to free the microscope for other purposes. This will be discussed in Section V. This minimizes the size of the carefully controlled environment in the microscopy rooms. It does, however, necessitate providing space for these other tasks.

B. HVAC/Ducting

The microscope systems employed for studying GFP tend to generate a lot of heat and ozone. These by-products must be properly handled to maintain a safe and comfortable working environment for the users and the experiment. Ozone (O_3) is generated by energy such as electricity and UV light passing through air. It is a common industrial pollutant created by power supplies, high-voltage sources (such as lasers), and high-energy light, all of which are present in a fluorescence microscopy laboratory. Ozone is an irritant that can cause fatigue and headaches and is generally not pleasant to work in for prolonged periods. Ozone is handled by exchanging the air in the room at a sufficient rate as to prevent ozone accumulation. In general, we use about ten air exchanges per hour.

In addition to exchanges of air, heat must also be dissipated from the laboratory. Heat is often measured in British thermal units (Btus) per hour. Because many fluorescence microscopy laboratories have a variety of instruments, many of which do not specify heat production, it may be necessary to make these calculations yourself. A BTU can be approximated by the following formula:

$$\text{Btu/h} = [\text{Room area (ft}^2) \times 10] + (\text{Number of occupants} \times 25) + \text{Equipment heat.}$$

On average, electrical systems run at about 30% of the power maximum reported on the faceplate so Btus of the equipment heat can be estimated by

$$\text{BTU/h} = 3.412\ \text{Btu/h} \times (0.30 \times \text{volts} \times \text{amps}) \text{ for each faceplate in the lab.}$$

These air flow and cooling requirements have the potential of creating air currents that can severely effect long-time-lapse experiments and that can lead to focus drifts in microscope systems. A couple of things can be done to prevent these effects. First of all, if possible, design the duct work such that the air exhaust is closest to the microscope and the return is furthest from the microscope. It is also a good idea to choose a diffuser for the return that maximizes air exchanges away from the microscope. This can be accomplished either by choosing a diffuser that limits air mixing to the ceiling or by creating a cowling to fit over the return to divert the air away from the microscope.

C. Ergonomics

> "The best position for microscopical observations is when the observer is lying on his back. . . . The worst of all positions is that in which we look downwards vertically"
>
> Sir David Brewster, 1835.

Scientists often confuse ergonomics with luxury and extravagance. In reality, paying attention to the design of the work area is an effective way to minimize repetitive strain injuries, prevent fatigue, and improve performance. Microscopy can often require sitting in a single position for hours at a time. Doing this in a poorly designed work space can not only reduce the mental acuity of the scientist but can also lead to career-limiting injuries such as carpal tunnel syndrome, thoracic duct syndrome, and lower back injuries. These phenomena have been well documented from the 19th century (Brewster, 1835). Clearly the cost of injury far outweighs the cost of carefully selecting a proper chair, work surface, and other working conditions. Many institutions now employ industrial hygienists to assist in designing a proper work environment. The reader is also encouraged to read Helander *et al.* (1991), Haines and McAtamney (1993), and Kwan and Humphreys (1985). My own experience is that I am most comfortable in a stool or chair with a rigid back that provides good support and that can be raised or lowered to force me to sit with proper posture in order to see into the oculars. The work surface should be kept clear of clutter, and commonly needed items (paper, writing instrument, cotton swabs, cleaning solution, immersion oil, etc.) should be kept at hand. As a side note, it is very important that the scientist take the time during experiments to stretch regularly, to rotate the shoulders, neck, and back, and to get out of the chair whenever possible in order to move around.

III. Maintaining Cells

A. Media

Cells and tissues are, of course, sensitive to their environment. Specimens that are not maintained in a proper environment may die or, worse yet, behave differently than they would in a more suitable environment. The medium that bathes the specimen must not only transport gases and nutrients, it is also the primary means by which temperature, osmolality, and pH are maintained as well as being the primary method by which factors such as hormones and growth factors are delivered to the cells. Each cell or tissue type has its own set of requirements for homeostasis, and it is not within the scope of this chapter to address each of these. Suffice it to say that the microenvironment must be carefully maintained in order to assure that the cells both survive and behave in a physiologically relevant manner. A couple of cautions are worth noting. Many growth media contain phenol red for a pH indicator to assure that the media is not contaminated. Unfortunately, phenol red can interfere with GFP fluorescence. It is generally a good idea to replace the medium with a phenol-red-free medium before attempting to image the cells with fluorescence microscopy. The other suggestion deals with maintaining pH. Most biologists maintain pH in culture by growing cells in the presence of 5% carbon dioxide. Unfortunately, this can be hard to maintain. A simple solution is to use a HEPES-buffered medium (Boehringer Ingelheim Bioproducts). This removes the need to maintain a 5% carbon dioxide environment, and it is generally well tolerated by most cells. If cells are to be maintained for a long period of time (hours), then some method will generally be needed to assure either that adequate medium exists in the viewing chamber or that some sort of perfusion method exists to replace the medium throughout the experiment.

B. Enclosed Systems

One way to maintain a homeostatic viewing environment is to fully enclose the entire microscope in a sealed compartment in which gas mixtures and temperature can be maintained. If the researcher has adequate space to accommodate such a system and if the experiments demand very long viewing times (days), then these systems are relatively easy to set up and maintain. One common method is to contract with a local Plexiglas manufacturer to design a closed compartment that will fully encompass the microscope. Temperature and humidity in this compartment can be maintained by controlling a common hair dryer with a thermostat and blowing the warm air over a distilled water reservoir. Gases can be delivered to the enclosure by having the manufacturer attach nozzle fittings onto the exterior of the enclosure. Enclosed chambers do introduce their own set of problems. The elevated temperatures and humidity can lead to oxidation problems on microscope and electronics components. The temperature in the chamber should be held constant all of the time or a small oven should be available to store objectives at constant temperature. If the temperature of the objective is changed excessively, stress can build in the objective, which may hamper strain-sensitive microscopy methods such as differential interference contrast (DIC). Worse yet, temperature cycling can cause the adhesives in the objectives to fail, which will permanently damage the objectives.

C. Chamber Systems

Most laboratories cannot afford the luxury of fully enclosing their microscope, or they wish to use the system in more normal conditions. These researchers may prefer to use one of a number of chamber systems to maintain just the environment of the imaging area of the microscope. Several home-made and commercial systems for this exist; however, there are a couple that I have found to be most useful. If temperature is not a critical consideration, a simple solution is to use the NUNC chamber slides (LabTek Chambered Coverglasses). These slides work well with inverted microscope systems. They are available as chambers mounted on top of #1 coverslips. They are relatively inexpensive and are disposable. The thin coverslip bottom of the NUNC chamber makes it particularly well suited for high-resolution viewing of live cells. These chambers have simple lids that can be easily removed for adding fluids (like agonists or drugs) to cells or for micromanipulators.

The most complete and effective systems that I have seen for imaging chambers come from Dan Focht of Bioptechs, Inc. These systems make use of optical glass surfaces that are coated with indium tin oxide. This metal is virtually clear within visible wavelengths and it heats when a current is applied across it. By making use of thermisters and a tightly coupled feedback mechanism, the temperature in the chamber can be controlled to very tight tolerances. The chambers come in either an open design without flow (the ΔT) or in a flow chamber configuration (FCS2). Both systems can effectively control the temperature at the surface of the cells or the tissue to about 0.2°C. The glass used in these chambers is an alkaline glass that most cells can grow on, even cells that don't normally do well on other glass surfaces.

IV. Imaging Systems

The very presence of this monograph is testament to the varied assays to which GFP is being applied in biomedical research. These assays each have their own requirements for imaging. In some cases, relatively low magnification is needed, such as plaque and expression screening (Ross-Macdonald *et al.,* 1997). In other assays, the limits of resolution with light microscopy are tested, such as protein trafficking experiments (Htun *et al.,* 1996). In some cases only a static image is needed; in others, time-lapse is needed. In some cases, only a single overview is needed, sometimes the investigator needs to follow fluorescence in time-resolved three-dimensional imaging. Each of these applications has different instrumentation demands. I give here a quick overview of the available technologies. For further study see Wang and Taylor (1989), Shotton (1992), Pawley (1995), and Inoué and Spring (1997).

A. Macroscopic Imaging

In a number of assays, including plaque screening (Ross-Macdonald *et al.,* 1997), viral infection assays, and migration assays, single-cell imaging is not needed. In these cases, macroscopic imaging provides much higher throughput and may be more statistically valid than trying to sample with a microscope-based system. Some of these macroscopic imaging systems are self-contained and are sold as complete systems. These include microtiter plate fluorescence readers such the Biolumin 960 from Molecular Dynamics and the FLIPR from Molecular Devices. For plaque-forming assays and clonal assays, good results have been achieved with a dissecting microscope equipped with a fluorescence attachment such as the Leica fluorescent dissection microscope. Dissection microscopy is simple to use and can generate excellent images of both GFP fluorescence and darkfield microscopy of plaques and colonies. These systems can be equipped with either 35-mm cameras or digital imaging systems for documentation and quantitation. In colony-forming assays, gene transfer assays, viral infection assays, and the like, dissection microscope imaging holds a distinct advantage in that the culture dish, almost irrespective of its shape, can be easily inspected without violating the sterile environment. This makes it a superb solution for monitoring the progress of ongoing experiments.

B. Video Microscopy

Video-enhanced microscopy was suggested by Zworykin in the first paper announcing the successful creation of a vacuum-tube device for converting light into an electronic image in 1934. His early dream had been the creation of an instrument that could "[observe] phenomena at present completely hidden from the eye, as in the case of the ultra-violet microscope." For a comprehensive review, see Inoué and Spring (1997). Today's digital video microscopes are the embodiment of Zworykin's vision. In these systems, video cameras are used to convert light from the microscope into an electric signal, which is, in turn, converted into a two-dimensional array of discrete numbers by an analog to digital converter (A/D converter). This array of numbers can then be stored digitally in computer memory (RAM) and processed or analyzed by software that resides in the computer. The array of digital numbers can be converted back into an analog electrical signal (D/A converter) and displayed on a video monitor (or computer monitor) and/or saved onto an analog medium (e.g., video tape). Alternatively, the stored digital array or image can be saved directly using a digital medium such as a hard disk or recordable CD-ROM.

These systems have taken on many different forms. They range in cost from less than $10,000 to over $100,000. They can be based on virtually any computer platform from an Intel/Windows-based PC, to an Apple Macintosh, to complex systems built on UNIX workstations. There are commercially produced systems and home grown systems. For a list of vendors, see Ted Inoué's Web site at VideoMicroscopy.com. Of special note is a free package available from the National Institutes of Health called NIH-Image. This is a freeware package written mostly by Wayne Rasband of the National Institute of Mental Health (NIMH). This package has two main components: a user interface and a sophisticated Pascal-like macro language. The source code is also freely available. NIH-Image is available for both PC and Macintosh platforms, but it should be noted that at this time, the PC version is a couple of years behind the Macintosh version in development. NIH-Image works with a variety of image capture devices and can use Adobe Photoshop plug-ins. This makes it very easy to use a large variety of input devices for image acquisition. Another excellent feature of NIH-Image is that there is a very large worldwide user base that communicate through a listserv. This connects users together so that they can share ideas, get help, and codevelop custom imaging applications. NIH-Image is available by anonymous ftp:

ftp server: zippy.nimh.nih.gov
login: anonymous
password: your e-mail (e.g., yourname@your.org)
directory: /pub/nih-image

Here you can find the program for both platforms, the source code, documentation, user-contributed macros and variants of NIH-Image, and helpful suggestions. To join the listserv send electronic mail to:

listserve@soils.umn.edu

in the body of the e-mail, type:

subscribe nih-image Your Realname

You then will be notified that you have been added to the listserv.

There is not room here to go through every permutation possible in assembling a video microscope system. In general, the components that need to be evaluated are the microscope, the camera, the framegrabber (in most cases), the computer, and the software.

These systems generally have the advantage that they are relatively inexpensive to assemble and they can generally collect images in "real time" (i.e., 30 frames/s). The disadvantage is that they generally have a fairly limited dynamic range and the signal-to-noise ratio is low. The limited dynamic range hampers the ability to use background subtraction to correct for illumination and camera errors. The limited signal-to-noise ratio means that frame averaging is often needed to achieve a satisfactory image which in turn limits the speed of image acquisition. These systems are most often limited to two-dimensional imaging, and the resolution is limited by out-of-plane blurring.

C. Confocal Microscopy

The confocal principle was first described by Minsky in 1957 (Minsky 1957, 1988). In practice, the first commercial systems became available in 1988. In general, if we think of fluorescence as coming from a point source smaller than a light microscope can resolve, say a single fluorochrome molecule, the light from that object will spread out in all three dimensions to create what can be described as concentric cones with their vertices focused on the point source. These conelike images are referred to as the point-spread function (PSF) because they describe how a single point spreads in three-dimensional image space. The difficulty of imaging a three-dimensional object is that objects above or below the plane of focus contribute intensity to the plane of focus. The PSFs of the objects in all planes can interact. If enough PSFs interact, the image becomes difficult to interpret and there is considerable loss of resolution. In a confocal microscope, the PSFs from out-of-focus planes are optically removed from the plane of focus by placing pinholes at the appropriate locations in the optical path resulting in highly focused excitation and emission light paths. Most of the light that makes it through this aperture is from the plane of focus. Light that initiates from other planes is mostly rejected by this aperture. By moving this aperture across a plane of focus, the instrument is able to construct an image of that plane within the three-dimensional object. By moving the specimen axially (e.g., in the z-dimension) and scanning each plane (e.g., the x- and y-dimensions), the confocal microscope is able to create a three-dimensional image of the object. For a thorough review of confocal microscopy, see Pawley (1995) and Shotton (1992). There are two types of commercial systems, Nipkow disks and scanning laser confocal microscopes. In a Nipkow disk system, the aperture is scanned across the plane by a spinning disk with an array of tiny holes in it. The holes are carefully designed so that within a revolution of the disk the entire plane is scanned (Sheppard and Wilson, 1981; Boyde, 1985; Lemp *et al.,* 1985–1996). This has the advantage of permitting direct viewing of the confocal image; unfortunately, the light loss through these apertures is so great that these systems have somewhat limited utility for live cell imaging. The other method of confocal imaging is by laser scanning confocal microscope (Valkenburg *et al.,* 1985). In this method, a laser beam is used for exciting the fluorochrome. The physics of the coherent laser beam is such that is approximates an excitation aperture. In most commercial systems, this beam is scanned through the image plane in a raster pattern by a pair of mirrors driven by galvanometers. Because fluorescence is a fast event, the mirrors are in virtually the same position when the emission light returns from the sample, so the same mirrors can be used to collect the raster pattern back into a single emission beam. That beam is then passed through a pinhole and is collected by a photomultiplier tube (PMT). The PMT transduces the light energy into an analog signal, which is then converted to a digital array through a framegrabber. This digital image can then be displayed, stored to disk, or manipulated in memory.

Laser scanning microscopes in general allow more light to be delivered to the specimen than Nipkow disk systems, which makes them more sensitive for moderate to weak fluorescent imaging. Because the raster scanning of the plane is generally accomplished by mirrors under the control of galvanometers, the positioning errors are usually less in laser scanning systems, which improves the resolution of the imaging system compared to Nipkow disk systems. There are intermediate solutions that increase acquisition speed (up to video rates) by using scanning a slit of light across the image plane instead of a point. These systems are generally good for fast dynamic studies that do not require the resolution obtained by the scanning point methods. For a thorough review of confocal microscopy, the reader is strongly encouraged to read Pawley (1995).

The advantage of confocal microscopy is the ability to optically section through relatively thick specimens (up to 100–200 μm), because the image that is displayed on the screen is the optical section, the research gets quick feedback as to the status of the experiment. There are factors that limit the efficacy of laser scanning microscopes. In point scanning systems, the acquisition times can be quite protracted. Confocal systems, by design, must reject light and are further limited by the quantum efficiencies of the PMTs. The dependence on lasers for excitation limits the flexibility; for example, to add the ability to image Hoechst staining, most confocal users would have to add a laser and safety features to the confocal to isolate the user from direct exposure to the UV laser. This can easily add $70,000 to the cost of a system. Most confocal systems rely on 8-bit (256 discrete digital values) data. Pawley argues that the number of quanta generated within the dwell of the beam through the sample is limited to less than 256, so 8-bit acquisition is adequate. In my experience, I have often seen cases in which 8-bit data is constraining. Most of this is due to the relatively high noise levels of confocal systems. The relatively low signal-to-noise ratio of confocal systems limits the amount of postprocessing that can be done on the images, though we have demonstrated some success by applying deconvolution algorithms to confocal data. The limited signal-to-noise ratio of confocal systems and a heavy reliance on spherical aberration free optics limits the quantifiability of confocal data sets. Some of the new confocal systems coming onto the market are starting to offer 12-bit acquisition systems which could improve the quality of the data sets. A more severe limitation on confocal microscopy is caused by the flux density of a highly focused laser beam (White *et al.,* 1987). Vigers *et al.* (1988) have shown that high flux densities (i.e., light power focused into a small volume) can cause significant heating in microenvironments within the cell, which can have profound effects on microtubule assembly and presumably other aspects of live cell biology as well as on photobleaching of the fluorochrome. Some of these effects are insidious and may be difficult to detect. This may limit the application of high-resolution laser scanning confocal in live cell imaging.

D. Multiphoton Microscopy

Webb and others (see Denk *et al.,* 1990; Piston *et al.,* 1995) have demonstrated that very short bursts of high-intensity relatively long wavelength light could excite a fluorochrome that would normally excite at one-half of that wavelength. They demonstrated that in a highly focused optical system, the energy of multiple photons striking a fluorochrome at the same time can generate a summing of the energies such that the effective wavelength is one-half of the source. So, if a laser could generate a short pulse (e.g., 20 fs) of light at 720 nm and that light could be highly focused, then there was a statistical likelihood that at the point of best focus multiple photons would collide to produce the equivalence of a single excitation at 360 nm. Webb showed that using such a laser system on a scanning laser confocal microscope would produce a system with considerable advantages over normal confocal systems. The depth of penetration of light into a sample is proportional to the wavelength, so a longer wavelength will allow deeper visualization into the sample. Longer wavelengths have less energy than shorter wavelengths. This means that there is much less photodamage to the specimen by 720-nm light than there is by the same power at 360 nm. The original two-photon system from Webb was built on a confocal microscope with the pinhole wide open, in effect using the system to merely deliver and detect light as a scanning laser microscope. Current commercially available multiphoton systems from BioRad and Zeiss are built on confocal microscopes and offer both single-photon confocal and multiphoton capabilities. Early adoption of multiphoton systems required considerable expertise in laser design and maintenance. Commercial multiphoton systems are now available that utilize off-the-shelf titanium–sapphire tunable laser systems. These systems hold great promise in cell biology and developmental biology for examining live samples in three dimensions. The costs and reliability of the laser systems will have to be significantly improved before these systems will become commonplace. Whether the laser manufacturers can meet that challenge remains to be seen.

E. Deconvolution Microscopy

As early as 1968, Goodman suggested the use of mathematical methods to improve the resolution and contrast of optical microscopy. Goodman suggested that the image generated by an optical system was the product of the actual image confused with (convolved with) the blurring errors introduced by the optical system itself. If the blurring could be measured, he argued, then the actual image could be regenerated by deconvolving the generated image with the PSF:

$$\text{Actual image} = \text{Generated image} \otimes \text{PSF}.$$

Goodman (1968) argued that in such a system, the resolution at which the actual image could be reconstructed was limited by the accuracy of the OTF and the signal-to-noise ratio of the generated image. Sedat and Agard (Agard and Sedat, 1983; Agard, 1984; Agard, *et al.,* 1989; Hiraoka, *et al.,* 1990; Scalettar *et al.,* 1996) and Fay (Coggins *et al.,* 1986, Fay *et al.,* 1986) pioneered the use of mathematical methods for three-dimensional microscopy. Others have followed with their own implementations of these methods, including blind deconvolution, which attempts to create an optical transform function (OTF) from the measured image itself (Holmes, 1992; Bell and Sejnowski, 1995). The most successful implementations of these methods are those systems that fully optimize the quality of the generated image and the OTF prior to implementing the algorithm. Because the ability to generate the actual image is limited by the OTF, careful measurement of the OTF is critical for success. This places a requirement on the microscopy itself to be as good as possible with proper Kohler illumination of the fluorescence system. This, of course, is also true in confocal microscopy. The second requirement from Goodman is that the signal-to-noise ratio must be as high as possible. The best deconvolution systems are using 12-bit or higher research-grade cooled CCD cameras for detectors. These cameras have very little internal noise, excellent quantum efficiencies, and tremendous linearity over more than three orders of magnitude.

Properly implemented, these systems are proving to be fabulous tools for imaging live cells. By using nonconfocal (wide-field) illumination, the flux density is greatly reduced, perhaps by as much as four logs, over confocal illumination. As mentioned earlier, this will greatly reduce the photodamage and photobleaching of the sample. Because these systems use arc lamps for illumination, there is considerable flexibility in excitation and emission wavelengths (from UV to near-IR). Adding additional wavelengths costs a fraction of what it would cost in confocal systems. The improved quantum efficiency of the cooled CCD detector over a PMT and the lack of a pinhole result in improved light detection and sensitivity. The improved signal-to-noise ratio of these systems produces data sets that are much more quantitative than their confocal counterparts (Goodwin and Brown, 1996). These systems have demonstrated superior planar (x and y) and axial (z) resolution as compared to confocal or video microscopy applications. We have showed that the measured intensities of deconvolution images are more quantitative than those of confocal images (Goodwin and Brown, 1996). There is a common misconception that these systems are slow. In fact, the improved signal-to-noise ratio results in generally faster image acquisition than laser scanning confocal systems. Deconvolution times have reduced dramatically over the past 2 years, being driven by tremendous improvements in workstation performance. To give an example, when I first started using a Delta Vision system (Applied Precision, Inc., Issaquah, WA) 3 years ago, a 512- × 512- × 64-section deconvolution (15 iterations of a the Sedat and Agard reiterative constrained algorithm) took over 2 h. The current record on a multiprocessor system is about 6 min. These improvements will continue as workstation performance increases. Surprisingly, these systems are considerably less expensive than their confocal counterparts. I have found deconvolution microscopy to be the method of choice for most live cell applications. Deconvolution microscopy currently works best for relatively thin sections, up to about 150 μm, being limited mostly by spherical aberrations in the sample beyond that point. Improved methods of dealing with spherical aberration should improve deconvolution microscopy out to the full working distance of the microscope objectives being used.

V. Computer Systems

Computer imaging systems can be very costly, with some two-photon systems costing well in excess of a half of a million dollars. Part of the challenge of a fluorescence imaging laboratory is finding ways to maximize the use of the acquisition system for acquiring data and, as much as possible, using other computer systems for processing, visualization, analysis, and publication of the images. By investing in computer support resources and efficient networking, image acquisition systems can be used as efficiently as possible. There are certainly costs associated with developing these computer resources, but they are generally a fraction of the costs of additional imaging systems. In general, an efficient computing environment is composed of a number of components. The laboratory needs to meet all of these needs to perform efficiently, and establishing an imaging facility without meeting these needs will lead to restrictions in the flow of data. The computer environment is made of acquisition, networking, fileserving, user workstations, and data archiving and backup. Some computer systems are sophisticated enough to meet multiple needs simultaneously. Other systems can meet only one need at a time. In general, the more flexible a given computer system is, the more complex it also is. We now explore these functions in detail.

A. Acquisition

The computer resources at the acquisition system can limit the sampling rate and length of the experiment. The key elements of the acquisition system are image memory, hard disk size, and the speed at which data can be moved from memory to the hard disk. In some systems, the image memory is contained in the frame grabber. In other systems, image memory is the same as system memory. In either case, images generally need to be moved efficiently through the system bus and hard disk controller to the hard disk. If the acquisition system has sufficient hard disk resources and the data can be moved there efficiently, then there is generally not a huge requirement for system memory. If the system cannot efficiently move data to the hard disk, then more memory is needed to buffer the image data. Hard disk performance and size is considerably cheaper to purchase than image memory, so the features to look for in the acquisition are a fast system speed, fast hard disk transfer rates, and a relatively large hard disk. The hard disk size determines the size of the experiment and how much data can be acquired before it is necessary to archive files or move them to a fileserver. Efficient computer systems can transfer files without substantially hindering data acquisition, which can place less of a requirement on the hard disk.

B. Fileserver

One of the most efficient ways to share data between computer systems is to centralize file storage resources in the form of a fileserver. These computer systems are specialized to store large amounts of data and to make that data available to multiple users simultaneously. For laboratories with large computing needs, these systems can also serve as compute servers. These servers are often UNIX or Windows NT workstations with large amounts of memory, large hard disks, and fast efficient network resources. A typical choice for this purpose is a Silicon Graphics or Sun Workstation with 200+ MB of RAM, 10+ GB of hard disk, and a 100-T network.

C. Client

Networked computer environments are referred to as client/server. The server portion was discussed earlier. The client is the desktop computer system that is used for data visualization, data analysis, and graphics. In a properly designed environment, the client can sit on the investigator's desk and provide efficient distributed access to data on the server. These systems are usually inexpensive compared to the server and the acquisition computers. This is a rapidly evolving field and is likely to continue to change substantially over the next few years. Currently, Macintosh PowerPC computers and Intel-processor-based Windows computers are the systems of choice, but the advent of Web-based computing may have profound effects on the choices available over the next few years. The volume of desktop computers enables software vendors to sell software for less on desktop systems than they can on workstations or other platforms. Software on a desktop computer usually costs about one half of what it costs on servers, and desktop computers are usually the first to get software improvements. The hardware needs for the client depend largely on the visualization, analysis, and graphics needs. If the need is simply looking at still images, then the software and memory demands may be small. If the user needs to analyze large movie files, then substantial amounts of memory may be needed. If the purpose is to create multimedia graphics on the desktop system, then there may be tremendous needs for hard disks, memory, and system speed.

D. Archival and Backup

One consistent aspect of the imaging laboratory is the tremendous amount of data that is generated. While at the Hutchinson Cancer Center, the Image Analysis Laboratory generated and stored more data than any other resource. We archived more than four gigabytes of data per week and maintained data for several hundred users. The task was daunting. Over a thousand worker hours a year were spent just maintaining the data. There is often confusion between two very different aspects of data management: data archival and data backup. Backup is routine copying of the entire computer system for the purpose of disaster recovery. This is usually a rotating tape library that is used to recover the whole computer system or large blocks of data in the case of catastrophic loss of a computer or a disk system. Archiving, on the other hand, is the systematic migration of data from relatively fast and expensive disk to less expensive and slower media. The basic assumption of archiving is that the data that is most recently acquired or accessed is more likely to be accessed again in the near future than data that has been left untouched on the computer system for some length of time. By migrating the older data onto inexpensive media, the faster and more expensive medium is freed for use by new data and frequently accessed data. Historically, this process has been time-consuming and expensive. It required manually moving data to archive media (say tape, erasable optical disks, or CD-ROM Recordable), verifying the archived files, cataloging the media, and then deleting the original files. This method can be made efficient on UNIX systems by the creation of some simple scripts for listing the entire contents of a archive volume into a file and another script to search for files within the volumes. For a medium sized multiuser facility, this can require 10 to 20 h/week of labor. Over the past few years, new schemes have been developed for automating the file migration process. Hierarchical storage management (HSM) and variants of HSM have become available at fairly reasonable costs. In HSM, protocols are set (say, any confocal file that has not been opened in 2 weeks is to be archived). The HSM system constantly monitors the computer or network systems to find files that meet or exceed the protocol condition. Those files are migrated on to alternative media and verified. The original file on the faster media is replaced with a pointer file of 0 size that remains in the user's directory. Whenever a user tries to access this file, the HSM finds the archived version of the files and replaces the pointer file with a copy of the original file. If HSM is used in conjunction with a jukebox for the archive media, then this whole process can proceed virtually unattended. After some point, another protocol can be set to handle removing files from the jukebox on to a removable media that is also cataloged. Some HSM systems can also automate system backups for disaster recovery. For larger multiuser facilities, these systems can usually be cost justified on labor savings alone, and the added security of automating the process makes this a robust and extremely useful file management system.

The decision of which medium to choose for archival is critical and difficult to make. At issue are the costs of the original medium, hardware costs, longevity, and access speed. Table I lists many of the more common archive media and rates them relative to each other based on these criteria. Longevity is divided into two components: the longevity of the medium and the longevity of the

Table I
Archive Media Options

Type of medium	Medium cost	HW cost	Medium longevity	Hardware longevity	Access speed
Removable Disk	++++	+	++	++	++++
Erasable Optical Disk	+++	++	++++	++	+++
Exabyte	++	+++	++	++	+
Digital Audio Tape	++	+++	++	+++	++
Digital Linear Tape	+	+++	++++	++	++
CD-ROM Recordable	++	+	++++	++++	+++
Digital Video Disc Recordable	+	+++	++++	++++(?)	++++

hardware to read the medium. Of those in the table, the one that I have found most useful is CD-ROM Recordable (CD-R), because it offers the best compromise on costs and performance. In the next year or so, the hardware costs of Digital Video Disc Recordable (DVD-R) should drop dramatically and questions of how generally accepted it is will have been answered. DVD-R, if it does indeed become widely accepted as a commodity appliance, holds the best promise for fast access speed, high density, and overall costs.

VI. Output

A. Data Analysis

One of the significant improvements in microscope instruments that has occurred over the past 10 years is that most systems now use standard computer components instead of proprietary systems. This makes it much easier to choose systems and gives the researcher the freedom to choose analysis software independent of the acquisition hardware. By using separate computer systems for acquisition and analysis, the most expensive component, the microscope system, is available full time for acquisition and the analysis can be moved off-line to a less expensive computer system. The basic philosophy that I have followed in the lab is that the microscope is a specialized data entry system. Once the image is in the computer, it doesn't matter where the analysis is done. The computer system requirements can be dictated by the processing needs and the software tools rather than having to live with whatever tools a given microscope vendor provides. The computer also enables the researcher to add additional visualization and analysis stations for a fraction of the cost of a new microscope system and allows for the possibility of delivering images directly to the investigator's desktop. This is especially important in a shared instrumentation environment, in which the instrument may very well not be in a given investigator's laboratory. The net effect of this is that the investigator can generally make this decision independent of the instrument purchase as long as the system vendor has an adequate export facility that can export image files as a common file format such as TIFF.

Selecting an analysis system starts with thinking through the types of questions that one hopes to answer with the system. Are you analyzing static two-dimensional data? Are you analyzing time-resolved data? Are you visualizing three-dimensional data sets? Are you quantitating time-resolved, multiwavelength, three-dimensional data sets? Are you measuring lengths and areas or volumes and velocities? Are you measuring mean intensity of a whole field, or are you attempting to automate the analysis of many individual cells? Do you need to be able to do image math (e.g., image ratioing) prior to analysis? All of these questions drive the decision of which tools are necessary and what software system is best suited for the task. When I managed a shared resource facility, I felt that it was necessary to have a variety of tools available. We settled on two levels of solutions. One very inexpensive and quite good environment is NIH Image by Wayne Rasband at the National Institute of Mental Health (NIMH). Currently, this system is best developed on the Macintosh platform. Versions are being developed on the Windows/Intel platform (PC Image or Scion Image from Scion, Corp.) and under Java (Image/J), but they lag behind the Macintosh version by many years. The most exciting of the new versions is Image/J because it will offer platform-independent application building. Java is a Web-aware device-independent programming environment. Code written on, say, a Macintosh would be 100% compatible on a Windows/Intel platform or on a UNIX workstation. This flexibility is going to be interesting to follow over the next several years. NIH Image has a number of considerable advantages for simple to moderately complex iamge analysis needs. First of all, it is free and the source code for the program is also freely available. Second, the listserv for Image mentioned earlier is an active helpful group of users from around the world in virtually every discipline. This gives broad depth in the expertise available and fosters interdisciplinary communications that often yield exciting and new ways of solving imaging problems.

The other level of solution that we embraced was computationally intensive analysis systems for multidimensional data sets. These programs run on Silicon Graphics, Inc., workstations. They are designed to handle multidimensional data sets. The tools are specifically designed to generate time-resolved volume renderings. They also have tools designed to segment features within three-dimensional data sets and to extract data from these sets. Although some of these tools are excellent, it should be noted that it is this segment of software development that has the greatest room for improvement. Unfortunately, much of what is needed is a whole new paradigm for image segmentation. Exciting new methods such as adaptive contours (snakes, worms, and balloons) and Markov random fields are coming out of the defense industry, but these have yet to make it into commonly available systems for cell biologists, so the best systems that now exist are semiautomated tools that help investigators model and analyze their data sets.

B. Data Reduction and Statistics

There are many very sophisticated programs for data reduction and statistical analysis. Over the years, however, I have found that for most purposes a good spreadsheet program capable of macro programming is sufficient for most applications. I generally recommend Excel from Microsoft for a spreadsheet program. It is by far the best-selling spreadsheet program, it is sold and understood worldwide, it is available on multiple desktop computer platforms, it is relatively inexpensive, and it has a very sophisticated programming language (Visual Basic) for macro programming. Although it is written specifically for business applications, a number of good statistical and engineering functions are available within the program. The graphing functions, although somewhat primitive by some standards, are adequate for most purposes.

C. File Formats

One of the key elements of an imaging laboratory is the ability to share the data with individuals outside of the laboratory. If the images and data generated in the laboratory cannot be published or shared, their value is greatly diminished. The breadth of media choices for data publishing have been greatly enhanced over the past 5 years with the advent of multimedia standards and publishing on the World Wide Web. However, the primary source of peer reviewed publication for most disciplines is still publication in the print media such as scientific journals. This, then, places a requirement on the suppliers of imaging software either to have a large array of graphics editing software as an integral part of the imaging package or to have convenient and accurate tools for exporting images into universally standard image formats. Most of the good commercially available graphics programs were not created specifically for the scientific community but rather for commercial applications such as print prepress and page layout. This commerical market has also driven the graphics file formats. In the commercial world, there has been little demand for a universal file format that is sufficiently flexible for the demands of the scientific community. Until recently, commercial programs had little use for more than 256 levels of red, green, or blue in a color image (24-bit RGB). Scientific applications may well require 64,000 levels (16-bit gray scale) in any number of wavelengths, in three dimensions, and all time resolved. This requirement has rendered most commercial formats inadequate for primary data storage. However, to be able to use the scientific images in commercial software, the images must be in a standard format that the graphics program understand. The most sensible solution is for vendors of scientific imaging software to provide accurate and easy to use tools for exporting images into standard commercial formats. The most flexible and commercially accepted of these formats is the Tag Image File Format (TIFF). TIFF was created by a consortium of the desktop publishing community headed then by Aldus, Corp. (now Adobe, Corp.). This consortium realized the value of an extensible format that could handle a variety of data types and that would be made freely available to the programming community to foster the free exchange of image and graphical data between different computer platforms and software programs. In its purest form, TIFF has the capacity to handle most of the requirements of the scientific imaging community; however, the difficulty of implementing the full flexibility of the format in commercial software products has led to a pared-down version of TIFF, which is now the most accepted. In general, TIFF is now generally recognized as the most general standard for image file exchange. The most successful TIFF types that are generally useable are 8-bit grayscale and 24-bit RGB. Most recently, 16-bit grayscale and 48-bit RGB are gaining popularity, but they still compose a small fraction of the TIFF market. Compression is one of the greatest sources of file incompatibility and should be used with caution. JPEG, created by the Joint Photographic Experts Group, is one of the best of the lossy compression methods, and it has gained wide acceptance on the World

Wide Web for continuous tone images like those generated by digital imaging systems. The other common image format on the Web is the Graphics Interchange Format or GIF. GIF is much more lossy than JPEG and is usually best suited for line art and other applications that do not require a large range of colors or grayscales.

For multiple frame data, I have found Quicktime from Apple Computer, Inc., to be the best interchangeable format. Compression/Decoders (CODECs) for Quicktime range from no compression to MPEG, JPEG, video, animation, and others. This gives a great deal of flexibility for compression, compatibility, and file size. MoviePlayer is a free software viewer for Quicktime available for almost every computer platform from Apple. Versions of NIH Image 1.6 or later can directly import Quicktime movies. This provides a convenient method for viewing and analyzing time-resolved data. On the Windows/Intel platform, Quicktime can be used. Microsoft/Intel have also created AVI for these purposes, but I have found Quicktime to be more generally useful.

D. Graphics Programs

There are many good programs on the market for creating professional quality graphics from digital images. These fall into a few broad categories: painting programs, drawing programs, and page-layout programs. Painting programs refer to a class of programs that treat graphics as an array of values, like a digital image. Painting programs historically started with a blank canvas. The user painted from a palette of colors using tools common to a painter. Examples include Microsoft Paint on the Windows platform and Claris MacPaint on the Macintosh program. The premier program of this type is Adobe Photoshop. This excellent program sets the standard for image manipulation. Historically, one of the big problems with paint programs was that the changes made to the image were made by overwriting the image data with the changes so that even simple items like arrows and text overwrote the original image. This made it very difficult to use these programs and required saving multiple copies of the original and modified image files. In Photoshop version 3 and later, layers were added to the imaging model. This makes it possible to place items such as text and arrows into a separate layer without modifying the underlying image. One of the problems that remains with paint programs is that the resolution of text layers is still tied to the resolution of the original image. This requires that the image resolution be increased to high resolution before overlaying text; otherwise, the text resolution is too low to be useful for publication.

Another class of graphics programs is the drawing program. These programs use mathematical descriptors for generating graphics. For example, a circle is not a pattern of darker colored pixels on a light array, but a formula for a circle, given a centroid, a radius, and a color. A text object refers to a font, which is a table of formulas that describe the curves of the font, the size, the color, and the location. This enables drawing programs like Macromedia Freehand and

Adobe Illustrator to define objects of essentially infinite resolution with small file sizes, because the printing or display of the graphics of any size is merely a matter of scaling the math. These vector-based drawing programs are preferable for line drawings and labeling figures; however, many of the drawing programs are clumsy at handling paint objects such as images within them.

The class of programs that I prefer for actually creating final artwork are programs that integrate drawing and painting. Programs like Deneba Canvas and Corel Draw provide tools for working around and with paint objects such as images and may even include some of the functions of programs like Photoshop. In addition, they have excellent drawing and text tools, like the better drawing programs. In this way, users can use vector objects such as line drawings, arrows, and text and get the advantage of scalability without having to change the resolution of the images.

E. Slides and Transparencies

Because 35-mm slides and transparencies use similar color schemes that a computer monitor does, generating 35-mm slides is generally pretty easy. Most color film recorders work on a relatively simple principle. The driver software analyzes each image to be recorded for color content (red, green, and blue). The film recorder itself is a high-resolution white phosphor monitor with a filter wheel between the camera lens and the white phosphor screen. The driver directs the film recorder to place the red filter in front of the camera lens, the shutter is opened, and the red portion of the image is displayed on the screen. This is then repeated for the green and blue portion of the image. The net effect is a very high-resolution image recorded onto film with better color registration than could be achieved by using a color monitor. This relatively simple process can easily produce film images that surpass that of any printer. Film recorders of this type are available from Lasergraphics, Polaroid, Agfa, and many other manufacturers.

F. Printing Images

One of the greatest potential sources of frustration in setting up a fluorescence imaging facility is the difficulty of getting publication-quality prints from digital images. Part of the difficulty can be associated with learning a new graphics program and dealing with file formats. In most cases, however, the greatest difficulty is associated with trying to understand the subtleties of color image formation on a computer screen, on slide film, and on a printed page. The difficulty is created by a fundamental difference in the way color is created on light-based imaging, such as computer screens and transparencies, and the way it is crated on absorptive-based imaging systems, such as printing. Absorptive imaging systems are said to be subtractive. This is the system were taught in grade school. It starts with the primary colors. All of the other colors can be created by combining the primary colors, and the sum of all of the colors is black. On the other hand, illumination-based systems such as slide transparencies and computer screens are said to be additive. The colors add to each other, and the sum of all colors together is white. This fundamental difference between the two causes confusion and makes it difficult to go from one imaging scheme to the other. This is one of the main sources for the difference between the way an image looks on a computer screen and the way it prints. The other source of the difference between what is on the computer screen and what prints is related to the monitor. Each monitor manufacturer uses slightly different phosphors to generate colors on the screen. This produces differences in the actual color generated on the screen. Calibration systems can now be purchased that greatly aid in normalizing the print of an image to a given computer display. The most difficult colors to print are the pure colors, particularly red and blue. Part of the problem with printing pure red and blue is that common sources of lighting do not have much red or blue in them. For example, incandescent lights have almost no blue in them. When you look for a pure blue image against a black background, you are really looking at an area that absorbs red and green light and reflects back only blue light against a background of areas that absorb red, green, and blue. If the light that you are looking at the picture with contains no blue, then there is no difference between the black and blue areas. Similarly, some fluorescent lights produce virtually no red light. Under these lighting conditions, it is very difficult to see red against a black background. One method that I have employed to overcome this problem is to make use of the green light that is almost always in the lighting regardless of its source. This has the further advantage that the human visual system is most sensitive to green light (546 nm). If some green color is added to the pure blue, it shifts the printed color closer to cyan. If green is added to red, it shifts it closer to orange. In both cases, the resulting colors print against a black background in a way that it easier to see. The easiest way that I have found to generate these intermediate colors is to use a program like Adobe Photoshop. Photoshop has utilities for performing math between images, including between color channels of the same image. Using the Image . . . Calculations . . . command, one can take the blue channel and add it to the green channel. I have found the best amount to add to be about 25% for the printers that I normally use. In Fig. 1 (see color plate), the image on the left is the original image. Using Photoshop 4.0, the image on the right has had all of the color ranges adjusted and the 25% of the blue channel was added to the green channel.

For printing images for publication, I prefer to use a dye-sublimation printer, such as the Tektronix Phaser 440. Dye-sublimation printers use a dye ribbons that have inks on them (usually cyan, magenta, and yellow). The dyes are formulated such that they sublimate from solid to gaseous states upon heating. The amount of dye that sublimates is related to how high above the sublimation temperature a point is heated. The inks have very high affinities for esters such as polyester. Dye-sublimation printers then use papers that have a polyester layer, usually under a semiporous layer. As these papers are drawn close to a given ink in the gaseous phase, the ink passes through the pores and adheres to the polyester layer. By controlling the temperature at each individual point on the image (300 to 600 dots per inch) a carefully controlled amount of ink can be deposited on the paper. By successively laying down the cyan, magenta, and yellow dyes, each dot on the paper can be printed in one of several million of colors. This enables these printers to generate prints that are almost continuous tones without dithering patterns. These prints can be rescanned if necessary by the publisher for creating the final galleys for publication. Many publishers are now permitting authors to submit images in digital form. This can greatly reduce the requirements for printing in the lab; however, it is still recommended that authors submit at least one copy of the figures as they want them printed so that publishers can color balance to what the author had in mind. The process from digital image to publication is continually evolving, and it is much improved over what existed even 5 years ago.

VII. Conclusions

The fluorescence imaging laboratory best functions as a carefully planned system of tools for optimal image generation, display, analysis, and output. These systems are supported by the physical plant in which the lab exists. The interactions of these tools within the laboratory will largely determine the success of the facility. Any weak component will drive the quality of the final product and failure to attend to the details of the facility will lead to frustration on the part of the users. Certainly few facilities exists that have every imaging tool. Fewer still are the facilities that have had the luxury of designing the laboratory at one time. Most facilities have evolved from a series of components that have been carefully assembled and fought for by the hard work of the facility director or manager. Because every component of the laboratory is expensive, each piece must be carefully considered and thought of in terms of its relationship to the other components in the lab and, even more important, in light of the scientific questions that need to be answered. Any facility, no matter how well equipped, is of little actual value if it fails to perform and solve real scientific problems. When the tools of the imaging laboratory are carefully chosen in light of the science as well as their interactions with the entire system, then science is enabled and the very presence of the imaging laboratory becomes almost a nonissue. When scientists are constantly up against inappropriate instrumentation or have to struggle with systems that are not carefully integrated with each other, then the science is hindered by the imaging laboratory. In many ways, the goal of the imaging facility is to so enable the science while not become an ends of itself. The mark of a great imaging laboratory is the science that is enabled and not the assembling of technology.

Above all, the successful imaging laboratory is not the assembly of pieces and technology; rather, it is gifted individuals that manage, maintain, and develop the facility. No system can overcome severe weaknesses in the staff of the facility and many clever scientist and technologists have succeeded in taking barely adequate facilities and converted them in to vital resources by their determination, skill, and hard work. It is the combination of appropriate technology, carefully integrated by gifted scientists, and directed toward relevant scientific inquiry that brings about the realization of Zworykin's dream of seeing beyond what the eye can see into the mechanisms of life itself.

References

Agard, D. A. (1984). Optical sectioning microscopy: cellular architecture in three dimensions. *Annu. Rev. Biophys. Bioeng.* **13,** 191–219.

Agard, D. A., Hiraoka, Y., Shaw, P. J., and Sedat, J. W. (1989). Fluorescence microscopy in three dimensions. *Methods Cell Biol.* **30,** 353–377.

Agard, D. A., and Sedat, J. S. (1983). Three-dimensional architecture of a polytene nucleus. *Nature* **302,** 676–681.

Bell, A. J., and Sejnowski, T. J. (1995). An information-maximization approach to blind separation and blind deconvolution. *Neural Computation* **7,** 1129–1159.

Boyde, A. (1985). Stereoscopic images in confocal (tandem scanning) microscopy. *Science* **230,** 1270–1272.

Brewster, D. (1835). "Treatise on Optics," 2nd American ed. Philadelphia, Eavey, Lea, & Blanchard.

Coggins, J. M., Fay, F. S., and Fogarty, K. E. (1986). Development and application of a three-dimensional artificial visual system. *Computer Methods Programs Biomed.* **22,** 69–77.

Denk, W., Strickler, J. H., and Webb, W. W. (1990). Two-photon laser scanning fluorescence microscopy. *Science* **248,** 73–76.

Fay, F. S., Fagarty, K. E., and Coggins, J. M. (1986). Analysis of molecular distribution in single cells using a digital imaging microscope. *Soc. Gen. Physiol. Ser.* **40,** 51–63.

Goodman, J. W. (1968). "Introduction to Fourier Optics." McGraw-Hill, New York.

Goodwin, P. C., and Brown, C. (1996). Wide-field deconvolution vs. confocal microscopy of living cells. *Scanning* **18,** 144–145.

Haines, H., and McAtamney, L. (1993). Applying ergonomics to improve microscopy work. *USA Microsc. Analy.* **July.**

Helander, M. G., Grossmith, E. J., and Prabhu, P. (1991). Planning and implementation of microscope work. *Appl. Ergon.* **22,** 36–42.

Hiraoka, Y., Sedat, J. W., and Agard, D. A. (1990). Determination of three-dimensional imaging properties of a light microscope system: partial confocal behavior in epifluorescence microscopy. *Biophys. J.* **57,** 325–333.

Holmes, T. J. (1992). Blind deconvolution of quantum limited coherent imagery: maximum likelihood approach. *J. Optical Soc. Am. A Optics Image Sci.* **9,** 1052–1061.

Htun, H., Barsony, J., Renyi, I., Gould, D. L., and Hager, G. L. (1996). Visualization of glucocorticoid receptor translocation and intranuclear organization in living cells with a green fluorescent protein chimera. *PNAS* **93,** 484–504.

Inoué, S., and Spring, K. R. (1997). "Video Microscopy: The Fundamentals." 2nd ed. Plenum Press, NY.

Kwan, S. L., and Humphreys, L. A. (1985). "Physical Stress Reduction of Microscopic Operators." Proceedings of the Human Factors Society, 29th Annual Meeting, Baltimore, MD.

Lemp, M. A., Dilly, P. N., and Boyde, A. (1985–1986). Tandem-scanning (confocal) microscopy of the full-thickness cornea. *Cornea* **4,** 205–209.

Minsky, M. (1957). U.S. Patent #3013467, Microscopy Apparatus.

Minsky, M. (1988). Memoir on inventing the confocal scanning microscope. *Scanning* **10,** 128–138.

Pawley, J. B. (Ed.) (1995). "Handbook of Biological Confocal Microscopy." 2nd ed. Plenum Press, NY.

Piston, D. W., Masters, B. R., and Webb, W. W. (1995). Three-dimensionally resolved NAD(P)H cellular metabolic redox imaging of the *in situ* cornea with two-photon excitation laser scanning microscopy. *J. Microsc.* **178,** 20–27.

Ross-Macdonald, P., Sheehan, A., Roeder, G. S., and Byder, M. (1997). A multipurpose transposon system for analyzing protein production, localization, and function in *Saccharomyces cerevisiae. PNAS* **94,** 190–195.

Scalettar, B. A., Swedlow, J. R., Sedat, J. W., and Agard, D. A. (1996). Dispersion, aberration and deconvolution in multi-wavelength fluorescence images. *J. Microsc.* **182,** 50–60.

Sheppard, C. J., and Wilson, T. (1981). The theory of the direct-view confocal microscope. *J. Microsc.* **124,** 107–117.

Shotton, D. M. (Ed). "Electronic Light Microscopy." John Wiley and Sons, Inc., 1992.

Vigers, G. P., Coue, M., and McIntosh, J. R. (1988). Fluorescent microtubules break up under illumination. *J. Cell Biol.* **107,** 1011–1124.

Wang, Y. L., and Taylor, D. L. (Eds.). (1989). "Fluorescence Microscopy in Living Cells. Methods in Cell Biology," vols. 29 and 30, San Diego, Academic Press.

White, J. G., Amos, W. B., and Fordham, M. (1987). An evaluation of confocal versus conventional imaging of biological structures by fluorescence light microscopy. *J. Cell Biol.* **105,** 41–48.

Willis, R. A. (1969). Designing a biological electron microscopy laboratory. Medical & Biological Illustration. **19(2):**82–88.

Valkenburg, J. A., Woldringh, C. L., Brakenhoff, G. J., van der Voot H. T., Nanninga, N. (1985). Confocal scanning light microscopy of the Escherichia coli nucleoid: comparison with phase-contrast and electron microscope images. Journal of Bacteriology. **161(2):**478–483.

Zworykin, V. K. (1934): The iconoscope—a modern version of the electric eye. Proc. IRE **22,** 16–32.

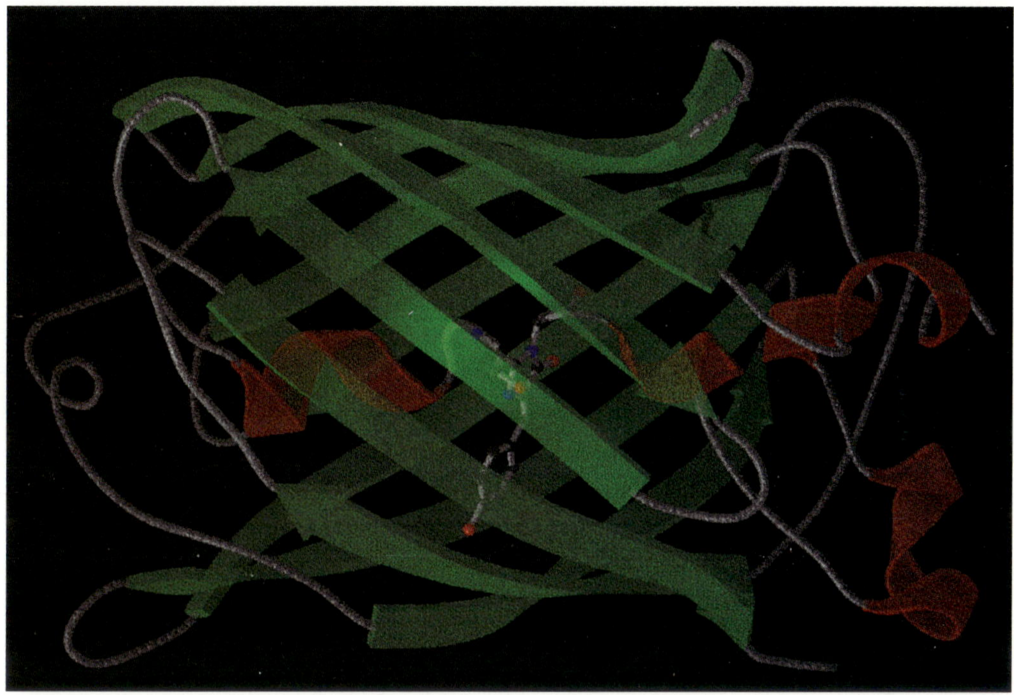

Ch.2, Fig. 1 Ribbon diagram of the wild-type GFP structure. The α-helices are shown in red, the β-strands are shown in green, and the chromophore is shown as a ball-and-stick model. The figure is reproduced with permission from Brejc *et al.* (1997).

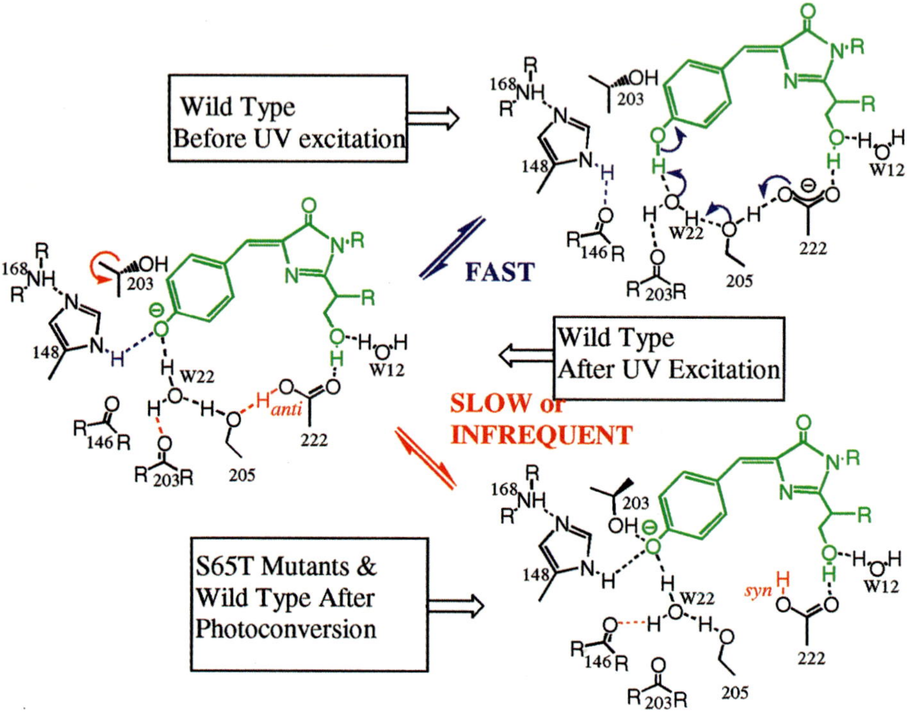

Ch.2, Fig. 2 Proposed mechanism for the photoisomerization of wild-type GFP based on the structural data and spectroscopic work (Chattoraj *et al.*, 1996). Rapid changes in structure caused by UV excitation are shown in blue. Slower changes associated with photoconversion are shown in red. The side chains of Arg-96 and Gln-94, which make hydrogen bonds with the carbonyl of the imidazolinone ring, do not show major conformational changes and are not shown in the figure. The figure is adapted from Brejc *et al.* (1997) with permission.

Ch.2, Fig. 3 Normalized relative brightness of GFP mutants. The excitation and emission spectra of the brightest and best folding GFP mutants are shown normalized for relative to the GFP mutant topaz (red). Spectra were calculated by multiplying normalized excitation and emission spectra by the quantum yield and molar extinction of each mutant. The relative brightness of each mutant was then normalized with respect to the brightest mutant topaz.

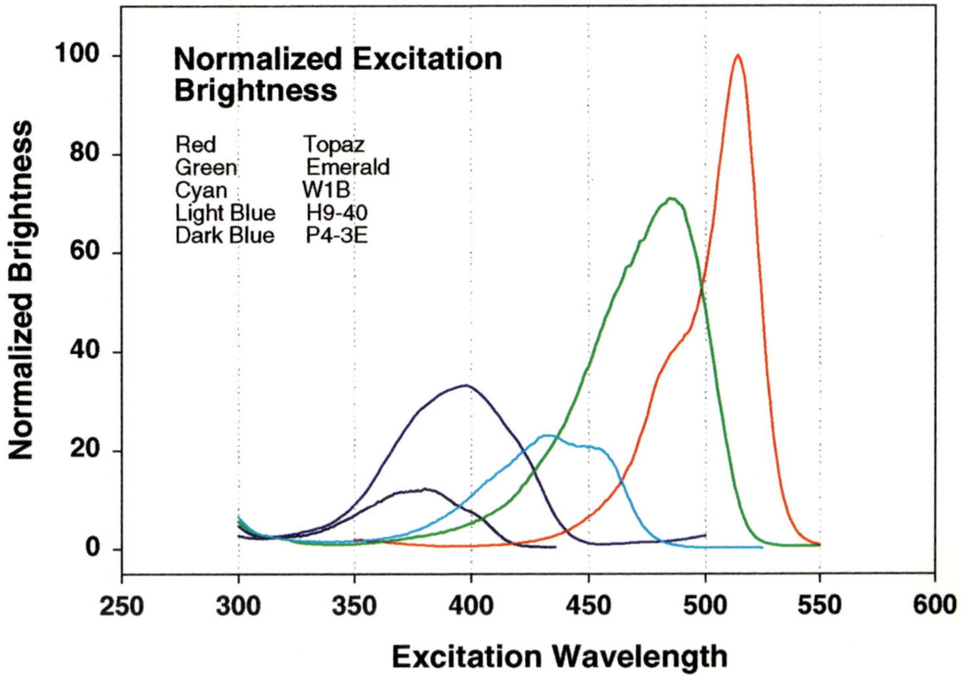
Normalized Excitation Brightness
Red Topaz
Green Emerald
Cyan W1B
Light Blue H9-40
Dark Blue P4-3E
Normalized Brightness
0
20
40
60
80
100
250
300
350
400
450
500
550
600
Excitation Wavelength

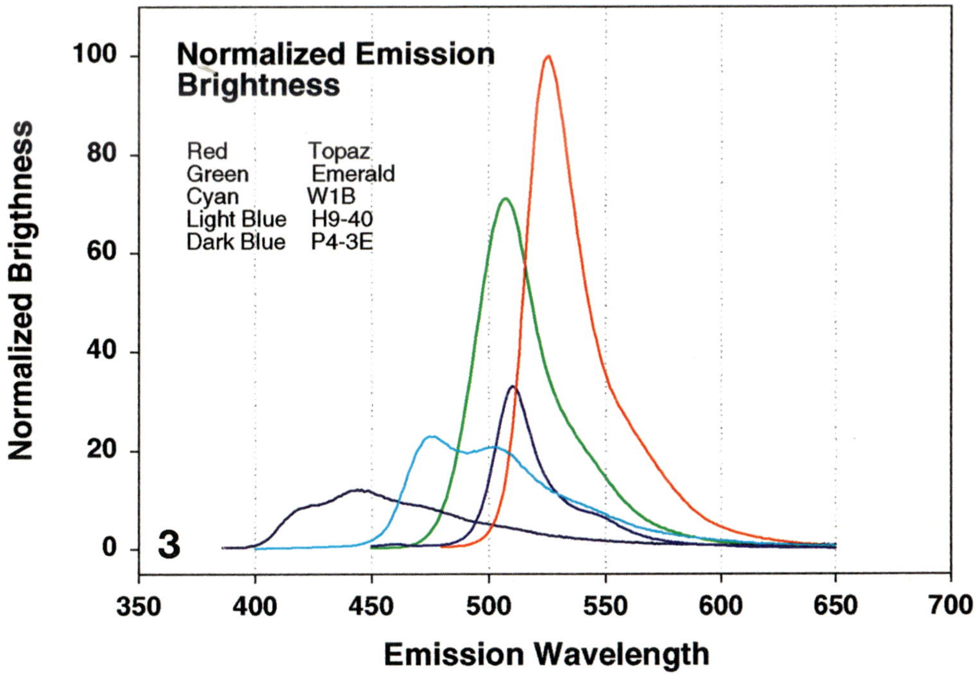
Normalized Emission Brightness
Red Topaz
Green Emerald
Cyan W1B
Light Blue H9-40
Dark Blue P4-3E
Normalized Brigthness
0
20
40
60
80
100
350
400
450
500
550
600
650
700
Emission Wavelength
3

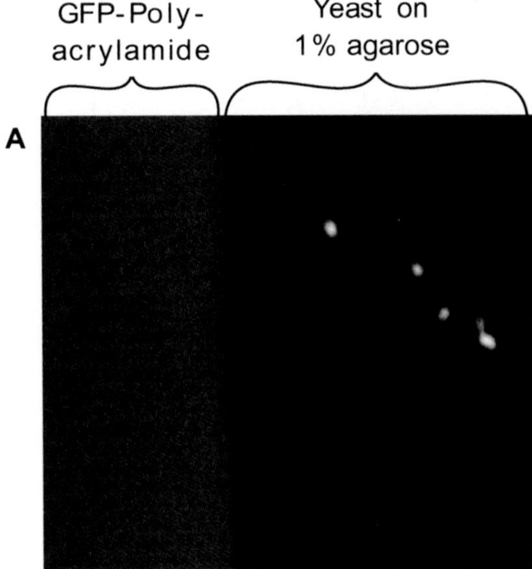

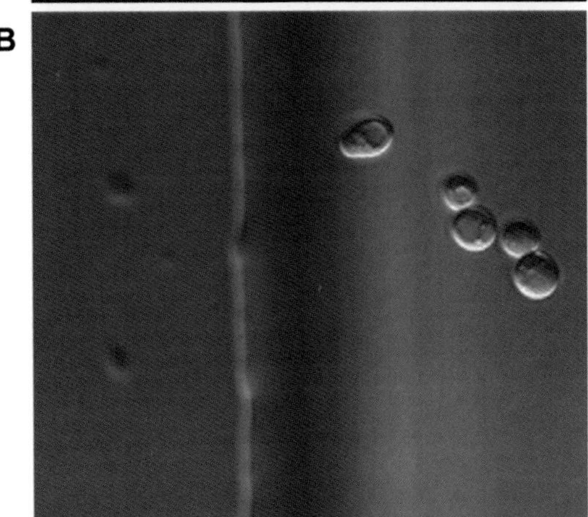

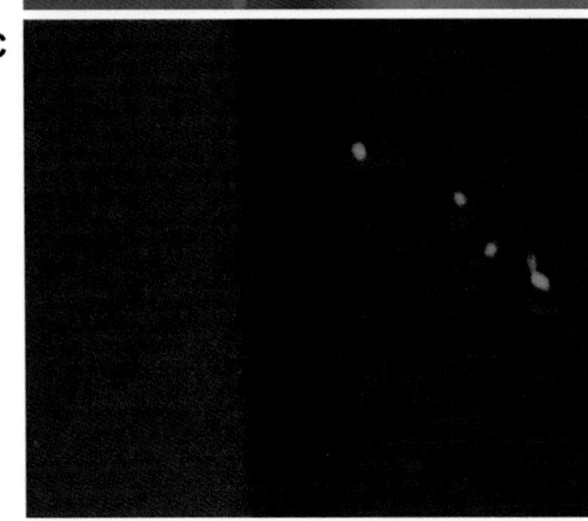

Ch.3, Fig. 2 Optical layout of the TIR microscope constructed in this laboratory. This diagram is intended for schematic purposes only, and not all the fold mirrors for the illumination pathway are shown; all optics present between the sample and the camera are indicated. The objective lens, telen lens, dichroic mirror 2, and barrier filter are housed within a Nikon nosepiece and epifluorescence attachment; the binocular head and lens L2 are from a Zeiss infinity-corrected inverted microscope. All other components are mounted by standard or custom optical bench mounts. The sample is held on a Leitz X–Y stage using a custom-made low-profile slide holder so as not to constrain motion of the prism relative to the slide; the prism, slide, and stage are moved as a unit by the focus micrometer (Newport ESA-CSA electrostrictive actuator). The lasers and ICCD camera are discussed in Appendix I. In brief, the other components are these:
Shutters: Newport 846 HP.
Interference filters: CVI Laser F10-490-4, F03514.5-4, and F10-632.8-4 for the 488-, 514-, and 632-nm laser lines, respectively. 1/4 plates: for 632-nm, CVI laser QWPM-632.8-05-4; for 488-, and 514-nm, CVO QWP0-514.5-05-4 (this is a zero-order waveplate with sufficient bandwidth to circularly polarize both 488-, and 514-nm light).
ND 1.0 filter: Melles–Griot, quartz substrate.
Dichroic Mirror 1: Chroma 530 DCLP.
ND Filter Wheel: New Focus 5214A.
Mirrors 1 and 2 (and other mirrors not shown): New Focus 5101-VIS.
Power Meter: Melles–Griot 13PEM001.
Lens L1: 100-mm doublet achromat.
Prism: CVI laser custom fabricated from UV-grade fused silica with CVI BBar antireflection coating for 488–633 nm on the two diagonal surfaces only.
Slides: fused silica, from Matsunami Trading Company, Japan.
Objective lens: Nikon PlanApo100/1.4.
Lens L3: CVI laser-fused silica PLCX-25.4-257.5-425-675, with the last two numbers indicating the approximate range of the antireflection coating. The magnification can be changed (without changing the objective) by substitution of this lens and movement of the camera (mounted on an optical rail) to the new focal plane.
The telen lens is supplied with the Nikon nosepiece. The dichroic mirror 2 and barrier filter sets for GFP, Cy3, and Cy5 are from Chroma (see text and Fig. 6). Mirror 3 is a custom optic from CVI coated with visible-enhanced aluminum.

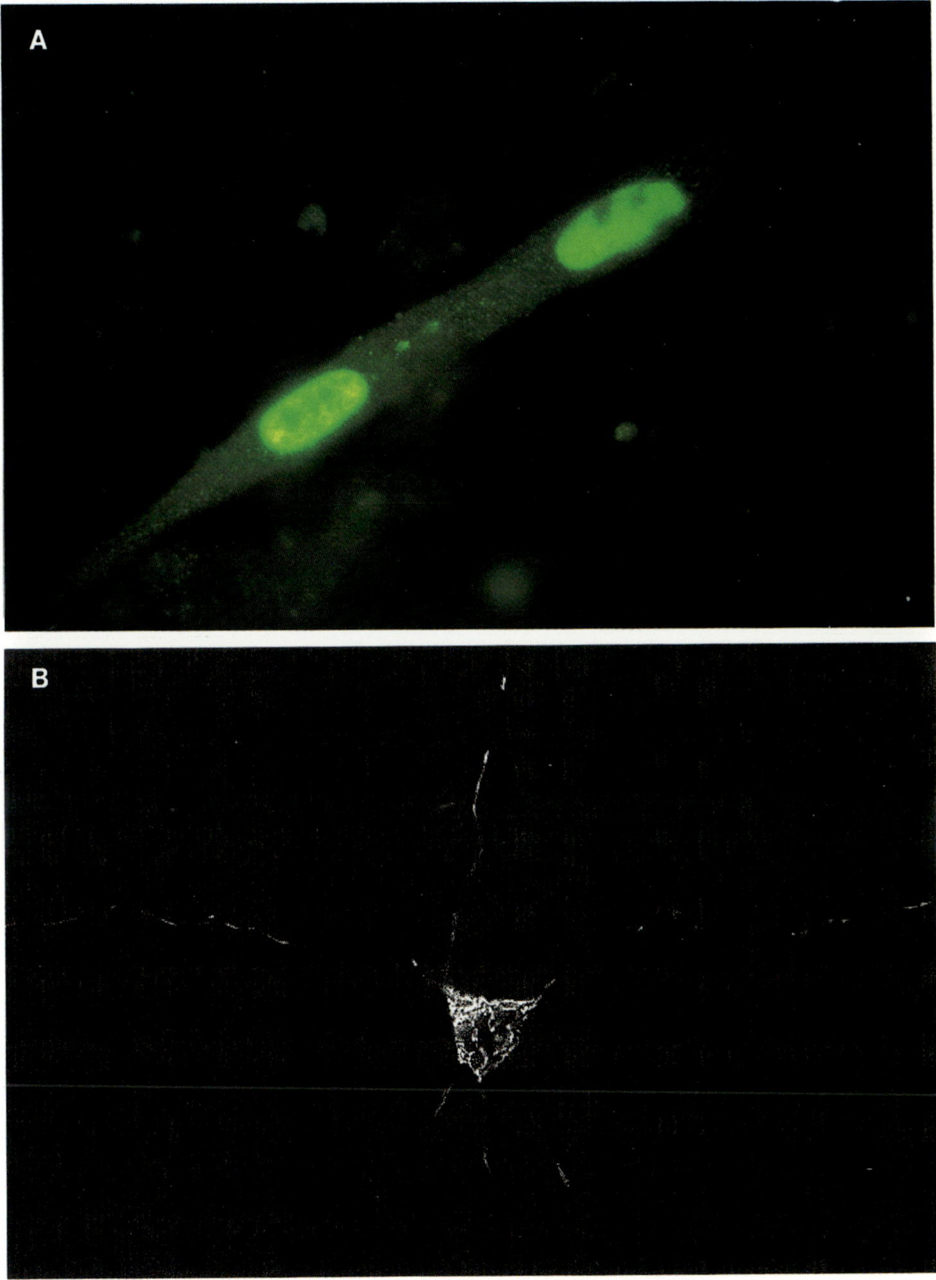

Ch.5, Fig. 4 Expression of organelle-targeted GFP in primary cultures: (A) rat skeletal myotubes transfected with nuGFP. Cells have been treated for 4 h with dexamethasone to induce the translocation of the chimera into the nucleus. (B) Cortical neuron expressing mtGFP.

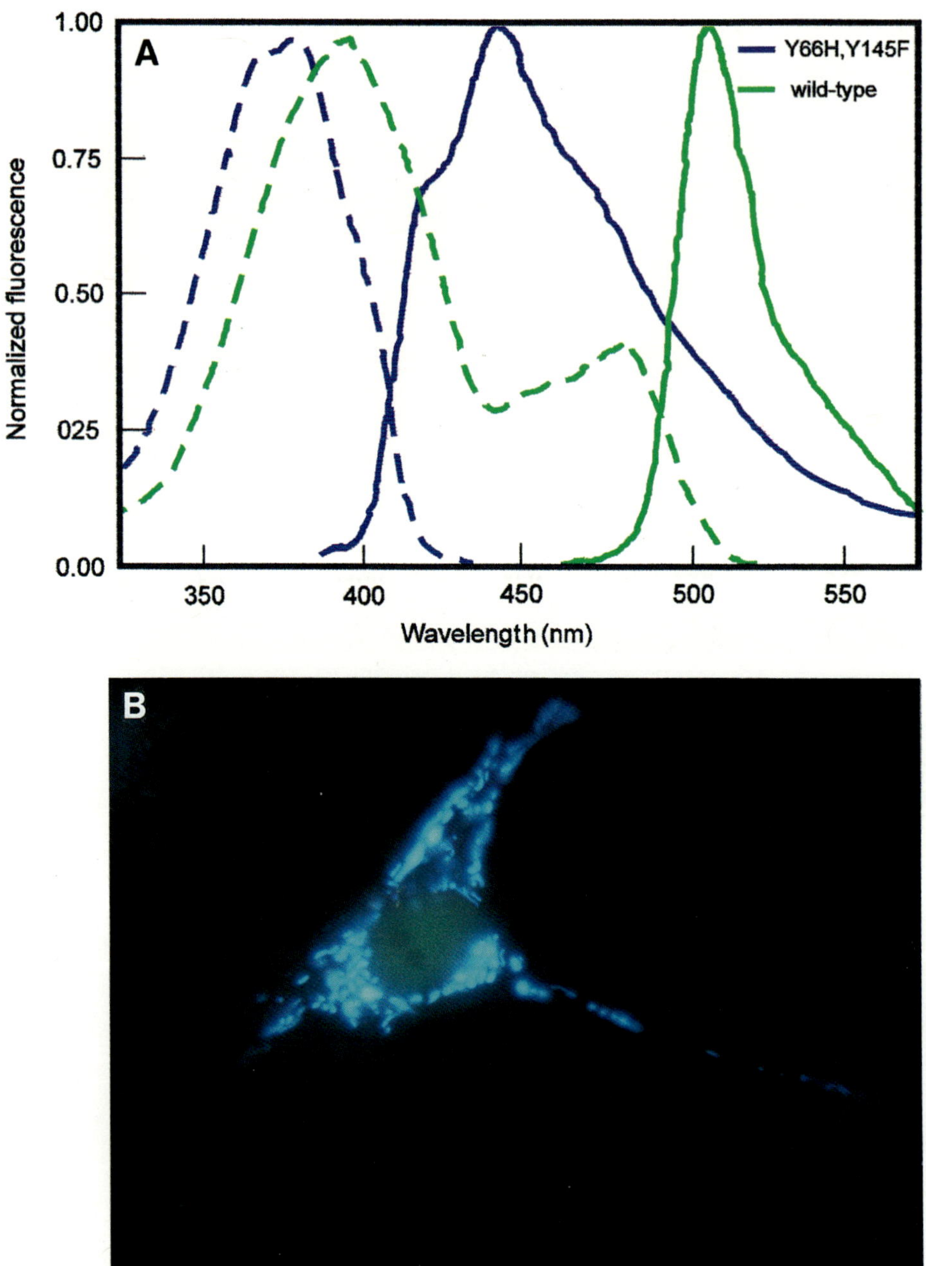

Ch.5, Fig. 5 (A) Excitation and emission spectra of GFPwt and GFP(Y66H, Y145F), respectively. (B) Cos7 cell cotransfected with nuGFP and mtGFP(Y66H, Y145F) and illuminated with UV light.

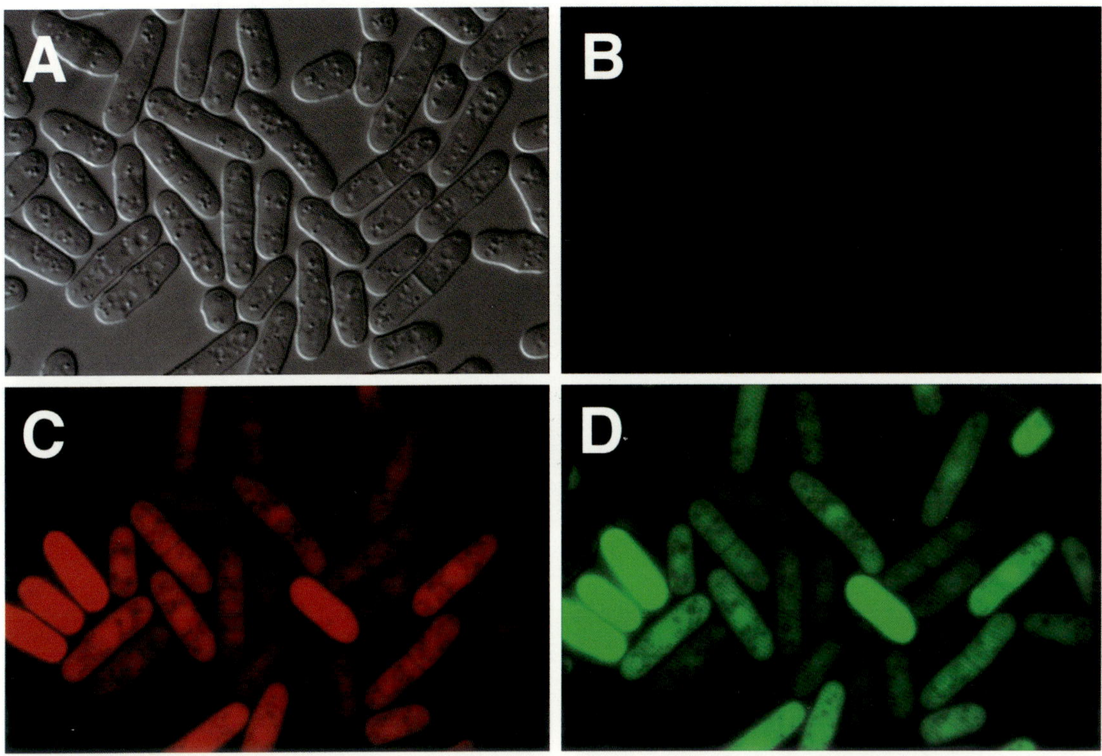

Ch.8, Fig. 5 Photoactivation of GFP. (A) DIC image before photoactivation (0.04-s exposure). (B) The same field in the Texas Red channel, before photoactivation (2-s exposure). (C) The same field in the Texas Red channel after 5-s photoactivation with blue light (2-s exposure). (D) The same field in the fluorescein channel, after photoactivation (0.5- exposure).

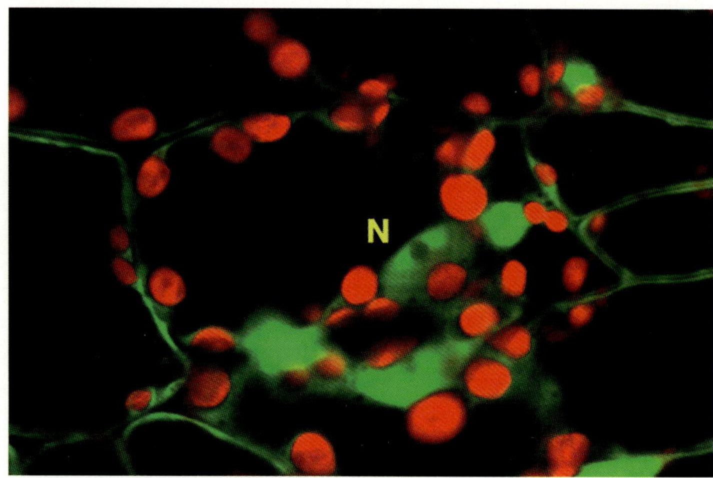

Ch.9, Fig. 1 Confocal optical sectioning of intact *Arabidopsis* plants. An *Arabidopsis* seedling that expressed a 35S promoter-driven GFP gene was subjected to microscopic examination. The image was collected using a Nikon Optiphot microscope equipped with a Biorad MRC-600 confocal scan head, and a Nikon planapo 60× (NA 1.2) water-immersion lens. GFP and chlorophyll were excited using the 488- and 568-nm lines, respectively, of a 25-nm krypton–argon ion laser. The green and red emissions were collected in separate channels and combined and pseudocolored using Adobe Photoshop. Individual cells and subcellular structure can be clearly resolved. Nuclear accumulation of GFP is seen in these cells (N), with exclusion from the nucleolus. The cytoplasm is pressed to the cell wall in these highly vacuolate cells. Chloroplasts are red autofluorescent.

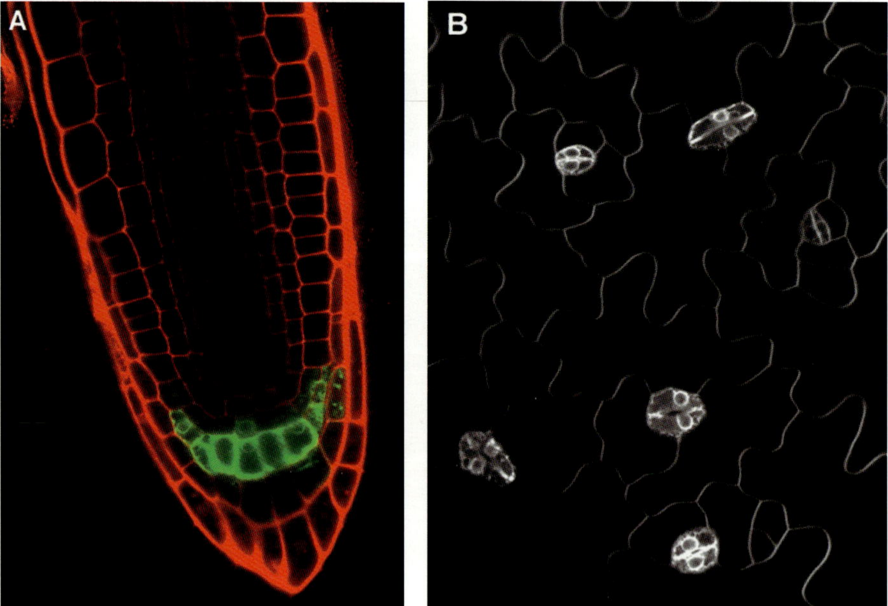

Ch.9, Fig. 2 Vital fluorescent dyes for counterstaining. (A) Transverse longitudinal section of a 5-day-old *Arabidopsis* root tip that has been stained with propidium iodide. Propidium iodide fluorescence is seen in the red channel, and reveals the outlines of cells in the root tip. GFP expression in this *Arabidopsis* transgenic line Q1630 (Sarah Hodge and J. Hasselhoff, unpublished results) is seen in part of the root cap, and is shown in the green channel. (B) Optical section through the epidermal layer of an FM 1-43 stained cotyledon of a 5-day-old seedling (Line J0991; J. Hasselhoff, unpublished results). FM 1-43 selectively stains the plasma membrane, to provide an outline of these cells. GFP fluorescence was collected in the same channel, and can be seen to effectively "fill" guard cells, which express the protein.

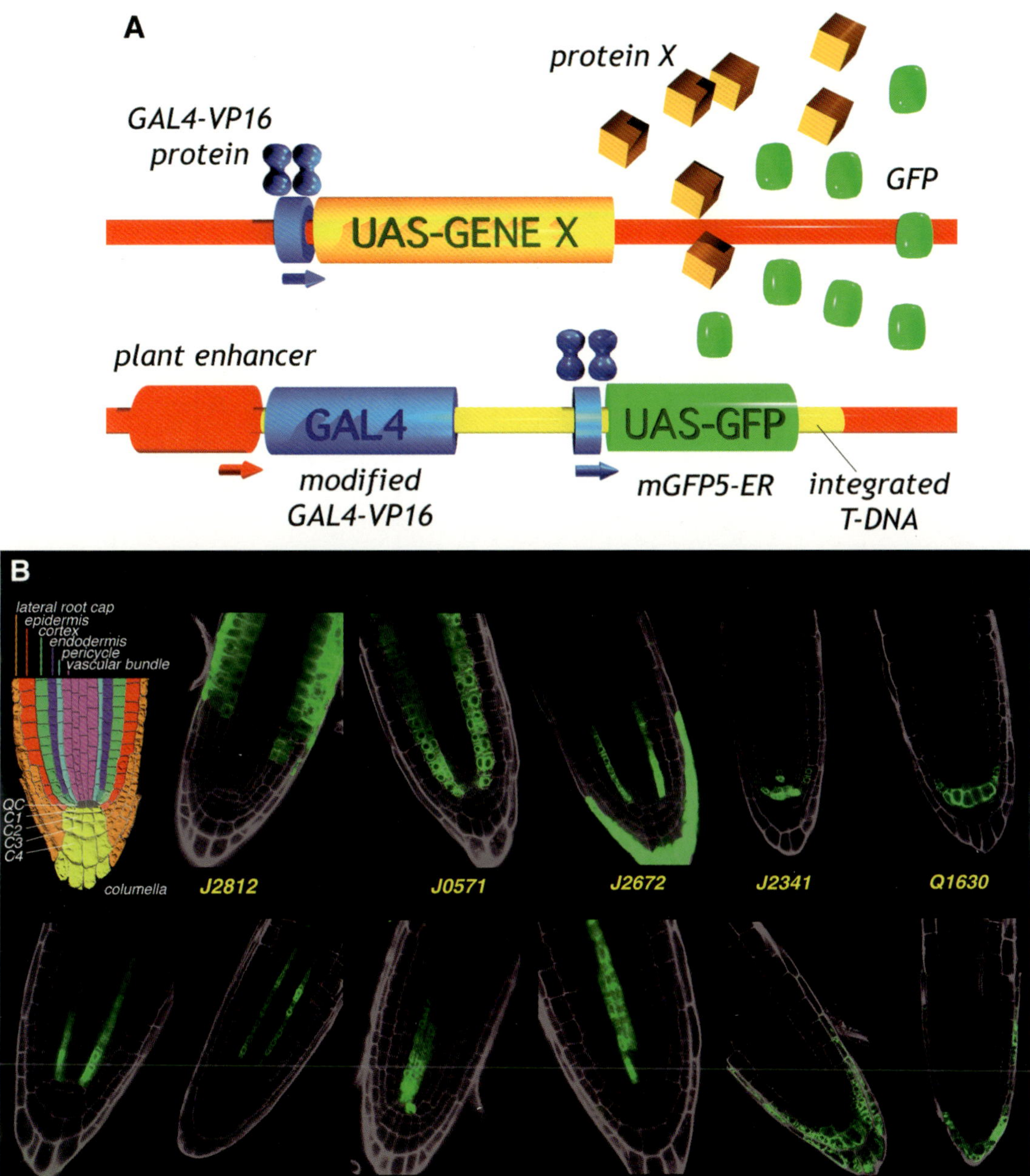

Ch.9, Fig. 3 *Arabidopsis* GAL4-VP16 enhancer trap lines. (A) *Agrobacterium*-mediated plant transformation was used to randomly integrate an engineered T-DNA vector into *Arabidopsis*. The vector contains a *GAL4-VP16* gene with modified codon usage adjacent to a naive promoter at a border of the transferred DNA, a kanamycin resistance selection marker and a GAL4-responsive *mGFP5-ER* gene. The modified GFP has improved fluorescent properties and is targeted to the endoplasmic reticulum. Cell-specific activation of the *GAL4-VP16* gene by a cellular enhancer results in the expression of the GFP marker gene, allowing the simple characterization of expression patterns. Targeted expression of another gene (X) can be induced by a genetic cross with a *GAL4-VP16* line. (B) Confocal micrographs of selected enhancer trap lines showing root-tip-specific expression. The roots were counterstained with propidium iodide, which outlines all live cells. The *GAL4-VP16*-driven GFP signal is shown superimposed. A schematic diagram of the different types of cell in the root tip is shown (upper left).

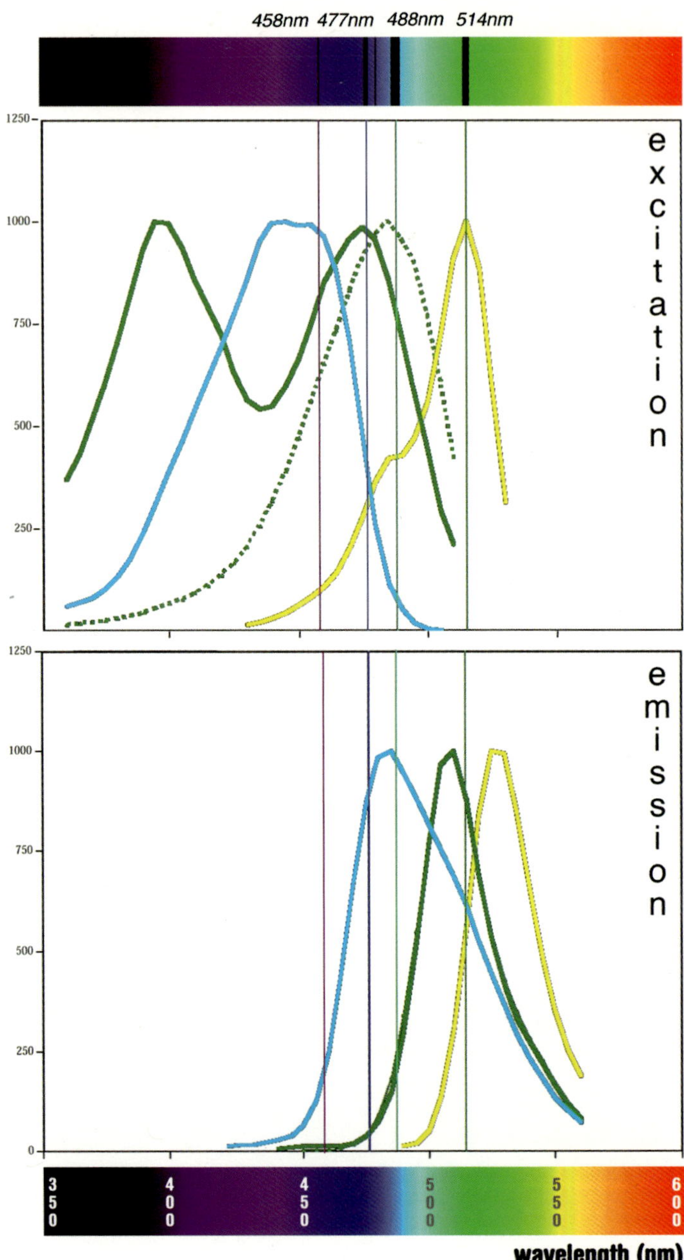

Ch.9, Fig. 4 Fluorescence spectra for variant GFPs. The excitation and emission spectra of cyan, green, and yellow fluorescent protein variants are indicated in their respective colors. The spectra of mGFP5 shown in solid green lines, and those of a green fluorescent variant containing the widely used S65T mutation (mGFP6; J.Hasselhoff, unpublished results) are shown as green dashed lines. The main laser lines for an argon-ion laser are indicated by vertical tracing.

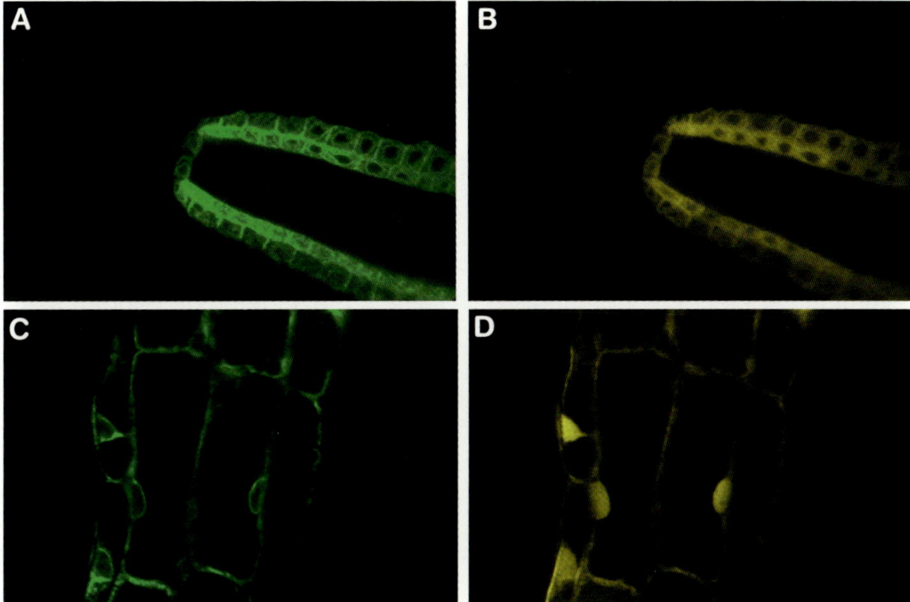

Ch.9, Fig. 5 Dual-channel confocal imaging of GFP variants. The GAL4-VP16 enhancer trap line J0571 was used to drive expression of a KNAT3–yellow fluorescent protein fusion in the endodermis and cortex of the *Arabidopsis* root. The KNAT3 fusion protein is localized in the cytoplasm of meristematic cells (B), but accumulates within the nuclei of older cells (D). The perinuclear and endomembrane distribution of ER-localized green fluorescent protein is also shown (A, C).

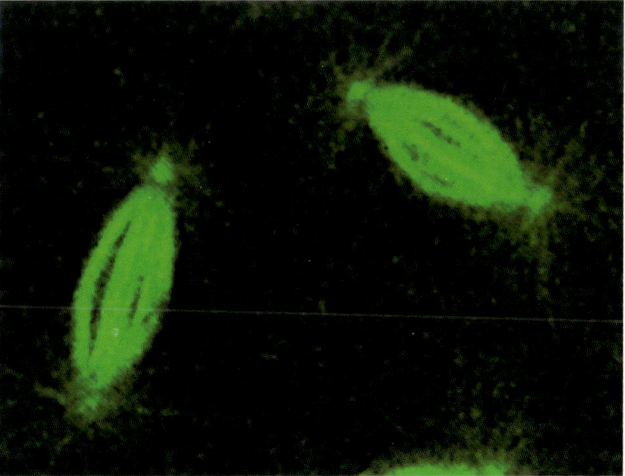

Ch.10, Fig. 2 Mitotic spindles labeled with Ncd-GFP. Ncd-GFP forms pole-to-pole fibers along the spindle microtubules, which are especially prominent in metaphase and early anaphase. The distribution of Ncd-GFP differs from that of tubulin during metaphase and early anaphase in that the fibers of Ncd-GFP extend across the chromosomes rather than terminating at the chromosomes positioned on or just moving off the metaphase plate. The spindles are in early anaphase of mitotic cycle 9. The centrosomes and astral microtubules are also labeled with Ncd-GFP.

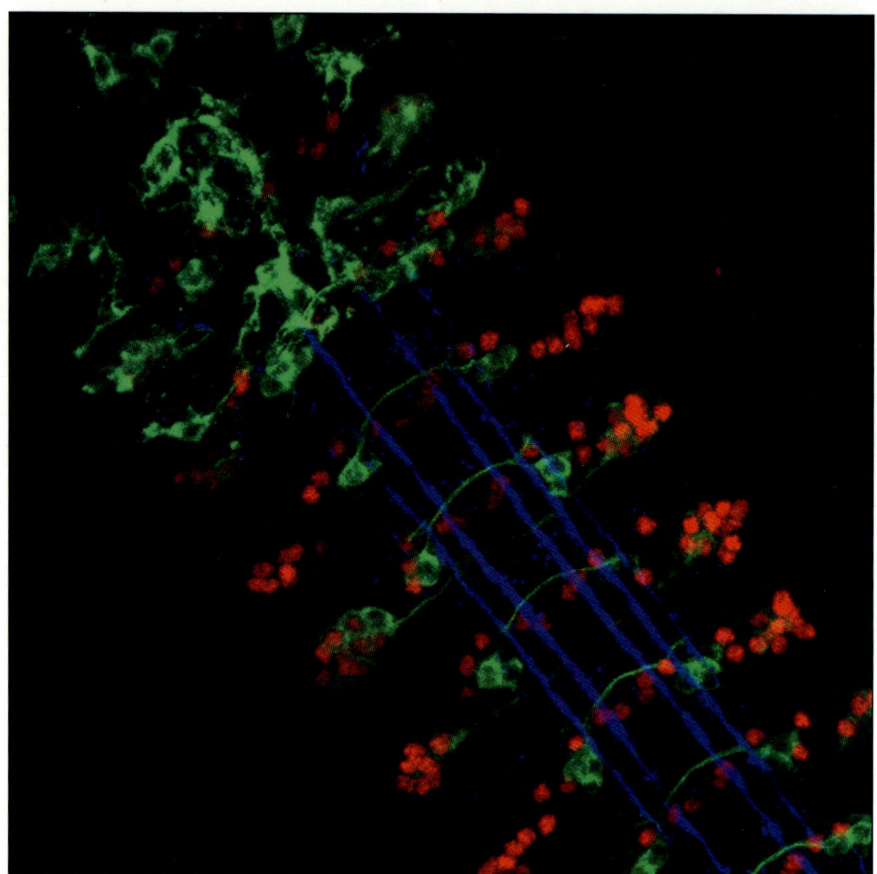

Chapter 11, Figure 5 Targeted expression of tau–GFP (green) labels individual neurons as they extend their axons in the central nervous system of living embryos. Tau–GFP continues to fluoresce after fixation, which allows immunolabeling of other proteins, such as Even-skipped (red) and Fasciclin II (blue).

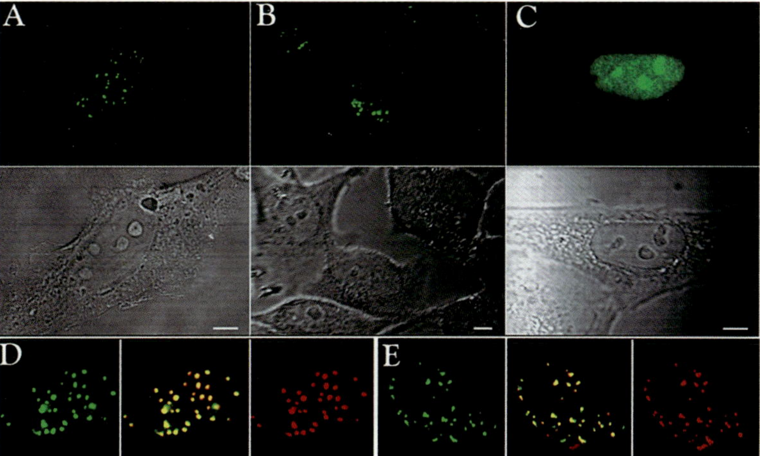

Ch.12, Fig. 1 Characterization of CENP-B–GFP in HeLa cells. Plasmids were transfected into cells using cationic lipids and visualized 48 h after transfection. (A) CENP-B–GFP in formaldehyde (4%)-fixed cells and (B) live HeLa cells; (C) CENP-B–GFP D4–16 in fixed cells. This experiment demonstrates that the localization of CENP-B–GFP is a function of its DNA-binding activity. Immunofluorescence using (D) human anti-centromere autoantibodies or (E) monoclonal mACA-1 against the COOH-terminus of CENP-B, which recognizes only endogenous CENP-B. In each panel the GFP signal (green) is on the left and the antibody signal, detected with a rhodamine-coupled secondary antibody (red), is on the right. The center panel shows the two signals superimposed to evaluate co-distribution. Reprinted with permission from Shelby *et al.* (1996).

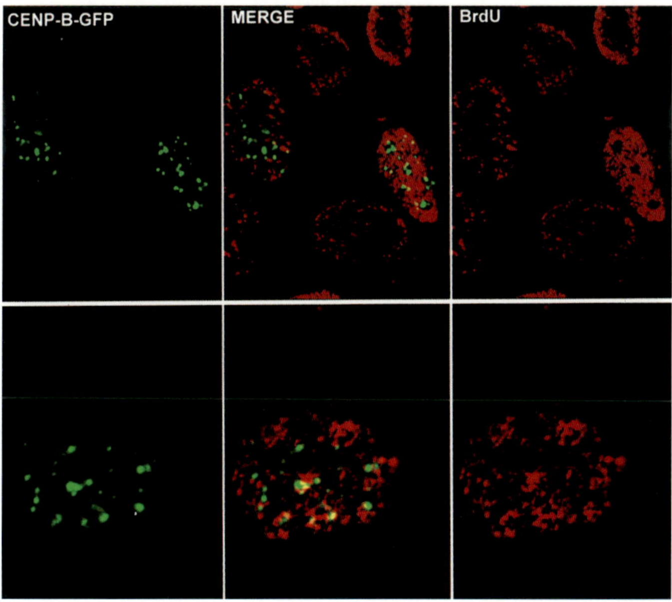

Ch.12, Fig. 2 Bromodeoxyuridine labeling in cells expressing CENP-B–GFP. HeLa cells were transfected with CENP-B–GFP as in Fig. 1. After 36 h, the cells were placed in 2 m*M* thymidine and incubated for 16 h. The cells were lableled with BrdU for 1 h at different times following release and then fixed in 70% formaldehyde, 50 m*M* glycine, pH 2.0, overnight at –20°C. The cells were then subjected to immunofluorescence using an anti-BrdU monoclonal antibody according to the manufacturer's instructions (Boehringer-Mannheim), and the antibody signal was detected with a rhodamine-coupled secondary antibody. GFP (green) fluorescence is on the left and the BrdU signal (red) is on the right, with a superimposition in the center.

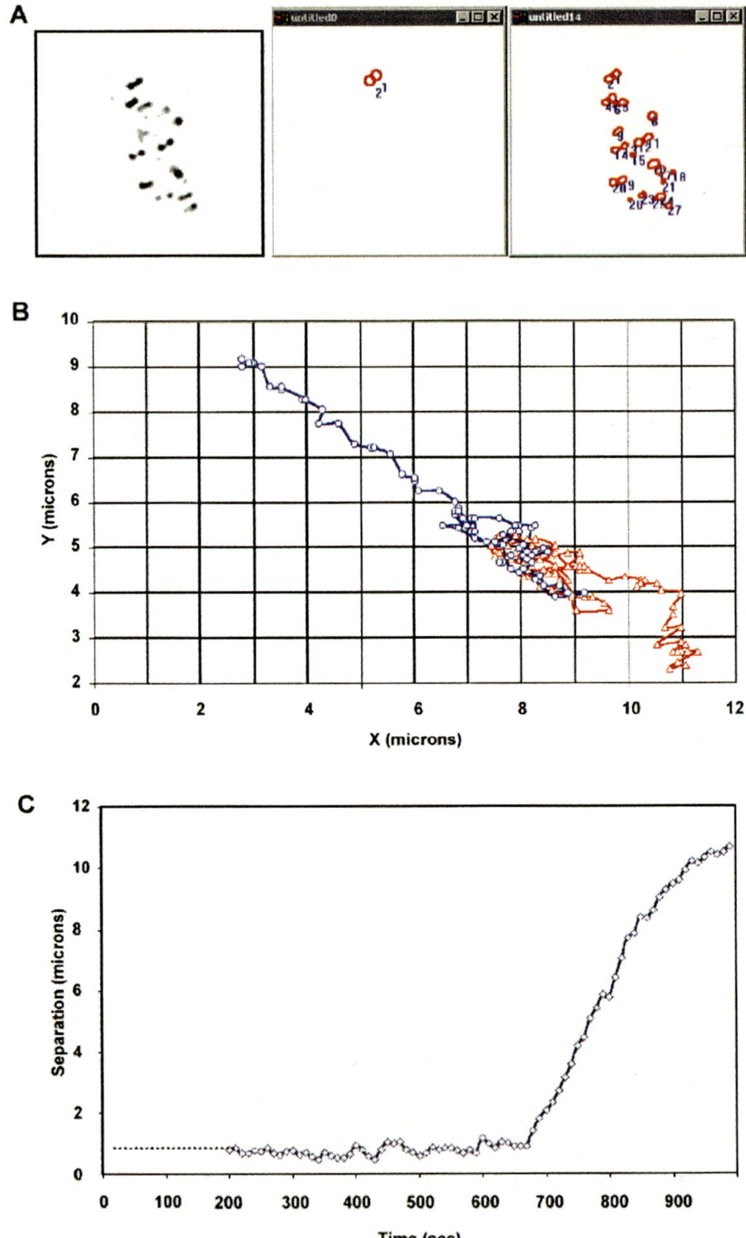

Ch.12, Fig. 3 Illustration of image analysis technique. (A) The left panel shows a single frame from a time-lapse image of a cell in mitosis. The image has been reversed for clarity. The two approaches for quantitating position discussed in the text are illustrated. Each panel shows the position of individual centromeres determined by the automatic object identification algorithm of the count/measure routine (center) or by the manual tag technique (left). Numerical data derived from the measurements are imported into a spreadsheet for analysis. (B) Raw motility data for a pair of sister centromeres. These data are replotted in (C) as the center–center distance between sisters.

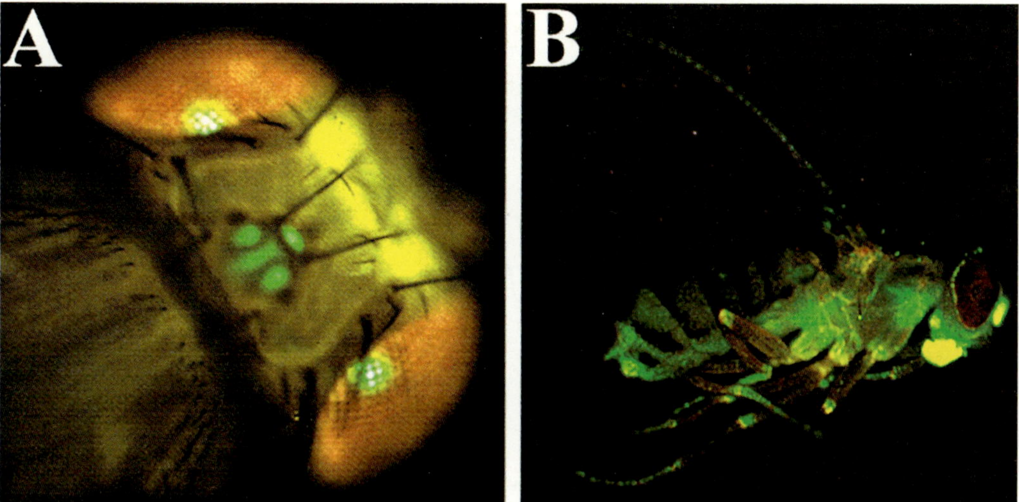

Ch.17, Fig. 1 GFP fluorescence in *Drosophila*. Variants of GFP can be detected in living adult *Drosophila*; confounding effects of autofluorescence are minimized with color detection. (A) Fluorescence from a *glass*-responsive element, which drives wild-type GFP in the eyes and ocelli. This image was captured with a 35-mm camera and a Chroma endogenous GFP filter cube. (B) A transgenic fly in which the *per* promoter indirectly drives S65T GFP. Note the fluorescence in the individual cells in the legs and wing. The image was captured with a Hamamatsu C5810 color camera, a Chroma HQ:GFP filter cube, and Inovision ISEE software.

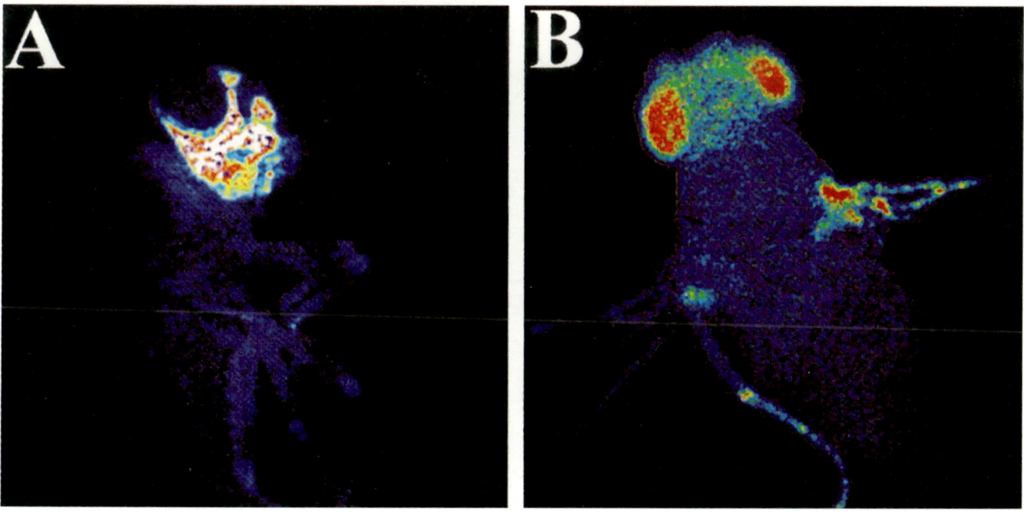

Ch.17, Fig. 2 Luciferase-driven bioluminescence in *Drosophila*. Luciferase-mediated bioluminescence can be visualized in living adult *Drosophila*. In these cases, bioluminescence is driven from transcriptional fusions using either the heat-shock (A) or *per* (B) promoter. Both images were captured with a Hamamatsu VIM camera system and pseudocolored using Adobe Photoshop.

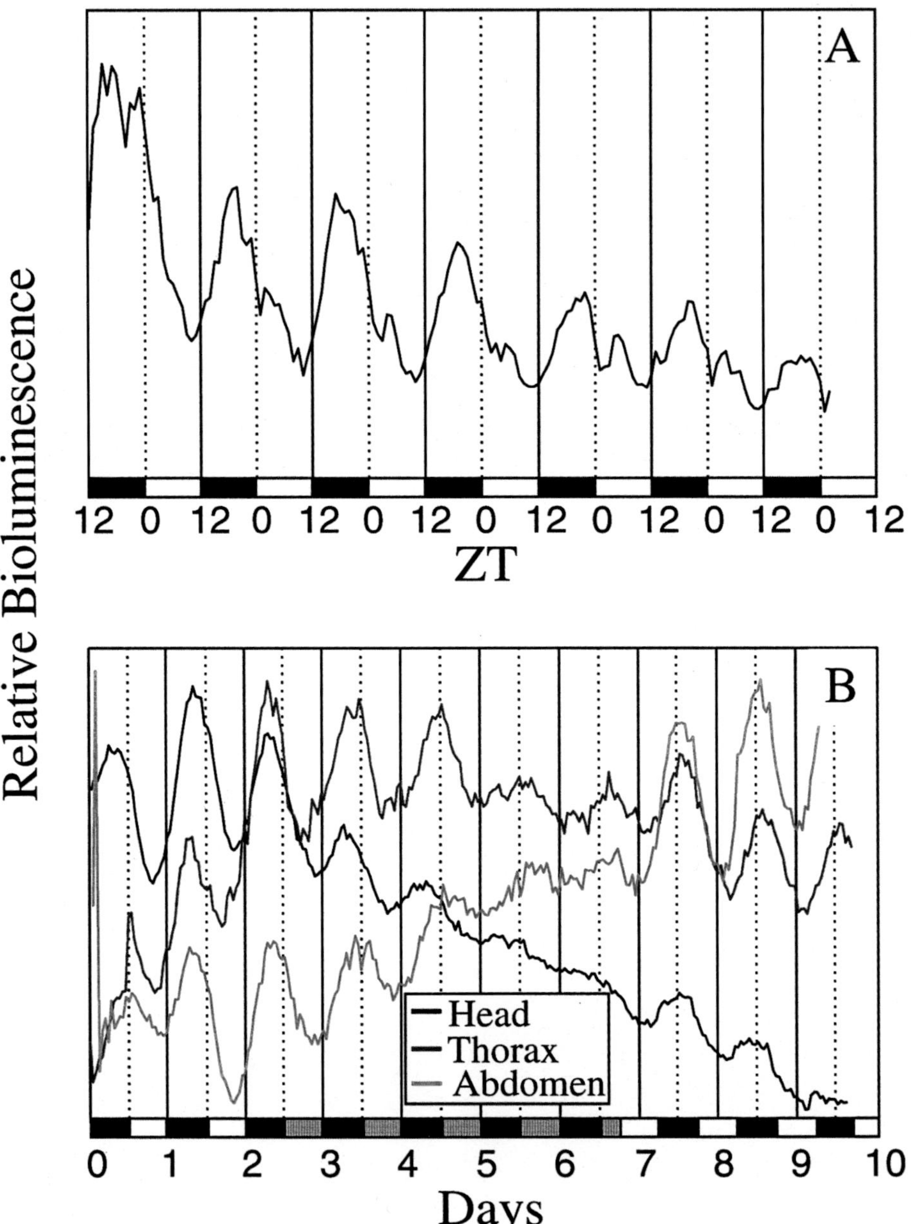

Ch.17, Fig. 3 Rhythmic bioluminescence in transgenic *per–luc Drosophila*. Bioluminescence can be quantitatively measured over several days in both living adult (A) or cultured pieces (B) of transgenic *per–luc Drosophila*. Note that the pieces in panel B show a period and phase similar to that of the whole animals in panel A. Also, the pieces are responsive to an imposed light cycle.

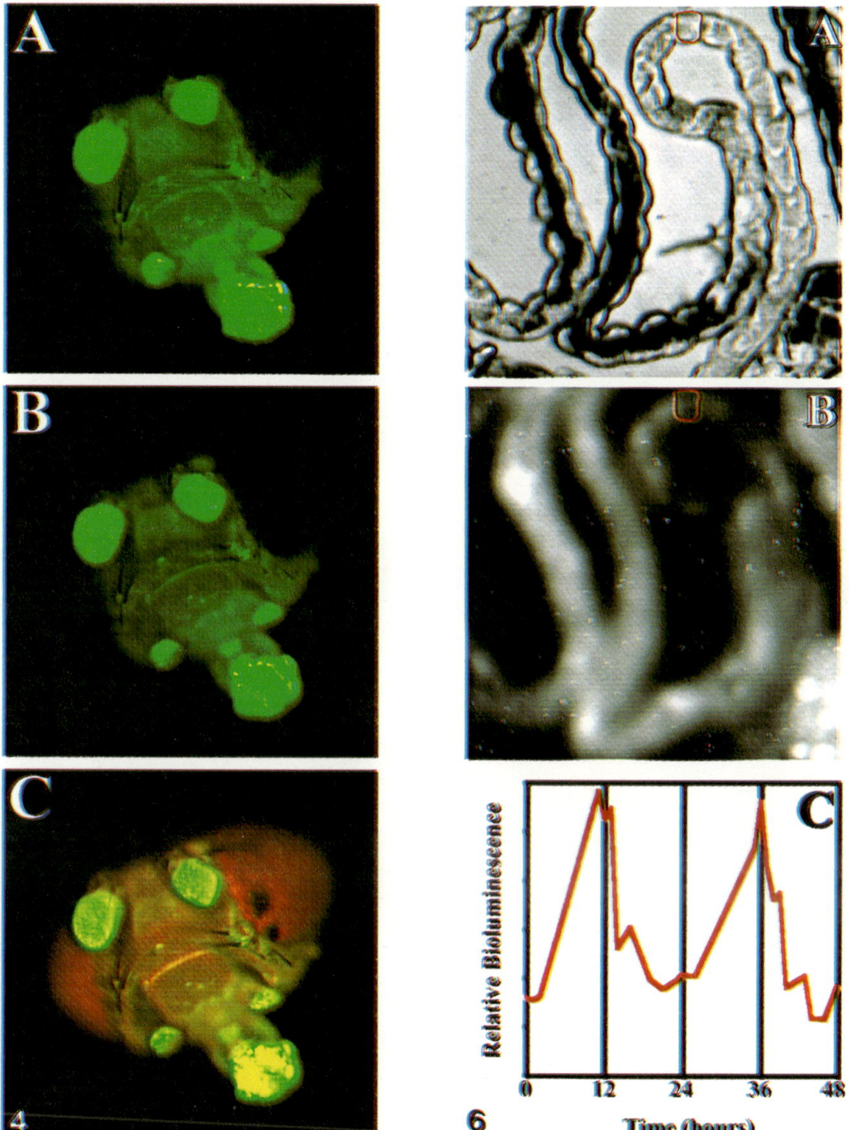

Ch.17, Fig. 4 Different filter sets for S65T GFP detection. The spectral characteristics of different filter sets determine their overall usefulness in fluorescence imaging. Here, three different, but similar, filter cubes are compared with a single sample emitting S65T GFP fluorescence (a *per*–GAL4; UAS–GFP adult *Drosophila* as described in Plautz *et al.*, 1997b). (A) Chroma HQ:FITC filter cube; (B) Chroma Endow GFP filter cube (optimized for detecting wild-type GFP); (C) Chroma HQ:GFP filter cube (optimized for detecting S65T GFP). In all cases, the images were exposed for 3 s; brightness was adjusted to the point where the antennae and proboscis were saturated to approximately the same extent (which was the same in panels A and B; panel C did not require as much adjustment).

Ch.17, Fig. 6 Bioluminescence can be quantitated in single cells. This figure shows data from a single biolumninescent *Drosophila* Malpighian tubule. (A) A transmitted-light image of a single Malpighian tubule (with a single cell outlined in red), (B) biolumninescence from the same tubule (again, with the same cell outlined in red). (C) Two cycles of quantitated bioluminescence; the average brightness of all the pixels in the area was determined at each timepoint, revealing a circadian rhythm in biolumninescence similar to that seen in whole animals or cultured body segments (see Fig. 3).

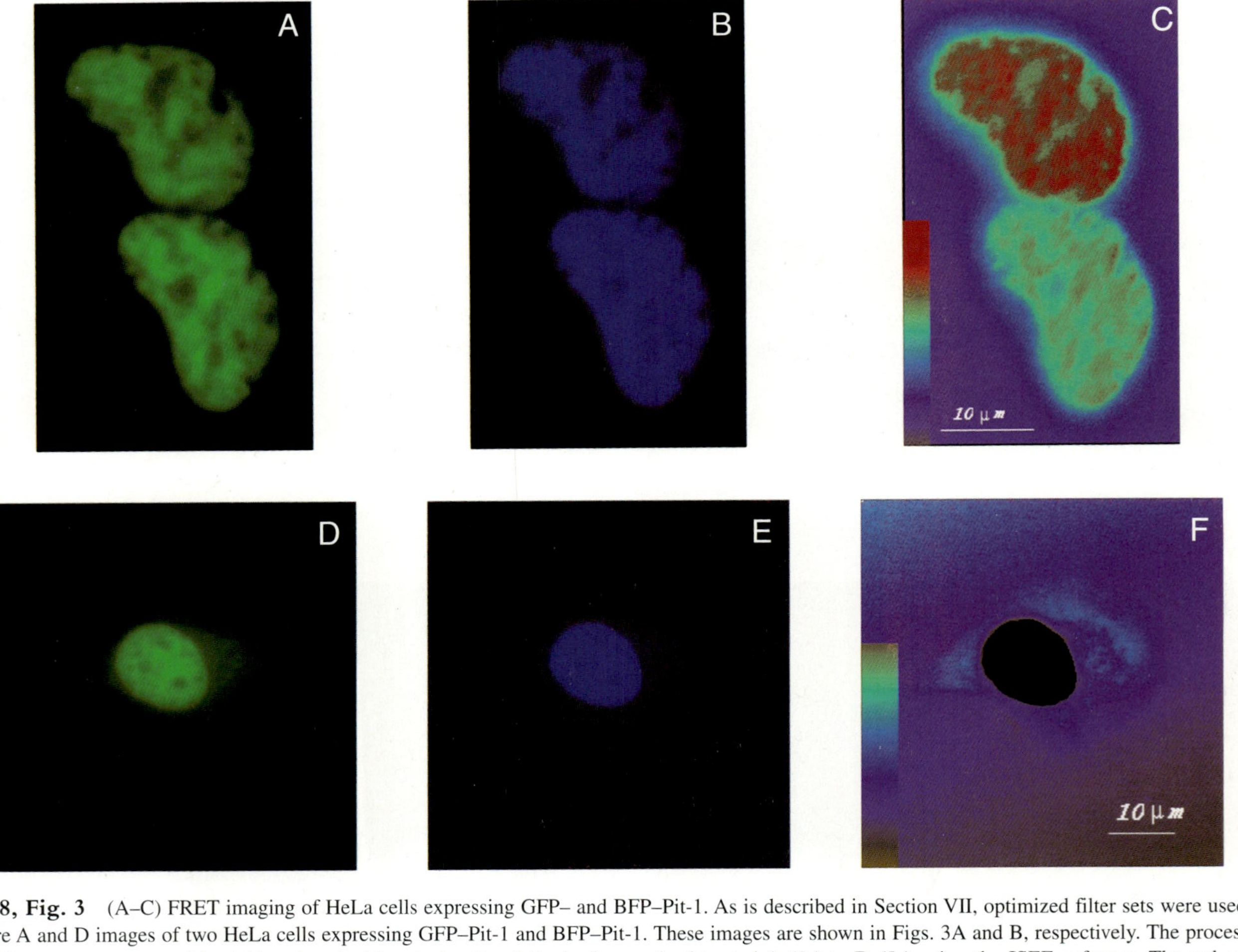

Ch.18, Fig. 3 (A–C) FRET imaging of HeLa cells expressing GFP– and BFP–Pit-1. As is described in Section VII, optimized filter sets were used to acquire A and D images of two HeLa cells expressing GFP–Pit-1 and BFP–Pit-1. These images are shown in Figs. 3A and B, respectively. The processed FRET image was generated by taking the pixel-by-pixel ratio of the background-subtracted A (I_A) to D (I_D) using the ISEE software. The red color indicates that the spatial distribution of the energy transfer signal in the nuclei of these cells at the focal plane and is consistent with energy transfer from BFP–Pit-1 to GFP–Pit-1. (D–F) FRET imaging of a HeLa cell expressing GFP–NLS and BFP–Pit-1. To demonstrate that simple colocalization of D and A fluorophores was not sufficient for FRET, I_A and I_D images were acquired from a single cell expressing nuclear localized GFP–NLS and BFP–Pit-1. Levels of the fluorophores in the nuclear compartment were similar to those shown above. The processed FRET image (I_A/I_D) is shown in Fig. 3F. These results demonstrate a decreased I_A/I_D ratio within the nuclear compartment, consistent with the absence of energy transfer between the noninteracting proteins.

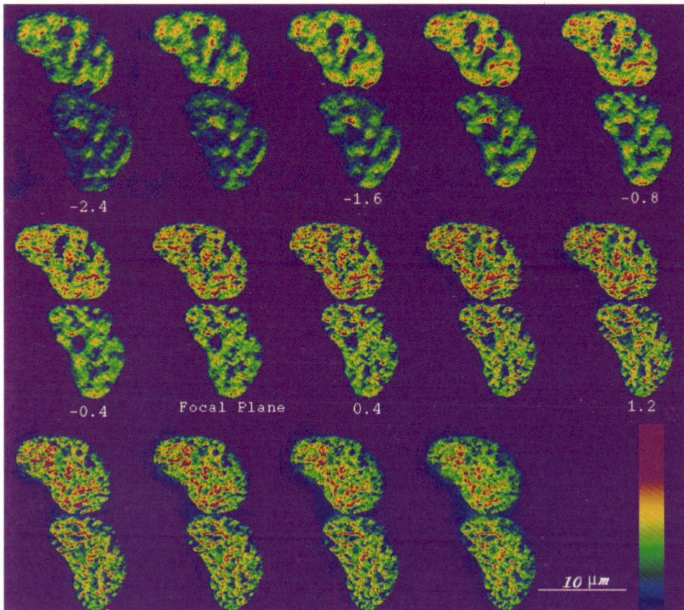

Ch.18, Fig. 4 Digital deconvolution of the FRET signal at different optical sections from cells expressing GFP–Pit-1 and BFP–Pit-1. The D and A images were acquired at the indicated optical sections using the computer driven stage and a motorized emission filter wheel. Fourteen optical sections were acquired at 0.4-µm steps starting at the bottom of the nucleus (–2.4 µm to 2.4 µm). These optical sections were then digitally deconvolved using DeltaVision software. When compared to Fig. 3C, the digitally deconvolved image at the focal plane provides much higher resolution of the FRET signals. Further, optical sectioning resolves a complex pattern of localized Pit-1 dimers within the two nuclei.

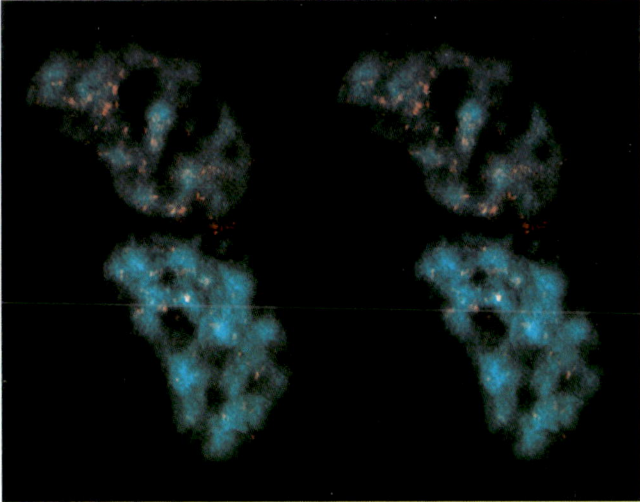

Ch.18, Fig. 5 Three-dimensional projection of the localized Pit-1 dimerization. Using the volume rendering software provided by the DeltaVision, the digitally deconvolved optical sections of A, D, and FRET were reconstructed to create a movie of 15 projections. One set of these projections (the same as that shown in Fig. 4) was selected, and the images were superimposed to generate a volume rendering of that optical section. The FRET signal from Pit-1 dimers (red or white color spots) can be localized at different depths within this volume view using 3-D glasses.

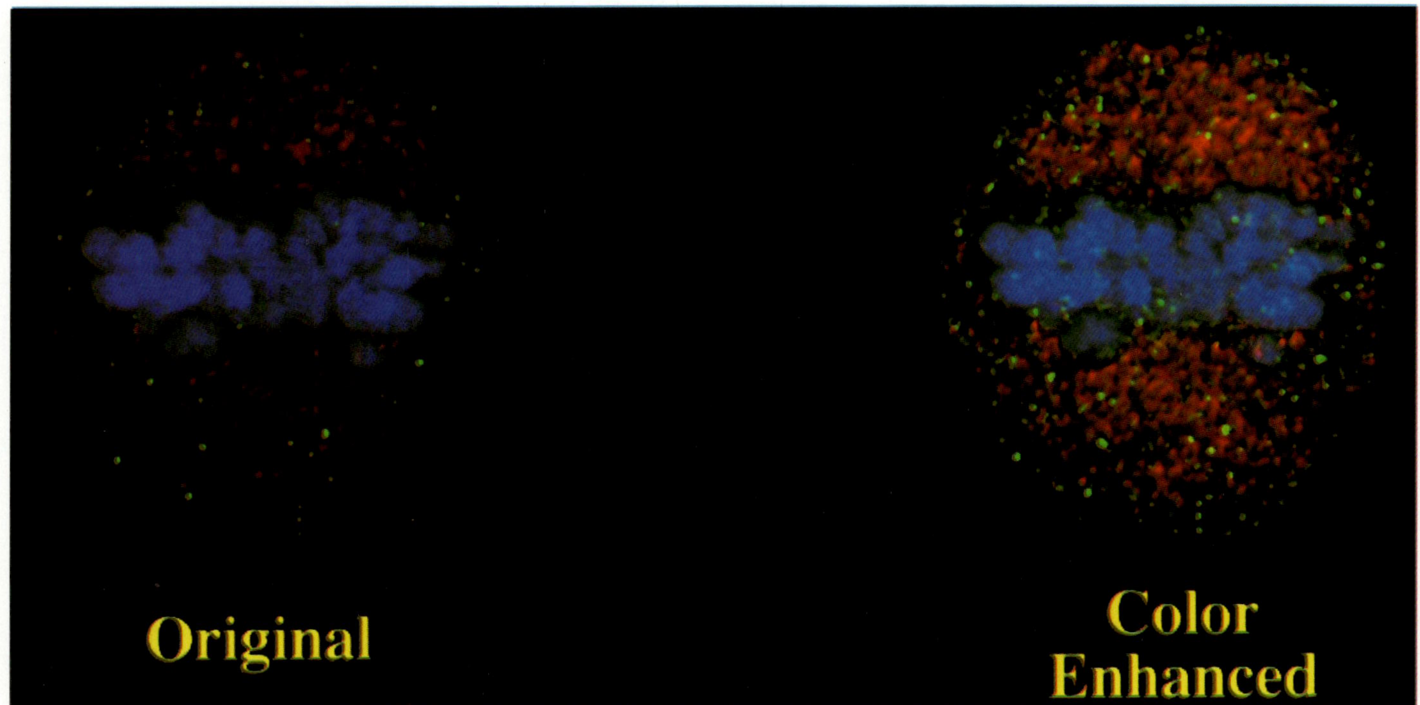

Ch.20, Fig. 1 Prints of a digital fluorescence image before (left) and after (right) color adjustment. The color adjustment was done in Photoshop v.4 (Adobe Systems, San Jose, CA). First, each channel (red, green, and blue) was optimized such that the full range of the data was utilized. To do this, each channel had the minimum value mapped to a level of 0 and the maximum mapped to 255. Next, 25% of the blue channel was added to the green channel using the Image . . . Calculations command. The image is of a mitotic spindle with two different antibodies labeled with FITC (green) and Texas Red (red) and the chromosomes stained with DAPI (blue). Image acquired and processed with a DeltaVision wide-field deconvolution microscope system.

INDEX

D

E

F

M

N

Q

R

S